Sturm-Liouville Problems

Theory and Numerical Implementation

Monographs and Research Notes in Mathematics

Series Editors:
John A. Burns, Thomas J. Tucker, Miklos Bona, Michael Ruzhansky

Actions and Invariants of Algebraic Groups, Second Edition
Walter Ricardo Ferrer Santos, Alvaro Rittatore

Lineability
The Search for Linearity in Mathematics
Richard M. Aron, Luis Bernal-Gonzalez, Daniel M. Pellegrino, Juan B. Seoane Sepulveda

Iterative Methods and Preconditioning for Large and Sparse Linear Systems with Applications
Daniele Bertaccini, Fabio Durastante

Monomial Algebras, Second Edition
Rafael Villarreal

Matrix Inequalities and Their Extensions to Lie Groups
Tin-Yau Tam, Xuhua Liu

Elastic Waves
High Frequency Theory
Vassily Babich, Aleksei Kiselev

Difference Equations
Theory, Applications and Advanced Topics, Third Edition
Ronald E. Mickens

Sturm-Liouville Problems
Theory and Numerical Implementation
Ronald B. Guenther, John W. Lee

For more information about this series please visit: https://www.crcpress.com/Chapman–HallCRC-Monographs-and-Research-Notes-in-Mathematics/book-series/CRCMONRESNOT

Sturm-Liouville Problems

Theory and Numerical Implementation

Ronald B. Guenther
John W. Lee
Department of Mathematics
Oregon State University Corvallis

CRC Press
Taylor & Francis Group
Boca Raton London New York

CRC Press is an imprint of the
Taylor & Francis Group, an **informa** business
A CHAPMAN & HALL BOOK

CRC Press
Taylor & Francis Group
6000 Broken Sound Parkway NW, Suite 300
Boca Raton, FL 33487-2742

Printed on acid-free paper

International Standard Book Number-13: 978-1-138-34543-0 (Hardback)

Library of Congress Cataloging-in-Publication Data

Names: Guenther, Ronald B., author. | Lee, John W., 1942- author.
Title: Sturm Liouville problems : theory and numerical implementation / R.B. Guenther, J.W. Lee (Department of Mathematics, Oregon State University, Corvallis, OR).
Description: Boca Raton, Florida : CRC Press, 2018. | Series: Monographs and research notes in mathematics | Includes bibliographical references and index.
Identifiers: LCCN 2018035973| ISBN 9781138345430 (hardback : alk. paper) | ISBN 9780429437878 (ebook)
Subjects: LCSH: Sturm-Liouville equation. | Differential equations. | Eigenvalues.
Classification: LCC QA372 .G84 2018 | DDC 515/.352--dc23
LC record available at https://lccn.loc.gov/2018035973

Visit the Taylor & Francis Web site at
http://www.taylorandfrancis.com

and the CRC Press Web site at
http://www.crcpress.com

Contents

Preface

This book on Sturm-Liouville problems is written for scientists and engineers, for applied mathematicians, and for advanced undergraduate and graduate students in these fields. We have endeavored to keep the level of mathematical precision at an accessible level for all readers. A reader with the logical reasoning skills acquired in a course in Euclidean geometry, a beginning course in matrix and linear algebra, and a good course in calculus can profitably read this book.

Scientists and engineers may find this book most useful as a reasonably-comprehensive, one-stop reference for the main results about Sturm-Liouville boundary value problems and eigenvalue problems that are needed for real-world applications. The book also can serve as a text for a capstone course in applied mathematics for advanced undergraduate and beginning graduate students. It gives these students valuable insight into how abstract theorems of analysis and linear algebra, that are typically covered in isolation, were motivated and discovered by the need to carefully analyze significant applied problems.

We have endeavored to choose topics that will be most useful to scientists, engineers, and applied mathematicians who encounter Sturm-Liouville problems in their modeling work. Some readers will want to use results from the book, such as existence theorems, continuous dependence, convergence of eigenfunction expansions, and error analyses of numerical methods, but are not especially interested in the proofs. Other readers will want the proofs. The book is organized to accommodate both groups, with some guidance given along the way.

Chapters 4, 5, 6, and 7 that cover regular Sturm-Liouville problems, two types of singular problems, and the numerical approximation of eigenvalues and eigenfunctions by shooting methods are written to substantially reduce the dependence of one of the chapters on its predecessors. Thus, a reader primarily interested in a topic in Chapter 6, say, will find little need to consult prior chapters for essential background material.

Sturm-Liouville problems arise naturally in engineering, physics, and, more recently, in biology and the social sciences. These problems lead to eigenvalue problems for ordinary and partial differential equations. This book addresses, in a unified way, the key issues that must be faced in science and engineering applications when separation of variables, variational methods, or other considerations lead to Sturm-Liouville eigenvalue problems and boundary value problems. In addition, effective numerical procedures for the approximation of eigenvalues and eigenfunctions of both regular and singular Sturm-Liouville problems are presented. Such procedures are essential because explicit evaluation of the eigenvalues and eigenfunctions is rarely possible.

Both regular and singular problems are treated with a high level of rigor and an emphasis on the types of problems that actually arise in mathematical modeling. Our treatment often follows familiar lines but also contains new results, especially in the chapters dealing with singular problems and in the careful justification of shooting methods that can be used to find accurate numerical approximations to eigenvalues and eigenfunctions of both regular and singular problems. A significant feature of the book is its treatment of singular problems: properties of solutions established for singular problems that involve Bessel functions or other special functions are shown to hold for two classes of singular Sturm-Liouville problems that embrace the special cases.

Regular and singular problems are treated with a high level of rigor for more than mathematical reasons: it is an essential element of good mathematical modeling. The quality and accuracy of predictions obtained from a mathematical model depend upon the precision with which physical concepts are incorporated in the model, the effectiveness of those concepts at capturing the principal physical attributes of the real-world situation, and a clear understanding of the likely impact of the simplifying assumptions made to derive the governing equations in the model. Solid physical reasoning and rigorous mathematics lead to better models and better models to better predictions. A rigorous physical model solved by rigorous mathematical methods enables the user to determine under what conditions the model holds and to have confidence in the predictions it makes. All of the models in Chapter 1 and later chapters can be derived from a careful interplay of underlying physical laws and accompanying mathematical precision. This book concentrates on the properties (predictions, behavior) of solutions to such models when Sturm-Liouville boundary value and eigenvalue problems arise. For example, a model for the vibrations of a violin string or a drum head leads to such Strum-Liouville problems. It must be proved that the model equations predict the oscillatory behavior observed in the real world. Moreover, the mathematical analysis of the model should lead to a better understanding of the physical situation being studied by adding precision to our understanding of expected behaviors and, ideally, by predicting behavior not previously observed. Without a rigorously derived mathematical theory to support the modeling none of the foregoing can be accomplished.

During the 19th century and the first half of the 20th century a number of analytical techniques and an extensive theory for dealing with Sturm-Liouville problems were developed. As long as the equations had constant coefficients or were of a special form such as a Bessel or hypergeometric equation, the problems could be dealt with by hand, in the sense that the eigenfunctions could be expressed in terms of special functions and the eigenvalues could be expressed in terms of the zeros of such functions. Asymptotic expansions, series representations, and other specialized methods made it possible to calculate numerical approximations to the first few eigenvalues and eigenfunctions. Eigenfunction expansions, whose coefficients involved integrals, were often hard to find explicitly and laborious to evaluate numerically. Moreover, if the medium was not homogenous, explicit solutions could rarely be found and one had to revert to approximations which were computationally intensive. All in all, numerical calculations were quite challenging.

The availability of digital computers began to change the picture dramatically by the middle of the 20th century, with increasing effect in subsequent decades. The advent of the modern computer seemed to be a quick and easy way out of all the difficulties mentioned above. One could reduce the differential equation to a system of difference equations and let the computer solve the resulting linear system of equations. The eigenvalues of a matrix eigenvalue problem approximated the (first few) eigenvalues of the Sturm-Liouville problem. It turned out that finding the eigenvalues in this way could be difficult, especially because after finding the first two or three eigenvalues, the approximation of subsequent eigenvalues tended to decreases noticeably in accuracy. Often this was the result of use of a numerical orthogonalization procedure. An alternative approach is to use the Ritz-Galerkin method. This technique yields an approximation to the first eigenvalue and gives an approximation to a corresponding eigenfunction. Finding the second eigenvalue can already be rather painful. Fortunately, it is usually the first eigenvalue that is of most interest.

We make the foregoing observations to contrast the more commonly used finite difference schemes for approximation of some eigenvalue, eigenfunction pairs of a Sturm-Liouville problem with the shooting methods developed in Chapter 7. The shooting methods determine each pair independently of the others, involve no orthogonalization procedure, and in principle can be used to find any eigenvalue as accurately as desired.

A precis of each chapter follows.

Chapter 1 Setting the Stage

Typical problems of applied mathematics that lead to Sturm-Liouville problems are formulated to make it clear what types of differential equations and boundary conditions arise in real-world applications and what the natural assumptions on the data are. The chapter starts with the first pure eigenvalue problem to arise historically, Euler buckling. Separation of variables in initial boundary value problems and steady-state problems for partial differential equations lead to Sturm-Liouville eigenvalue and boundary value problems for ordinary differential equations: examples arising from vibration problems (e.g., mechanical and acoustical waves), diffusion problems (e.g., heat flow and chemical transport), and steady-state phenomena are given. Connections with variational methods are discussed briefly. Green's functions are introduced and motivated by physical considerations. The chapter concludes with a discussion of the mathematical problems that must be faced to establish that Sturm-Liouville eigenvalues and eigenfunctions have the qualitative properties that physical intuition suggests and to fully justify the formal solutions that arise from eigenfunction expansions.

Chapter 2 Preliminaries

This chapter is designed as a convenient reference for background material needed for a rigorous treatment of Sturm-Liouville problems. Putting the background material here has three purposes. First, it enables us not to interrupt the treatment of Sturm-Liouville problems to develop background material on the spot in the midst of other reasoning unique to the problems at hand. Second, it frees the reader already familiar with the background material from an unnecessary distraction. Third, the chapter introduces most of the notational conventions used throughout the book. We recommend that readers familiar with the results collected here just skim the chapter to become familiar with the notation that is used later. We hope other readers will find it convenient to have proofs of some essential background results available in one place.

Chapter 3 Integral Equations

The unified approach to Sturm-Liouville eigenvalue and boundary value problems that we follow emphasizes the interplay between the mathematical assumptions made and their connection to the characteristic oscillation properties that physical systems leading to such problems exhibit. This is accomplished by recasting a problem expressed in terms of a differential equation and related boundary conditions as a single integral equation. Of course, this leads us to a presentation of the principal results of Hilbert-Schmidt theory.

The unified treatment also includes material on Kellogg kernels that has essentially vanished from the current literature on Sturm-Liouville problems. This is unfortunate because use of Kellogg kernels adds considerable insight into the interplay between mathematical properties of eigenvalues and eigenfunctions of Sturm-Liouville problems and the well-documented behavior of physical systems that lead to such problems. Our treatment of *singular* Sturm-Liouville problems leads naturally to *singular* Kellogg kernels. We believe the material on singular Kellogg kernels is new.

The development we follow also involves extensions of properties of integral operators with suitably positive kernels (due originally to Jentzsch) and some pioneering work of Schur on the relationship between the eigenvalues and eigenfunctions of a kernel and those of its compound kernels. The work of Jentzsch and especially of Schur is not as well known as it should be.

The field of integral equations is vast. We have restricted the chapter to just those results needed for a unified treatment of Sturm-Liouville problems.

Chapter 4 Regular Sturm-Liouville Problems

A Sturm-Liouville eigenvalue is *regular* on an interval $a < x < b$ if the coefficients in the differential equation $-(p(x)y')' + q(x)y = \lambda r(x)y$ are continuous on the corresponding closed interval and p and r are positive there. Likewise for the Sturm-Liouville boundary value problems where λry is replaced by f, a given continuous function. We treat regular Sturm-Liouville problems in depth when the associated boundary conditions are linear and separated and also in considerable detail when they are mixed. Green's functions are introduced and characterized in a standard way. A glance at the table of contents indicates the scope of topics treated.

Since $p(x)$ is only assumed to be continuous, which may be the natural assumption corresponding to the underlying physical situation, we look closely into what it means for y to be a solution to such an equation and to the consequences of the definition adopted for a solution. Many treatments of Sturm-Liouville differential equations assume further smoothness on $p(x)$. For that reason, the existence, uniqueness, and continuous dependence results established here are more general than those often encountered in the literature.

Chapter 5 Singular Sturm-Liouville Problems - I

The treatment of the singular problems in this chapter and the next was motivated by an appendix in [43] by Tychonoff and Samarski, that seems to have had much less influence that it should have. We flesh out and extend the important developments of the appendix. The appendix sets out the qualitative properties of *bounded* solutions to singular Sturm-Liouville differential equations of the form $(p(x)y')' + (\lambda r(x) - q(x))y = 0$ where p and/or q are singular and $a < x < b$. The nature of the singularities are chosen to cover all the classical Sturm-Liouville differential equations and special functions that arise from practical applications. In particular, the equations of Bessel, Legendre, Hermite, and Laguerre are included.

We assume in Chapter 5 that p, q, and r are continuous on $a \leq x \leq b$, $p > 0$ on $a < x \leq b$, and p has a simple zero at $x = a$. The weight function $r(x)$ is positive on $a < x \leq b$ but can have a zero of finite order at $x = a$. Bessel's equation of order 0 is a prototype for such problems.

The conversion of a singular Sturm-Liouville problem to an integral equation is done via a mildly singular Green's function. (See the precis of Appendix A.) As far as we know, a careful treatment of mildly singular Green's functions and their compound kernels is new as is the introduction of singular Kellogg kernels.

The topics covered in this setting are given in the table of contents for the chapter. The bottom line is that all the familiar oscillation behavior exhibited by regular Sturm-Liouville problems extends to this setting.

Chapter 6 Singular Sturm-Liouville Problems - II

Just as for Chapter 5, the treatment of the singular problems in this chapter was motivated by an appendix in [43] by Tychonoff and Samarski: see the precis for Chapter 5.

In Chapter 6, the coefficient q is allowed to have a simple pole at $x = a$ in addition to the singular behavior allowed in Chapter 5. This is the situation for Bessel's equations of order $n > 0$. To accommodate the additional singularity in q some additional smoothness is needed.

Just as in Chapter 5, a singular Sturm-Liouville problem is converted into an equivalent singular integral equation whose kernel (Green's function) is mildly singular. As far as we know, a careful treatment of the mildly singular kernels of Chapter 6 and their compound kernels is new.

The full scope of topics covered in Chapter 6 is given in the table of contents. The bottom line is that all the familiar oscillation behavior exhibited by regular Sturm-Liouville problems extends to this setting.

Chapter 7 Approximation of Eigenvalues and Eigenfunctions

Shooting methods are presented for the regular and singular Sturm-Liouville eigenvalue problems studied in Chapters 4, 5, and 6. The shooting method used for the regular problems of Chapter 4 extends naturally to the singular problems of Chapter 5 with the aid of the boundary condition, $(q(a) - \lambda r(a))y(a) - p'(a)y'(a) = 0$, that is forced by the singularity at $x = a$ and by the fact that $y(a) \neq 0$ for any bounded nontrivial solution to the singular differential equation in the eigenvalue problem. In sharp contrast, in Chapter 6 any bounded solution to the singular differential equation satisfies $y(a) = 0$ and $y'(a)$ may not exist. Consequently, a completely different shooting method must be used. A careful convergence analysis is given for all three methods. Many examples are given with output from each shooting method and effective strategies for choosing initial guesses are discussed.

Chapter 8 Concluding Examples and Observations

The results of preceding chapters are applied to a few particularly important problems of both historical and current importance. The main examples are the vibrations of a hanging chain (or cable), the oscillations of a string, and the vibrations of a bar (beam). We also indicate how some of the shooting methods in Chapter 7 can be extended to higher order, self-adjoint problems.

Appendix A Mildly Singular Compound Kernels

The Green's functions for the singular Sturm-Liouville problems of Chapters 5 and 6 are mildly singular in different senses. This appendix develops the properties of two types of mildly singular kernels and their compound kernels that include the Green's functions encountered in Chapters 5 and 6 and that are needed to establish the oscillation properties of singular Kellogg kernels and the singular Sturm-Liouville eigenvalue problems in Chapters 5 and 6.

Appendix B Iteration of Mildly Singular Kernels

This appendix develops the properties of kernels that arise by iteration of mildly singular kernels and of their compound kernels that are needed to establish the properties of the singular Sturm-Liouville eigenvalue problems in Chapters 5 and 6.

Appendix C The Kellogg Conditions

The defining conditions of a Kellogg kernel are shown to be essentially equivalent to five simple physical properties of elastic strings. The treatment closely follows that given in Gantmacher and Krein [16] and makes this important link between mathematical assumptions and observed physical behavior available to a wider audience.

We extend our sincere thanks to the individuals of the acquisitions, editorial and publication staffs at CRC Press, Taylor & Francis Group, and Nova Techset who assisted in the publication of our book. We especially appreciated the friendly, helpful attitudes and quick responses to our queries from Sarfraz Khan, Editor, Mathematics; Callum Fraser, Editorial Assistant; Suzanne Lassandro, Production Editorial Manager; Teena Lawrence, Manager (Project Management); and Jeanne Washington, Freelancer (proofreader) whose combined efforts have resulted in a much-improved final product.

Finally, the authors would be grateful to receive feedback from readers about misprints, mistakes, misconceptions, and so on that you find. Thank you. We can be reached at guenth@math.orst.edu and jwlee@math.oregonstate.edu.

Chapter 1

Setting the Stage

There are many reasons for seeking eigenvalues and their corresponding eigenfunctions. Here are a few of them. First, solutions to many problems modeled by ordinary and partial differential equations can often be given explicitly in terms of eigenfunction expansions. (For the problems we shall treat, such eigenfunction expansions are strictly analogous to the representation of a vector in terms of its **i**, **j**, and **k** components in 3-space or the corresponding representations in n-space.) Second, eigenvalues are often of independent interest. They may tell us where bifurcations can occur in nonlinear models by looking at their linearizations. Eigenvalues give sharp estimates about the rates of decay (or growth) of solutions arising in heat conduction, concentration analyses, flow in porous media, and so on. In vibration problems, they give fundamental frequencies and overtones of musical instruments. Eigenvalues are important in determining the critical mass for nuclear reactions in a given geometry. Finally, eigenvalues arise naturally in optimization and in the calculus of variations.

Since most eigenvalue problems cannot be solved explicitly, we will take a hard look at the qualitative behavior of both the eigenvalues and the eigenfunctions, and analyze both regular and singular problems. For the same reason, we present an effective numerical technique for the practical evaluation of eigenvalues and corresponding eigenfunctions.

To further motivate the types of Sturm-Liouville problems that are the subject of this book, we present, without detailed derivations, a few important problems of mathematical physics and the Sturm-Liouville eigenvalue and boundary value problems to which they lead, usually via separation of variables in a partial differential equation. However, eigenvalue problems arise in many contexts involving ordinary and partial differential equations as well as in matrix theory and more general operator settings. It seems likely that Euler considered the first eigenvalue problem when he discussed the buckling of a beam. We start our survey with that problem.

In the survey of problems that follows, we assume that all functions are real-valued and all constants are real numbers, which is natural for the scenarios presented. See the final section of this chapter concerning complex-valued functions and data.

1.1 Euler Buckling

A straight elastic bar (beam, rod) of length l is positioned vertically upward and is anchored at its base. Experimentally one can take a thick metal wire. A small compressive force of magnitude K acts vertically downward on the free end of the bar as in Figure 1.1.

The equations governing the shape of the bar are

$$\begin{cases} EIy'' = -Ky, & 0 < x < l \\ y(0) = 0, y(l) = 0 \end{cases}$$

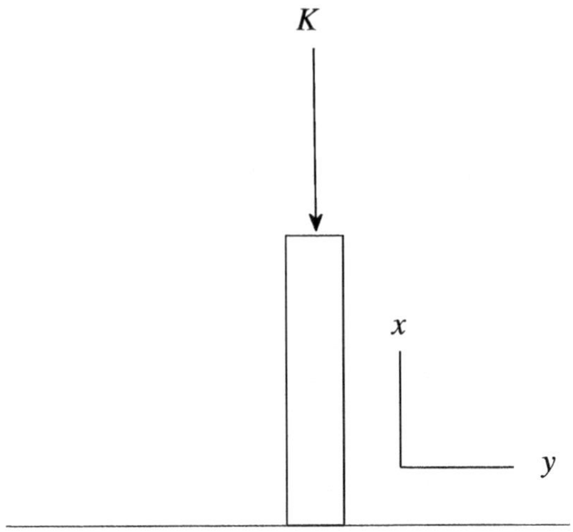

FIGURE 1.1: The Euler Beam

where $y = y(x)$ is the transverse deflection of the midline of the bar from its vertical equilibrium position. The physical constants E and I are determined by the elastic and geometric properties of the bar. The governing equations always have the solution $y(x) = 0$ for $0 \leq x \leq l$. Experiments confirm that this is the shape of the bar when K is small but that the bar will buckle when K is increased to a critical value. Buckling means the bar will deflect from the vertical into a new equilibrium shape. Do the governing equations predict buckling? Euler answered this question in the affirmative in 1757.

Express the governing equations as

$$y'' + \lambda y = 0, \; y(0) = y(l) = 0,$$

where $\lambda = K/EI > 0$. If buckling occurs, it must be possible to find a solution (or solutions) to the governing equations different from the obvious solution $y(x) = 0$ for $0 \leq x \leq l$, the so-called trivial solution. Other solutions, if any exist, are called nontrivial. The differential equation has general solution

$$y = A\cos(\sqrt{\lambda}x) + B\sin(\sqrt{\lambda}x).$$

Since $y(0) = 0$, $A = 0$. If nonzero deflections (solutions) are possible, we must have $B \neq 0$ and $y(l) = B\sin(\sqrt{\lambda}l) = 0$. Thus, nonzero solutions y exist if and only if $\lambda = \lambda_n = (n\pi/l)^2$ for n a positive integer. The corresponding nontrivial solutions are $y = y_n(x) = B_n \sin(n\pi x/l)$ with $B_n \neq 0$. The values of λ (hence, K) that permit nontrivial deflections are now called eigenvalues and the corresponding nontrivial solutions are called eigenfunctions. The problem we have just solved is called an eigenvalue problem.

Buckling in the Euler beam first occurs at $\lambda_1 = \pi^2/l^2$; that is when $K = EI\pi^2/l^2$ and the bent beam takes the new equilibrium state

$$y = B\sin(\pi x/l)$$

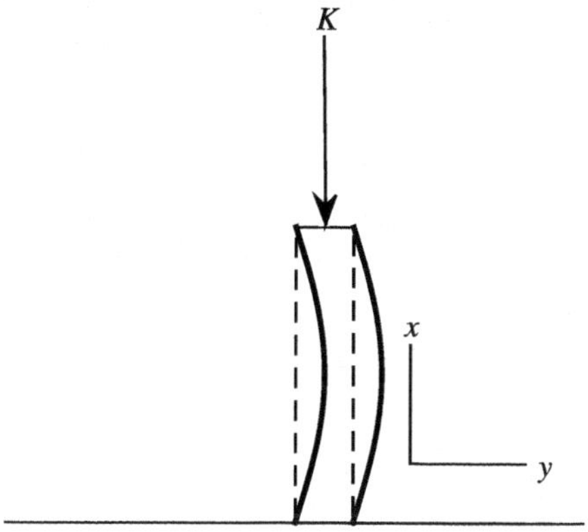

FIGURE 1.2: The Buckled Beam

for some $B \neq 0$. Figure 1.2 illustrates the case with $B > 0$. Since $\lambda_1 = (K/EI)^{1/2}$, the smallest eigenvalue determines the minimum compressive force needed to buckle a beam of given flexural rigidity EI.

The Euler model predicts that buckling can occur and does occur only at the eigenvalues λ_n and that the corresponding buckled equilibrium states are multiples of $\sin(n\pi x/l)$. Actually, once the bar has buckled, a new model is needed because the physical situation has become nonlinear. Nevertheless, even in the nonlinear regime the linear problem above, which is the linearization of an appropriate nonlinear model, still predicts the values of K at which buckling can occur. Problems of this sort are called bifurcation (branching) problems because nonlinear states branch from a stable linear state (usually $y = 0$) at certain critical values, the eigenvalues of the linearized problem. The eigenfunction y_n corresponding to the branch point determined by the eigenvalue λ_n approximates the shape of the nonlinear buckled responses of small amplitude that occur near the branch point.

The governing equations for Euler buckling,

$$y'' + \lambda y = 0, \; y(0) = y(l) = 0,$$

where $\lambda = K/EI > 0$ comprise a regular Sturm-Liouville eigenvalue problem. Regular means that the differential equation is regular. A detailed study of regular Sturm-Liouville boundary value problems and eigenvalue problems is presented in Chapter 4.

1.2 Hanging Chain

In contrast to Euler buckling, the problem of determining the normal modes of a hanging chain leads to a singular Sturm-Liouville differential equation and corresponding eigenvalue problem. This problem was solved by Daniel Bernoulli (1700–1782) in 1732 and involved the first use of a Bessel function. Small transverse displacements from equilibrium were assumed. The problem was discussed further by Euler in 1781. F. W. Bessel (1784–1846) investigated the functions that now bears his name.

In Bernoulli's treatment the chain is a one-dimensional continuum that has constant density. We formulate a slightly more general model that permits variable density. Suppose the length of the chain is l. Set up coordinates so that the x-axis is directed vertically upward with the origin at the free end of the chain when the chain hangs in its vertical equilibrium position. Let $\rho_0(x)$ be the density of the chain when it is hanging in equilibrium and $u(x,t)$ be the transverse displacement at time t of the point on the chain that is located at position x when the chain hangs in equilibrium. The only external force acting on the chain is gravity, with constant acceleration g, and the tension at a cross section of the chain acts tangentially and is due to the part of the chain that lies below the cross section.

Under these assumptions the initial boundary value problem for the chain is

$$\begin{cases} \rho_0(x)u_{tt} = (p(x)u_x)_x, & 0 < x < l, \quad t > 0, \\ |u(0,t)| < \infty, & u(l,t) = 0, \quad t \geq 0, \\ u(x,0) = f(x), & u_t(x,0) = v(x) \quad 0 \leq x \leq l, \end{cases} \tag{1.1}$$

where

$$p(x) = g\int_0^x \rho_0(\xi)\, d\xi$$

for $0 \leq x \leq l$, $f(x)$ specifies the initial shape of the chain, and $v(x)$ is its initial velocity profile. Observe that the differential equation is singular because $p(0) = 0$. Typically such equations can have both bounded and unbounded solutions. Physically realistic solutions for the displacement $u(x,t)$ must be bounded. This leads to the boundary condition $|u(0,t)| < \infty$ which means that the displacement is bounded for $x > 0$ and near 0 for all time t. It follows that $u(x,t)$ is bounded in space and time.

The normal modes of the chain are the motions where each point of the chain vibrates at the same frequency. Such motions have the form $u(x,t) = T(t)X(x)$, so-called separated solutions of the partial differential equation. A separated solution will satisfy the wave equation in (1.1) if and only if

$$\rho_0(x)\,T''X = (p(x)\,TX')',$$

$$\frac{T''}{T} = \frac{(p(x)X')'}{\rho_0(x)}X = -\lambda,$$

where $-\lambda$ is a separation constant. Thus,

$$T'' + \lambda T = 0$$

and

$$(p(x)X')' + \lambda\rho_0(x)X = 0,\ |X(0)| < \infty,\ X(l) = 0,$$

where the boundary conditions on X follow from those in (1.1). The normal modes $u(x,t) = T(t)X(x)$, apart from the trivial solution, are determined by those values of λ (eigenvalues) for which the X-problem has nontrivial solutions (eigenfunctions). For such λ the frequency of oscillation of each point on the chain, $\sqrt{\lambda}/2\pi$, is determined by the T-equation.

In Bernoulli's original problem $\rho_0(x) = \rho_0$ a given positive constant, $p(x) = g\rho_0 x$, the wave equation for the chain is

$$u_{tt} = g(xu_x)_x,$$

and the separated solutions $u(x,t) = T(t)X(x)$ are determined by

$$T'' + \lambda T = 0$$

and

$$g(xX')' + \lambda X = 0, \; |X(0)| < \infty, \; X(l) = 0.$$

It turns out that the X-equation is reducible to a Bessel's equation of order 0 and, hence, that the spatial component of a normal mode is a multiple of a bounded solution of that equation. We will discuss Bernoulli's problem further in Chapter 8 together with numerical results. For the moment, we mention that Bessel's equation of order 0 is a prototype for the singular Sturm-Liouville boundary value and eigenvalue problems that are the subject of Chapter 5.

If the density is $\rho_0(x) = \rho_0 x^n$ where ρ_0 is a positive constant and $n \geq 0$, then the wave equation is

$$x^n u_{tt} = g\left(\frac{x^{n+1}}{n+1} u_x\right)_x.$$

The normal modes $u(x, t) = T(t)X(x)$ are determined from

$$x^n T'' X = \frac{g}{n+1}(x^{n+1} TX')'.$$

Hence,

$$\frac{T''}{T} = \frac{g}{(n+1)} \frac{1}{x^n X}(x^{n+1} X')' = -\lambda,$$

where λ is a separation constant,

$$T'' + \lambda T = 0,$$

and

$$\left(\frac{g}{n+1} x^{n+1} X'\right)' + \lambda x^n X = 0, \; |X(0)| < \infty, \; X(l) = 0.$$

The X-equation is reducible to a Bessel's equation of order n and, hence, the spatial component of a normal mode is expressible in terms of a bounded solution of that equation. We will discuss this generalized Bernoulli problem further in Chapter 8 together with numerical results. For the moment, we mention that Bessel's equation of order n with $n > 0$ is a prototype for the singular Sturm-Liouville boundary value and eigenvalue problems that are the subject of Chapter 6.

1.3 Separation of Variables

The method of separation of variables seeks to solve a linear partial differential equation, such as the wave equation or heat equation, together with given side conditions by ultimately reducing the problem to finding solutions to a family (or families) of related ordinary differential equations with corresponding side conditions. The separated solutions are superposed appropriately so as to satisfy any remaining conditions of the original problem. This solution strategy leads naturally to many interesting and important eigenvalue problems. Most of the physical situations described in this chapter involve either the homogeneous wave equation

$$bu_{tt} = \operatorname{div}(p\nabla u) - cu$$

or the homogeneous heat equation (also known as the diffusion equation)

$$bu_t = \operatorname{div}(p\nabla u) - cu$$

with $u = u(x,t)$ where x varies in real Euclidean n-space and $b(x) > 0$, $p(x) \geq 0$, and $c(x) \geq 0$ are given functions of the spatial variables in a domain of interest. A natural first step in the method of separation of variables is to seek separated solutions of the form $u = w(x)T(t)$. Such a separated solution satisfies the wave equation or the heat equation if and only if

$$\frac{T''}{T} = \frac{\operatorname{div}(p\nabla w) - cw}{bw}$$

or

$$\frac{T'}{T} = \frac{\operatorname{div}(p\nabla w) - cw}{bw}$$

holds for all relevant times t and positions x. The left member of each equation must be constant in time because the right member does not change as time varies. Likewise, the right member must be constant in space because the left member does not vary as the spatial variable changes. Thus, both sides of each equation must be one and the same constant; that is,

$$\frac{T''}{T} = -\lambda \quad \text{and} \quad \frac{\operatorname{div}(p\nabla w) - cw}{bw} = -\lambda,$$

or

$$\frac{T'}{T} = -\lambda \quad \text{and} \quad \frac{\operatorname{div}(p\nabla w) - cw}{bw} = -\lambda,$$

for some **separation constant**, here $-\lambda$. In typical applications, λ is positive: for the wave equation this is equivalent to separated solutions $u(x,t) = T(t)w(x)$ being periodic in time while for the heat equation it is equivalent to separated solutions that decay in time. For either type of problem

$$\operatorname{div}(p\nabla w) - cw + \lambda bw = 0$$

in the interior of the spatial domain of interest. If p is constant, the differential equation has the form

$$\Delta w - cw + \lambda bw = 0,$$

where Δw is the Laplacian of w. For separated solutions to be useful they must satisfy some of the homogeneous side conditions of the problem and they must be nontrivial, not identically zero. This is how eigenvalue problems emerge.

Separation of variables was first used by Euler (1748) in an isolated case to find a solution of the one-dimensional wave equation together with boundary and initial data and to determine the fundamental frequencies of a vibrating violin or piano string. D'Alembert gave the general solution to the one-dimensional wave equation in 1746. A heated controversy arose between Euler and D'Alembert about the meaning of a solution to the wave equation. Lagrange sided with Euler in the debate. Later Fourier (1805) significantly extended the method of separation of variables in his pioneering studies on heat conduction. Dirichlet put Fourier's method on a firm foundation about 25 years later. These developments led to the boundary value problems and eigenvalue problems now called Sturm-Liouville problems. In the 1820s and 1830s Sturm and Liouville initiated the systematic study of such problems and in the process initiated the study of qualitative properties of solutions to differential equations when explicit solutions were not available.

We shall deal with problems that have one spatial dimension or can be reduced to one spatial dimension. Such problems include higher-dimensional spatial situations where the geometry and symmetry lead to an initial boundary value problem with only one spatial dimension.

1.4 Vibration Problems

Problems of this kind typically lead to an initial boundary value problem for the wave equation.

1.4.1 Vibrations of a String

Consider a taut, homogeneous string, such as a piano or violin string, stretched between two posts. Pluck the string so that it experiences small transverse vibrations. If $u = u(x,t)$ is the deflection of the string from its rest (unplucked) position, then u satisfies the wave equation, initial conditions, and boundary conditions that follow:

$$\begin{cases} u_{tt} = c^2 u_{xx} & \text{for } 0 < x < l,\ t > 0, \\ u_t(x,0) = f(x),\, u_t(x,0) = g(x), & \text{for } 0 \le x \le l, \\ u(0,t) = 0,\ u(l,t) = 0, & \text{for } t \ge 0, \end{cases} \tag{1.2}$$

where l is the length of the string, $c = \sqrt{\tau/\rho}$ is the speed of wave propagation, τ is the (constant) horizontal component of tension in the string, and ρ is the linear density of the string. The functions f and g specify the initial displacement and velocity of the string.

If external forces act on the homogeneous string the wave equation becomes $u_{tt} = c^2 u_{xx} + F(x,t)$, where $F(x,t)$ models the external transverse forces, such as those that arise when the string is bowed as in Figure 1.3.

If the string is inhomogeneous, then $\rho = \rho(x) > 0$ and a careful derivation of the wave equation shows that $\tau > 0$ is either a constant or a function of time t. In the more general case, the wave equation in the model becomes

$$\rho(x) u_{tt} = \left(\tau(t) u_x\right)_x.$$

If the string is homogeneous so that $\rho(x) = \rho_0$ and $\tau = \tau_0$ where ρ_0 and τ_0 are constants and $c = \sqrt{\tau_0/\rho_0}$, then

$$u(x,t) = X(x)\,T(t)$$

will satisfy the wave equation and the boundary conditions in (1.2) if $X(x)$ and $T(t)$ satisfy

$$\begin{aligned} &X''(x) + \lambda X(x) = 0,\ X(0) = 0,\ X(l) = 0, \\ &T''(t) + \lambda c^2 T(t) = 0, \end{aligned} \tag{1.3}$$

where $-\lambda$ is the separation constant. Furthermore, a separated solution $u(x,t) = X(x)\,T(t)$ will only be useful if it is not identically zero. The equation for $X(x)$ always has the trivial

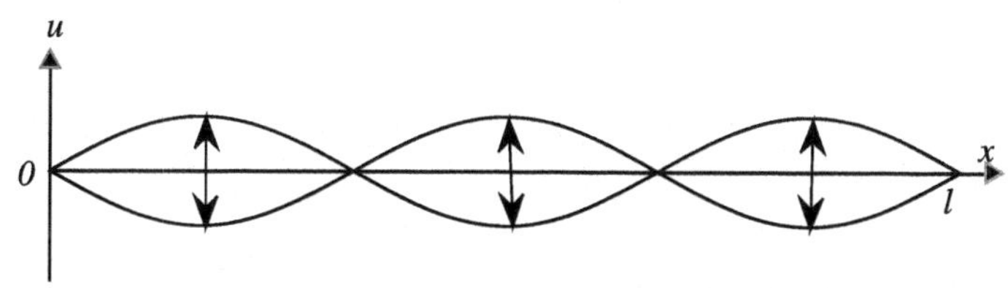

FIGURE 1.3: Vibrations of a String

solution $X(x) = 0$. Thus, separated solutions will only be useful if there are values of λ such that the problem (1.3) has nontrivial solutions. Thus, separation of variables has led to an eigenvalue problem for X. It is the same eigenvalue problem that we encountered in Euler buckling.

If the string is inhomogeneous and the horizontal component of the tension is time dependant, then (1.3) becomes

$$X''(x) + \lambda\rho(x)X(x) = 0, \; X(0) = 0, \; X(l) = 0. \tag{1.4}$$

and the temporal factor of a separated solution satisfies

$$T''(t) + \lambda\tau(t)T = 0.$$

We will discuss the vibrations of a string more fully in Chapter 8.

1.4.2 Vibrations of a Circular Membrane

Consider the membrane of a circular drum, assumed homogeneous. Suppose the membrane is set in motion by a displacement and velocity that is radially symmetric. The transverse vibrations u that result are radially symmetric; that is, $u = u(r, t)$. Use of the 2-D wave equation in polar coordinates with pole at the center of the membrane leads to the initial boundary value problem

$$\begin{cases} u_{tt} = c^2 \dfrac{1}{r}(ru_r)_r & \text{for } 0 \le r < b, \; t > 0 \\ u(r,0) = f(r), \; u_t(r,0) = g(r) & \text{for } 0 \le r \le b, \\ u(b,t) = 0, & \text{for } t \ge 0, \end{cases} \tag{1.5}$$

where b is the radius of the membrane. If separated solutions $u = R(r)T(t)$ are required to satisfy the wave equation and the homogeneous boundary condition, the eigenvalue problem which arises for $R(r)$ is

$$R'' + \frac{1}{r}R' + \lambda R = 0, \; R(b) = 0.$$

The differential equation has a singularity at $r = 0$. Solutions to such equations can become unbounded. Unbounded solutions are not physically meaningful for the vibrating membrane. So there is an unstated (hidden) boundary condition that R must satisfy. It must be bounded. This is often expressed by stating the eigenvalue problem for R as

$$R'' + \frac{1}{r}R' + \lambda R = 0, \; |R(0)| < \infty, \; R(b) = 0, \tag{1.6}$$

because any unbounded behavior of R must occur at the singularity $r = 0$.

Separation of variables succeeds in this example because polar coordinates were used. Separation of variables fails in rectangular coordinates because there are no nontrivial separated solutions in that coordinate system. Use of polar coordinates facilitates the solution process but brings in a singularity that is somewhat fake. The vibrating drum has no intrinsic physical singularity that would lead to the singular term, $1/r$, in the mathematical model or its solution. This singularity is an artifact of expressing the wave equation in polar coordinates: the transformation $x = r\cos\theta$, $y = r\sin\theta$ is not 1-1 at $r = 0$ and its Jacobian $\partial(x, y)/\partial(r, \theta) = r$ is zero at $r = 0$. This problem with the choice of coordinates is overcome by introducing the hidden boundary condition $|R(0)| < \infty$ in the eigenvalue problem. Similar situations arise when cylindrical or spherical polar coordinates are used.

The differential equation in this eigenvalue problem is a Bessel equation. One of the motivations for investigating Bessel functions and other higher transcendental functions was to solve problems arising in mathematical physics.

1.4.3 Spherically Symmetric Vibrations in a Ball

If a ball with radius b experiences radially symmetric vibrations $u = u(r, t)$, then u satisfies the 3-dimensional wave equation

$$u_{tt} = c^2 \frac{1}{r^2}(r^2 u_r)_r, \quad 0 \le r < b, \ t > 0$$

and the eigenvalue problem that arises from separation of variables is

$$R''(r) + \frac{2}{r}R'(r) + \lambda R(r) = 0, \ |R(0)| < \infty, \ R(b) = 0. \tag{1.7}$$

Problem (1.4) is typical of a regular Sturm-Liouville eigenvalue problem: the coefficients in the differential equation are smooth functions and no coefficient multiplying the highest order derivative in the equation is zero at any point. Problems (1.6) and (1.7) are typical singular Sturm-Liouville eigenvalue problems: at least one condition of a regular problem is violated. The singular problems point out that the more general treatment of singular eigenvalue problems in Chapters 5 and 6 must allow singularities that behave like $1/r$.

1.5 Diffusion Problems

Many diffusion problems in one spatial dimension are governed by a diffusion equation of the form

$$u_t = (p(x)u_x)_x + b(x)u_x - c(x)u \tag{1.8}$$

together with appropriate initial and boundary conditions. Here x is the spatial variable and t is time, $0 \le x \le l$, $t > 0$, $p \ge 0$ is a diffusion coefficient and b and $c \ge 0$ are given. The physical meaning of b, c, and $u = u(x, t)$ depends on the problem at hand. Depending on the field, the names associated with (1.8) are Fourier, Fick, Darcy, and Nerust among others. In typical applications $p > 0$, except possibly at $x = 0$ or $x = l$ and the diffusion equation is satisfied for $0 < x < l$ and $t > 0$.

It is beneficial to express the diffusion equation, and other such equations that have the term $b(x)u_x$, in what is called formally self-adjoint form by means of the change of variable $u(x, t) = g(x)v(x, t)$ where

$$g(x) = \exp\left(-\int^x b(\xi)/p(\xi)\, d\xi\right).$$

It is easy to check that diffusion equation for $v(x, t)$ has the form

$$g(x)v_t = (p(x)g(x)v_x)_x - q(x)v,$$

so that the change of variables preserves $p(x)g(x) > 0$, except possibly at $x = 0$ or $x = l$. Replacing $p(x)g(x)$ by $p(x)$, the transformed equation has the form

$$g(x)v_t = (p(x)v_x)_x - q(x)v.$$

Separated solutions $v(x, t) = X(x)T(t)$ of this partial differential equation must satisfy the pair of ordinary differential equations

$$\begin{aligned} (p(x)X'(x))' - q(x)X(x) + \lambda g(x)X(x) &= 0, \\ T'(t) + \lambda T(t) &= 0, \end{aligned}$$

where the separation constant is $-\lambda$.

1.5.1 Chemical Transport

A simple model for the flow of a chemical through a partially saturated soil column in which the ground water is at rest and scrubbers reduce the concentration of chemical to zero at the ends of the column is

$$\begin{cases} u_t = (p(x)u_x)_x & \text{for } 0 < x < l,\ t > 0, \\ u(x,0) = f(x) & \text{for } 0 \le x \le l, \\ u(0,t) = 0,\ u(l,t) = 0 & \text{for } t \ge 0. \end{cases} \tag{1.9}$$

Here $u(x,t)$ is the concentration of the chemical or pollutant at position x at time t and $p \ge 0$ is the diffusion coefficient. The coefficient b in the general diffusion equation is zero because the ground water is at rest, and c also is zero when only the diffusion effect is modeled. The initial concentration is given by the function $f(x)$.

The eigenvalue problem for X that arises when separated solutions are required to satisfy the boundary conditions is

$$(p(x)X'(x))' + \lambda X(x) = 0,\ X(0) = X(l) = 0.$$

1.5.2 Heat Conduction in a Rod

A laterally insulated rod of length l has an initial temperature distribution $g(x)$ and is surrounded by a medium held at temperature zero. Each end of the rod communicates heat to its surroundings according to Newton's law of cooling. A reasonable model in this situation for the temperature $u(x,t)$ in the rod at position x and time t is

$$\begin{cases} u_t = (p(x)u_x)_x - q(x)u, & 0 < x < l,\ t > 0, \\ u(x,0) = g(x), & 0 \le x \le l, \\ \alpha u(0,t) - \beta u_x(0,t) = 0, & t \ge 0, \\ \gamma u(l,t) + \delta u_x(l,t) = 0, & t \ge 0, \end{cases} \tag{1.10}$$

In this model, $q(x) \ge 0$ is a heat loss coefficient that allows for imperfect lateral insulation along the lateral surface of the rod, and α, β, γ, and δ are positive constants determined by the characteristics of the rod and Newton's law of cooling.

The eigenvalue problem for X that arises when separated solutions are required to satisfy the boundary conditions is

$$\begin{aligned} &(p(x)X'(x))' - q(x)X(x) + \lambda X(x) = 0,\ 0 < x < l, \\ &\alpha X(0) - \beta X'(0) = 0,\ \gamma X(l) + \delta X'(l) = 0. \end{aligned}$$

Explicit solutions for the eigenvalues λ and the corresponding eigenfunctions X do not exist except for a few simple but important choices of $p(x)$, $q(x)$, and the boundary conditions. However, the basic heat equation $u_t = au_{xx}$ with a a positive constant and with homogeneous Dirichlet boundary conditions $u(0,t) = 0$ and $u(l,t) = 0$ leads to the eigenvalue problem

$$\begin{aligned} &X''(x) + \lambda X(x) = 0,\ 0 < x < l, \\ &X(0) = 0,\ X(l) = 0, \end{aligned}$$

whose solution yields eigenvalues $\lambda_n = (n\pi/l)^2$ and corresponding eigenfunctions $X_n(x) = \sin n\pi x/l$ for $n = 1, 2, \ldots$.

1.5.3 Heat Conduction in a Disk

Consider a heat conducting disk of radius b with insulated top and bottom, constant thermal conductivity 1, initial temperature distribution $f(r,\theta)$, and with its bounding circle held at temperature zero. The temperature $u(r,\theta,t)$ is given by

$$\begin{cases} u_t = \Delta u & \text{for } 0 \le r < b,\ 0 \le \theta \le 2\pi,\ t > 0, \\ u(r,\theta,0) = f(r,\theta) & \text{for } 0 \le r \le b,\ 0 \le \theta \le 2\pi, \\ u(b,\theta,t) = 0 & \text{for } 0 \le \theta \le 2\pi,\ t > 0. \end{cases}$$

Separation of variables in space and time with separation constant $-\lambda$ via $u = T(t)v(r,\theta)$ leads to

$$T' + \lambda T = 0 \quad \text{and} \quad \Delta v + \lambda v = 0.$$

So $T(t)$ is a multiple of $e^{-\lambda t}$. A second separation of variables via $v = R(r)\Theta(\theta)$ yields

$$r^2R'' + rR' + \lambda r^2 R - \mu R = 0,$$
$$\Theta'' + \mu\Theta = 0.$$

Since (r,θ) and $(r,\theta+2\pi)$ mark the same point in the plate, Θ must be 2π periodic

$$\Theta'' + \mu\Theta = 0,\ \Theta(0) = \Theta(2\pi),\ \Theta'(0) = \Theta'(2\pi).$$

The condition on the derivative follows from Fourier's law of heat flow.

The eigenvalue problem for Θ has eigenvalues $\mu = \mu_n = n^2$ for $n = 0,\ 1,\ 2, \ldots$ and corresponding eigenfunctions $\Theta = \Theta_n = a_n \cos n\theta + b_n \sin n\theta$ where $a_n^2 + b_n^2 \neq 0$. The eigenvalue $\mu_n = n^2$ has multiplicity 2 because two linearly independent eigenfunctions correspond to it, unlike all the foregoing examples where each eigenvalue has multiplicity 1.

Since $\mu = \mu_n = n^2$, the differential equation for $R = R_n(r)$ becomes

$$r^2R'' + rR' + (\lambda r^2 - n^2)R = 0,$$

which is Bessel's equation of order n with parameter λ. Equivalently,

$$-(rR')' + \left(\frac{n^2}{r} - \lambda r\right)R = 0,$$

which reveals a singularity in the highest derivative term because $r = 0$ at the origin and a singularity in the coefficient of R that becomes positively infinite as $r \to 0$. We will deal with singular problems of this type in more generality in Chapter 6.

Each separated solution $u_n(r,\theta,t) = e^{-\lambda t}\Theta_n(\theta)R_n(r)$ will satisfy the condition that the temperature on the boundary of the disk is 0 and be physically realistic (remain bounded) if the separation constant λ can be chosen so that $R = R_n(r)$ satisfies

$$-(rR')' + \left(\frac{n^2}{r} - \lambda r\right)R = 0,\ 0 < r < b,$$

$$|R(0)| < \infty,\ R(b) = 0,$$

which is an eigenvalue problem for a Bessel's equation for each $n = 0,\ 1,\ 2 \ldots.$

1.6 Steady State Regimes

If a wave or diffusion phenomenon involves significant external forcing or sources (sinks), the homogeneous wave and heat equations must be modified to include such effects. The resulting equations are the inhomogeneous wave equation,

$$bu_{tt} = \operatorname{div}(p\nabla u) - cu - f,$$

and the inhomogeneous heat equation,

$$bu_t = \operatorname{div}(p\nabla u) - cu - f,$$

where $u = u(x,t)$ and x varies in some domain in Euclidean n-space, t is time, $b > 0$, $p \geq 0$, and $c \geq 0$ are given functions of the spatial variables in a domain of interest, and f describes the external influences. If f is independent of t or $f(x,t) \to f(x)$, a time independent limit, and the boundary conditions are time independent, then it is natural to expect that $u(x,t)$ will converge to a steady-state, that is, time independent, solution $u(x)$ to the wave or heat equation. In either case, the steady-state solution $u = u(x)$ satisfies the differential equation

$$\operatorname{div}(p\nabla u) - cu = f$$

or

$$p\Delta u + \nabla p \cdot \nabla u - cu = f$$

and any prescribed boundary conditions. If p is a constant, the steady-state equation has the form

$$\operatorname{div}(\nabla u) - cu = f$$

or

$$\Delta u - cu = f.$$

If $c = 0$ the steady-state equation becomes

$$\Delta u = f,$$

which is the Laplace equation when $f = 0$ and the Poisson equation when $f \neq 0$.

1.6.1 Heat Conduction in a Rectangular Plate

If the plate with insulated top and bottom has dimensions $0 \leq x \leq l_1$, $0 \leq y \leq l_2$, its horizontal side $0 \leq x \leq l_1$, $y = 0$ is held at temperature $f(x)$, and the other three sides are held at temperature zero, the thermal diffusivity $p = p(x)$ is independent of y, and only diffusion effects are modeled, then the steady-state temperature $u = u(x,y)$ satisfies

$$\begin{cases} (pu_x)_x + (pu_y)_y = 0 & \text{for } 0 < x < l_1,\ 0 < y < l_2, \\ u(x,0) = f(x) & \text{for } 0 \leq x \leq l_1, \\ u = 0 & \text{on the other sides.} \end{cases}$$

The initial data $f(x)$ must satisfy the compatibility conditions $f(0) = f(l_1) = 0$ because the temperature on the boundary must be continuous.

Separated solutions $u(x,y) = X(x)Y(y)$ that satisfy the differential equation and the homogeneous boundary conditions lead to the eigenvalue problem

$$(p(x)X')' + \lambda p(x)X = 0,\ X(0) = X(l_1) = 0$$

and the companion problem

$$Y'' - \lambda Y = 0,\ Y(l_2) = 0.$$

The companion problem has for solutions multiples of $Y(y) = \sinh\sqrt{\lambda}(l_2 - y)$. So once the eigenvalues λ_n and eigenfunctions $X_n(x)$ are determined the separated solutions are constant multiples of

$$X_n(x) \sinh\sqrt{\lambda_n}(l_2 - y).$$

1.6.2 Heat Conduction in a Circular Plate

If the heat conducting body with insulated top and bottom is a circular disk with radius b, it is natural to use polar coordinates r and θ with origin at the center of the disk. Assume the thermal conductivity is independent of θ so that $p = p(r)$ and the circular boundary of the disk is held at temperature $f(\theta)$. In this situation, the steady-state temperature $u = u(r, \theta)$ satisfies

$$\begin{cases} p(r)\left(u_{rr} + \dfrac{1}{r}u_r + \dfrac{1}{r^2}u_{\theta\theta}\right) + p'(r)u_r = 0 & \text{for } 0 \le r < b,\ 0 \le \theta \le 2\pi, \\ u(b, \theta) = f(\theta) & \text{for } 0 \le \theta \le 2\pi \end{cases}$$

The initial data $f(\theta)$ must satisfy the compatibility condition $f(0) = f(2\pi)$ because the temperature cannot have two values at the same point.

Separated solutions $u(r, \theta) = R(r)\Theta(\theta)$ that satisfy the partial differential equation must satisfy the ordinary differential equations

$$r^2R'' + r(1 + rp'/p)R' - \lambda R = 0$$

and

$$\Theta'' + \lambda\Theta = 0.$$

Since Θ must be 2π periodic, $\lambda = \lambda_n = n^2$ for $n = 0, 1, 2, \ldots$ and

$$\Theta_n(\theta) = a_n \cos n\theta + b_n \sin n\theta$$

with $a_n^2 + b_n^2 \neq 0$.

The differential equation and "hidden" boundary condition for $R = R_n$ are

$$r^2R'' + r(1 + rp'/p)R' - n^2R = 0,\ |R(0)| < \infty.$$

The R problem reduces to

$$r^2R'' + rR' - n^2R = 0,\ |R(0)| < \infty$$

when the thermal conductivity is constant. This is an Euler equation and its bounded solutions are the constant multiples of

$$R_n(r) = r^n.$$

Euler equations also are obtained if $p(r)$ is proportional to any power of r.

1.7 On Models

The models presented earlier arise in other settings, for example in biological systems and in acoustics. Diffusion of molecules in a fluid at rest satisfies the partial differential equation

$$\frac{\partial C}{\partial t} = \frac{\partial}{\partial x}\left(D\frac{\partial C}{\partial x}\right) + \frac{\partial}{\partial y}\left(D\frac{\partial C}{\partial y}\right) + \frac{\partial}{\partial z}\left(D\frac{\partial C}{\partial z}\right)$$

where $C(x, y, z, t)$ is the concentration of the substance and D is the diffusion coefficient. The equation modeling sound waves in the atmosphere,

$$\frac{\partial^2 p}{\partial t^2} = c^2\left(\frac{\partial^2 p}{\partial x^2} + \frac{\partial^2 p}{\partial y^2} + \frac{\partial^2 p}{\partial z^2}\right),$$

where p is the pressure and c is the speed of sound, is fundamental to acoustics theory. The vibrations of a circular membrane also satisfies an equation of this type.

The fact that the same equations arise over and over in different contexts is one of the fundamental strengths of mathematical modeling: consider

$$\frac{\partial u}{\partial t} = \frac{\partial^2 u}{\partial x^2},$$

a non-dimensionalized partial differential equation. But what does it mean? The meaning depends on the context in which it was derived. One researcher might say that u is a concentration of molecules and the equation describes their diffusion. Another might say u is temperature and the equation describes heat conduction. A third might say u is pressure and the equation describes flow through a porous medium. A fourth might say u is the signal in a fiber optic cable when the leakage to ground is negligible, and so on. The same equation holds in all cases. The time and length scales differ as do the interpretations of the solution, but, after introducing dimensionless coordinates, the partial differential equation is the same.

From the mathematical standpoint this means that the solution techniques developed in one field can be applied in another field in which the approach may not naturally suggest itself. From a practical standpoint, intuition developed from the study of, say, heat conduction can be applied to the study of molecular diffusion, Brownian motion, pressure waves, and so on as long as the underlying partial differential is the same.

1.8 Sturm-Liouville Boundary Value Problems

Return to heat conduction in the laterally insulated rod and the initial boundary value problem (1.10). If we assume that there are distributed heat sources and sinks along the rod, then the heat equation contains a new term $f(x, t)$ that describes the heat generation due to the sources and sinks. The modified problem is

$$\begin{cases} u_t = (p(x)u_x)_x - q(x)u + f(x,t), & 0 < x < l,\ t > 0, \\ u(x,0) = g(x), & 0 \le x \le l, \\ \alpha u(0,t) - \beta u_x(0,t) = 0, & t \ge 0, \\ \gamma u(l,t) + \delta u_x(l,t) = 0 & t \ge 0. \end{cases} \tag{1.11}$$

If $f(x, t)$ is independent of t or tends to a time independent limit, say $f(x)$, as $t \to \infty$, then we expect that the temperature will tend to a time independent limit, $u = u(x)$, that will satisfy the heat equation with $\partial u/\partial t = 0$ and the (time independent) boundary conditions

$$\begin{cases} -(p(x)u')' + q(x)u = f(x), & 0 < x < l, \\ \alpha u(0) - \beta u'(0) = 0,\ \gamma u(l) + \delta u'(l) = 0 \end{cases} \tag{1.12}$$

A problem of this form consisting of a Sturm-Liouville differential equation and certain boundary conditions is called a **Sturm-Liouville boundary value problem**.

We note as a matter of convenience that if we replace $q(x)$ by $q(x) - \lambda r(x)$ in the differential equation, the Sturm-Liouville boundary value problem becomes

$$\begin{cases} -(p(x)u')' + (q(x) - \lambda r(x))u = f(x) & 0 < x < l, \\ \alpha u(0) - \beta u'(0) = 0,\ \gamma u(l) + \delta u'(l) = 0 \end{cases}$$

This problem reduces to the general Sturm-Liouville eigenvalue problem when $f = 0$ and to the general Sturm-Liouville boundary value problem when $\lambda = 0$.

1.9 Calculus of Variations

The calculus of variations is another source of Sturm-Liouville boundary value problems and eigenvalue problems. For example, according to the principle of minimum mechanical energy, a one-dimensional continuum, modeled by a curve $y = y(x)$ that extends from $y(a) = c_a$ to $y(b) = c_b$ assumes the shape that minimizes (more properly, makes stationary) the integral

$$I(y) = \int_a^b \left(\frac{1}{2}p(x)y'^2 + \frac{1}{2}q(x)y^2 - yf(x)\right) dx$$

among all continuously differentiable functions y that satisfy the boundary conditions. In the context of an elastic string suspended between two posts, $p(x)$ is the mass density along the string, $q(x)$ is a coefficient of elasticity, and $f(x)$ is an external force. The first variation of the integral is

$$\delta I(y)(\zeta) = \frac{d}{d\varepsilon} I(y + \varepsilon\zeta)\bigg|_{\varepsilon=0} = \int_a^b (p(x)y\zeta' + q(x)y\zeta - \zeta f(x))\, dx$$

where ζ is any continuously differentiable function satisfying $\zeta(a) = 0$ and $\zeta(b) = 0$. The conditions on ζ guarantee that $y + \varepsilon\zeta$ satisfies the boundary conditions and, hence, determines a potential shape assumed by the continuum. If this derivative is zero, that is, if the first variation $\delta I(y)(\zeta) = 0$ for some function y and all ζ, then y has a continuous second derivative by the Theorem of Du-Bois Reymond and the fundamental lemma of the calculus of variations implies that

$$-(p(x)y')' + q(x)y = f(x).$$

Thus, the problem of "minimizing" $I(y)$ is equivalent to solving the Sturm-Liouville boundary value problem

$$-(p(x)y')' + q(x)y = f(x), \;\; y(a) = c_a, \;\; y(b) = c_b.$$

The problem of finding the eigenvalues and eigenfunctions of the Sturm-Liouville eigenvalue problem

$$-(p(x)y')' + q(x)y = \lambda y, \;\; y(a) = 0, \;\; y(b) = 0$$

is equivalent to "minimizing" the integral

$$J(y) = \int_a^b \left(\frac{1}{2}p(x)y'^2 + \frac{1}{2}q(x)y^2\right) dx$$

over continuously differentiable functions y satisfying $y(a) = 0$ and $y(b) = 0$ subject to the normalizing constraint

$$\int_a^b y^2\, dx = 1.$$

In this case, the eigenvalue arises essentially as a Lagrange multiplier. There must be a constant λ such that the "minimizing" y is also a stationary value of

$$K(y) = \int_a^b \left(\frac{1}{2}p(x)y'^2 + \frac{1}{2}q(x)y^2 - \frac{1}{2}\lambda y^2\right) dx.$$

1.10 Green's Functions

Since the solution to a Sturm-Liouville boundary value problem involves two integrations, it is natural to expect that there is a solution formula that involves integration. There is such a formula when the problem has a unique solution. It expresses the solution to the boundary value problem in terms of a Green's function, or influence function, that is determined by the Sturm-Liouville differential operator and the boundary conditions of the problem, and which has an important physical interpretation.

In this section, we use physical arguments to motivate the existence of a Green's function, to point out some of its important properties, and to find the solution formula. A fully rigorous mathematical treatment of these topics will be given in later chapters.

We carry out the physical reasoning in the context of steady-state heat flow in a rod, as described in (1.11) and (1.12). It is convenient to denote the steady-state temperature by $y(x)$ instead of $u(x)$ as in (1.12) so $y(x)$ satisfies

$$\begin{cases} -(py')' + q(x)y = f(x), & 0 < x < l, \\ \alpha y(0) - \beta y'(0) = 0, \ \gamma y(l) + \delta y'(l) = 0. \end{cases} \tag{1.13}$$

Let $Ly = -(py')' + q(x)y$ so the differential equation in (1.13) is $Ly = f$.

The solution formula for the Sturm-Liouville problem (1.13) is

$$y(x) = \int_0^l g(x, s) f(s)\, ds, \tag{1.14}$$

where $g(x, s)$ is called the **Green's function** or **influence function** for the Sturm-Liouville differential operator Ly and the given boundary conditions. More precisely, $g(x, s)$ is called a Green's function for (1.13) if it is continuous on $0 \le x,\ s \le l$ and *uniquely* solves the given Sturm-Liouville problem for all continuous right-hand sides $f(x)$.

We argue as follows to understand why the solution formula (1.14) is reasonable: the right member $f(x)$ is the given rate at which heat is generated per unit length per second by sources and sinks along the rod. Let $\varepsilon > 0$ be fixed and $f_s(x)$ specify a unit rate of heating per second concentrated near the point $x = s$ in the rod. That is, $f_s(x)$ is continuous, zero outside the interval $s - \varepsilon < x < s + \varepsilon$, and

$$\int_0^l f_s(x)\, dx = 1.$$

Let $y = g_\varepsilon(x, s)$ be the steady-state temperature in the rod produced by the input $f_s(x)$. Analytically, $g_\varepsilon(x, s)$ is the solution to (1.13) when $f = f_s$. It is plausible that, as ε tends to zero, the temperature distribution $g_\varepsilon(x, s)$ will converge to a limiting continuous temperature distribution $g(x, s)$ that corresponds to a heat source of unit intensity at the point s. Thus, $g(x, s)$ is the temperature at x due to a unit heat source applied at location s in the rod.

Now, let $f(x)$ be an arbitrary continuous rate of heat generation per unit length per unit time along the rod. Imagine the rod decomposed into n nonoverlapping segments each of length Δs and centered at s_k. In the kth segment, the heat input from the continuous distribution $f(s)$ is closely approximated by $f(s_k)\Delta s$, with the approximation improving as $\Delta s \to 0$. Consequently, the contribution to the temperature $y(x)$ at point x due to the heating in the kth segment is approximately $g(x, s_k)f(s_k)\Delta s$, and this approximation should improve as $\Delta s \to 0$. Since the differential equation governing heat flow in the rod is linear, the temperature that arises in the rod due to the combined effect of all the inputs $f(s_k)\Delta s$ is

$$\sum_{k=1}^{n} g(x, s_k) f(s_k) \Delta s$$

and this should closely approximate the temperature $y(x)$ in the rod produced by the continuous distribution $f(x)$, with the approximation improving as $\Delta s \to 0$. This suggests that

$$y(x) = \lim_{n\to\infty} \sum_{k=1}^{n} g(x, s_k) f(s_k) \Delta s = \int_0^l g(x, s) f(s)\, ds,$$

which is just the solution formula (1.14).

The foregoing reasoning also leads to several important properties of the Green's function. The solution $g_\varepsilon(x, s)$ to (1.13) with right member $f_s(x)$ satisfies

$$Lg_\varepsilon(x, s) = f_s(x) = 0 \quad \text{for } |x - s| > \varepsilon,$$

where the differential operator L acts on functions of x. Since $g_\varepsilon(x, s)$ converges to $g(x, s)$ as ε tends to zero, this suggests that

$$Lg(x, s) = 0 \quad \text{for } x \neq s,$$

where L acts on functions of x. Furthermore, each solution $g_\varepsilon(x, s)$ satisfies the boundary conditions of the problem so passing to the limit as ε tends to zero it follows that, as a function of x for fixed s,

$$g(x, s) \text{ satisfies the boundary conditions of the problem.}$$

Next, we look at the effect of the infusion of a unit of energy at the point $x = s$. This infusion suggests that some type of singular behavior must occur in $g(x, s)$ when $x = s$. Integrate $Lg_\varepsilon(x, s) = f_s(x)$ from 0 to l to obtain

$$-pg'_\varepsilon(x, s)\Big|_{x=s-\varepsilon}^{x=s+\varepsilon} + \int_{s-\varepsilon}^{s+\varepsilon} q(x) g_\varepsilon(x, s)\, dx = 1,$$

where the prime indicates differentiation with respect to x. As ε tends to 0, the integral tends to zero because the temperature $g_\varepsilon(x, s)$ must remain bounded. Thus,

$$-pg'(x, s)\Big|_{x=s-}^{x=s+} = 1,$$

$$g'(x, s)\Big|_{x=s-}^{x=s+} = -\frac{1}{p(s)}$$

because $p(x)$ is continuous; hence,

$$\frac{\partial g(x, s)}{\partial x}\Bigg|_{x=s-}^{x=s+} = -\frac{1}{p(s)};$$

that is, the derivative of the Green's function with respect to x is discontinuous at $x = s$ and has a jump of $-1/p(s)$ there. We will show later that the Green's function is characterized by the foregoing properties.

It is informative to think about the solution formula (1.14) in a slightly different way, in terms of inverse processes. The Green's function $g(x, s)$, called a kernel in this context, defines an integral operator G that transforms a continuous function f into another continuous function Gf defined by

$$Gf(x) = \int_0^l g(x,s)f(s)\,ds.$$

Since (1.13) is uniquely solved by

$$y(x) = \int_0^l g(x,s)f(s)\,ds = Gf(x),$$

the Sturm-Liouville differential operator L together with its boundary conditions and integral operator G are related by

$$Ly = f \text{ if and only if } y = Gf.$$

The integral operator G is the inverse of the differential operator L.

Finally, the Green's function enables us to express a Sturm-Liouville eigenvalue problem as an eigenvalue problem for an integral operator: simply replace f by λry in (1.13) and (1.14) to find that the Sturm-Liouville eigenvalue problem $Ly = \lambda ry$ can be expressed as

$$y(x) = \lambda \int_0^l g(x,s)r(s)y(s)\,ds.$$

The eigenvalues λ of (1.13), that is, the values of λ for which (1.13) has a nontrivial solution y, are also called the eigenvalues of the kernel $g(x,s)r(s)$. This conversion to an integral equation eigenvalue problem will be our principal means for studying Sturm-Liouville eigenvalue problems in Chapters 4, 5, and 6.

1.11 The Path Ahead

The theoretical results that stand behind applications of Sturm-Liouville boundary value and eigenvalue problems divide roughly into two threads. The first thread concerns the existence and basic properties of eigenvalues, the orthogonality of the eigenfunctions, and the convergence of eigenfunction expansions. The second thread concerns the oscillatory behavior of the eigenfunctions, the nature of their zeros, and the approximation properties they possess that are very much like those of ordinary polynomials. Both threads are approached most naturally by expressing a Sturm-Liouville eigenvalue problem as a corresponding integral equation eigenvalue problem. Indeed, much of the theory of integral equations was motivated and developed in order to analyze eigenvalues problems that arose in the realm of differential equations. The conversion to an equivalent integral equation is advantageous for two primary reasons – the integral operator defined by the Green's function is better behaved than the differential operator in the original problem and the boundary conditions are built into the Green's function, which avoids the need to deal separately with side conditions defined by additional equations.

Once the theoretical results of both threads are established, questions of practical implementation arise. Effective numerical procedures for the approximation of eigenvalues and eigenfunctions of both regular and singular Sturm-Liouville problems are needed because explicit evaluation of the eigenvalues and eigenfunctions is rarely possible.

We describe the main problems addressed by the two threads and state a few key results in the sections, Thread I and Thread II, that follow. All of the issues raised will be addressed more fully later in the book.

1.11.1 Thread I

The vast majority of eigenvalue problems that come up in practice are self-adjoint. For now it is enough to know that the Green's function $g(x, s)$ associated with a self-adjoint problem with all real-valued data is symmetric: $g(x, s) = g(s, x)$. Sturm-Liouville eigenvalue problems with real-valued data and separated boundary conditions are self-adjoint. (Separated means that each boundary condition only involves the function and its first derivative at one end-point.) Periodic boundary conditions also determine self-adjoint problems. The discussion that follows is restricted to the case of separated boundary conditions because they occur most frequently in applications.

Our overall approach is as follows. An eigenvalue problem will be reduced to Sturm-Liouville form: a differential equation of the form

$$-(p(x)y')' + (q(x) - \lambda r(x))y = 0$$

for $a < x < b$ together with appropriate boundary conditions. In simple cases, the eigenvalues and eigenfunctions can be found explicitly. In general, explicit solutions are not available and to obtain theoretical properties of the eigenvalues and eigenfunctions, we replace the eigenvalue problem by an equivalent integral equation

$$y(x) = \lambda \int_a^b g(x, s)r(s)y(s)\, ds,$$

where $g(x, s)$ is the Green's function corresponding to the Sturm-Liouville differential operator and its associated boundary conditions.

To make clearer the issues to be faced, we introduce them via the diffusion problem (1.10) with $f(x, t) = 0$ (no sources or sinks along the rod), homogeneous Dirichlet boundary conditions, and initial temperature distribution $g(x)$ now relabeled $f(x)$:

$$\begin{cases} u_t = (p(x)u_x)_x - q(x)u, & 0 < x < l,\ t > 0, \\ u(x, 0) = f(x), & 0 \le x \le l, \\ u(0, t) = 0,\ u(l, t) = 0, & t \ge 0. \end{cases} \tag{1.15}$$

When separation of variables is used to seek nontrivial separated solutions $u(x, t) = X(x)T(t)$ that satisfy the diffusion equation and the homogeneous boundary conditions, one is led to the eigenvalue problem

$$\begin{cases} (p(x)X')' - q(x)X + \lambda X = 0, & 0 < x < l, \\ X(0) = 0,\ X(l) = 0; \end{cases} \tag{1.16}$$

and the companion equation $T' + \lambda T = 0$ for the time factor. A key step in separation of variables is to superpose the separated solutions with the aim of satisfying all the remaining conditions in the initial boundary value problem at hand. For this to work, one almost always needs an infinite superposition of the separated solutions. That is, the eigenvalue problem must have an infinite number of eigenvalues and corresponding eigenfunctions. We will establish these properties for general regular and singular Sturm-Liouville eigenvalue problems. In (1.15), as for most problems with physically realistic boundary conditions, the eigenvalues are all real, simple,

$$0 < \lambda_1 < \lambda_2 < \cdots < \lambda_n < \cdots \quad \text{and} \quad \lambda_n \to \infty \text{ as } n \to \infty.$$

The corresponding eigenfunctions $\phi_1(x), \ldots, \phi_n(x), \ldots$ can be chosen real and orthonormal,

$$\int_0^l \phi_n(x)\phi_m(x)\,dx = \delta_{nm}$$

where δ_{nm} is the Kronecker delta with value 1 if $n=m$ and value 0 otherwise. For (1.15), the corresponding functions $T(t)$ in the separated solutions are multiples of $T_n(t) = e^{-\lambda_n t}$.

Since any (finite) linear combination of the separated solutions will satisfy the diffusion equation and the homogeneous boundary conditions, it is reasonable to expect that an infinite superposition

$$u(x,t) = \sum_{n=1}^{\infty} \alpha_n e_n^{-\lambda_n t}\phi_n(x)$$

will too. This is true if the infinite series is suitably convergent. This is another of the issues we must face. Finally we want the series to satisfy any remaining conditions imposed by the model. Here we want

$$u(x,0) = \sum_{n=1}^{\infty} \alpha_n\phi_n(x) = f(x).$$

That is we need to know that any reasonable function $f(x)$ can be represented by an eigenfunction expansion. So two more issues emerge. What do we mean by a reasonable function? In what sense does the series converge?

The questions raised above are addressed in the Hilbert-Schmidt theorem and its corollaries, which are among the principal results of Chapter 3. Applications to Sturm-Liouville problems are given in Chapters 4, 5, and 6.

1.11.2 Thread II

We continue to assume that the Sturm-Liouville eigenvalue problem has separated boundary conditions, just as in Thread I. In this case, each eigenvalue has a uniquely determined corresponding eigenfunction up to nonzero constant multiples and the eigenvalues can be listed as

$$\lambda_0 < \lambda_1 < \cdots < \lambda_n < \cdots$$

with $\lambda_n \to \infty$ as $n \to \infty$. The corresponding eigenfunctions are denoted by

$$\phi_0(x), \phi_1(x), \phi_2(x), \ldots, \phi_n(x), \ldots$$

and are continuous and orthogonal in the underlying interval, say $0 \le x \le 1$. The eigenfunctions $\phi_0(x), \phi_1(x), \phi_2(x), \ldots, \phi_n(x)$ have oscillatory and approximation properties analogous to those possessed by ordinary polynomials of degree n. Several of these properties are listed later in this section and established in later chapters. A unified approach to the properties we have in mind began with O. D. Kellogg in 1916–1918, [26] and [27], when he introduced what are now called Kellogg kernels. He wanted to determine what properties of the Green's function of a Sturm-Liouville eigenvalue problem, or more generally of a real-valued symmetric kernel, would imply all the familiar oscillatory properties of the eigenfunctions. Kellogg discovered the properties from a purely mathematical perspective. Later, beginning in the mid 1930s, Gantmacher and Krein [16] significantly extended Kellogg's pioneering work and added an important physical perspective: a few simple physical properties of an elastic continuum imply that its influence function must be a Kellogg kernel and, hence, have the properties Kellogg discovered. The investigations of Gantmacher and Krein also extended Kellogg's results to include certain nonsymmetric kernels. See Pincus

[32] for a much deeper analysis of the contributions of Kellogg and of Gantmacher and Krein than is given here. Just as in Thread I, the unified treatment we will give later of the oscillatory and approximation properties of Sturm-Liouville eigenvalue problems is made possible by converting the Sturm-Liouville eigenvalue problem into an equivalent integral equation eigenvalue problem.

Kellogg starts his 1916 paper with an example of three continuous, piecewise linear, orthogonal functions on $[0, 1]$, say $\psi_0(x)$, $\psi_1(x)$, and $\psi_2(x)$ such that $\psi_0(x)$ has no zero in the interval and $\psi_1(x)$ and $\psi_2(x)$ each have exactly one zero in $[0, 1]$ that occurs at an interior point of the interval. His point is that the familiar properties of the orthogonal eigenfunctions of a real, symmetric kernel, say $k(x, s)$, cannot all be a consequence only of their orthogonality. Kellogg goes on to show that the familiar oscillatory and approximation properties of the eigenfunctions $\phi_0(x)$, $\phi_1(x)$, $\phi_2(x), \ldots$ hold if

$$\det\left[\phi_i(x_j)\right]_{n\times n} > 0 \tag{1.17}$$

for all $0 < x_1 < \cdots < x_n < 1$, $i = 0, \ldots, n-1$, and $n = 1, 2, \ldots$. In particular, Kellogg showed that the determinantal inequalities imply:

- Given $n+1$ distinct points in $(0, 1)$ and given $n+1$ values, there is a unique ϕ-polynomial of the form $\sum_{k=0}^{n} a_k\phi_k(x)$ that takes on the given values at the given points.
- If a nonzero ϕ-polynomial $\sum_{k=0}^{n} a_k\phi_k(x)$ vanishes at n distinct points, then it changes sign at those points.
- $\phi_n(x)$ has exactly n zeros in $(0, 1)$ and changes its sign at each of these zeros.
- The zeros of $\phi_{n-1}(x)$ and $\phi_n(x)$ in $(0, 1)$ strictly interlace.
- For $m \le n$, a nonzero ϕ-polynomial $\sum_{k=m}^{n} a_k\phi_k(x)$ changes sign at least m times and at most n times on $(0, 1)$.

Kellogg concluded his 1916 paper by stating that it would be desirable to find conditions on the kernel $k(x, s)$ that imply the inequalities (1.17). He did just that in his 1918 paper. Kellogg's conditions from 1918 are:

$$\begin{aligned} &\text{K1.} \quad \det\left[k(x_i, x_j)\right]_{n\times n} > 0 \quad \text{for } 0 < x_1 < \cdots < x_n < 1, \\ &\text{K2.} \quad \det\left[k(x_i, s_j)\right]_{n\times n} \ge 0 \quad \text{for } \begin{matrix} 0 \le x_1 \le \cdots \le x_n \le 1, \\ 0 \le s_1 \le \cdots \le s_n \le 1, \end{matrix} \end{aligned}$$

for $n = 1, 2, 3, \ldots$ and all choices of $x_1, x_2, \ldots, x_n$ and $s_1, s_2, \ldots, s_n$ that satisfy the given conditions. As noted above, Gantmacher and Krein significantly extended Kellogg's work, establishing the existence of an infinite sequence of positive eigenvalues and corresponding eigenfunctions for nonsymmetric kernels that satisfy Kellogg's conditions and explaining the physical meaning of the Kellogg conditions. Reference [16], which concentrates on the symmetric case for ease of exposition, gives a rich account of the interplay between the Kellogg conditions and the oscillatory behavior of discrete and continuous mechanical systems. See Appendix C for a physical motivation of the Kellogg conditions.

1.11.3 Finding Eigenvalues and Eigenfunctions

We continue in the context of the initial boundary value problem (1.15) and the corresponding eigenvalue problem (1.16)

$$\begin{cases} (p(x)y')' - q(x)y + \lambda y = 0, \quad 0 < x < l, \\ y(0) = 0, \ y(l) = 0, \end{cases}$$

where $X(x)$ has been replaced by $y(x)$ for convenience. The eigenvalues and eigenfunctions for this problem can only be found explicitly for simple choices of $p(x)$ and $q(x)$. So we must face a basic question. How can we actually construct the eigenvalues λ_n and the eigenfunctions $\phi_n(x)$, both theoretically and numerically?

We will use the following procedure to address both issues. Although the basic idea we are about to describe is not new, the proofs needed to justify it for both regular and singular Sturm-Liouville problems are new, as far as we know. The basic idea is this: To solve the eigenvalue problem

$$\begin{cases} (p(x)y')' - q(x)y + \lambda y = 0, & 0 < x < l, \\ y(0) = 0, \ y(l) = 0 \end{cases}$$

consider the initial value problem

$$\begin{cases} (p(x)u')' - q(x)u + \lambda u = 0, \\ u(0) = 0, \ u'(0) = 1 \end{cases}$$

and denote its solution by $u(x, \lambda)$ for $0 \le x \le l$. The solution $u(x, \lambda)$ will be an eigenfunction of the eigenvalue problem and λ will be its corresponding eigenvalue if

$$u(l, \lambda) = 0.$$

So more issues arise that we must face. How do we establish theoretically the global solvability of the initial value problem? How do we know that $u(l, \lambda) = 0$ has an infinite number of solutions? And, once these questions are answered, how do we compute accurate numerical approximations of the eigenvalues and eigenfunctions? The answers to these questions involve basic existence, uniqueness, and continuous dependence results for ordinary differential equations, the Newton-Raphson method, and initial value problems solvers for ordinary differential equations. The existence, uniqueness, and continuous dependence results needed are standard for regular Sturm-Liouville problems when p' is continuous. They are either new or not well known for singular Sturm-Liouville problems or when p is merely continuous for regular problems. All of these matters are addressed in Chapters 4, 5, 6, and 7.

1.12 Intrinsic Interest of Eigenvalues

Eigenvalues are often of interest in their own right. In the discussion of the heat conduction model (1.10) we noted that the eigenvalues and eigenfunctions for the basic diffusion or heat conduction model, $u_t = au_{xx}$ with $a > 0$ a constant and with Dirichlet boundary conditions, are $\lambda_n = (n\pi/l)^2$ and $X_n(x) = \sin n\pi x/l$. For this problem, the series for the temperature $u(x, t)$ given in Section 1.11.1 is

$$u(x, t) = \sum_{n=1}^{\infty} c_n \exp(-a\lambda_n t) \sin \sqrt{\lambda_n}/l.$$

The first term in the expansion of the solution contains the factor $\exp(-a\lambda_1 t)$ that determines the overall rate at which the solution decays in time. In particular, it determines how soon the transient effects due to the initial conditions can be neglected and when a steady state is reached, or in the case of forcing, how soon only the forcing terms have an appreciable effect on the solution.

In the case of acoustics, or vibration problems, $c\sqrt{\lambda_n}$, where c is the speed of propagation of the disturbance, gives the frequency of the vibrations. In the case of a piano string, those values are basis for tuning the piano. The frequency $c\sqrt{\lambda_1}$ is called the fundamental tone and the ratio, $c\sqrt{\lambda_n}/c\sqrt{\lambda_1} = n$ is an integer, a fact discovered empirically by the Pythagoreans.

In studying chain nuclear reactions, one is led to an equation of the form $u_t = a\Delta u + ku$, where u represents the number of neutrons per unit volume. Separation of variables with $u = T(t)v(x)$ leads to the series solution $u = \sum_{n=1}^{\infty} \exp\big((k - a\lambda_n)t\big) v_n(x)$ where λ_n and $v_n(x)$ are the eigenvalues and eigenfunctions for $-\Delta v$ together with appropriate boundary conditions. The eigenvalues depend upon the geometry of the container. If k is greater than $a\lambda_n$ for some value of n, the result is a reaction out of control, if $a\lambda_n$ is greater than k for all n, the reaction damps out, and if $a\lambda_1 = k$, the reaction is critical and one has a controlled reaction which can be used to generate electric power.

In quantum mechanics, the eigenvalues of the Schrödinger equation yield the energy levels of, say, electrons (see [43] or [45]).

Eigenvalues occur in many other contexts. We have seen a simple example in the case of Euler buckling, but they arise also in more general buckling problems. They arise in mathematical biology, in particular in the study of populations.

In applied problems, it is often the case that only the first eigenvalue is of critical interest because it determines how the system will behave for large values of the time. In such instances, the first eigenvalue and eigenfunction or a small number of eigenvalues and eigenfunctions can be used to accurately represent the solution as it evolves in time.

1.13 Real Versus Complex Solutions

There are occasions when it is desirable or necessary to deal with complex-valued solutions to differential equations. Most applied problems lead to differential equations and side conditions that involve only real data and the solutions of interest are real-valued or must be real-valued. The catalog of problems just discussed are typical examples.

Perhaps the most important situation involving Sturm-Liouville problems and where complex-valued functions enter is when separation of variables is used in the Schrödinger equation

$$i\hbar\Psi_t = -\frac{\hbar^2}{2m}\Delta\Psi + V(x,t)\Psi,$$

where i is the imaginary unit, $\hbar$ is Planck's constant divided by 2π, Ψ is a wave function, and V is a real-valued potential energy function. Complex-valued solutions must be considered in any situation in which the differential equation or side conditions involve complex-valued data.

In the chapters that follow we often allow the coefficients in a differential equation to be complex-valued and likewise any constants in the problem may be complex numbers. The results obtained about solutions and their properties apply to any complex-valued or real-valued solutions that may exist. Frequently, if the equations and data involve only real quantities, it is natural to expect that the solutions must be real-valued. We establish such results for initial value problems, for boundary value problems, and for Green's functions in sufficient generality to cover scenarios in typical applications. Corresponding results are established for eigenfunctions of eigenvalue problems whose eigenvalues are known to be real.

A final observation is in order. If a problem is expressed in terms of linear differential and linear boundary condition equations any of which can be either homogeneous or inhomogeneous but involve only real-valued data, then the real part of any complex-valued solution of the problem is a real-valued solution and the imaginary part is a real-valued solution of the corresponding homogeneous problem.

Chapter 2

Preliminaries

We collect together in this chapter background results from calculus, analysis, and linear algebra that play a prominent role later, as we study Sturm-Liouville boundary value and eigenvalue problems. The chapter serves as a convenient reference and avoids the obligation to develop background material in the midst of arguments in later chapters that are focused on differential and integral equations.

All readers should at least skim through the chapter to become familiar with the notation that we use. The notation is standard, for the most part. Readers who are familiar with the topics in the chapter can move on quickly to later chapters, perhaps never needing to refer back. For other readers, we have endeavored to present the material as a focused, readable introduction to essential background results that can be consulted as needed.

We emphasize that although solutions and other functions may sometimes assume complex values, the domains of all solutions and other functions are either sets of real numbers or sets in real n-dimensional space.

2.1 Euclidean Spaces

We use standard notation and denote the real numbers equipped with the usual algebraic operations by $\mathbb{R}$ and the complex numbers equipped with the usual operations by $\mathbb{C}$.

2.1.1 Real Euclidean Spaces

Real n-dimensional Euclidean space is denoted by $\mathbb{R}^n$,

$$\mathbb{R}^n = \{x = (x_1, \ldots, x_n) : x_j \text{ a real number for each } j\}.$$

The elements x of $\mathbb{R}^n$ are called points or more often vectors, when we identify x with the position vector from the origin to the point x. Vectors in $\mathbb{R}^n$ are added and multiplied by scalars (real numbers) componentwise. The norm (length or magnitude) of a vector x is

$$\|x\| = \left(\sum_{j=1}^{n} |x_j|^2 \right)^{1/2}.$$

For any vectors x and y and scalar α, the norm satisfies

$$\begin{aligned} \|x\| &\geq 0 \text{ with equality only if } x = 0, \\ \|\alpha x\| &= |\alpha| \, \|x\|, \\ \|x + y\| &\leq \|x\| + \|y\| \quad \text{(triangle inequality)}. \end{aligned}$$

The usual inner product in $\mathbb{R}^n$ enables us to define angles between vectors x and y in $\mathbb{R}^n$. It is defined by

$$\langle x, y \rangle = \sum_{j=1}^{n} x_j y_j$$

and related to the norm through

$$\|x\| = \sqrt{\langle x, x\rangle}$$

and the Cauchy-Schwarz inequality

$$|\langle x, y\rangle| \leq \|x\|\|y\|.$$

The usual inner product is linear in its first variable and is symmetric: for any vectors x, y, z and scalars α and β

$$\langle \alpha x + \beta y, z\rangle = \alpha\langle x, y\rangle + \beta\langle y, z\rangle,$$
$$\langle x, y\rangle = \langle y, x\rangle.$$

Consequently, the inner product is linear in its second variable as well.

Fix $a < b$. We use the following notation for a simplex in $\mathbb{R}^n$

$$\Delta_n = \{x = (x_1, x_2, \ldots, x_n) \in \mathbb{R}^n : a \leq x_1 \leq x_2 \leq \cdots \leq x_n \leq b\}.$$

So Δ_1 is the closed interval $[a, b]$ on the real line, Δ_2 is the triangle with vertices (a, ca) , (a, b) and (b, b) in the Euclidean plane, Δ_3 is tetrahedron in 3-space. The set of points inside the simplex, its interior, is

$$\overset{\circ}{\Delta}_n = \{x = (x_1, x_2, \ldots, x_n) \in \mathbb{R}^n : a < x_1 < x_2 < \cdots < x_n < b\}.$$

2.1.2 Complex Euclidean Spaces

Apart from simplices, corresponding language and results hold in complex Euclidean spaces. Complex n-dimensional Euclidean space is denoted by $\mathbb{C}^n$,

$$\mathbb{C}^n = \{z = (z_1, \ldots, z_n) : z_j \text{ a complex number for each } j\}.$$

If $n = 1$, each complex number c can be expressed as $c = a + ib$ where a and b are real numbers. The real number a is the real part of c and is denoted by $\mathrm{Re}(c)$. The real number b is the imaginary part of c and is denoted by $\mathrm{Im}(c)$. The complex conjugate of c is $\bar{c} = a - ib$. The absolute value of a complex number is $|c| = \sqrt{a^2 + b^2}$.

The elements z of $\mathbb{C}^n$ are called points or more often vectors, when we identify z with the position vector from the origin to the point z. Vectors in $\mathbb{C}^n$ are added and multiplied by scalars (complex numbers) componentwise. The norm (length or magnitude) of a vector z is

$$\|z\| = \left(\sum_{j=1}^{n} |z_j|^2\right)^{1/2}.$$

For any vectors z and w and scalar α, the norm satisfies

$$\|z\| \geq 0 \text{ with equality only if } z = 0,$$
$$\|\alpha z\| = |\alpha|\|z\|,$$
$$\|z + w\| \leq \|z\| + \|w\| \quad \text{(triangle inequality)}.$$

The usual inner product in $\mathbb{C}^n$ is defined by

$$\langle z, w\rangle = \sum_{j=1}^{n} z_j \bar{w}_j$$

and related to the norm through

$$\|z\| = \sqrt{\langle z, z\rangle}$$

and the Cauchy-Schwarz inequality

$$|\langle z, w\rangle| \le \|z\| \, \|w\|.$$

The usual inner product is linear in its first variable and is complex symmetric: for any vectors z, w, u and scalars a and β

$$\langle \alpha z + \beta w, u\rangle = \alpha\langle z, u\rangle + \beta\langle w, u\rangle,$$

$$\langle z, w\rangle = \overline{\langle w, z\rangle}.$$

Consequently, the inner product in $\mathbb{C}^n$ is **conjugate linear** in its second variable

$$\langle u, \alpha z + \beta w\rangle = \bar{\alpha}\langle u, z\rangle + \bar{\beta}\langle u, w\rangle.$$

2.1.3 Elements of Convergence

A **sequence** $\{x_n\}_{n=1}^{\infty}$ in real or complex Euclidean space is an ordered list of points in that space. Other notations for a sequence are $\{x_n\}$ or simply x_n, with the range of the index n understood from the context. A sequence $\{x_n\}_{n=1}^{\infty}$ **converges** to x, if $\|x_n - x\| \to 0$ as $n \to \infty$; that is, given any $\varepsilon > 0$ there is an integer N, dependent on ε, such that $\|x_n - x\| < \varepsilon$ whenever $n > N$.

A **subsequence** of a sequence $\{x_n\}_{n=1}^{\infty}$ is an ordered sub-list $\{x_{n_k}\}_{k=1}^{\infty}$ where $1 \le n_1 < n_2 < \cdots$.

If a sequence x_n in a Euclidean space converges, say with limit x, then given $\varepsilon > 0$ there is an integer N such that $\|x_n - x\| < \varepsilon/2$ whenever $n > N$. Consequently,

$$m, n \ge N \text{ implies that } \|x_m - x_n\| \le \|x_m - x\| + \|x - x_n\| < \varepsilon.$$

A sequence with this property is called a **Cauchy sequence**. So convergent sequences in Euclidean spaces are Cauchy (sequences). The converse is true: every Cauchy sequence in a Euclidean space converges. This so-called **Cauchy criterion** for convergence gives a means of establishing convergence in Euclidean spaces without knowing the limit in advance.

2.1.4 Upper Bounds and Sups

Let X be a nonempty set of real numbers. A real number b is an **upper bound** for X if $x \le b$ for all x in X, in which case X is said to be **bounded above**. A real number l is the **least upper bound** of X, also called the **supremum** of X, if l is an upper bound for X and $l \le b$ whenever b is an upper bound for X. The supremum of X is usually denoted by sup X.

A fundamental property of the real number system, which is equivalent to the Cauchy criterion in $\mathbb{R}$, is:

every nonempty set in $\mathbb{R}$ that is bounded above has a supremum.

We will need the following results that follow directly for the definition of a supremum.

Lemma 1 *Let X be a nonempty set of real numbers that is bounded above. A real number l is the supremum of X if and only if l is an upper bound for X and given any $\varepsilon > 0$ there is an element x in X such that $l - \varepsilon < x \le l$.*

Corollary 2 *If $l = \sup X$ then there is a sequence x_n of elements in X that converge to l.*

The terms lower bound, bounded below, and greatest lower bound (or infimum) are defined by reversing the inequalities in the earlier definitions.

2.1.5 Closed and Compact Sets

There are several equivalent ways to define closed sets and compact sets in Euclidean spaces. We prefer sequential definitions. A set S in $\mathbb{R}^n$ or $\mathbb{C}^n$ is **closed** if whenever a sequence $\{x_n\}$ with elements in S converges its limit lies in S. A set S in $\mathbb{R}^n$ or $\mathbb{C}^n$ is **compact** if every sequence of elements in S contains a convergent subsequence whose limit lies in S.

Theorem 3 (*Heine-Borel*) *A set S in $\mathbb{R}^n$ or $\mathbb{C}^n$ is compact if and only if it is closed and bounded.*

A proof can be found in any advanced calculus book.

2.2 Calculus and Analysis

Sturm-Liouville problems involve analyzing solutions $y(x)$ of a differential equation of the form

$$-(p(x)y'(x))' + q(x)y(x) = f(x) \quad \text{for } a < x < b,$$

where $y(x)$ is subject to certain boundary conditions at $x = a$ and $x = b$ or is subject to initial conditions. Since $y(x)$ is required to satisfy the differential equation, it is subject to certain smoothness (differentiability) requirements on $a < x < b$. It is important for applications to know if, and to what extent, that smoothness extends to the behavior of $y(x)$ at $x = a$ and $x = b$. To answer this question among others, we will make use of standard concepts and results from advanced calculus on continuity, differentiability, and integrability. The most important of them are stated here for convenient reference. Occasionally, a proof is given, especially for less familiar results. Missing arguments can be found in almost any book on advanced calculus. Reliable references include [15], [29], [30], [34] and [39].

2.2.1 Continuity

A real or complex-valued function f defined on a set S in Euclidean space is continuous at x_0 in S if given any $\varepsilon > 0$ there is a $\delta > 0$, dependent on x_0 and on ε, such that $|f(x) - f(x_0)| < \varepsilon$ whenever $x \in S$ satisfies $|x - x_0| < \delta$. A function f is **continuous** on S if it is continuous at every point of S. A function f is **uniformly continuous** on S if given any $\varepsilon > 0$ there is a $\delta > 0$, dependent only on ε, such that $|f(x) - f(x')| < \varepsilon$ whenever x, $x' \in S$ satisfy $|x - x'| < \delta$. Uniform continuity on a set S means that there is a single $\delta > 0$ in the definition of continuity of f at x_0 that works simultaneously for all x_0 in S.

Theorem 4 (Maximum Minimum Value Theorem) *A real-valued continuous function defined on a closed and bounded set in finite dimensional Euclidean space assumes its maximum and minimum values at points in the set.*

Theorem 5 *A real or complex-valued continuous function defined on a closed and bounded set in finite dimensional Euclidean space is uniformly continuous on the set.*

Theorem 6 (Intermediate Value Theorem) *If $f(x)$ is a real-valued continuous function on an interval $[a, b]$ and C is a value strictly between $A = f(a)$ and $B = f(b)$, then there is a point c with $a < c < b$ such that $f(c) = C$.*

Proposition 7 *If a function $f(x)$ with values in a Euclidean space is uniformly continuous on an open interval (a, b), then $f(x)$ has a unique extension by continuity to a continuous function on the closed interval $[a, b]$.*

Proof. Let $\varepsilon > 0$. Since $f(x)$ is uniformly continuous on (a, b), there is a $\delta > 0$ such that $|f(x) - f(x')| < \varepsilon$ when x and x' in (a, b) satisfy $|x - x'| < \delta$. Let x_n be a sequence in (a, b) with $x_n \to a$ as $n \to \infty$. Given $\delta/2$ there is an index N such that $n > N$ implies that $|x_n - a| < \delta/2$. Consequently, m, $n > N$ implies that $|x_m - x_n| < \delta$ and $|f(x_m) - f(x_n)| < \varepsilon$. Thus, $f(x_n)$ is a Cauchy sequence and, hence, converges, say to A. Let $m \to \infty$ to obtain

$$|A - f(x_n)| \leq \varepsilon$$

for $n > N$. Define $f(a) = A$. The extended function $f(x)$ is continuous at $x = a$. Indeed, fix $n > N$. Then for $|x - a| < \delta/2$,

$$|x - x_n| \leq |x - a| + |a - x_n| < \delta$$

and

$$|f(x) - f(a)| \leq |f(x) - f(x_n)| + |f(x_n) - A| \leq \varepsilon + \varepsilon = 2\varepsilon,$$

which establishes the continuity of $f(x)$ at $x = a$. The continuous extension to $x = b$ is done in the same way. ∎

Corollary 8 *If $f(x)$ is defined on an open interval (a, b) with values in a Euclidean space and $f'(x)$ is bounded on (a, b), then $f(x)$ has a unique extension by continuity to the closed interval $[a, b]$.*

Proof. If $f(x) = (f_1(x), f_2(x), \ldots, f_n(x))$, then each component function $f_j(x)$ is differentiable on (a, b) and has a bounded derivative there, say $|f_j'(x)| < M_j$ for x in (a, b). By the mean value theorem for derivatives there is a point c_j in (a, b) such that,

$$|f_j(x) - f_j(x')| = |f_j'(c_j)(x - x')| \leq M_j|x - x'|$$

for x and x' in (a, b). It follows that each $f_j(x)$ is uniformly continuous on (a, b); hence, each $f_j(x)$ and $f(x)$ has a unique extension by continuity to $[a, b]$. ∎

The next result seems obvious on geometric grounds. An elementary proof uses the maximum minimum value theorem and involves several nearly identical cases. We omit the details; see [1].

Theorem 9 *If a real-valued continuous function defined on an interval of any type is not strictly increasing or strictly decreasing on the interval, then the function assumes either a local maximum or a local minimum at an interior point of the interval.*

2.2.2 Differential Calculus

When we say a function is differentiable on an interval (of any type, open, closed, half-open) we mean it has an ordinary two-sided derivative at each point in the interval that is not an endpoint of the interval and has the appropriate one-sided derivative at an endpoint of the interval that belongs to the interval.

Theorem 10 (Mean Value Theorem for Derivatives) *If a function $y(x)$ is continuous on $a \leq x \leq b$ and differentiable on $a < x < b$, then there is a point ξ (strictly) between a and b such that $y(b) - y(a) = y'(\xi)(b - a)$.*

An important observation about functions that satisfy a differential equation and boundary conditions follows from the mean value theorem.

Lemma 11 *If a function $y(x)$ is continuous on $a \le x < b$, differentiable on $a < x < b$, and $\lim_{x\to a} y'(x)$ exists, then $y(x)$ is differentiable on $a \le x < b$ and its derivative is continuous at $x = a$. (The corresponding result holds with the roles of the endpoints interchanged.)*

Proof. By the mean value theorem

$$\frac{y(x) - y(a)}{x - a} = y'(\xi_x)$$

for some ξ_x between a and x. Hence, there exists

$$y'(a) = \lim_{x\to a} \frac{y(x) - y(a)}{x - a} = \lim_{\xi_x \to a} y'(\xi_x) = \lim_{x\to a} y'(x)$$

and y' is continuous at $x = a$. ■

Limits involving indeterminate forms often can be evaluated by an appropriate form of l'Hôpital's rule.

Theorem 12 *Let c be a real number or $\pm\infty$ and $\lim$ stand for one of $\lim_{x\to c}$, $\lim_{x\to c+}$, or $\lim_{x\to c-}$. If either*

(i) $\lim f(x) = 0$ and $\lim g(x) = 0$ or
(ii) $\lim g(x) = \pm\infty$, then

$$\lim \frac{f(x)}{g(x)} = \lim \frac{f'(x)}{g'(x)}$$

provided the limit on the right exists, finite or infinite.

It is implicit in the statement of l'Hôpital's rule that f and g satisfy the minimal hypothesis for the statements to make sense. For example, if lim means $\lim_{x\to c+}$, then $c < \infty$ or $c = -\infty$ and there is an open interval (c, d) in the domain of both f and g and $g' \neq 0$ there because the limit on the right is assumed to exist.

A much simpler form of l'Hôpital's rule suffices for our purposes: if c is real, $f(c) = 0$, $g(c) = 0$, $f'(c)$ exists, $g'(c)$ exists, and $g'(c) \neq 0$, then

$$\lim \frac{f(x)}{g(x)} = \frac{f'(c)}{g'(c)}.$$

The proof of this simple form of the rule follows immediately from

$$\frac{f(x)}{g(x)} = \frac{(f(x) - f(a))/(x - a)}{(g(x) - g(a))/(x - a)}$$

and a limit passage.

2.2.3 Integral Calculus

Theorem 13 (Fundamental Theorem of Calculus I) *If $f(x)$ is continuous on $[a, b]$ and c is a point in $[a, b]$, then the following integral is differentiable at each x in $[a, b]$ and has the indicated derivative*

$$\frac{d}{dx} \int_c^x f(s)\, ds = f(x).$$

We will often use Theorem 13 as follows: with f as in the theorem, there exists

$$\lim_{x\to c}\frac{1}{x-c}\int_c^x f(s)\,ds = f(c)$$

directly from the definition of a derivative because

$$\frac{1}{x-c}\int_c^x f(s)\,ds = \frac{1}{x-c}\left(\int_c^x f(s)\,ds - \int_c^c f(s)\,ds\right).$$

Alternatively, the existence and value of the limit can be established by using l'Hôpital's rule.

Theorem 14 (Fundamental Theorem of Calculus II) *If $f'(x)$ is Riemann integrable on $[a, b]$, then*

$$\int_a^b f'(x)\,dx = f(b) - f(a).$$

In particular the equality holds if $f'(x)$ is continuous on $[a, b]$.

Proof. Let $a = x_0 < x_1 < \cdots < x_n = b$ be a partition of $[a, b]$ such that

$$\max_{1\le j\le n}\left(x_j - x_{j-1}\right) \to 0$$

as $n \to \infty$ and

$$R_n = \sum_{j=1}^{n} f'(\xi_j)(x_j - x_{j-1})$$

be the Riemann sum determined by the partition and the mean value theorem for derivatives through

$$f(x_j) - f(x_{j-1}) = f'(\xi_j)(x_j - x_{j-1})$$

for some ξ_j with $x_{j-1} < \xi_j < x_j$. Then

$$\int_a^b f'(s)\,ds = \lim_{n\to\infty} R_n = \lim_{n\to\infty}(f(b) - f(a)) = f(b) - f(a).$$

■

Theorem 15 (Mean Value Theorem for Integrals) *If $f(x)$ is continuous on $[a, b]$ and $p(x)$ is continuous and nonnegative there, then*

$$\int_a^b f(x)p(x)\,dx = f(\xi)\int_a^b p(x)\,dx$$

for some ξ between a and b.

The reader may wish to consult the next section where uniform convergence is defined.

Theorem 16 *If $f_n(x)$ and $f(x)$ are Riemann integrable on $[a, b]$ and if f_n converges uniformly to f on $[a, b]$, then*

$$\lim_{n\to\infty}\int_a^b |f_n(s) - f(s)|\,ds = 0.$$

Consequently,

$$\lim_{n\to\infty} \int_a^b f_n(s)\, ds = \int_a^b f(s)\, ds$$

and, for any c in $[a, b]$, $F_n(x) = \int_c^x f_n(s)\, ds$ converges uniformly on $[a, b]$ to $F(x) = \int_c^x f(s)\, ds$.

If $f(s)$ is Riemann integrable on $[c, b]$ for all c with $a < c < b$, is not Riemann integrable on $[a, b]$, and

$$\lim_{c\to a} \int_c^b f(s)\, ds$$

exists, finite or infinite, then

$$\int_a^b f(s)\, ds = \lim_{c\to a} \int_c^b f(s)\, ds$$

is called an **improper (Riemann) integral**. If the limit is finite, the improper integral is called **convergent** or is said to converge. Otherwise, it diverges to $+\infty$ or $-\infty$, as the case may be. If $\int_c^b f(s)\, ds$ does not have a limit as $c \to a$, no value is assigned to the improper integral. This language is consistent with that used for infinite series.

The following version of the fundamental theorem of calculus deserves mention: if the improper integral $\int_a^x f(s)\, ds$ converges for $x > a$ and if f is continuous on $a < x \leq b$, then

$$\frac{d}{dx} \int_a^x f(s)\, ds = f(x) \quad \text{for } a < x \leq b.$$

To verify this simply observe that given any x in $a < x \leq b$ there is an x_1 with $a < x_1 < x$ such that $\int_a^{x_1} f(s)\, ds$ is a convergent improper integral and

$$\int_a^x f(s)\, ds = \int_a^{x_1} f(s)\, ds + \int_{x_1}^x f(s)\, ds.$$

The desired conclusion follows from the usual fundamental theorem of calculus.

The following convergent improper integrals will be important later when we study singular Sturm-Liouville problems:

$$\int_0^1 |\ln s|\, ds = \lim_{c\to 0} \int_c^1 -\ln s\, ds = -\lim_{c\to 0} [s \ln s - s]_c^1 = 1$$

because $c \ln c - c \to 0$ as $c \to 0$. Likewise, since $s\ln s - s$ is an antiderivative of $\ln s$ or equivalently $d(s \ln s - s) = \ln s\, ds$ integration by parts gives

$$\begin{aligned}\int_0^1 |\ln s|^2\, ds &= \lim_{c\to 0} \int_c^1 \ln s \ln s\, ds = \lim_{c\to 0} \int_c^1 \ln s\, d(s \ln s - s)\\ &= \lim_{c\to 0}\left([\ln s\,(s \ln s - s)]_c^1 - \int_c^1 (\ln s - 1)\, ds\right) = 2\end{aligned}$$

because $\ln c\,(c \ln c - c) \to 0$ as $c \to 0$. Continuing in this way establishes that

$$I_p = \int_0^1 |\ln s|^p\, ds$$

is a convergent improper integral for $p = 1, 2, 3, \ldots$ and that $I_p = p!$ for $p \geq 1$. The fact that $(\ln c)^{p-1}(c \ln c - c) \to 0$ as $c \to 0$ can be established by applying l'Hôpital's rule to each ratio in $\big((\ln c)^{p-1}/c^{-1/2}\big)((\ln c - 1)/c^{-1/2})$.

It follows, by a change of variable, that

$$\int_a^{a+1} |\ln(s-a)|^p\, ds \quad \text{and, hence,} \quad \int_a^b |\ln(s-a)|^p\, ds$$

converges for $p \geq 1$ and any finite limits with $b > a$.

The following comparison test, the analogue of the basic comparison test for infinite series, is often useful.

Proposition 17 *If* $|f(s)| \leq g(s)$ *for* $a < s \leq b$, *both* f *and* g *are continuous on* $[c, b]$ *for any* $c > a$ *in* $[a, b]$, *and* $\int_a^b g(s)\, ds$ *converges, then* $\int_a^b |f(s)|\, ds$ *and* $\int_a^b f(s)\, ds$ *both converge.*

For example, if $b > a$ and $f(s)$ is a continuous function on $a < s \leq b$ and

$$|f(s)| \leq A|\ln(s-a)|^p + B$$

for $a < s \leq b$ and some constants A, B, and $p \geq 1$, then both improper Riemann integrals $\int_a^b f(s)\, ds$ and $\int_a^b |f(s)|\, ds$ converge.

Integrals involving $\ln(\max(x, s) - a)$ for (x, s) in $[a, b] \times [a, b] \backslash \{(a, a)\}$ occur in our treatment of singular Sturm-Liouville problems in Chapter 5. The following results will be needed there. A quick glance at the graph of $|\ln(t-a)|$ for $a < t < \infty$ helps to confirm the following observation,

$$|\ln(\max(x, s) - a)| \leq |\ln(s-a)| + |\ln(b-a)| \tag{2.1}$$

for all (x, s) in $[a, b] \times [a, b]$ with $s > a$. Indeed, if $\max(x, s) = s$ the inequality is clear. If $\max(x, s) = x$, then $x > a$ and there are two cases to consider. If $a < x \leq a + 1$, then $a \leq s \leq x \leq a + 1$, $|\ln(s-a)| \geq |\ln(x-a)|$, and

$$|\ln(\max(x, s) - a)| = |\ln(x-a)| \leq |\ln(s-a)|.$$

If $\max(x, s) = x$ and $a + 1 < x \leq b$, then $|\ln(b-a)| \geq |\ln(x-a)|$, and

$$|\ln(\max(x, s) - a)| = |\ln(x-a)| \leq |\ln(s-a)| + |\ln(b-a)|$$

as asserted.

For t, u, and p positive,

$$(t+u)^p \leq (2\max(t, u))^p \leq 2^p \max(t^p, u^p) \leq 2^p(t^p + u^p).$$

Consequently, (2.1) and the basic comparison test for improper integrals implies that for (x, s) in $[a, b] \times [a, b]$ with $s > a$,

$$|\ln(\max(x, s) - a)|^p \leq 2^p|\ln(s-a)|^p + 2^p|\ln(b-a)|^p$$

and

$$\int_a^b |\ln(\max(x, s) - a)|^p\, ds \leq 2^p \int_a^b (|\ln(s-a)|^p + |\ln(b-a)|^p)\, ds = M_p < \infty,$$

where M_p is a constant independent of x in $[a, b]$. Another application of the basic comparison test implies that

$$\int_a^b h(x, s)\big(\ln(\max(x, s) - a)\big)^p \, ds \quad \text{and} \quad \int_a^b |h(x, s)||\ln(\max(x, s) - a)|^p \, ds$$

both converge for any function $h(x, s)$ that is continuous on $[a, b] \times [a, b]$. Moreover,

$$\int_a^b |h(x, s)||\ln(\max(x, s) - a)|^p \, ds \le HM_p < \infty \tag{2.2}$$

where $H = \max_{a \le x, s \le b} |h(x, s)|$.

Proposition 18 *Let S be a closed bounded subset in $\mathbb{R}^m$ whose m-dimensional volume $\int_S 1\, ds = |S|$ exists. If $k(x, s)$ is continuous on $X \times S$, where X is a closed bounded set in $\mathbb{R}^n$, and $k(x, s)$ is integrable over S for each x in X, then $\int_S k(x, s)\, ds$ is a continuous function for x in X. (Here ds is shorthand for $ds_1\, ds_2 \cdots ds_m$.)*

Proof. Since S is closed and bounded, $|S| < \infty$. Since (x, s) varies in a closed, bounded set $X \times S$ in $\mathbb{R}^{n+m}$, $k(x, s)$ is uniformly continuous there. So given $\varepsilon > 0$, there is a $\delta > 0$ such that

$$|k(x, s) - k(x_0, s)| < \varepsilon$$

for $|x - x_0| < \delta$ with x and x_0 in X and all s in S. Consequently,

$$\left| \int_S k(x, s)\, ds - \int_S k(x_0, s)\, ds \right| \le \int_S |k(x, s) - k(x_0, s)|\, ds \le \varepsilon |S|.$$

Since $\varepsilon|S|$ can be made arbitrarily small, it follows that $\int_S k(x, s)\, ds$ is continuous on X. ■

Proposition 19 *If $f(x)$ is a nonnegative continuous function on $[a, b]$ and $\int_a^b f(x)\, dx = 0$, then $f(x) = 0$ for all x in $[a, b]$.*

Proof. Suppose $f(c) > 0$ for some c with $a < c < b$. By continuity of f there is a $\delta > 0$ such that $f(x) > f(c)/2$ for $a < c - \delta < x < c + \delta < b$ and

$$\int_a^b f(x)\, dx \ge \int_{c-\delta}^{c+\delta} f(x)\, dx \ge \frac{f(c)}{2}(2\delta) > 0.$$

This contradiction proves that $f(x) = 0$ for $a < x < b$ and by continuity it also is 0 at $x = a$ and $x = b$. ■

Corollary 20 *If $f(x)$ is a continuous real or complex-valued function on $[a, b]$ and $\int_a^b f(x) g(x)\, dx = 0$ for all continuous functions $g(x)$ on $[a, b]$, then $f(x) = 0$ for all x in $[a, b]$.*

Proof. Let $g(x) = \overline{f(x)}$ and apply the proposition. ■

Versions of Proposition 19 and Corollary 20, whose proof are virtually the same, hold for improper integrals: if $f(x)$ is nonnegative and continuous on $a < x \le b$ and the improper integral $\int_a^b f(x)\, dx = 0$, then $f(x) = 0$ on $a < x \le b$. The corresponding corollary asserts that $f(x) = 0$ on $a < x \le b$. The extended versions are used to establish uniqueness of Green's functions for singular Sturm-Liouville boundary value problems.

Corollary 21 *If $k(x, s)$ is a continuous function on $[a, b] \times [a, b]$ and*

$$\int_a^b \int_a^b k(x, s) g(x) h(s)\, dx ds = 0$$

for all continuous functions $g(x)$ on $[a, b]$ and all continuous functions $h(s)$ on $[a, b]$, then $k(x, s) = 0$ for all (x, s) in $[a, b] \times [a, b]$.

Proof. Since the inner integral in

$$\int_a^b \left(\int_a^b k(x, s) h(s)\, ds \right) g(x)\, dx = 0$$

is a continuous function of x, by the previous corollary,

$$\int_a^b k(x, s) h(s)\, ds = 0$$

for all x in $[a, b]$ and all continuous functions $h(s)$ on $[a, b]$. Apply the previous corollary again to conclude that $k(x, s) = 0$ for all x in $[a, b]$ and all s in $[a, b]$. ■

Sometimes it is important to know when equality holds in the triangle inequality for integrals.

Proposition 22 *If $w: [a, b] \to \mathbb{C}$ is continuous, then*

$$\left| \int_a^b w(s)\, ds \right| \leq \int_a^b |w(s)|\, ds$$

with equality if and only if $w = e^{i\theta_0} p$ for some real number θ_0 and nonnegative continuous function p.

Proof. Express the integral on the left in polar form

$$\int_a^b w(s)\, ds = r_0 e^{i\theta_0}$$

so that

$$\left| \int_a^b w\, ds \right| = r_0 = \int_a^b e^{-i\theta_0} w\, ds$$

Since the integral on the right is a real number

$$\left| \int_a^b w\, ds \right| = r_0 = \mathrm{Re}\left(\int_a^b e^{-i\theta_0} w\, ds \right) = \int_a^b \mathrm{Re}(e^{-i\theta_0} w)\, ds.$$

Since

$$\int_a^b |w|\, ds = \int_a^b |e^{-i\theta_0} w|\, ds,$$

$$\int_a^b |w|\, ds - \left| \int_a^b w\, ds \right| = \int_a^b |e^{-i\theta_0} w|\, ds - \int_a^b \mathrm{Re}(e^{-i\theta_0} w)\, ds.$$

For a complex number z, $|z| \geq \mathrm{Re}(z)$ with equality if and only if z is real and nonnegative. Hence

$$\left|\int_a^b w(s)\, ds\right| \leq \int_a^b |w(s)\,|ds$$

with equality if and only if $e^{-i\theta_0} w = p$ with $p \geq 0$. ∎

The proposition holds by the same proof if the simplex $\Delta_1 = [a, b]$ is replaced by the simplex Δ_n in $\mathbb{R}^n$. (More generally the region of integration can be any set in $\mathbb{R}^n$ for which the indicated integrals exist.)

2.2.4 Sequences and Series of Functions

A sequence of real or complex-valued functions $\{f_n(x)\}$ **converges pointwise** to a function $f(x)$ on a set S in Euclidean space if for each (fixed) x in S, the $\lim_{n\to\infty} f_n(x) = f(x)$. That is, given any $\varepsilon > 0$ there is an integer N, dependent on ε and x, such that $|f_n(x) - f(x)| < \varepsilon$ whenever $n > N$.

A sequence of real or complex-valued functions $\{f_n(x)\}$ **converges uniformly** to a function $f(x)$ on a set S in Euclidean space if given any $\varepsilon > 0$ there is a positive integer N, dependent only on ε and the set S, such that $|f_n(x) - f(x)| < \varepsilon$ for all x in S when $n > N$. The distinction between pointwise convergence and uniform convergence is that when the convergence is uniform once ε is given a single N can be found that works simultaneously for all x in S.

Theorem 23 *If a sequence of real or complex-valued continuous functions converges uniformly on a set to a limit function, then the limit function is continuous (on the set).*

If $\{f_n(x)\}$ is a sequence of real or complex-valued functions defined on a set S in Euclidean space, then its associated infinite series $\sum_{n=1}^{\infty} f_n(x)$ has partial sums $\{s_n(x)\}$ where $s_n(x) = \sum_{j=1}^{n} f_j(x)$. The series is said to converge pointwise, converge uniformly, or have any other property related to convergence if and only if its sequence of partial sums has the corresponding property.

Theorem 24 (Weierstrass M-test) *If $\{f_n(x)\}_{n=1}^{\infty}$ is a sequence of real or complex-valued functions defined on a set S in Euclidean space and there are constants M_n such that $|f_n(x)| \leq M_n$ for all x in S and all $n = 1, 2, \ldots$ and if $\sum_{n=1}^{\infty} M_n$ converges, then $\sum_{n=1}^{\infty} f_n(x)$ is absolutely and uniformly convergent on S.*

Let z be a real or complex-variable. The **Geometric series** $\sum_{n=0}^{\infty} z^n$ converges if and only if $|z| < 1$, in which case

$$\sum_{n=0}^{\infty} z^n = \frac{1}{1 - z}.$$

It follows from the Weierstrass M-test that the geometric series converges absolutely and uniformly on the set $|z| \leq r$ for any $0 \leq r < 1$.

Theorem 25 (Dini) *If a sequence $\{f_n(x)\}$ of continuous functions is nondecreasing, $f_n(x) \leq f_{n+1}(x)$, and converges pointwise to a continuous function $f(x)$ on a closed bounded set S in Euclidean space, then the convergence is uniform.*

Proof. Denote the pointwise limit function by $f(x)$ for x in S. If the convergence is not uniform, there is an $\varepsilon_0 > 0$ such that no N exists such that $n \geq N$ implies that $|f_n(x) - f(x)| < \varepsilon_0$ for all x in S. Consequently, if $N = 1$ there must be a function f_{n_1} in the sequence $\{f_n(x)\}$ and a point x_1 in S such that $|f_{n_1}(x_1) - f(x_1)| \geq \varepsilon_0$. If $N = n_1 + 1$ there must be a function $f_{n_2}(x)$ in the sequence $\{f_n(x)\}$ and a point x_2 in S such that $|f_{n_2}(x_2) - f(x_2)| \geq \varepsilon_0$. Repeat this reasoning with $N = n_k + 1$ for $k = 2, 3, \ldots$ to obtain a subsequence $\{f_{n_k}(x)\}_{k=1}^{\infty}$ of $\{f_n(x)\}$ and a sequence of points $\{x_k\}$ in S such that

$$|f_{n_k}(x_k) - f(x_k)| \geq \varepsilon_0.$$

Equivalently,

$$f(x_k) - \varepsilon_0 \geq f_{n_k}(x_k)$$

because the sequence is increasing.

Now for each fixed positive integer m, $n_k \geq m$ for all k sufficiently large. Hence, $f_{n_k}(x_k) \geq f_m(x_k)$ and

$$f(x_k) - \varepsilon_0 \geq f_m(x_k)$$

for all k sufficiently large. Since a closed bounded set in Euclidean space is compact, there is a subsequence of $\{x_k\}$ that converges to c in S. By replacing the full sequence by the convergent subsequence and relabeling, we can assume that the full sequence converges to c. Let k tend to infinity in the inequality above to obtain

$$f(c) - \varepsilon_0 \geq f_m(c)$$

because f and f_m are continuous. But this inequality cannot hold for all positive integers m because of the pointwise convergence of f_m to f. This contradiction establishes that the convergence is uniform. ∎

Corollary 26 *If $\{f_n(x)\}$ is a sequence of continuous nonnegative functions on a closed bounded set S in Euclidean space and the series $\sum_{n=1}^{\infty} f_n(x)$ converges to a continuous function on S, then the convergence is uniform.*

2.3 Matrix and Linear Algebra

We use standard matrix notation and assume the reader is familiar with the elements of matrix algebra, determinants (mainly of low order), and linear algebra as they commonly occur in a first course in differential equations or earlier.

An $m \times n$ matrix $A = [a_{ij}]_{m \times n}$ is a rectangular array of elements a_{ij} with m rows and n columns. If all the a_{ij} are real numbers the matrix is called real. If the a_{ij} may be complex numbers the matrix is called complex. An $m \times 1$ matrix is a column vector and a $1 \times n$ matrix is a row vector. The matrix A^T, the transpose of A, is the $n \times m$ matrix obtained from A by interchanging its rows and columns. The matrix $\bar{A} = [\bar{a}_{ij}]_{m \times n}$ is the conjugate of A.

If $v_1, v_2, \ldots, v_m$ are vectors in a vector space and $c_1, c_2, \ldots, c_m$ are scalars, then

$$c_1 v_1 + c_2 v_2 + \cdots + c_m v_m$$

is called a linear combination of the given vectors, and the set of all such linear combinations as the scalars vary is called the span of $v_1, v_2, \ldots, v_m$, often denoted $\text{span}(v_1, v_2, \ldots, v_m)$.

We emphasize here aspects of linear algebra that are especially relevant to the treatment of Sturm-Liouville boundary value problems and eigenvalue problems given later. Useful references for topics in matrix and linear algebra are [11], [12], [14], [20], and [40].

2.3.1 Determinants

We summarize here the properties of determinants of real or complex square matrices that are needed later. It is convenient to use the following notation for a square matrix A

$$A = [a_{ij}]_{n\times n} = \begin{bmatrix} a_{11} & a_{12} & \cdots & a_{1n} \\ a_{21} & a_{22} & \cdots & a_{2n} \\ \cdots & \cdots & \cdots & \cdots \\ a_{n1} & a_{n2} & \cdots & a_{nn} \end{bmatrix} = \begin{bmatrix} a_1 & a_2 & \cdots & a_n \end{bmatrix}$$

where

$$a_j = \begin{bmatrix} la_{1j} \\ a_{2j} \\ \cdots \\ a_{nj} \end{bmatrix}$$

is the jth column of the matrix A for $j = 1, 2, \ldots, n$. A square matrix can be regarded as a function of its n^2 elements a_{ij} or as a function of the n column vectors a_j according to the convenience of the moment.

The following properties of the determinant were first developed for 2×2 and 3×3 matrices and later generalized to the $n \times n$ case. Each $n \times n$ matrix A, regarded as a function of its columns $a_1, \ldots, a_n$, has associated to it a real or complex number called its **determinant**, denoted by

$$\det A = \det\begin{bmatrix} a_1 & a_2 & \cdots & a_n \end{bmatrix},$$

and characterized by the following four properties:
(i) If B is the matrix obtained from A by interchanging any two of its columns, then

$$\det B = -\det A;$$

(ii) If B is the matrix obtained by multiplying each element in the first column of A by the same number c, then

$$\det B = c \det A;$$

(iii) If b and c are any two column vectors with n components, then

$$\det\begin{bmatrix} b+c & a_2 & \cdots & a_n \end{bmatrix} = \det\begin{bmatrix} b & a_2 & \cdots & a_n \end{bmatrix} + \det\begin{bmatrix} c & a_2 & \cdots & a_n \end{bmatrix}$$

(iv) If e_j is the column vector with 1 for its jth component and all other components 0, then

$$\det\begin{bmatrix} e_1 & e_2 & \cdots & e_n \end{bmatrix} = 1;$$

that is, the determinant of the identity matrix is 1.

Any rearrangement $j_1, j_2, \ldots, j_n$ of the natural numbers $1, 2, \ldots, n$ is called a **permutation** of $1, 2, \ldots, n$. By a rearrangement we mean that $j_1, j_2, \ldots, j_n$ is a list of the numbers $1, 2, \ldots, n$ but not necessarily in natural order. In other words, a permutation of the natural numbers $1, 2, \ldots, n$ is just a function $\sigma : \{1, 2, \ldots, n\} \to \{1, 2, \ldots, n\}$ that maps each element in its domain to a different element of the set $\{1, 2, \ldots, n\}$. The rearrangement $j_1, j_2, \ldots, j_n$ is the permutation σ defined by

$$\sigma(1) = j_1,\ \sigma(2) = j_2, \ldots, \sigma(n) = j_n.$$

Any permutation, such as the permutation 4, 5, 6, 1, 2, 3 of 1, 2, 3, 4, 5, 6, can be put back in natural order by a finite number of successive transpositions (interchanges of pairs of its element) in many ways. For example, the successive interchanges of the pairs 4 and 1, 5 and 2, and 6 and 3 returns the permutation 4, 5, 6, 1, 2, 3 to its natural order. Alternatively, the successive transpositions 6 and 1, 5 and 1, 4 and 1, 6 and 2, 5 and 2, 4 and 2, 6 and 3, 5 and 3, 4 and 3 also return 4, 5, 6, 1, 2, 3 to its natural order. In the first case 3 transpositions were used and in the second case 9 transpositions were used. Notice that 3 and 9 differ by an even integer. This illustrates an important property of permutations: if a permutation σ of $1, 2, \ldots, n$ can be put into natural order by t transpositions and also by s transpositions, then t and s differ by an even integer. Hence,

$$(-1)^t = (-1)^s,$$

a number ± 1 that is called the **sign** (or **signature**) of the permutation σ. The sign of a permutation σ is denoted by sgn σ. Thus,

$$\operatorname{sgn} \sigma = (-1)^t$$

if t transpositions return σ to its natural order. In particular, the sign of the permutation 4, 5, 6, 1, 2, 3 is $(-1)^3 = (-1)^9 = -1$.

It follows from properties (i)–(iv) above that

$$\det A = \sum_{\sigma} (\operatorname{sgn} \sigma) a_{1\sigma(1)} a_{2\sigma(2)} \cdots a_{n\sigma(n)}$$

where the sum is over all $n!$ permutations of the numbers $1, 2, \ldots, n$. Alternatively,

$$\det A = \sum (-1)^{|j_1, j_2 \ldots, j_n|} a_{1j_1} a_{2j_2} \cdots a_{nj_n}$$

if the permutation $j_1, j_2, \ldots, j_n$ can be put in natural order by $|j_1, j_2, \ldots, j_n|$ transpositions.

For example, if A is 2×2, there are two permutations of 1 and 2, namely 1, 2 with $|1, 2| = 0$ and $2, 1 = 0$ with $|2, 1| = 1$ and

$$\det A = (-1)^{|1,2|} a_{11} a_{22} + (-1)^{|2,1|} a_{12} a_{21} = a_{11} a_{22} - a_{12} a_{21}.$$

The definition of a determinant given above is primarily useful for theoretical purposes. Practical evaluation of determinants uses a process called Gauss elimination that takes advantage of some of the following well-known properties of determinants, each an elementary consequence of properties (i)–(iv):

1. $\det A = \det A^T$.
2. The determinant is a linear function of any one of its rows (columns).
3. A determinant changes it sign when any two of its rows (columns) are interchanged.
4. $\det A = 0$ if any two rows (columns) are linearly dependent.
5. The value of a determinant is unchanged if a multiple of one row (column) is added to a different row (column).

There is an important geometric interpretation of determinants that deserves mention. If the vectors $a_1, a_2, \ldots, a_n$ drawn from a common point are the sides of parallelepiped and A is the matrix whose columns (rows) are the given vectors, then $|\det A|$ is the n-dimensional volume of the parallelepiped. In 3-space this is the familiar result

$$V = |\mathbf{a} \cdot \mathbf{b} \times \mathbf{c}| = \left| \det \begin{pmatrix} \mathbf{a} \\ \mathbf{b} \\ \mathbf{c} \end{pmatrix} \right|,$$

where **a**, **b**, and **c** are row vectors.

2.3.2 Systems of Linear Algebraic Equations

The system of linear algebraic equations

$$\begin{aligned} a_{11}x_1 + a_{12}x_2 + \cdots + a_{1n}x_n &= b_1 \\ a_{21}x_1 + a_{22}x_2 + \cdots + a_{2n}x_n &= b_2 \\ &\cdots \\ a_{n1}x_1 + a_{n2}x_2 + \cdots + a_{nn}x_n &= b_n \end{aligned}$$

can be expressed compactly in matrix form as

$$Ax = b$$

where $A = [a_{ij}]$ is the coefficient matrix, $x = [x_1\, x_2 \ldots x_n]^T$ is the vector of unknowns, and $b = [b_1\, b_2 \ldots b_n]^T$ is the vector of right-hand sides. The system $Ax = 0$ is the corresponding homogeneous system. A solution x to a homogeneous system is trivial (the trivial solution) if $x = 0$ and nontrivial if $x \neq 0$; that is, at least one component of x is not zero.

If the system is square (has as many equations as unknowns), then we express the determinant of the matrix A either by $\det A$ or $\det(A)$ or $|A|$ as seems most convenient. The basic facts concerning solving such a system are:

$$\begin{aligned} &Ax = b \text{ has a unique solution } \Leftrightarrow \det A \neq 0. \\ &Ax = 0 \text{ has nontrivial solutions } \Leftrightarrow \det A = 0. \end{aligned}$$

Another useful equivalent way to express the last result is

$$Ax = 0 \text{ has only the trivial solution } \Leftrightarrow \det A \neq 0.$$

If $a_1, a_2, \ldots, a_n$ are column vectors in Euclidean n-space, and $x_1, x_2, \ldots, x_n$ are scalars, then

$$x_1 a_1 + x_2 a_2 + \cdots + x_n a_n = Ax$$

where $A = [\, a_1 \quad a_2 \quad \cdots \quad a_n \,]$. It follows that the vectors $a_1, a_2, \ldots, a_n$ are linearly dependent if and only if $\det A = 0$. Equivalently, $a_1, a_2, \ldots, a_n$ are linearly independent if and only if $\det A \neq 0$.

The most efficient way to solve an $n \times n$ linear system of algebraic equations that has a unique solution is to use systematic elimination of unknowns, Gaussian elimination. Nevertheless, for some theoretical purposes **Cramer's rule** is useful and may be useful in solving some 2×2 and 3×3 systems. Write the system in matrix form as $Ax = b$, where $x = [x_1, \ldots, x_n]^T$. The system has a unique solution if and only if its determinant $|A| \neq 0$. Cramer's rule is the assertion that the solution to the system is

$$x_j = \frac{|A_j|}{|A|} \quad \text{for } j = 1, 2, \ldots, n$$

and where A_j is the $n \times n$ matrix obtained from A by replacing its jth column by the column vector b.

2.3.3 Linear Dependence and Linear Independence

Linear dependence and linear independence can be thought of informally as distinguishing when a set of vectors contains redundant information from when it does not. The set of vectors $v_1, v_2, \ldots, v_m$ in a vector space is linearly dependent if there are scalars $c_1, c_2, \ldots, c_m$ not all zero such that

$$c_1 v_1 + c_2 v_2 + \cdots + c_m v_m = 0.$$

If, for example, $c_1 \neq 0$ then

$$v_1 = -c_1^{-1}(c_2 v_2 + \cdots + c_m v_m)$$

and any information in v_1 can be obtained from the other vectors. The set of vectors $v_1, v_2, \ldots, v_m$ is linearly independent if the relation

$$c_1 v_1 + c_2 v_2 + \cdots + c_m v_m = 0$$

holds only for $c_1 = c_2 = \cdots = c_m = 0$.

In a differential equations setting, the vector space is a space of suitably differentiable functions defined on an interval I and the vectors are functions, say $f_1, \ldots, f_m$ defined on I. The functions are linearly dependent if there are scalars $c_1, c_2, \ldots, c_m$ not all zero such that

$$c_1 f_1 + c_2 f_2 + \cdots + c_m f_m = 0,$$

which means

$$c_1 f_1(x) + c_2 f_2(x) + \cdots + c_n f_m(x) = 0 \text{ for all } x \text{ in } I.$$

The functions are said to be linearly dependent on I, when it is helpful to emphasize the interval on which the functions are defined. Likewise, the functions are linearly independent if the relation

$$c_1 f_1 + c_2 f_2 + \cdots + c_m f_m = 0$$

holds only for $c_1 = c_2 = \cdots = c_m = 0$. That is, if

$$c_1 f_1(x) + c_2 f_2(x) + \cdots + c_m f_m(x) = 0 \text{ for all } x \text{ in } I$$

holds only for $c_1 = c_2 = \cdots = c_m = 0$. The functions also are said to be linearly independent on I.

2.3.4 Eigenvalues and Eigenvectors

Let A be a real or complex $n \times n$ matrix. A real or complex number λ is an eigenvalue of A if there is a nonzero vector e such that $Ae = \lambda e$. The vector e is an eigenvector and is said to belong to its eigenvalue.

The matrix $I = [\delta_{ij}]_{n\times n}$, where δ_{ij} is the Kronecker delta, $\delta_{ij} = 0$ if $i \neq j$ and $\delta_{ii} = 1$, is the $n \times n$ identity matrix. The eigenvalue, eigenvector relation can be expressed as

$$(A - \lambda I)e = 0$$

and interpreted as a linear homogeneous system of equations for the components of e. Thus, λ will be an eigenvalue of A precisely when the system has a nontrivial (not identically zero) solution for e. This happens if and only if

$$|A - \lambda I| = 0,$$

which is the characteristic equation of the matrix. The left member is a polynomial of degree n. Its roots (zeros) are the eigenvalues of the matrix. If λ is an eigenvalue, the nontrivial solutions e of $(A-\lambda I)e=0$ are its corresponding eigenvectors.

The algebraic multiplicity m of an eigenvalue λ is the multiplicity of λ as a root of the characteristic equation. The geometric multiplicity of λ, m', is the number of linearly independent eigenvectors corresponding to λ. It is always the case that $m' \leq m$.

2.3.5 Self-Adjoint and Symmetric Matrices

The properties of eigenvalues and eigenfunctions of self-adjoint Sturm-Liouville eigenvalue problems are strictly analogous to those of self-adjoint matrices and in particular to real symmetric matrices. It is informative for that reason to summarize some key properties of such matrices here. The adjoint of A is the matrix $A^* = [a^*_{ij}]$ where $a^*_{ij} = \bar{a}_{ji}$ for all i,j; that is, $A^* = \bar{A}^T$. A is **self-adjoint** if $A = A^*$. A real, self-adjoint matrix is called **symmetric**.

The adjoint matrix arises from the following calculation:

$$\langle Ax, y\rangle = \sum_{i=1}^{n} (Ax)_i \bar{y}_i = \sum_{i=1}^{n}\left(\sum_{j=1}^{n} a_{ij}x_j\right)\bar{y}_i = \sum_{j=1}^{n} x_j \overline{\sum_{i=1}^{n} \bar{a}_{ij} y_i}$$

$$= \sum_{j=1}^{n} x_j \overline{\sum_{i=1}^{n} a^*_{ji} y_i} = \sum_{j=1}^{n} x_j \overline{(A^*y)_j} = \langle x, A^*y\rangle.$$

Hence,

$$\langle Ax, y\rangle = \langle x, A^*y\rangle$$

for all vectors x and y in $\mathbb{C}^n$. This result is more important and useful than first meets the eye. As an example, we use it to prove

Lemma 27 *All the eigenvalues of a self-adjoint matrix are real and eigenvectors belonging to distinct eigenvectors are orthogonal.*

Proof. If $Ax = \lambda x$ and $x \neq 0$, then

$$\lambda\langle x, x\rangle = \langle \lambda x, x\rangle = \langle Ax, x\rangle = \langle x, Ax\rangle = \langle x, \lambda x\rangle = \bar{\lambda}\langle x, x\rangle;$$

hence, $\lambda = \bar{\lambda}$ and λ is real. If $Ay = \mu y$ and $y \neq 0$, then

$$\lambda\langle x, y\rangle = \langle \lambda x, y\rangle = \langle Ax, y\rangle = \langle x, Ay\rangle = \langle x, \mu y\rangle = \mu\langle x, y\rangle$$

and $\langle x, y\rangle = 0$ if $\lambda \neq \mu$. ■

It follows from the principal axis theorem in the next section, that a self-adjoint matrix A has n real eigenvalues, not necessarily distinct, $\lambda_1, \ldots, \lambda_n$, and n corresponding real eigenvectors, $e_1, \ldots, e_n$, that are mutually orthogonal (hence, linearly independent); that is,

$$\langle e_i, e_j\rangle = \sum_{k=1}^{n} e_{ik} e_{jk} = c_i \delta_{ij},$$

where $c_i = \| e_i \|^2$ and δ_{ij} is the Kronecker delta. The eigenvectors are a basis for $\mathbb{R}^n$. Such a basis of eigenvectors provides the most natural basis for dealing with computational and theoretical problems related to the matrix A. (It is strictly analogous to the standard basis **i**, **j**, **k** in ordinary three space.) Each vector x in the space can be expressed as

$$x = x_1 e_1 + \cdots + x_n e_n$$

and taking inner (dot) product with e_j gives

$$\langle x, e_j\rangle = x_j c_j.$$

To illustrate the utility of this representation we solve the matrix equation $Ax = b$, where b is a given n-vector: take the inner product of $Ax = b$ with e_j to find $\langle x, e_j\rangle$,

$$\langle b, e_j\rangle = \langle Ax, e_j\rangle = \langle x, Ae_j\rangle = \lambda_j\langle x, e_j\rangle = \lambda_j c_j x_j,$$

$$x_j = \lambda_j^{-1} c_j^{-1}\langle b, e_j\rangle.$$

Hence,

$$x = \sum_{j=1}^{n} x_j e_j = \sum_{j=1}^{n} \frac{\langle b, e_j\rangle}{\lambda_j c_j} e_j.$$

If $Ax = b$ is the 2×2 system

$$\begin{bmatrix} 2 & 1 \\ 1 & 2 \end{bmatrix}\begin{bmatrix} x_1 \\ x_2 \end{bmatrix} = \begin{bmatrix} 5 \\ 1 \end{bmatrix},$$

then the matrix A has eigenvalue, eigenvector pairs

$$\lambda_1 = 3 \quad \text{and} \quad e_1 = \begin{bmatrix} 1 \\ 1 \end{bmatrix},$$

$$\lambda_2 = 1 \quad \text{and} \quad e_2 = \begin{bmatrix} 1 \\ -1 \end{bmatrix},$$

$c_1 = c_2 = 2$, $\langle b, e_1\rangle = 6$, $\langle b, e_2\rangle = 4$, and the system has solution

$$x = \begin{bmatrix} x_1 \\ x_2 \end{bmatrix} = \frac{6}{3\cdot 2}\begin{bmatrix} 1 \\ 1 \end{bmatrix} + \frac{4}{1\cdot 2}\begin{bmatrix} 1 \\ -1 \end{bmatrix} = \begin{bmatrix} 3 \\ -1 \end{bmatrix}.$$

Solutions to Sturm-Liouville boundary value problems have convenient analogous solution formulas, which are infinite series expansions in terms of orthonormal eigenfunctions.

2.3.6 Principal Axis Theorem

The principal axis theorem gets its name from the fact that the procedure we are about to describe determines the semi-axes of conic sections in the plane and of the corresponding surfaces in space via the eigenvalues and eigenvectors of a related matrix. When generalized to higher dimensions the result is:

Theorem 28 (Principal Axis Theorem) *Let A be an $n \times n$ symmetric matrix. Then A has n real eigenvalues, counted to multiplicity, and n corresponding pairwise orthogonal real eigenvectors, which may be chosen to be orthonormal.*

The function $f(x, y) = ax^2 + 2bxy + cy^2$, where a, b, and c are real numbers, is a quadratic form in the real variables x and y. The form has global extreme values on the unit circle $x^2 + y^2 = 1$. They can be found by the method of Lagrange multipliers: There is a constant λ, called a Lagrange multiplier, such that the global extreme values are taken on at points (x, y) on the circle where the function

$$g(x, y) = ax^2 + 2bxy + cy^2 - \lambda(x^2 + y^2)$$

is stationary; that is, its partial derivatives are zero. Since

$$g_x(x, y) = 2ax + 2by - \lambda 2x,$$
$$g_y(x, y) = 2bx + 2cy - \lambda 2y,$$

the extreme values occur at points (x, y) that satisfy the 2×2 system

$$ax + by = \lambda x,$$
$$bx + cy = \lambda y.$$

Rather than solve this system directly it is more informative to express it in matrix form as

$$Av = \lambda v$$

where

$$A = \begin{bmatrix} a & b \\ b & c \end{bmatrix}, \quad \text{and} \quad v = \begin{bmatrix} x \\ y \end{bmatrix}.$$

Thus, the global extreme values (that must exist) occur at eigenvectors of the matrix A. (Recall (x, y) is a point on the unit circle.)

This approach generalizes to $n \times n$ symmetric matrices and, of even more importance for us, to integral operators with symmetric kernels. Here is (most of) the story for $n \times n$ symmetric matrices.

A function $f(x) = \sum_{i,j=1}^{n} c_{ij} x_i x_j$, where c_{ij} are real numbers, is a real quadratic form in the variable(s) $x = (x_1, \ldots, x_n)$, where each x_j is real. Since

$$c_{ij} x_i x_j + c_{ji} x_j x_i = (c_{ij} + c_{ji}) x_i x_j$$

and $c_{ij} + c_{ji}$ is symmetric in i and j, by replacing each c_{ij} by $(c_{ij} + c_{ji})/2$ the quadratic form can be expressed as

$$f(x) = \sum_{i,j=1}^{n} a_{ij} x_i x_j$$

where $a_{ij} = (c_{ij} + c_{ji})/2$ is symmetric in i and j. Furthermore,

$$f(x) = \sum_{i=1}^{n} \left(\sum_{j=1}^{n} a_{ij} x_j \right) x_i = \sum_{i=1}^{n} (Ax)_i x_i = \langle Ax, x \rangle$$

where $(Ax)_i$ is the ith component of Ax and $\langle \cdot \, , \, \cdot \rangle$ is the usual inner product on $\mathbb{R}^n$. Thus, every real quadratic form can be expressed as

$$f(x) = \langle Ax, x \rangle$$

where A is a real symmetric matrix and any such matrix defines a quadratic form.

As in the case $n = 2$, $f(x)$ achieves both its global maximum and global minimum at points on the unit sphere $\|x\| = 1$, equivalently $\langle x, x \rangle = 1$. Thus, there is a Lagrange multiplier λ such that the global extreme values of f occur at points x where $g(x) = \langle Ax, x \rangle - \lambda \langle x, x \rangle$ is stationary. By the product rule

$$\frac{\partial g}{\partial x_j} = \langle Ae_j, x \rangle + \langle Ax, e_j \rangle - \lambda 2 \langle x, e_j \rangle,$$

where e_j is the jth standard basis vector in $\mathbb{R}^n$. Since Ae_j is the jth column of A and $\langle x, e_j \rangle = x_j$,

$$\frac{\partial g}{\partial x_j} = 2\Big((Ax)_j - \lambda x_j \Big).$$

Consequently, the global extrema of f occur at points x on the unit sphere where

$$(Ax)_j - \lambda x_j = 0$$

for $j = 1, \ldots, n$; that is, at points x on the unit sphere where

$$Ax = \lambda x.$$

That is, the global maximum and minimum occur at points x on the unit sphere that are eigenvectors of the symmetric matrix A. This shows two things: first, every symmetric matrix has at least one eigenvalue (and by pushing this line of reasoning a little harder n eigenvalues counted to multiplicity and n corresponding orthonormal eigenvectors). Second,

$$\max_{\|x\|=1} |\langle Ax, x\rangle|$$

occurs at an eigenvector of A.

In finite dimensions, we are guaranteed that the quadratic form assumes its extreme values on the unit sphere, $\|x\| = 1$. In the infinite dimensional setting of integral operators, we will need to show both that the extreme values exist and that they are taken on only at eigenfunctions of the integral operator.

2.3.7 Matrices as Linear Transformations

An $m \times n$ matrix A is often regarded as a transformation or operator that takes an n-vector x into the m-vector Ax gotten by matrix multiplication. If A is a real matrix it usually acts on vectors with real components. In this case, $A : \mathbb{R}^n \to \mathbb{R}^m$, where we use customary function notation to indicate that the domain of A, viewed as a transformation, is real n-dimensional Euclidean space and its range is in real m-dimensional Euclidean space. For example, the matrix $A : \mathbb{R}^3 \to \mathbb{R}^3$

$$A = \begin{bmatrix} 0 & 1 & 0 \\ -1 & 0 & 0 \\ 0 & 0 & 0 \end{bmatrix}$$

transforms the vector $v = [x, y, z]^T$ into the vector $Av = [y, -x, 0]^T$. In geometric terms, A projects v orthogonally onto the xy-plane and rotates the projection counterclockwise by $90°$ when viewed from $(0, 0, 1)$. If A is complex and acts on complex Euclidean space we write $A : \mathbb{C}^n \to \mathbb{C}^m$.

Linear integral operators which transform input functions into output functions by integration are continuous analogues of matrix operators. We will use them to study Sturm-Liouville boundary value and eigenvalue problems.

2.4 Interpolation and Approximation

The elements of interpolation and the results related to approximation theory and total positivity reviewed here are directly related to establishing corresponding properties possessed by the eigenfunctions and Green's functions of Sturm-Liouville eigenvalue problems. To anticipate why the sign properties of the determinants that appear below are related to

Sturm-Liouville problems see the discussion in Section 1.11.2. For comprehensive treatments of approximation theory and total positivity theory see [16], [24], [25], and [31].

It is sufficient for our purposes to assume throughout this section that all functions are real-valued and defined on intervals of real numbers. I denotes an interval with positive length. If the interval is not specified explicitly it is the entire real line. All constants and exponents that appear are real numbers.

2.4.1 Tchebycheff Systems

The following properties of ordinary polynomials of degree n or less, that is linear combinations of the power functions $1, x, \ldots, x^n$, are well known:

(A) *(Interpolation)* For $n \geq 1$, given any $n+1$ distinct real numbers $x_0, x_1, \ldots, x_n$ and any $n+1$ values $b_0, b_1, \ldots, b_n$, there is a *unique* polynomial p of degree n or less that assumes the given values at the given points; that is, $p(x_j) = b_j$ for $j = 0, 1, \ldots, n$.

(B) *(Zeros)* A nonzero (not identically zero) polynomial of degree $n \geq 1$ has at most n distinct real zeros.

A nonzero polynomial of degree n can be expressed as $p(x) = \sum_{k=0}^{n} a_k x^k$ with $a_n \neq 0$. One proof of (B) follows: suppose that $p(x)$ has $n+1$ distinct zeros, say $x_0, x_1, \ldots, x_n$. By Rolle's theorem $p'(x)$ at least one zero between each zero of $p(x)$; hence, $p'(x)$ has at least n distinct zeros. Likewise, $p''(x)$ has at least $n-1$ distinct zeros and $p^{(n)}(x) = n!a_n$ has at least one zero. Since $a_n \neq 0$ this is impossible, a contradiction that establishes (B).

This same line of reasoning can be used to prove the following evident but important fact: if two polynomials $p(x) = \sum_{k=0}^{n} a_k x^k$ and $\tilde{p}(x) = \sum_{k=0}^{n} \tilde{a}_k x^k$ are equal on an interval with endpoints a and b with $b - a > 0$, then $a_k = \tilde{a}_k$ for $k = 0, 1, \ldots, n$. Consequently, a polynomial $p(x) = \sum_{k=0}^{n} a_k x^k$ is identically equal to zero in an interval of positive length if and only if all its coefficients are equal to zero.

It will be useful for a point of view we will use shortly to observe that the properties (A) and (B) are equivalent:

(A) $\Rightarrow$ (B): if a polynomial of the form $p(x) = \sum_{k=0}^{n} a_k x^k$ with $a_n \neq 0$ has $n+1$ distinct real zeros, then this polynomial and the zero polynomial $z(x)$ both take on the value 0 at the $n+1$ points where $p(x)$ has zeros. By the uniqueness assertion in (A), it follows that $p(x) = z(x)$ on the interval containing the $n+1$ distinct zeros of $p(x)$. Thus, $p(x) = 0$ on that interval and all its coefficients are zero, which contradicts $a_n \neq 0$. This contradiction establishes the desired implication.

(B) $\Rightarrow$ (A): let distinct real numbers $x_0, x_1, \ldots, x_n$ be given. The system of equations

$$\sum_{k=0}^{n} a_k x_j^k = 0, \ j = 0, 1, \ldots, n,$$

for the unknowns $a_0, a_1, \ldots, a_n$ has a nontrivial solution if and only if its determinant $\det [x_j^k]_{(n+1)\times(n+1)} = 0$. (See Section 2.3.2.) That is, there is a nonzero polynomial $p(x) = \sum_{k=0}^{n} a_k x^k$ with $n+1$ distinct zeros if and only if $\det [x_j^k]_{(n+1)\times(n+1)} = 0$. It follows that $\det [x_j^k]_{(n+1)\times(n+1)} \neq 0$ because (B) holds. Consequently, the system

$$\sum_{k=0}^{n} a_k x_j^k = b_j, \ j = 0, 1, \ldots, n,$$

has a unique solution for $a_0, a_1, \ldots, a_n$ for any choice of $b_0, b_1, \ldots, b_n$. Equivalently, there is a unique polynomial $p(x)$ of degree n or less that assumes the values b_k at the points x_k. Thus, (B) $\Rightarrow$ (A).

(A) is now established because we have already proved (B). Equally interesting for us is the following point of view: the discussion above shows that det $[x_j^k]_{(n+1)\times(n+1)} \neq 0$ for all choices of $n+1$ distinct points $x_0, x_1, \ldots, x_n$. Now, in (A) and (B) we can relabel the points $x_0, x_1, \ldots, x_n$ so they appear in increasing order

$$x_0 < x_1 < \cdots < x_n.$$

We assume this is the case from now on.

The determinant det $[x_j^k]_{(n+1)\times(n+1)}$ is called a **Vandermonde determinant** and we denote it by

$$V(x_0, x_1, \ldots, x_n) = \begin{vmatrix} 1 & x_0 & \cdots & x_0^n \\ 1 & x_1 & \cdots & x_1^n \\ \vdots & & \vdots & \vdots \\ 1 & x_n & \cdots & x_n^n \end{vmatrix}$$

for all choices of points $x_0 < x_1 < \cdots < x_n$. We have already observed that

$$V(x_0, x_1, \ldots, x_n) \neq 0$$

for any choice of $x_0 < x_1 < \cdots < x_n$. Moreover, as $x_0, x_1, \ldots, x_n$ vary over the simplex $x_0 < x_1 < \cdots < x_n$, the determinant maintains a fixed sign; it is always positive or always negative. To see this we use a continuity argument. Let $x_0', x_1', \ldots, x_n'$ be any point in the simplex different from $x_0, x_1, \ldots, x_n$ and consider the function

$$f(t) = V(x_0 + tx_0', x_1 + tx_1', \ldots, x_n + tx_n')$$

for t in $[0, 1]$. The function $f(t)$ is clearly continuous on $[0, 1]$ and $x_0 + tx_0', x_1 + tx_1', \ldots, x_n + tx_n'$ is a point in the simplex for each t. So $f(t) \neq 0$ for t in $[0, 1]$. If $f(0)$ and $f(1)$ were to have opposite signs, then by the intermediate value theorem $f(t)$ would have a zero in $[0, 1]$, a contradiction. Thus $f(0)$ and $f(1)$ have the same sign. That is, V has the same sign, always positive or always negative, at every point in the simplex $x_0 < x_1 < \cdots < x_n$.

Finally, we show that the fixed sign of the Vandermonde determinant is positive. The following argument makes unnecessary the reasoning used in the last paragraph to show that the Vandermonde determinant has a fixed sign. The importance of that reasoning will be apparent later in the section. Define

$$D(x) = \begin{vmatrix} 1 & x_0 & \cdots & x_0^n & x_0^{n+1} \\ 1 & x_1 & \cdots & x_1^n & x_1^{n+1} \\ \vdots & & \vdots & & \vdots \\ 1 & x_n & \cdots & x_n^n & x_n^{n+1} \\ 1 & x & \cdots & x^n & x^{n+1} \end{vmatrix},$$

a polynomial of degree $n+1$ and with $n+1$ distinct zeros $x_0, x_1, \ldots, x_n$ where $x_0 < x_1 < \cdots < x_n$. Expand the determinant by its last row to see that the polynomial has leading coefficient, the coefficient of x^n, $V(x_0, x_1, \ldots, x_n)$. Since the polynomial has factors $(x - x_k)$ for $k = 0, 1, \ldots, n$, it follows that

$$D(x) = V(x_0, x_1, \ldots, x_n) \prod_{k=0}^{n} (x - x_k).$$

For any $x_{n+1} > x_n$, $D(x_{n+1}) = V(x_0, x_1, \ldots, x_{n+1})$ so

$$V(x_0, x_1, \ldots, x_{n+1}) = V(x_0, x_1, \ldots, x_n) \prod_{k=0}^{n} (x_{n+1} - x_k)$$

for $n = 0, 1, 2, \ldots$. Applying this recursion formula for a few values of $n = 0, 1, 2, \ldots$, more precisely by mathematical induction, it follows that

$$V(x_0, x_1, \ldots, x_n) = \prod_{0 \le j < k \le n} (x_k - x_j) > 0.$$

The foregoing considerations suggest the importance of the following in which the powers $1, x, \ldots, x^n$ are replaced by continuous functions $\phi_0(x), \phi_1(x), \ldots, \phi_n(x)$. The continuous functions $\phi_0(x), \phi_1(x), \ldots, \phi_n(x)$ are called a **Tchebycheff system**[1] on an interval I if

$$\det [\phi_k(x_j)] > 0$$

for all choices of $x_0, x_1, \ldots, x_n$ in I with $x_0 < x_1 < \cdots < x_n$. The functions in a Tchebycheff system are linearly independent on I, as we will establish shortly. If $\phi_0(x), \phi_1(x), \ldots, \phi_n(x)$ are linearly independent on I and

$$\det [\phi_k(x_j)] \ge 0$$

for all choices of $x_0, x_1, \ldots, x_n$ in I with $x_0 < x_1 < \cdots < x_n$, they form a **weak Tchebycheff system** on I.

An expression of the form $\sum_{k=0}^{n} c_k \phi_k(x)$, where $c_0, c_1, \ldots, c_n$ are real numbers, is called a **ϕ-polynomial** (or, for short, just a polynomial if the context is clear). It is nontrivial if it is not the zero function on I.

Proposition 29 *If $\phi_0(x), \phi_1(x), \ldots, \phi_n(x)$ is a Tchebycheff system on I, then:*
(*A*) (Interpolation) *For $n \ge 1$, given any $n+1$ distinct real numbers $x_0, x_1, \ldots, x_n$ in I and any $n+1$ values $b_0, b_1, \ldots, b_n$, there is a unique ϕ-polynomial ϕ that assumes the given values at the given points; that is, $\phi(x_j) = b_j$ for $j = 0, 1, \ldots, n$.*
(*B*) (Zeros) *A nonzero (not identically zero) ϕ-polynomial has at most n distinct real zeros.*

Proof. As usual we can label $x_0, x_1, \ldots, x_n$ so that $x_0 < x_1 < \cdots < x_n$. The ϕ-polynomial $\phi(x) = \sum_{k=0}^{n} c_k \phi_k(x)$ assumes the values b_j at x_j if and only if $c_0, c_1, \ldots, c_n$ satisfy the linear system

$$\sum_{k=0}^{n} c_k \phi_k(x_j) = b_j$$

for $j = 0, 1, \ldots, n$. Since $\det [\phi_k(x_j)] > 0$, the system has a unique solution, which establishes (A).

If the ϕ-polynomial $\phi(x) = \sum_{k=0}^{n} c_k \phi_k(x)$ has $n+1$ distinct zeros $x_0, x_1, \ldots, x_n$, then the system above is solvable with $b_j = 0$ for $j = 0, 1, \ldots, n$. Since $\det [\phi_k(x_j)] > 0$, the unique solution is $c_k = 0$ for $k = 0, 1, \ldots, n$. Thus, ϕ is the zero function and property (B) follows. ■

Property (B) in the proposition implies that a ϕ-polynomial $\sum_{k=0}^{n} c_k \phi_k(x)$ is nontrivial if and only if $\sum_{k=0}^{n} c_k^2 > 0$. This implies that the functions in a Tchebycheff system are linearly independent on I. Property (B) also implies that the zeros of a nontrivial ϕ-polynomial are isolated.

[1] Tchebycheff is one of several transliterations from the Cyrillic alphabet. We use this spelling because the polynomials associated with the name are denoted by $T_n(x)$.

Proposition 30 *The following are equivalent:*
(a) If $\phi_0(x), \phi_1(x), \ldots, \phi_n(x)$ are continuous functions on I and each nonzero ϕ-polynomial has at most n distinct zeros, then either $\phi_0(x), \phi_1(x), \ldots, \phi_n(x)$ is a Tchebycheff system on I or $\phi_0(x), \phi_1(x), \ldots, \phi_{n-1}(x), -\phi_n(x)$ is.
(b) If $\phi_0(x), \phi_1(x), \ldots, \phi_n(x)$ or $\phi_0(x), \phi_1(x), \ldots, \phi_{n-1}(x), -\phi_n(x)$ is a Tchebycheff system on I, then each nonzero ϕ-polynomial has at most n distinct zeros in I.

Proof. (a) If there were points $x_0, x_1, \ldots, x_n$ with $x_0 < x_1 < \cdots < x_n$ such that the linear system

$$\sum_{k=0}^{n} c_k \phi_k(x_j) = 0$$

for $j = 0, 1, \ldots, n$ had $\det[\phi_k(x_j)] = 0$, then the system would have a nontrivial solution for $c_0, c_1, \ldots, c_n$ and $\sum_{k=0}^{n} c_k \phi_k(x)$ would be a nonzero ϕ-polynomial with $n+1$ distinct zeros, a contradiction. Thus, $\det[\phi_k(x_j)] \neq 0$ for all choices of $x_0, x_1, \ldots, x_n$ with $x_0 < x_1 < \cdots < x_n$. It follows that either $\det[\phi_k(x_j)] > 0$ on the simplex $x_0 < x_1 < \cdots < x_n$ or $\det[\phi_k(x_j)] < 0$ there by the same argument used for the Vandermonde determinant. If the sign is positive, then $\phi_0(x), \phi_1(x), \ldots, \phi_n(x)$ is a Tchebycheff system and if it is negative, then $\phi_0(x), \phi_1(x), \ldots, \phi_{n-1}, -\phi_n(x)$ is.
(b) If $\phi_0(x), \phi_1(x), \ldots, \phi_n(x)$ is a Tchebycheff system on I or if $\phi_0(x), \phi_1(x), \ldots, \phi_{n-1}(x), -\phi_n(x)$ is, then any nontrivial φ-polynomial has at most n distinct zeros by by Proposition 29. ∎

The fundamental theorem of algebra states that a polynomial of degree n has exactly n zeros when each zero is counted to its multiplicity. A zero c of a polynomial $p(x)$ has multiplicity m if $p(c) = \cdots = p^{(m-1)}(c) = 0$ and $p^{(m)}(c) \neq 0$. A zero is simple if its multiplicity is 1 and is a double zero if its multiplicity is 2. A polynomial changes it sign at a simple zero and maintains a fixed sign near a double zero. Multiplicity in this sense does not apply to a ϕ-polynomial, unless it is sufficiently differentiable. But the zeros of a ϕ-polynomial can be counted in a way that distinguishes between zeros where a sign change occurs and those where no sign change occurs.

If $\phi_0(x), \phi_1(x), \ldots, \phi_n(x)$ is a Tchebycheff system on an interval I, then a zero c of a ϕ-polynomial is called a **nonnodal zero** of the polynomial if c is not an endpoint of I and the polynomial does not change sign at c. (c behaves like a double zero of an ordinary polynomial.) Any other zero of the ϕ-polynomial, including an endpoint of I that belongs to I, is called a **nodal zero (node)**. A nodal zero c that is not an endpoint of I behaves like a simple zero of an ordinary polynomial; the polynomial changes sign at c. We say a ϕ-polynomial changes sign at an interior zero c if every open interval that contains c also contains points where the polynomial is positive and points where it is negative. Proposition 29 asserts that a nontrivial ϕ-polynomial has at most n real zeros. The next proposition sharpens this result:

Proposition 31 *If $\phi_0(x), \phi_1(x), \ldots, \phi_n(x)$ is a Tchebycheff system on an interval I, then any nontrivial ϕ-polynomial has at most n zeros in I where each nodal zero is counted once and each nonnodal zero is counted twice.*

Proof. If the desired conclusion were false, there would be a ϕ-polynomial, say ϕ, with at least $n+1$ zeros when zeros are counted as in the proposition. The polynomial ϕ must have at least one nonnodal zero and at most $n-1$ nodal zeros by Proposition 29. Let the distinct zeros of ϕ in I be $t_1, \ldots, t_k$. Augment this set of zeros as follows: for each nonnodal zero t_j add the point $t_j + \varepsilon$ and to the first nonnodal zero t_{j_0} also add the point $t_{j_0} - \varepsilon$, where $\varepsilon > 0$ is chosen sufficiently small that the added points are all distinct from $t_1, \ldots, t_k$

and $\phi(t) \neq 0$ at each added point. The augmented set of points has at least $n+2$ points because ϕ has at least $n+1$ zeros. Put these points in increasing order and label the first $n+2$ of them as $x_0 < x_1 < \cdots < x_{n+1}$. The values $\phi(x_j)$ alternate in sign in the sense that $\phi(x_j)\phi(x_{j+1}) \leq 0$ for $j = 0, \ldots n$. Furthermore, not all of the $\phi(x_j)$ are zero because at least two of the first $n+2$ x_i must arise for the first nonnodal zero of ϕ. The determinant

$$\begin{vmatrix} \phi(x_0) & \cdots & \phi(x_{n+1}) \\ \phi_0(x_0) & \cdots & \phi_0(x_{n+1}) \\ \cdots & \cdots & \cdots \\ \phi_n(x_0) & \cdots & \phi_n(x_{n+1}) \end{vmatrix} = 0$$

because the first row is a linear combination of the following rows. Expand the determinant by its first row to get

$$\sum_{j=0}^{n+1} (-1)^j \phi(x_j) m_{1j} = 0$$

where m_{1j} is an $n+1$ by $n+1$ determinant of the form det $[\phi_i(x'_j)]$ with $\{x'_k\}$ the $n+1$ points $\{x_k\}_{k \neq j}$. Each $m_{1j} > 0$ because $\phi_0(x), \ \phi_1(x), \ldots, \phi_n(x)$ is a Tchebycheff system and $(-1)^j\phi(x_j) \geq 0$ for all j or satisfies the reverse inequality for all j by the alternating sign pattern of the $\phi(x_j)$. Thus, each summand in the displayed sum satisfies $(-1)^j\phi(x_j)m_{1j} \geq 0$ for all j or $(-1)^j\phi(x_j)m_{1j} \leq 0$ for all j. Since at least one of the $\phi(x_j) \neq 0$ this is a contradiction and the proposition is established. ■

2.4.2 Total Positivity

Consider approximation based not the consecutive power functions $1, x, \ldots, x^n$ and ordinary polynomials but rather on "polynomials" determined by the powers $x^{\alpha_0}, x^{\alpha_1}, \ldots, x^{\alpha_n}$ where $\alpha_0 < \alpha_1 < \cdots < \alpha_n$ and $x > 0$. We will show that $x^{\alpha_0}, x^{\alpha_1}, \ldots, x^{\alpha_n}$ is a Tchebycheff system on $(0, \infty)$ by means of Proposition 30 and the fact that all the Vandermonde determinants are positive. That is, first we show that any *nontrivial* polynomial

$$p(x) = \sum_{k=0}^{n} c_k x^{\alpha_k}$$

has at most n distinct zeros in $(0, \infty)$. The proof is by induction on the number of summands in a polynomial of the given form. If there is one summand, then $n = 0$, $c_0 \neq 0$, and the assertion is true. Assume by induction that any nontrivial polynomial of the given form with n summands has at most $n-1$ distinct zeros in $(0, \infty)$. Let $p(x)$ be a polynomial with $n+1$ summands. If $c_0 = 0$, then $p(x)$ has n summands and at most $n-1$ zeros by the induction hypothesis. If $c_0 \neq 0$,

$$(x^{-\alpha_0}p(x))' = \sum_{k=1}^{n} c_k(\alpha_k - \alpha_0)x^{\alpha_k - \alpha_0 - 1}$$

is a polynomial of the same form but with n summands, the latter polynomial is either identically zero or has at most $n-1$ positive zeros by the induction hypothesis. In the former case, $c_1 = 0, \ldots, c_n = 0$ and $p(x) = c_0x^{\alpha_0}$ has no zeros because $c_0 \neq 0$. In either case, the polynomial $(x^{-\alpha_0}p(x))'$ has at most $n-1$ positive zeros. If $p(x)$ had $n+1$ distinct positive zeros, then so would $x^{-\alpha_0}p(x)$ and, by Rolle's theorem, $(x^{-\alpha_0}p(x))'$ would have at least n distinct positive zeros, a contradiction. Hence, $p(x)$ has at most n distinct positive zeros, the induction step is advanced, and the original assertion that $p(x)$ has at most n

positive zeros is established. By Proposition 30 either $x^{\alpha_0}, x^{\alpha_1}, \ldots, x^{\alpha_n}$ or $x^{\alpha_0}, x^{\alpha_1}, \ldots, -x^{\alpha_n}$ is a Tchebycheff system on $(0, \infty)$.

A continuity argument shows that $x^{\alpha_0}, x^{\alpha_1}, \ldots, x^{\alpha_n}$ is a Tchebycheff system on $(0, \infty)$: the function

$$g(t) = \det [x_j^{(1-t)k+t\alpha_k}]_{(n+1)\times(n+1)}$$

for given points $x_0, x_1, \ldots, x_n$ with $0 < x_0 < x_1 < \cdots < x_n$ and real numbers $\alpha_0 < \alpha_1 < \cdots < \alpha_n$ is continuous for t in $[0, 1]$. Since $\alpha'_0 < \alpha'_1 < \cdots < \alpha'_n$ where $\alpha'_k = (1-t)k + t\alpha_k$, $g(t) \neq 0$ for each t by what was just established, $g(0) = V(x_0, x_1, \ldots, x_n) > 0$, and $g(1) = \det [x_j^{\alpha_k}] > 0$ because otherwise $g(t)$ would vanish somewhere in the interval $[0, 1]$. Thus, $x^{\alpha_0}, x^{\alpha_1}, \ldots, x^{\alpha_n}$ is a Tchebycheff system on $(0, \infty)$; equivalently

$$\det [x_j^{\alpha_k}] > 0$$

for any points $x_0, x_1, \ldots, x_n$ with $0 < x_0 < x_1 < \cdots < x_n$ and any real numbers $\alpha_0 < \alpha_1 < \cdots < \alpha_n$.

Now comes an additional important observation, that does not apply to the consecutive powers: if $m < n$ and $\beta_0 < \beta_1 < \cdots < \beta_m$ is a selection of $m+1$ the $\alpha_0 < \alpha_1 < \cdots < \alpha_n$, then by what we have just proved $x^{\beta_0}, x^{\beta_1}, \ldots, x^{\beta_m}$ is a Tchebycheff system on $(0, \infty)$. So if $y_0 < y_1 < \cdots < y_m$ is a selection of m of the $0 < x_0 < x_1 < \cdots < x_n$,

$$\det [y_r^{\beta_s}]_{m+1\times m+1} > 0.$$

Since $\det [y_r^{\beta_s}]_{m+1\times m+1}$ varies over all subdeterminants of the matrix $[x_j^{\alpha_k}]_{(n+1)\times(n+1)}$ as the β's and y's vary through all possible selections, we have established: the generalized Vandermonde matrix $[x_j^{\alpha_k}]_{(n+1)\times(n+1)}$ where $x_0, x_1, \ldots, x_n$ satisfy $0 < x_0 < x_1 < \cdots < x_n$ and $\alpha_0 < \alpha_1 < \cdots < \alpha_n$ are real numbers has the property that its determinant and the determinant of every square submatrix of $[x_j^{\alpha_k}]_{(n+1)\times(n+1)}$ is positive.

A square matrix with the property that its determinant and the determinants of all its square submatrices are nonnegative is called **totally positive** and if all the determinants are positive it is called **strictly totally positive.** In this terminology we have established:

Proposition 32 *The power functions $x^{\alpha_0}, x^{\alpha_1}, \ldots, x^{\alpha_n}$ where $\alpha_0 < \alpha_1 < \cdots < \alpha_n$ form a Tchebycheff system on the interval $(0, \infty)$. Consequently, a generalized Vandermonde matrix $[x_j^{\alpha_k}]_{(n+1)\times(n+1)}$ where $0 < x_0 < x_1 < \cdots < x_n$ and $\alpha_0 < \alpha_1 < \cdots < \alpha_n$ is strictly totally positive.*

Corollary 33 *Let $\sigma > 0$ and $g(s, t) = e^{-(s-t)^2/\sigma}$ be the Gauss kernel. If $s_1 < \cdots < s_n$ and $t_1 < \cdots < t_n$, then $\det [g(s_j, t_k)]_{n\times n} > 0$.*

Proof. Since

$$e^{-(s-t)^2/\sigma} = e^{-s^2/\sigma} e^{2st/\sigma} e^{-t^2/\sigma},$$

$$\det [e^{-(s_j-t_k)^2/\sigma}]_{n\times n} = \prod_{j=1}^{n} e^{-s_j^2/\sigma} \prod_{k=1}^{n} e^{-t_k^2/\sigma} \det [e^{2s_jt_k/\sigma}]_{(n+1)\times(n+1)}.$$

Let $x_j = \exp(2s_j/\sigma)$ and $\alpha_k = t_k$ to see that

$$\det [e^{2s_jt_k/\sigma}]_{n\times n} = \det [x_j^{\alpha_k}]_{n\times n} > 0$$

because $0 < x_1 < \cdots < x_n$ and $\alpha_0 < \alpha_1 < \cdots < \alpha_n$. Thus, $\det [g(s_j, t_k)] > 0$ as claimed. ∎

The Gauss kernel $g(x, s)$ is called strictly totally positive on $(-\infty, \infty) \times (-\infty, \infty)$ because it satisfies the determinantal inequalities in the corollary. Since the heat equation $u_t = au_{xx}$, $-\infty < x < \infty$, $t > 0$ has fundamental solution

$$k(x, t) = \frac{1}{\sqrt{4\pi at}} e^{-x^2/4at}$$

and the probability density of the normal probability distribution with mean μ and variance σ^2 is

$$\frac{1}{\sqrt{2\pi\sigma^2}} e^{-(x-\mu)^2/2\sigma^2},$$

the total positivity properties of the Gauss kernel have significant applications to diffusion problems and in probability theory.

Weierstrass used the fundamental solution to the heat equation with $4at = \sigma$ in his original proof of the Weierstrass approximation theorem. The primary step in Weierstrass' proof and a result we will need later is

Theorem 34 *If $f(x)$ is continuous on $[a, b]$, then*

$$\lim_{\sigma\to 0+} \frac{1}{\sqrt{\pi\sigma}} \int_{-\infty}^{\infty} e^{-(x-s)^2/\sigma} f(s)\, ds = f(x)$$

with uniform convergence for x in $[a, b]$. (Here f is extended to $(-\infty, \infty)$ by setting $f(x) = f(a)$ for $x < a$ and $f(x) = f(b)$ for $x > b$.)

Proof. Let $\sigma > 0$. In the proof we will use the result from calculus that

$$\frac{1}{\sqrt{\pi}} \int_{-\infty}^{\infty} e^{-t^2}\, dt = 1.$$

The change of variables $t = (x - s)/\sqrt{\sigma}$ with x fixed gives

$$\frac{1}{\sqrt{\pi\sigma}} \int_{-\infty}^{\infty} e^{-(x-s)^2/\sigma}\, ds = 1.$$

Since f is continuous on $[a, b]$ it is bounded, say by M. It is also uniformly continuous on $[a, b]$. For convenience extend f to a continuous function on $(-\infty, \infty)$ by setting $f(x) = f(a)$ for $x \le a$ and $f(x) = f(b)$ for $x \ge b$. Clearly f is bounded by M and is also uniformly continuous on $(-\infty, \infty)$. Consequently, given $\varepsilon > 0$ there is a δ, dependent only on ε, such that for s and x in $(-\infty, \infty)$

$$|f(s) - f(x)| < \varepsilon \text{ when } |s - x| < \delta.$$

With these preparations, let

$$\begin{aligned}
E(x) &= \frac{1}{\sqrt{\pi\sigma}} \int_{-\infty}^{\infty} e^{-(x-s)^2/\sigma} f(s)\, ds - f(x) \\
&= \frac{1}{\sqrt{\pi\sigma}} \int_{-\infty}^{\infty} e^{-(x-s)^2/\sigma} (f(s) - f(x))\, ds \\
&= \frac{1}{\sqrt{\pi\sigma}} \int_{x-\delta}^{x+\delta} e^{-(x-s)^2/\sigma} (f(s) - f(x))\, ds \\
&\quad + \frac{1}{\sqrt{\pi\sigma}} \int_{-\infty}^{x-\delta} e^{-(x-s)^2/\sigma} (f(s) - f(x))\, ds \\
&\quad + \frac{1}{\sqrt{\pi\sigma}} \int_{x+\delta}^{\infty} e^{-(x-s)^2/\sigma} (f(s) - f(x))\, ds \\
&= J_1 + J_2 + J_3
\end{aligned}$$

and make estimates as follows:

$$|J_1| < \varepsilon \frac{1}{\sqrt{\pi\sigma}} \int_{x-\delta}^{x+\delta} e^{-(x-s)^2/\sigma}\, ds < \varepsilon,$$

$$|J_2| \le 2M \frac{1}{\sqrt{\pi\sigma}} \int_{-\infty}^{x-\delta} e^{-(x-s)^2/\sigma}\, ds = \frac{2M}{\sqrt{\pi}} \int_{\delta/\sqrt{\sigma}}^{\infty} e^{-t^2}\, dt,$$

where the change of variable $t = (x-s)/\sqrt{\sigma}$ was used with x regarded as a parameter. Likewise,

$$|J_3| \le \frac{2M}{\sqrt{\pi}} \int_{-\infty}^{-\delta/\sqrt{\sigma}} e^{-t^2}\, dt.$$

Since

$$\frac{1}{\sqrt{\pi}} \int_{-\infty}^{\infty} e^{-t^2}\, dt = 1,$$

there exists $\sigma_\delta > 0$, not dependent on x, such that $|J_2| < \varepsilon$ and $|J_3| < \varepsilon$ for $0 < \sigma < \sigma_\delta$. Combine the estimates to find that $|E(x)| < 3\varepsilon$ when $0 < \sigma < \sigma_\delta$ for all x in $(-\infty, \infty)$. This establishes the asserted uniform convergence on $[a, b]$. ∎

Corollary 35 (Weierstrass Approximation Theorem) *Every continuous function $f(x)$ on a closed bounded interval $[a, b]$ can be uniformly approximated by a polynomial on $[a, b]$.*

Proof. We use the notation from the proof of the theorem and sketch the steps needed to complete proof. Use the change of variable $t = (x-s)/\sqrt{\sigma}$ to find that

$$\left| \frac{1}{\sqrt{\pi\sigma}} \int_{-\infty}^{\infty} e^{-(x-s)^2/\sigma} f(s)\, ds - \frac{1}{\sqrt{\pi\sigma}} \int_{-N}^{N} e^{-(x-s)^2/\sigma} f(s)\, ds \right|$$

$$= \frac{M}{\sqrt{\pi}} \left| \int_{-\infty}^{(x-N)/\sqrt{\sigma}} e^{-t^2}\, dt + \int_{(x+N)/\sqrt{\sigma}}^{\infty} e^{-t^2}\, dt \right|$$

can be made as small as desired uniformly for x in $[a, b]$ by choosing N suitably large. Once N is suitably fixed, let $T_m(u)$ be the mth Taylor polynomial about 0 of e^{-u}. Use the same change of variable to find that

$$\left| \frac{1}{\sqrt{\pi\sigma}} \int_{-N}^{N} e^{-(x-s)^2/\sigma} f(s)\, ds - \frac{1}{\sqrt{\pi\sigma}} \int_{-N}^{N} T_m\!\left(\frac{(x-s)^2}{\sigma}\right) f(s)\, ds \right|$$

$$= \frac{1}{\sqrt{\pi}} \left| \int_{(x-N)/\sqrt{\sigma}}^{(x+N)/\sqrt{\sigma}} (e^{-t^2} - T_m(t)) f(x - \sqrt{\sigma} t)\, dt \right|$$

$$\le \frac{M}{\sqrt{\pi}} \int_{(a-N)/\sqrt{\sigma}}^{(b+N)/\sqrt{\sigma}} \left| e^{-t^2} - T_m(t^2) \right| dt.$$

Since N and σ are fixed, there exists an m such the Taylor polynomial $T_m(u)$ approximates e^{-u} on $\left[0,\ \max\left((a-N)^2/\sigma,\ (b+N)^2/\sigma\right)\right]$ as accurately as desired and, hence, the right member above is as small as desired. Since

$$\frac{1}{\sqrt{\pi\sigma}} \int_{-N}^{N} T_m\!\left(\frac{(x-s)^2}{\sigma}\right) f(s)\, ds$$

is a polynomial in x the Weierstrass theorem is established. ■

If A is an $n \times n$ matrix, $1 \le i_1 < \cdots < i_p \le n$, and $1 \le j_1 < \cdots < j_p \le n$, then

$$A\begin{pmatrix} i_1, \ldots, i_p \\ j_1, \ldots, j_p \end{pmatrix} = \det\left[a_{i_r j_s}\right]_{r,s=1}^{p}$$

is the minor of A formed by the elements of A in rows $i_1, \ldots, i_p$ and columns $j_1, \ldots, j_p$.

The oscillatory behavior of the eigenfunctions of many Sturm-Liouville eigenvalue problems with separated boundary conditions follows from properties of its Green's function which are captured in the following lemma. An $n \times n$ matrix $G = [g_{ij}]_{n \times n}$ is a **Green's matrix** if its elements satisfy

$$g_{ij} = a_{\min(i,j)} b_{\max(i,j)} = \begin{cases} a_i b_j & \text{for } i \le j, \\ a_j b_i & \text{for } i \ge j, \end{cases}$$

where a_i and b_j are real numbers. A Green's matrix is symmetric.

Lemma 36 *Let $G = [g_{ij}]$ be an $n \times n$ Green's matrix so that $g_{ij} = a_{\min(i,j)} b_{\max(i,j)}$ and a_i and b_j are real numbers for $i, j = 1, \ldots n$. Fix $1 \le i_1 < \cdots < i_p \le n$ and $1 \le j_1 < \cdots < j_p \le n$. If*

$$1 \le i_1, j_1 < i_2, j_2 < \cdots < i_p, j_p \le n,$$

then

$$G\begin{pmatrix} i_1, \ldots, i_p \\ j_1, \ldots, j_p \end{pmatrix} = a_{k_1} \begin{vmatrix} a_{k_2} & a_{l_1} \\ b_{k_2} & b_{l_1} \end{vmatrix} \begin{vmatrix} a_{k_3} & a_{l_2} \\ b_{k_3} & b_{l_2} \end{vmatrix} \cdots \begin{vmatrix} a_{k_p} & a_{l_{p-1}} \\ b_{k_p} & b_{l_{p-1}} \end{vmatrix} b_{l_p},$$

where

$$k_\nu = \min(i_\nu, j_\nu) \quad \text{and} \quad l_\nu = \max(i_\nu, j_\nu).$$

If the condition $1 \le i_1, j_1 < i_2, j_2 < \cdots < i_p, j_p \le n$ does not hold, then

$$G\begin{pmatrix} i_1, \ldots, i_p \\ j_1, \ldots, j_p \end{pmatrix} = 0.$$

Proof. Assume for the moment that $a_i \ne 0$ for $i = 1, \ldots, n$. Suppose $i_1, j_1 < i_2, j_2$ does not hold. Since $i_1 < i_2$ and $j_1 < j_2$, either $i_1 < i_2 \le j_1$ or $j_1 < j_2 \le i_1$. If $i_1 < i_2 \le j_1$, then the first two rows of the determinant in question are

$$\begin{bmatrix} a_{i_1} b_{j_1} & a_{i_1} b_{j_2} & \cdots & a_{i_1} b_{j_p} \end{bmatrix}$$

and

$$\begin{bmatrix} a_{i_2} b_{j_1} & a_{i_2} b_{j_2} & \cdots & a_{i_2} b_{j_p} \end{bmatrix}.$$

Since these rows are proportional, the determinant is zero. Similarly, the first two columns are proportional if $j_1 < j_2 \le i_1$.

If $i_1, j_1 < i_2, j_2$ is satisfied, then since $i_1 < i_2$ and $j_1 < j_2$, either $i_1 < i_2 \le j_2$ or $j_1 < j_2 \le i_2$. If $i_1 < i_2 \le j_2$, then

$$k_2 = i_2 \quad \text{and} \quad l_2 = j_2$$

and

$$G\begin{pmatrix} i_1, \ldots, i_p \\ j_1, \ldots, j_p \end{pmatrix} = \begin{vmatrix} a_{k_1} b_{l_1} & a_{i_1} b_{j_2} & \cdots & a_{i_1} b_{j_p} \\ a_{j_1} b_{i_2} & a_{i_2} b_{j_2} & \cdots & a_{i_2} b_{j_p} \\ g_{i_3 j_1} & g_{i_3 j_2} & \cdots & g_{i_3 j_p} \\ & & \cdots & \end{vmatrix}.$$

Multiply row 2 by (a_{i_1}/a_{i_2}), subtract the result from row 1, and expand the determinant by row 1 to obtain

$$G\begin{pmatrix} i_1,\dots,i_p \\ j_1,\dots,j_p \end{pmatrix} = \left(a_{k_1}b_{l_1} - \frac{a_{i_1}}{a_{i_2}}a_{j_1}b_{i_2}\right)G\begin{pmatrix} i_2,\dots,i_p \\ j_2,\dots,j_p \end{pmatrix}.$$

Since $a_{i_1}a_{j_1} = a_{k_1}a_{l_1}$ and $i_2 = k_2$,

$$a_{k_1}b_{l_1} - \frac{a_{i_1}}{a_{i_2}}a_{j_1}b_{i_2} = \frac{a_{k_1}}{a_{k_2}}(a_{k_2}b_{l_1} - a_{l_1}b_{k_2}) = \frac{a_{k_1}}{a_{k_2}}\begin{vmatrix} a_{k_2} & a_{l_1} \\ b_{k_2} & b_{l_1} \end{vmatrix}$$

and

$$G\begin{pmatrix} i_1,\dots,i_p \\ j_1,\dots,j_p \end{pmatrix} = a_{k_1}\begin{vmatrix} a_{k_2} & a_{l_1} \\ b_{k_2} & b_{l_1} \end{vmatrix}\frac{1}{a_{k_2}}G\begin{pmatrix} i_2,\dots,i_p \\ j_2,\dots,j_p \end{pmatrix}.$$

If $j_1 < j_2 \le i_2$ instead of $i_1 < i_2 \le j_2$, similar reasoning yields the same result.

Continuing this line of reasoning step-by-step, either a $p \times p$ minor is 0 or

$$1 \le i_1, j_1 < i_2, j_2 < \cdots < i_p, j_p \le n,$$

and

$$G\begin{pmatrix} i_1,\dots,i_p \\ j_1,\dots,j_p \end{pmatrix} = a_{k_1}\begin{vmatrix} a_{k_2} & a_{l_1} \\ b_{k_2} & b_{l_1} \end{vmatrix}\begin{vmatrix} a_{k_3} & a_{l_2} \\ b_{k_3} & b_{l_2} \end{vmatrix}\cdots\begin{vmatrix} a_{k_p} & a_{l_{p-1}} \\ b_{k_p} & b_{l_{p-1}} \end{vmatrix}\frac{1}{a_{k_p}}G\begin{pmatrix} i_p \\ j_p \end{pmatrix}$$

Since

$$G\begin{pmatrix} i_p \\ j_p \end{pmatrix} = a_{k_p}b_{l_p}$$

the expansion of the minor is established when all the $a_i \ne 0$.

Since both members in the equality asserted in the lemma depend continuously on the a_i, the equality also holds when some of the a_i are 0. ■

Corollary 37 *Let G be an $n \times n$ Green's matrix with $a_ib_i \ne 0$ for $i = 1, 2, \dots, n$. Then G is totally positive if and only if the a_i and b_i have the same sign for $i = 1, \dots, n$ and*

$$\frac{a_1}{b_1} \le \frac{a_2}{b_2} \le \cdots \le \frac{a_n}{b_n}.$$

Moreover, for $1 \le i_1 < \cdots < i_p \le n$, $1 \le j_1 < \cdots < j_p \le n$,

$$G\begin{pmatrix} i_1,\dots,i_p \\ j_1,\dots,j_p \end{pmatrix} > 0$$

if and only if

$$1 \le i_1, j_1 < i_2, j_2 < \cdots < i_p, j_p \le n,$$

and

$$\frac{a_1}{b_1} < \frac{a_2}{b_2} < \cdots < \frac{a_n}{b_n}.$$

Proof. The 1×1 minors of G are $g_{ii} = a_ib_i$ and $g_{ij} = a_ib_j$ for $i \le j$. For $p = 2, \dots, n$, by Lemma 36 all $p \times p$ minors of G are 0 except possibly those with $1 \le i_1 < \cdots < i_p \le n$, $1 \le j_1 < \cdots < j_p \le n$, and $1 \le i_1, j_1 < i_2, j_2 < \cdots < i_p, j_p \le n$, equivalently $l_{\nu-1} < k_\nu$ for $\nu = 2, \dots, p$. For such minors, the 2×2 determinants in Lemma 36 are nonnegative if and only if

$$\frac{a_{l_{\nu-1}}}{b_{l_{\nu-1}}} \le \frac{a_{k_\nu}}{b_{k_\nu}} \quad \text{for } l_{\nu-1} = \max(i_{\nu-1}, j_{\nu-1}) < k_\nu = \min(i_\nu, j_\nu)$$

for $p = 1, 2, \dots, n$, which is equivalent to the first chain of inequalities in the corollary. The conclusions of the corollary follow at once from these observations. ■

2.5 Linear Spaces and Function Spaces

The geometric language that proves so useful and suggestive in Euclidean spaces is equally useful in function spaces or more general linear spaces.

2.5.1 Linear Spaces

A linear space (or vector space) is a set $\mathcal{M}$ together with a set of scalars $\mathcal{S}$ and two operations, addition and scalar multiplication such that $f+g$ and af belong to $\mathcal{M}$ for all f and g in $\mathcal{M}$ and all scalars a. The only scalar fields used in this book are $\mathbb{R}$ and $\mathbb{C}$. $\mathcal{M}$ is called a real, respectively complex, linear space according as the scalar field is $\mathbb{R}$ or $\mathbb{C}$. Addition and scalar multiplication in a linear space satisfy the following familiar rules. For all f, g, h in $\mathcal{M}$ and all scalars α and β,

$$\begin{aligned} f+g &= g+f, \\ (f+g)+h &= f+(g+h), \\ \alpha(f+g) &= \alpha f+\alpha g, \\ (\alpha+\beta)f &= \alpha f+\beta f, \\ 1f &= f. \end{aligned}$$

Finite dimensional linear spaces include $\mathbb{R}^n$ with scalar field $\mathbb{R}$, $\mathbb{C}^n$ with scalar field $\mathbb{C}$, and the space of $n \times m$ real or complex matrices with the usual algebraic operations and scalar fields $\mathbb{R}$ and $\mathbb{C}$, respectively. Infinite dimensional linear spaces that are functions spaces include the continuous function on an interval $[a, b]$, the differentiable functions on $[a, b]$, and the integrable functions on $[a, b]$.

We denote the linear space of real or complex-valued continuous functions on $[a, b]$ by $C[a, b]$. It will be clear from the context whether $C[a, b]$ is regarded as the real linear space with real-valued functions and real scalars or as the complex linear space with complex-valued functions and complex scalars.

The differentiable functions on $[a, b]$, denoted by $D[a, b]$, are a special subset of $C[a, b]$ because $D[a, b]$ is a linear space with the operations of addition and scalar multiplication it inherits from $C[a, b]$. We describe this relationship by saying that $D[a, b]$ is a subspace of $C[a, b]$. In general, a subset $\mathcal{N}$ of a linear space $\mathcal{M}$ is a **subspace** of $\mathcal{M}$ if $\mathcal{N}$ is a linear space in its own right with the addition and scalar multiplication it inherits from $\mathcal{M}$. It is routine to check that a subset $\mathcal{N}$ of $\mathcal{M}$ is a subspace of $\mathcal{M}$ if and only if it is closed under addition and scalar multiplication, which means $f+g$ belongs to $\mathcal{N}$ whenever f and g belong to $\mathcal{N}$ and, for any scalar a, af belongs to $\mathcal{N}$ whenever f belongs to $\mathcal{N}$.

It is convenient to use the following, mostly standard, notation. Let I be an interval of any type, open, closed, half-open, bounded, or unbounded. Then

$B(I)$ is the set of real or complex-valued bounded functions on I;

$C(I)$ is the set of real or complex-valued continuous functions on I;

$C^n(I)$ is the set of real or complex-valued functions whose nth derivative is continuous on I.

Here n is a positive integer. Sometimes $C^0(I)$ is an alternative notation for $C(I)$. Each of these spaces is a linear space with the usual operations of addition and scalar multiplication of functions. Each is a subspace of its predecessor.

If $f_1, f_2, \ldots, f_m$ are vectors in $\mathcal{M}$ and $c_1, c_2, \ldots, c_m$ are scalars,

$$c_1 f_1 + c_2 f_2 + \cdots + c_m f_m$$

is called a **(finite) linear combination** of the vectors $f_1, f_2, \ldots, f_m$. The notions of linear dependence and linear independence extend to a general linear space $\mathcal{M}$ in a natural way. For example, the vectors $f_1, f_2, \ldots, f_m, \ldots$ are linearly independent if the finite set of vectors $f_1, f_2, \ldots, f_m$ are **linearly independent** for every m. Otherwise the vectors $f_1, f_2, \ldots, f_m, \ldots$ are **linearly dependent**. The span of a set of vectors $\mathcal{V}$ in $\mathcal{M}$, denoted span $(\mathcal{V})$ is the set of all finite linear combinations of vectors in $\mathcal{V}$. A set of vectors $\mathcal{B}$ in $\mathcal{M}$ that is linearly independent and spans $\mathcal{M}$ is a basis for $\mathcal{M}$.

2.5.2 Normed Linear Spaces

Linear spaces with only the algebraic structure above are of limited value in applications outside algebra. They lack a notion of distance between points (or vectors) in $\mathcal{M}$ and of convergence in $\mathcal{M}$. Normed spaces bring in these missing elements. A linear space $\mathcal{M}$ is called a **normed (linear) space** if there is a function called a **norm**, $\|\cdot\|$, defined on $\mathcal{M}$ such that for all f, g in $\mathcal{M}$ and all scalars a

$$\begin{aligned}
&\text{(i)} \quad \|f\| \geq 0 \text{ with equality if and only if } f = 0,\\
&\text{(ii)} \quad \|\alpha f\| = |\alpha|\|f\|,\\
&\text{(iii)} \quad \|f + g\| \leq \|f\| + \|g\|.
\end{aligned}$$

$\|f\|$, the norm of f, is analogous to the length or magnitude of a vector in Euclidean spaces. The distance between two points f and g in a normed linear space is (by definition) $\|f - g\|$. A sequence $\{f_n\}_{n=1}^{\infty}$ **converges** to f in $\mathcal{M}$ and we write $f_n \to f$ if $\|f_n - f\| \to 0$ as $n \to \infty$. Once again, these definitions are motivated by the corresponding notions in Euclidean spaces, and, of course, $\mathbb{R}^n$ and $\mathbb{C}^n$ are normed spaces with the norm being the usual Euclidean distance of a vector from the origin.

If a sequence converges, it is usually important to know that its limit inherits some characteristics of the terms in the sequence. If the elements of the sequence lie in a closed set this is true. A set S in $\mathcal{M}$ is **closed** if every convergent sequence of elements in S has its limit in S. That is, S is closed if whenever s_n belong to S and $s_n \to x$ for some x in $\mathcal{M}$, then x belongs to S.

A set S in $\mathcal{M}$ is **bounded** if there is a constant M such that $\|s\| \leq M$ for all s in S.

The most important normed linear spaces for us are **function spaces**, linear spaces whose elements are functions and whose norms are chosen to induce the type of convergence that is needed to study a problem at hand. One such space is $C[a, b]$ equipped with the **maximum norm** (also known as the supremum norm or sup norm)

$$\|f\|_{\max} = \max_{a \leq x \leq b} |f(x)|.$$

We omit the easy check that the maximum norm is in fact a norm on $C[a, b]$. It follows immediately from the definitions that convergence in the maximum norm is uniform convergence on $[a, b]$. That is, $\|f_n - f\|_{\max} \to 0$ as $n \to \infty$ if and only if the sequence of continuous functions f_n converges uniformly on $[a, b]$ to f.

In problems in which an integral provides a more useful measure of the size (norm) of a continuous function than does the maximum norm, $C[a, b]$ is often equipped with one of the following norms

$$\|f\|_1 = \int_a^b |f(x)|\, dx \quad \text{or} \quad \|f\|_2 = \left(\int_a^b |f(x)|^2\, dx\right)^{1/2}.$$

It is easy to check that $\|f\|_1$ is a norm on $C[a, b]$. When $C[a, b]$ is equipped with the 1-norm the resulting normed space is denoted by $\mathcal{L}_1[a, b]$. When $C[a, b]$ is equipped with the 2-norm the resulting normed space is denoted by $\mathcal{L}_2[a, b]$. The check that $\|f\|_2$ is a norm is more involved than the check for $\|f\|_1$. We will return to it shortly from the point of view of inner product spaces.

Since the choice of a norm for a function space is dictated primarily by the type of convergence relevant to the situation at hand, it is important to realize that many different norms produce the same notion of convergence. Such norms are called equivalent and one of them may be more convenient to use than another. Two norms $\|\cdot\|_r$ and $\|\cdot\|_s$ on a normed space $\mathcal{M}$ are **equivalent** if there are constants $M \geq m > 0$ such that

$$m\|f\|_r \leq \|f\|_s \leq M\|f\|_r$$

for all f in $\mathcal{M}$. Equivalent norms induce the same notion of convergence in $\mathcal{M}$ because the relations

$$m\|f_n - f\|_r \leq \|f_n - f\|_s \leq M\|f_n - f\|_r$$

reveal that $f_n \to f$ in the r-norm if and only if $f_n \to f$ in the s-norm.

A norm that is equivalent to the maximum norm and is often more convenient to use when studying existence, uniqueness, and continuous dependence questions for initial value problems for differential equations is: fix $L > 0$ and for f in $C[a, b]$ define

$$\|f\|_L = \max_{a \leq x \leq b} e^{-L(x-a)}|f(x)|.$$

It is routine to check that $\|f\|_L$ is a norm on $C[a, b]$. It is equivalent to the maximum norm because

$$e^{-L(b-a)}|f(x)| \leq e^{-L(x-a)}|f(x)| \leq |f(x)|,$$

for all x in $[a, b]$, and, hence,

$$e^{-L(b-a)}\|f\|_{\max} \leq \|f\|_L \leq \|f\|_{\max}.$$

So convergence of functions in the L-norm is uniform convergence on $[a, b]$.

We mention in passing that none of the norms, the maximum norm, the 1-norm, or the 2-norm, are equivalent norms on $C[a, b]$.

2.5.3 Inner Product Spaces

Once again important notions from $\mathbb{R}^n$ and $\mathbb{C}^n$ serve as a guide. A real linear space $\mathcal{M}$ is a real inner product space if it is equipped with a real-valued function called an **inner product** (or dot product), denoted $\langle \cdot, \cdot \rangle$, that is defined for pairs of elements of $\mathcal{M}$ and assigns to them a value in the scalar field such that for any f, g, h in $\mathcal{M}$ and scalars α and β

$$\begin{aligned}
\langle f, f \rangle &\geq 0 \text{ with equality if and only if } f = 0, \\
\langle \alpha f + \beta g, h \rangle &= \alpha\langle f, h \rangle + \beta\langle g, h \rangle, \\
\langle f, g \rangle &= \langle g, f \rangle.
\end{aligned}$$

The inner product is linear in its first variable and, by the symmetry property, it is linear in its second variable as well.

Correspondingly, a complex linear space $\mathcal{M}$ is a complex inner product space if it is equipped with a complex-valued inner product, denoted $\langle \cdot, \cdot \rangle$, that is defined for pairs of elements of $\mathcal{M}$ and assigns to them a value in the scalar field such that for any f, g, h in $\mathcal{M}$ and scalars α and β

$$\begin{aligned} \langle f, f\rangle &\geq 0 \text{ with equality if and only if } f = 0, \\ \langle \alpha f + \beta g, h\rangle &= \alpha\langle f, h\rangle + \beta\langle g, h\rangle, \\ \langle f, g\rangle &= \overline{\langle g, f\rangle}. \end{aligned}$$

Consequently, the inner product is linear in its first variable and, by the complex symmetry property, it is conjugate linear in its second variable,

$$\langle f, \alpha g + \beta h\rangle = \bar{\alpha}\langle f, g\rangle + \bar{\beta}\langle f, h\rangle.$$

A real or complex inner product space is also a normed linear space with the (induced) norm $\|f\| = \sqrt{\langle f, f\rangle}$. Inner product spaces are normed spaces with additional structure. An inner product space is always equipped with the norm $\|f\| = \sqrt{\langle f, f\rangle}$, unless explicitly stated to the contrary. The confirmation that $\|f\| = \sqrt{\langle f, f\rangle}$ is a norm follows most easily from the following inequality:

Lemma 38 (Schwarz Inequality) *If f and g are elements of an inner product space $\mathcal{M}$, then*

$$|\langle f, g\rangle| \leq \|f\|\|g\|.$$

Proof. We may assume in the proof that $g \neq 0$ because if $g = 0$ the inequality is evident. Assume that $\mathcal{M}$ is a complex inner product space. For any complex scalar λ,

$$0 \leq \langle f + \lambda g, f + \lambda g\rangle = \langle f, f\rangle + \lambda\langle g, f\rangle + \bar{\lambda}\langle f, g\rangle + |\lambda|^2\langle g, g\rangle$$

or, since $\langle g, f\rangle = \overline{\langle f, g\rangle}$,

$$0 \leq \langle f + \lambda g, f + \lambda g\rangle = \langle f, f\rangle + 2\mathrm{Re}\lambda\overline{\langle f, g\rangle} + |\lambda|^2\langle g, g\rangle.$$

Set $\lambda = -\langle f, g\rangle/\|g\|^2$ to obtain

$$0 \leq \|f\|^2 - \frac{|\langle f, g\rangle|^2}{\|g\|^2}.$$

The desired conclusion follows. The same proof works for a real inner product space. ■

The choice for λ in the proof of the Schwarz inequality was not as arbitrary as it might have appeared. If the inner product space is real, for $g \neq 0$ and any real scalar λ,

$$0 \leq \langle f + \lambda g, f + \lambda g\rangle = \langle f, f\rangle + 2\lambda\langle f, g\rangle + \lambda^2\langle g, g\rangle.$$

The quadratic in the right member of this inequality is minimized for $\lambda = -\langle f, g\rangle/\|g\|^2$, the value for λ selected in the proof. It is also informative to observe that the graph of the parabola in λ never crosses the λ-axis. Consequently, its discriminant must be nonpositive; that is, $4\langle f, g\rangle^2 - 4\|f\|^2\|g\|^2 \leq 0$, which gives another proof of the Schwarz inequality for real inner product spaces.

Lemma 39 *The assignment $\|f\| = \sqrt{\langle f, f\rangle}$ defines a norm in any inner product space $\mathcal{M}$.*

Proof. For any f and g in $\mathcal{M}$ and any scalar a, clearly $\|f\| \geq 0$ with equality if and only if $f = 0$ and

$$\|\alpha f\| = \sqrt{\langle \alpha f, \alpha f \rangle} = \sqrt{\alpha \bar{\alpha} \langle f, f \rangle} = |\alpha| \|f\|.$$

The triangle inequality follows from the Schwarz inequality:

$$\begin{aligned} \|f+g\|^2 &= \langle f+g, f+g \rangle = \|f\|^2 + 2\mathrm{Re}\langle f, g \rangle + \|g\|^2 \\ &\leq \|f\|^2 + 2|\langle f, g \rangle| + \|g\|^2 \leq \|f\|^2 + 2\|f\|\|g\| + \|g\|^2 = \left(\|f\| + \|g\|\right)^2 \end{aligned}$$

Taking square roots establishes the triangle inequality and $\|f\| = \sqrt{\langle f, f \rangle}$ is a norm on $\mathcal{M}$. ■

The assignment

$$\langle f, g \rangle = \int_a^b f(x)\overline{g(x)}\, dx$$

defines an inner product on the complex linear space $C[a, b]$. The same assignment with the bar removed defines an inner product on the real linear space $C[a, b]$. We omit the routine check that this assignment is an inner product on $C[a, b]$. In either case, we denote the inner product space by $\mathcal{L}_2[a, b]$. The corresponding norm induced by the inner product is

$$\|f\|_2 = \left(\int_a^b |f(x)|^2\, dx \right)^{1/2},$$

the 2-norm. In this setting the Schwarz inequality $|\langle f, g \rangle| \leq \|f\|_2 \|g\|_2$ is

$$\left| \int_a^b f(x)\overline{g(x)}\, dx \right| \leq \left(\int_a^b |f(x)|^2\, dx \right)^{1/2} \left(\int_a^b |g(x)|^2\, dx \right)^{1/2}.$$

Although it is not obvious, we mention in passing that there is no inner product on $C[a, b]$ whose induced norm is the 1-norm introduced earlier. So $\mathcal{L}_1[a, b]$ is a normed space but is not an inner product space.

Other useful inner products on $C[a, b]$ are determined by weight functions. A **weight function** $r(x)$ is a nonnegative continuous function on $[a, b]$ such that $\int_c^d r(x)\, dx > 0$ for every subinterval $[c, d]$ of $[a, b]$. In typical applications $r(x) > 0$ on $[a, b]$, except perhaps for a finite number of points z where $r(z) = 0$. The associated inner product on $C[a, b]$ is defined by

$$\langle f, g \rangle_r = \int_a^b f(x)\overline{g(x)} r(x)\, dx$$

and the norm induced by the weighted inner product is $\|f\|_r = \sqrt{\langle f, f \rangle_r}$. Weighted inner products arise naturally when the behavior of a function for certain values of x in $[a, b]$ is more important than its values at other points in $[a, b]$, for the problem under study.

The standard basis in Euclidean 3-space, $\mathbf{i}$, $\mathbf{j}$, and $\mathbf{k}$, is especially convenient because the relations

$$\mathbf{i} \cdot \mathbf{i} = \mathbf{j} \cdot \mathbf{j} = \mathbf{k} \cdot \mathbf{k} = 1,$$

and

$$\mathbf{i} \cdot \mathbf{j} = \mathbf{i} \cdot \mathbf{k} = \mathbf{j} \cdot \mathbf{k} = 0,$$

imply that any vector $\mathbf{v}$ in 3-space can be expressed by

$$\mathbf{v} = v_1 \mathbf{i} + v_2 \mathbf{j} + v_3 \mathbf{k}$$

with components and norm easily evaluated by

$$v_1 = \mathbf{v} \cdot \mathbf{i},\ v_2 = \mathbf{v} \cdot \mathbf{j},\ v_3 = \mathbf{v} \cdot k$$

and

$$\|\mathbf{v}\| = \sqrt{v_1^2 + v_2^2 + v_3^2}.$$

In any n-dimensional inner product space $\mathcal{M}$ it is possible to construct an **orthonormal set** of elements

$$\phi_1,\ \phi_2, \ldots,\ \phi_n$$

such that $\langle \phi_j, \phi_j \rangle = 1$ and $\langle \phi_j, \phi_k \rangle = 0$ for $j \neq k$ and $j, k = 1, 2, \ldots, 77n$. (See the Gram-Schmidt process later in this section.) Normal in the word orthonormal means that each vector is normalized to have length 1 and ortho means that each pair of vectors is orthogonal. Two vectors ϕ and ψ are **orthogonal** in an inner product space if $\langle \phi, \psi \rangle = 0$. The elements in an orthonormal set are always linearly independent. Since the space is n-dimensional, they are a basis. Hence, given f in $\mathcal{M}$ there are constants $f_1, \ldots, f_n$ such that

$$f = f_1\phi_1 + f_2\phi_2 + \cdots + f_n\phi_n.$$

Take inner product of each side with ϕ_j to find that

$$f_j = \langle f, \phi_j \rangle$$

and that

$$\|f\| = \left(\sum_{j=1}^{n} |\langle f, \phi_j \rangle|^2 \right)^{1/2},$$

all in strict analogy with the situation in three space.

Inner product spaces $\mathcal{M}$ that are function spaces are typically infinite dimensional. In such spaces it is always possible to construct an orthonormal sequence of elements

$$\phi_1,\ \phi_2, \ldots,\ \phi_n, \ldots$$

with $\langle \phi_j, \phi_j \rangle = 1$ for $j = 1, 2, \ldots$ and $\langle \phi_j, \phi_k \rangle = 0$ for $j \neq k$ and $j, k = 1, 2, \ldots$. (See the Gram-Schmidt process later in this section.) If f is in $\mathcal{M}$, then

$$\left\| f - \sum_{n=1}^{N} \langle f, \phi_j \rangle \phi_j \right\|^2 = \|f\|^2 - \sum_{n=1}^{N} |\langle f, \phi_j \rangle|^2.$$

This is confirmed by a straightforward expansion of the inner product of $f - \sum_{n=1}^{N} \langle f, \phi_j \rangle \phi_j$ with itself. Since the left member is nonnegative,

$$\sum_{n=1}^{N} |\langle f, \phi_j \rangle|^2 \leq \|f\|^2$$

and, letting $N \to \infty$,

$$\sum_{n=1}^{\infty} |\langle f, \phi_j \rangle|^2 \leq \|f\|^2, \tag{2.3}$$

which is **Bessel's inequality**. It follows directly from these considerations that equality holds in Bessel's inequality for every f in $\mathcal{M}$ if and only if

$$f = \sum_{n=1}^{\infty} \langle f, \phi_j \rangle \phi_j,$$

for every f in $\mathcal{M}$ in which case $\phi_1, \phi_2, \ldots, \phi_n, \ldots$ is called an **orthonormal basis** for $\mathcal{M}$. The inner product $\langle f, \phi_j \rangle$ is called the jth **Fourier coefficient** of f with respect to the orthonormal system $\phi_1, \phi_2, \ldots$ because of connections with Fourier series.

2.5.3.1 Gram-Schmidt Process

In an inner product space $\mathcal{M}$, it is usually advantageous to replace a linearly independent set of vectors by an orthonormal set that spans the same subspace of $\mathcal{M}$. The Gram-Schmidt process is a natural way to find such an orthonormal set.

We describe the process for a sequence of linearly independent vectors $f_1, f_2, \ldots, f_n, \ldots$ in $\mathcal{M}$. Define vectors

$$g_1 = f_1,$$
$$g_2 = f_2 - \frac{\langle f_2, g_1 \rangle}{\langle g_1, g_1 \rangle} g_1,$$
$$g_3 = f_3 - \frac{\langle f_3, g_1 \rangle}{\langle g_1, g_1 \rangle} g_1 - \frac{\langle f_3, g_2 \rangle}{\langle g_2, g_2 \rangle} g_2,$$
$$\vdots$$

where g_n is f_n minus its projection on all the previously constructed vectors $g_1, g_2, \ldots, g_{n-1}$. It is routine to check step-by-step that the vectors $g_1, g_2, \ldots, g_n, \ldots$ are nonzero because $f_1, f_2, \ldots, f_n, \ldots$ are linearly independent, that f_n is a linear combination of $g_1, g_2, \ldots, g_n$ for each n, and that $g_1, g_2, \ldots, g_n, \ldots$ are orthogonal. Thus,

$$\phi_1 = \frac{g_1}{\|g_1\|}, \; \phi_2 = \frac{g_2}{\|g_2\|}, \; \ldots, \; \phi_n = \frac{g_n}{\|g_n\|}, \; \ldots$$

is an orthonormal set such that

$$\operatorname{span}(f_1, f_2, \ldots, f_n) = \operatorname{span}(\phi_1, \phi_2, \ldots, \phi_n)$$

for each positive integer n. The vectors $\phi_1, \phi_2, \ldots, \phi_n, \ldots$ are said to be obtained from $f_1, f_2, \ldots, f_n, \ldots$ by the Gram-Schmidt process.

The following simple observation is useful. If $f_1, f_2, \ldots, f_n, \ldots$ are real or complex-valued continuous functions on an interval $[a, b]$ and

$$\langle f, g \rangle = \int_a^b f(x)\overline{g(x)}\, dx$$

is the usual inner product, then the orthonormal sequence $\phi_1, \phi_2, \ldots, \phi_n, \ldots$ obtained by the Gram-Schmidt process consists of complex-valued functions. However, if the original sequence $f_1, f_2, \ldots, f_n, \ldots$ is a sequence of real-valued functions, then the orthonormal sequence $\phi_1, \phi_2, \ldots, \phi_n, \ldots$ also consists of real-valued functions. This is true because all the inner products and functions that occur in the orthogonalization process are real-valued.

2.6 Completeness and Completion

In typical applications involving normed spaces, sequences of successive approximations $\{f_n\}_{n=1}^{\infty}$ are constructed in order to solve a particular problem. This means that f_n is an approximate solution to a problem we want to solve and that the approximations improve in the sense that $f_n \to f$ as $n \to \infty$, where f is the exact solution. In most real-world applications, the exact solution f cannot be found by analytical methods. It is known only through the sequence of approximations $\{f_n\}$. So how can we test that $f_n \to f$, that is, that $\|f_n - f\| \to 0$ if we do not know f? Cauchy answered this question for the real number system in the nineteenth century. A sequence of real numbers $\{x_n\}$ converges, that is, there is a real number x such that $x_n \to x$ as $n \to \infty$, if and only if the following condition is satisfied: given any $\varepsilon > 0$ there is an integer $N > 0$ such that $|x_{n+p} - x_n| < \varepsilon$ for $n > N$ and all natural numbers p. Notice that Cauchy's test for convergence does not require advance knowledge of the limit.

The primary motivation for the development of the real number system was to fill in holes or gaps in the rational number line that exist because the characterization of convergence given by Cauchy does not hold in the rational number system. In that system, sequences that "appear to converge," that is, that satisfy the Cauchy condition of the last paragraph, can fail to converge. The real numbers are obtained from the rational numbers by a completion process that fills in the gaps and supplies the missing limits. Since the system of rational numbers $\mathbb{Q}$ is a normed linear space with norm the usual absolute value, there are normed linear spaces in which the Cauchy condition for convergence fails. This is a serious problem for the successive approximation approach. It is very important to know which spaces of interest behave like the real number system in that sequences that "appear to converge" really do converge and which spaces lack this property.

A sequence $\{f_n\}$ in a normed space $\mathcal{M}$ is called a **Cauchy sequence** if given any $\varepsilon > 0$ there is an integer $N > 0$ such that $\|f_{n+p} - f_n\| < \varepsilon$ for $n > N$ and all natural numbers p. (A Cauchy sequence is also called a fundamental sequence.) Convergent sequences in normed spaces are Cauchy sequences. Indeed, if $f_n \to f$, then given $\varepsilon > 0$ there is a positive integer N such that

$$\|f_n - f\| < \varepsilon/2 \quad \text{for } n > N.$$

By the triangle inequality

$$\|f_{n+p} - f_n\| \leq \|f_{n+p} - f\| + \|f - f_n\| < \varepsilon \quad \text{for } n > N \text{ and all natural numbers } p.$$

That is, convergent sequences always have the Cauchy property. The converse is true in the real number system as Cauchy discovered; however, the converse is not true in all normed spaces. (The rational number system is but one example. Examples in function spaces are coming.) A space in which every Cauchy sequence converges is said to be **complete**. A complete normed linear space is called a **Banach space**. Since inner product spaces are normed spaces the definitions of Cauchy and complete apply to inner product spaces. A complete inner product space is called a **Hilbert space.**

Among the three function spaces of particular interest for us, $C[a, b]$ with the maximum norm, $\mathcal{L}_1[a, b]$ and $\mathcal{L}_2[a, b]$ only the first is complete. We will address the lack of completeness of $\mathcal{L}_1[a, b]$ and $\mathcal{L}_2[a, b]$ shortly. First, we confirm that $C[a, b]$ with the maximum norm is complete and, hence, a Banach space.

Lemma 40 *$C[a, b]$ equipped with the maximum norm is a Banach space.*

Proof. We must show that every Cauchy sequence $\{f_n\}$ in $C[a, b]$ equipped with the maximum norm converges to a function f in $C[a, b]$. Since $\{f_n\}$ is Cauchy, given any $\varepsilon > 0$ there exist N such that $\|f_{n+p} - f_n\|_{\max} < \varepsilon$ for all $n > N$ and all positive integers p. It follows that for each x in $[a, b]$ and all $n > N$ and all positive integers p

$$|f_{n+p}(x) - f_n(x)| \leq \|f_{n+p} - f_n\|_{\max} < \varepsilon.$$

Thus, $\{f_n(x)\}$ is a Cauchy sequence in the real or complex numbers. Since these spaces are complete, $\lim_{n\to\infty} f_n(x)$ exists for each x in $[a, b]$. For each x denote the limit by $f(x)$. This defines a real or complex-valued function on $[a, b]$. Let $p \to \infty$ in the displayed inequality to obtain: for each x in $[a, b]$ and all $n > N$

$$|f(x) - f_n(x)| \le \varepsilon.$$

This establishes that the sequence of continuous functions f_n converges uniformly on $[a, b]$ to the function f. The limit function f is continuous on $[a, b]$ by Theorem 23. Finally, since $|f(x) - f_n(x)| \le \varepsilon$ for all x in $[a, b]$ it follows that

$$\|f - f_n\|_{\max} = \max_{a\le x\le b} |f(x) - f_n(x)| \le \varepsilon \quad \text{for all } n > N;$$

that is, $f_n \to f$ in $C[a, b]$ with the maximum norm. Thus, $C[a, b]$ equipped with the maximum norm is complete; it is a Banach space. ■

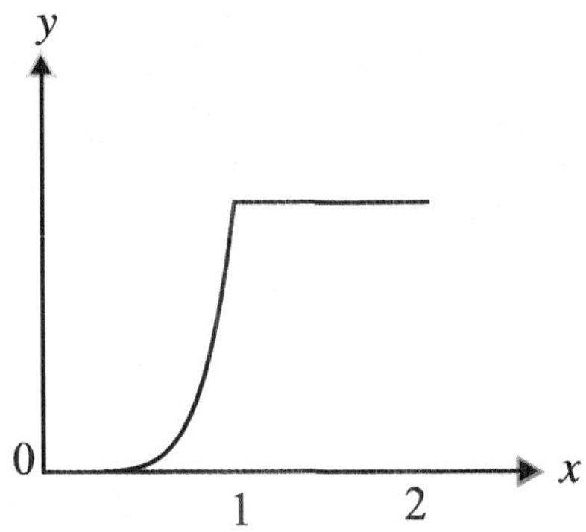

FIGURE 2.1: Graph of $f_n(x)$ for $n = 6$

We asserted that the spaces $\mathcal{L}_1[a, b]$ and $\mathcal{L}_2[a, b]$, the spaces $C[a, b]$ equipped with the 1-norm and the 2-norm respectively, are not complete. The following example confirms the assertion for $\mathcal{L}_2[a, b]$. Minor modifications in the argument confirm it for $\mathcal{L}_1[a, b]$. The sequence of continuous functions

$$f_n(x) = \begin{cases} x^n & \text{for } 0 \le x < 1 \\ 1 & \text{for } 0 \le x < 1 \end{cases}$$

is a Cauchy sequence in $\mathcal{L}_2[0, 2]$ because

$$\|f_{n+p} - f_n\|_2^2 = \int_0^1 \left(x^{n+p} - x^n\right)^2 dx = \int_0^1 \left(x^{2n+2p} - 2x^{2n+p} + x^{2n}\right) dx$$
$$= \frac{1}{2n+2p+1} - \frac{2}{2n+p+1} + \frac{1}{2n+1} < \frac{2}{2n+1}$$

can be made arbitrarily small for all p by taking n sufficiently large. See Figure 2.1.

If there were a function f in $\mathcal{L}_2[0, 2]$ such that $\|f_n - f\|_2 \to 0$, then f would be continuous on $[a, b]$ and

$$0 = \lim_{n\to\infty} \|f_n - f\|_2^2 = \lim_{n\to\infty} \left(\int_0^1 \left(x^n - f(x)\right)^2 dx + \int_1^2 \left(1 - f(x)\right)^2 dx \right)$$
$$= \lim_{n\to\infty} \int_0^1 \left(x^n - f(x)\right)^2 dx + \int_1^2 \left(1 - f(x)\right)^2 dx.$$

Consequently,

$$\lim_{n\to\infty}\int_0^1 \left(x^n - f(x)\right)^2 dx = 0 \quad \text{and} \quad \int_1^2 \left(1 - f(x)\right)^2 dx = 0.$$

Since f is continuous it follows that $f(x) = 1$ for $1 \le x \le 2$, and by the mean value theorem for integrals

$$\int_0^1 \left(x^n - f(x)\right)^2 dx = \frac{1}{2n+1} + \frac{2f(\xi_n)}{n+1} + \int_0^1 f(x)^2\, dx$$

for some ξ_n in $[0, 1]$. Since the continuous function f is bounded on $[a, b]$, letting $n \to \infty$ yields

$$\int_0^1 f(x)^2\, dx = 0$$

and $f(x) = 0$ for $0 \le x \le 1$. If there were a function f in $\mathcal{L}_2[a, b]$ to which the Cauchy sequence converged, then $f(1) = 0$ and $f(1) = 1$, which contradicts the fact that a function must be single-valued. Thus, the Cauchy sequence $\{f_n\}$ does not converge to a function in $\mathcal{L}_2[a, b]$.

There is a function f to which the Cauchy sequence above converges but it is not in $\mathcal{L}_2[a, b]$ because it is not continuous. In fact, it is not difficult to guess what the limit function is because the usual pointwise limit

$$\lim_{n\to\infty} f_n(x) = \begin{cases} 0 & \text{for } 0 \le x < 1 \\ 1 & \text{for } 1 \le x \le 2 \end{cases}$$

exists and the discontinuous function $f(x)$ defined by the two-part formula on the right is the missing limit function.

The situation in this example in $\mathcal{L}_2[a, b]$ is analogous to a similar situation that arose centuries ago and was confronted by the Pythagoreans. The natural numbers and their quotients, the positive rational numbers, were known to the ancient Greeks. It was thought (hoped) by the Pythagoreans that all geometric lengths could be expressed by one of these numbers. They discovered that the length of the diagonal of a square of side 1 was not given by such a rational number. The Pythagoreans expressed this by saying that the diagonal of the square is incommensurate with its side. Today we say the $\sqrt{2}$ is an irrational number. There is no point on the rational number line corresponding to $\sqrt{2}$. There is a gap or hole there. On the other hand, there are rational numbers that approximate this missing number as accurately as may be desired. For example, the familiar sequence 1.4, 1.41, 1.414, 1, 4142, 1.41423, ... The nth rational number q_n in this sequence is chosen so that $q_n^2 < 2$ and so that the rational number $q_n' = q_n + 10^{-n}$ satisfies $\left(q_n'\right)^2 > 2$. It follows that $\{q_n\}$ is a Cauchy sequence in $\mathbb{Q}$ that has no limit in $\mathbb{Q}$ but whose terms cluster about a "hole" in the rational number line that corresponds to the missing limit, the real number $\sqrt{2}$. The real number system $\mathbb{R}$ is constructed from the rational number system $\mathbb{Q}$ by adjoining to it all the missing limits of Cauchy sequences in $\mathbb{Q}$ that fail to converge to rational limits. The construction preserves the algebraic structure of $\mathbb{Q}$ and distances in $\mathbb{Q}$ while adding the missing limits and doing so in an economical way. Economical means $\mathbb{Q}$ is dense in its completion $\mathbb{R}$. That is, every real number is the limit of a sequence of rational numbers. In the same way, a set D is **dense** in a normed linear space $\mathcal{L}$ if every point in $\mathcal{L}$ is the limit of a sequence of elements in D.

It turns out that virtually the same construction, called completion, can be carried out in any normed linear space $\mathcal{L}$. The completion L of a normed linear space $\mathcal{L}$ is a Banach space that preserves the algebraic structure of $\mathcal{L}$, preserves distances between points in $\mathcal{L}$, and $\mathcal{L}$ is a dense subset of its completion L. When the completion process is carried out for

$\mathcal{L}_2[a, b]$ with the 2-norm the resulting Banach space is denoted by $L_2[a, b]$. It is the space of Lebesgue square integrable functions on $[a, b]$. The norm in $L_2[a, b]$ is still denoted by $\| \cdot \|_2$. Since $\mathcal{L}_2[a, b]$ is dense in $L_2[a, b]$, given a function f in $L_2[a, b]$ there is a sequence of continuous functions f_n such that $f_n \to f$ in $L_2[a, b]$; that is, $\|f_n - f\|_2 \to 0$. Corresponding remarks apply $L_1[a, b]$.

The completion process provides a strategy for proving theorems in the completion. For example, to establish a result in $L_2[a, b]$ it often suffices to establish it first in $\mathcal{L}_2[a, b]$, where all functions are continuous, and then to extend the result to $L_2[a, b]$ by a limiting argument using the fact that for any f in $L_2[a, b]$ there is a sequence of continuous functions f_n on $[a, b]$ such that $f_n \to f$ in $L_2[a, b]$.

2.7 Compact Sets in $C[a, b]$

Notions of compactness play essential roles in many parts of mathematics and its applications. In very general terms, compactness is a property of a set S (in a space with a notion of convergence) that makes it behave in many respects just as if it were a finite set. For example, a real-valued function f defined on a finite subset in $\mathbb{R}$ (is automatically continuous) and obviously takes on minimum and maximum values. A continuous function f defined on a compact set S in $\mathbb{R}$ has the same property.

Compact sets can be defined in settings much more general that we shall encounter. All the sets we shall deal with lie in normed linear spaces. In that setting, perhaps the most useful definition of compactness is suggested by the Bolzano-Weierstrass theorem in $\mathbb{R}$. A set S in a normed linear space is **compact** if every sequence of elements in S contains a subsequence that converges to a limit that lies in S. We already have defined compact sets in $\mathbb{R}^n$ and $\mathbb{C}^n$ in this way and stated the Heine-Borel theorem: a set S in $\mathbb{R}^n$ or $\mathbb{C}^n$ is compact if and only if it is closed and bounded.

For our purposes, we also need to identify the compact sets in $C[a, b]$, the space of continuous real or complex valued functions on $[a, b]$. The next theorem characterizes compact sets in $C[a, b]$ when it is equipped with the maximum norm. For a proof see [9] or [35].

Theorem 41 (Arzelà-Ascoli) *A set S of functions in $C[a, b]$ equipped with the maximum norm is compact if and only if it is uniformly bounded and equicontinuous on $[a, b]$.*

The terminology use in the Arzelà-Ascoli theorem needs some elaboration. A collection of real or complex-valued functions S defined on $[a, b]$ is **uniformly bounded on** $[a, b]$ if for all f in S, $|f(x)| \leq M$ for all x in $[a, b]$.

There are two notions of equicontinuity for families of functions, equicontinuity of the family on a set and equicontinuity of the family at a point of a set. Both concepts are useful.

A set of functions S in $C[a, b]$ is **equicontinuous on** $[a, b]$ if given any $\varepsilon > 0$ there is a $\delta > 0$, dependent only on ε, such that for all f in S, $|f(x) - f(x')| < \varepsilon$ whenever x and x' in $[a, b]$ satisfy $|x - x'| < \delta$.

If S consists of a single function f, equicontinuity is just uniform continuity of f on $[a, b]$. The "equi" in equicontinuity means that uniform continuity of f holds uniformly across all functions f in S. That is, the δ in the definition of uniform continuity that depends on ε and the function f in question can be chosen independently of the functions in an equicontinuous family.

If the common domain of a set of functions S is compact, then equicontinuity on the domain follows from equicontinuity at each point in the domain. This observation often makes it easier to verify equicontinuity on a domain. A set of functions S in $C[a, b]$ is **equicontinuous at** x_0 in $[a, b]$ if given any $\varepsilon > 0$ there is a $\delta > 0$ depending on ε and on x_0 such that for all f in S, $|f(x) - f(x_0)| < \varepsilon$ for x in $[a, b]$ with $|x - x_0| < \delta$.

The following table helps to distinguish between the indicated type of continuity: in the table f is a function, $\mathcal{F}$ is a family of functions, all functions are defined on a set D, and a positive ε is given.

Type of Continuity	δ(Depends On)
f is continuous at x_0 in D	$\delta(f, \varepsilon, x_0)$
f is uniformly continuous on D	$\delta(f, \varepsilon, D)$
F is equicontinuous continuous at x_0 in D	$\delta(\mathcal{F}, \varepsilon, x_0)$
$\mathcal{F}$ is equicontinuous continuous on D	$\delta(\mathcal{F}, \varepsilon, D)$

Proposition 42 *A set S of functions in $C[a, b]$ is equicontinuous on $[a, b]$ if it is equicontinuous at x_0 for every x_0 in $[a, b]$.*

Proof. We use proof by contradiction. If the lemma were false, then S would not be equicontinuous on $[a, b]$. Consequently, there must be an $\varepsilon_0 > 0$ such that no $\delta > 0$ exists such that for all f in S, $|f(x) - f(x')| < \varepsilon_0$ for x and x' in $[a, b]$ with $|x - x'| < \delta$. For $n = 1, 2, 3, \ldots$ and $\delta = 1/n$ this means that there is a function in S, say f_n, and points in $[a, b]$, say x_n and x'_n, such that

$$|f_n(x_n) - f_n(x'_n)| \geq \varepsilon_0 \quad \text{and} \quad |x_n - x'_n| < 1/n.$$

The sequence $\{x_n\}$ has a convergent subsequence with limit in $[a, b]$ because $[a, b]$ is compact. That is, there is c in $[a, b]$ and a subsequence $\{x_{n_k}\}$ of $\{x_n\}$ such that $x_{n_k} \to c$ as $k \to \infty$. By equicontinuity of S at c, given $\varepsilon_0/2$ there is a $\delta_c > 0$ such that for all f in S, $|f(x) - f(c)| < \varepsilon_0/2$ for x in $[a, b]$ with $|x - c| < \delta_c$. Since x_{n_k} and x'_{n_k} are in $[a, b]$, $x_{n_k} \to c$ as $k \to \infty$, and $|x_{n_k} - x'_{n_k}| < 1/n_k$, it follows that $x'_{n_k} \to c$ as $k \to \infty$. Hence, $|x_{n_k} - c| < \delta_c$ and $|x'_{n_k} - c| < \delta_c$ for all k sufficiently large and for such k,

$$|f_{n_k}(x_{n_k}) - f_{n_k}(c)| < \varepsilon_0/2 \quad \text{and} \quad |f_{n_k}(x'_{n_k}) - f_{n_k}(c)| < \varepsilon_0/2.$$

Consequently,

$$|f_{n_k}(x_{n_k}) - f_{n_k}(x'_{n_k})| \leq |f_{n_k}(x_{n_k}) - f_{n_k}(c)| + |f_{n_k}(c) - f_{n_k}(x'_{n_k})| < \varepsilon_0,$$

which contradicts $|f_{n_k}(x_{n_k}) - f_{n_k}(x'_{n_k})| \geq \varepsilon_0$. This contradiction establishes the lemma. ■

2.8 Contraction Mapping Theorem

The method of successive approximations works as follows. An initial guess, say x_0, is made of the solution to a difficult problem. A reasonable process is known (think Newton's method) that will produce an apparently better, updated guess, say x_1, to the solution. If the reasonable process is expressed as a function f, then $x_1 = f(x_0)$ and the successive approximations $x_0, x_1, x_2, \ldots$ are generated by

$$x_{n+1} = f(x_n)$$

for $n = 0, 1, 2, \ldots$. The hope is that as $n \to \infty$ the approximate solutions converge to the exact solution, say $\bar{x}$. If the hope is realized, that is if $x_n \to \bar{x}$, and if f is continuous, then letting $n \to \infty$ in the recursion formula yields an equation satisfied by the solution $\bar{x}$,

$$\bar{x} = f(\bar{x}).$$

A point $\bar{x}$ that satisfies $\bar{x} = f(\bar{x})$ is called a **fixed point** of the function f. A fixed point x of f is also a **zero** (root) of the equation

$$F(x) = 0,$$

where $F(x) = x - f(x)$, and conversely. Thus, solving equations and determining fixed points are two approaches to the same problem. Which point of view is taken is often of great importance.

The theorem that follows, in the generality stated here, was first proved by Caccioppoli and is often attributed to Caccioppoli and Banach. The underlying method was used earlier by Picard to establish the existence and uniqueness of solutions to initial value problems for rather general differential equations and has its roots in work of Kepler. It is now referred to as the contraction mapping theorem. A function (mapping, transformation) f from a subset S of a normed space into that space is a **contraction** (contraction mapping) if there is a constant ρ with $0 < \rho < 1$ such that

$$\|f(x') - f(x)\| \le \rho\|x' - x\| \text{ for all } x' \quad \text{and} \quad x \text{ in } S.$$

The constant ρ is called a **contraction constant**. Note that a contraction f on S is continuous; in fact, it is uniformly continuous on S.

Theorem 43 (Contraction Mapping Theorem) *If $f: C \to C$ is a contraction defined on a closed subset C of a Banach space $\mathcal{M}$, then f has a unique fixed point $\bar{x}$ in C. If x_0 is any point in C and successive approximations are defined by $x_{n+1} = f(x_n)$ for $n = 0, 1, 2, \ldots$, then $x_n \to \bar{x}$ as $n \to \infty$. Moreover, if f has contraction constant ρ, the error in x_n as an approximation to $\bar{x}$ satisfies*

$$\|x_n - \bar{x}\| \le \frac{\rho^n}{1 - \rho}\|x_1 - x_0\|.$$

Proof. With the notation as in the theorem, we first establish that $\{x_n\}$ is a Cauchy sequence: since

$$x_{n+p} - x_n = (x_{n+1} - x_n) + (x_{n+2} - x_{n+1}) + \cdots + (x_{n+p} - x_{n+p-1}),$$

$$\|x_{n+p} - x_n\| \le \sum_{k=0}^{p-1} \|x_{n+k+1} - x_{n+k}\|.$$

The difference of consecutive approximations can be estimated as follows,

$$\begin{aligned}\|x_{n+k+1} - x_{n+k}\| &= \|f(x_{n+k}) - f(x_{n+k-1})\| \le \rho\|x_{n+k} - x_{n+k-1}\| \\ &\le \rho(\rho\|x_{n+k-1} - x_{n+k-2}\|) \le \cdots \le \rho^{n+k}\|x_1 - x_0\|.\end{aligned}$$

Thus,

$$\|x_{n+p} - x_n\| \le \sum_{k=0}^{p-1} \rho^{n+k}\|x_1 - x_0\| \le \rho^n\|x_1 - x_0\| \sum_{k=0}^{\infty} \rho^k$$

and summing the geometric series yields

$$\|x_{n+p} - x_n\| \le \frac{\rho^n}{1-\rho}\|x_1 - x_0\|.$$

Since $0 < \rho < 1, \rho^n \to 0$ as $n \to \infty$ and the right member of the inequality can be made as small as desired for all n sufficiently large and for all p. Consequently, $\{x_n\}$ is a Cauchy sequence. Since $\mathcal{M}$ is a Banach space,

$$x_n \to \bar{x} \text{ for some } \bar{x} \text{ in } M.$$

Since C is closed and $x_n \to \bar{x}$ it follows that $\bar{x}$ belongs to C. Since f is continuous on C, passing to the limit in $x_{n+1} = f(x_n)$ gives

$$\bar{x} = f(\bar{x});$$

that is, $\bar{x}$ is a fixed point of f. Let $p \to \infty$ in the inequality above to conclude that

$$\|\bar{x} - x_n\| \le \frac{\rho^n}{1-\rho}\|x_1 - x_0\|.$$

It remains to prove that $\bar{x}$ is the only fixed point of f in C. If y were also a fixed point of f in C, then

$$\|y - \bar{x}\| = \|f(y) - f(\bar{x})\| \le \rho\|y - \bar{x}\|,$$

$\|y - \bar{x}\| = 0$ because $0 < \rho < 1$, and $y = \bar{x}$, which establishes uniqueness of the fixed point. ■

The following simple example illustrates the use of the contraction mapping theorem as a means for solving equations. It also illustrates that some ingenuity is required. The cubic equation $x^3 + 2x - 1 = 0$ has exactly one real root somewhere in the interval $[0, 1]$ because $f(x) = x^3 + 2x - 1$ satisfies $F'(x) = 3x^2 + 2 > 0$ so F is increasing, $f(0) = -1$, and $f(1) = 2$. One way to express the equation to be solved in fixed point form is

$$x = \frac{1}{x^2 + 2} = f(x).$$

Since $0 \le 1/(x^2 + 2) \le 1/2$ for all x, $f: C \to C$ for $C = [0, 1/2]$ and f is a contraction on C with contraction constant $1/4$ because

$$\left|\frac{1}{x^2+2} - \frac{1}{y^2+2}\right| = \frac{|x^2 - y^2|}{(x^2+2)(y^2+2)} \le \frac{|x-y|}{4}.$$

So the contraction mapping theorem applies with $\mathcal{M}$ the Banach space $\mathbb{R}$. Thus, there is a fixed point r in $[0, 1/2]$; it is the real root of the cubic. We use the successive approximations,

$$x_{n+1} = \frac{1}{x_n^2 + 2} \quad \text{with} \quad x_0 = 1/4,$$

to get accurate approximations to r. Since $x_1 = 16/33$ and $|x_1 - x_0| = 31/132$, the successive approximations satisfy

$$|x_n - r| \le \frac{(1/4)^n}{3/4}\frac{31}{132} = \frac{31}{396 \cdot 4^{n-1}}.$$

To estimate r to three place accuracy, that is to guarantee that $|x_n - r| < 5 \times 10^{-4}$, the error estimate implies it suffices to choose $n \ge 5$. To three places $x_5 = 0.453$.

In fact, three place accuracy is already achieved for $n = 4$. The error estimate in the contraction mapping theorem has to cover all possible situations. Therefore, it typically yields a conservative estimate for an x_n giving the required accuracy.

Since determining the range of a function or even useful qualitative information about its range can be difficult, the following corollary to the contraction mapping theorem is useful. If c is a point in a normed linear space and $r > 0$, the set of points x satisfying the inequality $\|x - x_0\| \leq r$ is called the **closed ball** of radius r and center x_0. The closed ball is denoted by $C_{x_0}(r)$. It is a closed set.

Corollary 44 (of the Contraction Mapping Theorem) *Let $\mathcal{M}$ be a Banach space and $f: C_{x_0}(r) \to M$ be a contraction with contraction constant ρ on the closed ball $C_{x_0}(r)$. If f moves the center of the ball a distance at most $(1 - \rho)r$, then f has a unique fixed point in the closed ball.*

Proof. If $x_1 = f(x_0)$, then $\|x_1 - x_0\| \leq (1 - \rho)r$. It follows that f maps $C_{x_0}(r)$ into itself: if x is in $C_{x_0}(r)$,

$$\begin{aligned}\|f(x) - x_0\| &\leq \|f(x) - x_1\| + \|x_1 - x_0\| = \|f(x) - f(x_0)\| + \|x_1 - x_0\| \\ &\leq \rho\|x - x_0\| + (1 - \rho)r \leq \rho r + (1 - \rho)r = r.\end{aligned}$$

So $f(x)$ lies in the closed ball, $f: C_{x_0}(r) \to C_{x_0}(r)$, and the contraction mapping theorem applies to f. ∎

In our study of Sturm-Liouville problems, we will need to know how a solution to a particular Sturm-Liouville differential equation changes when the coefficients in the differential equation are perturbed. A variant of the contraction mapping theorem will get us to the results we will need. Here is the setup. We have a family of contraction mappings f_s, one for each s in a set S. The contraction mapping theorem applies to each map f_s and yields a unique fixed point, x_s. If the maps vary continuously with s in a suitable sense, we should be able to conclude that the fixed points x_s also vary continuously with s. The next theorem establishes just such a conclusion.

Theorem 45 *Let C be a closed subset of a Banach space, S be a subset of a normed linear space, and $F: C \times S \to C$ be a function on pairs of element (x, s) with x in C and s in S that has the following properties.*

(i) For each s in S the function $f_s: C \to C$ is a contraction, where $f_s(x) = F(x, s)$.

(ii) There is a constant ρ, with $0 < \rho < 1$, such that ρ is a contraction constant for f_s for all s in S.

(iii) For each fixed x in C, the function $F(x, s)$ from S into C is continuous.

Then the contraction f_s has a unique fixed point x_s in C and this fixed point varies continuously with s; that is, the function $g: S \to C$ given by $g(s) = x_s$ is continuous.

Proof. The existence of the unique fixed point x_s follows immediately from the contraction mapping theorem. To establish continuity of g, fix s_0 in S and let s vary in S. Then

$$\begin{aligned}\|x_s - x_{s_0}\| &= \|f_s(x_s) - f_{s_0}(x_{s_0})\| = \|f_s(x_s) - f_s(x_{s_0}) + f_s(x_{s_0}) - f_{s_0}(x_{s_0})\| \\ &\leq \|f_s(x_s) - f_s(x_{s_0})\| + \|f_s(x_{s_0}) - f_{s_0}(x_{s_0})\| \\ &\leq \rho\|x_s - x_{s_0}\| + \|F(x_{s_0}, s) - F(x_{s_0}, s_0)\|.\end{aligned}$$

Hence,

$$\|x_s - x_{s_0}\| \leq \frac{1}{1 - \rho}\|F(x_{s_0}, s) - F(x_{s_0}, s_0)\|.$$

The continuity of $g(s) = x_s$ follows because $F(x_{s_0}, s)$ is continuous on S by (iii). ∎

2.9 Bisection and Newton-Raphson Methods

A shooting method for the numerical calculation of eigenvalues and eigenfunctions of regular and singular Sturm-Liouville problems is given in Chapter 7. The shooting method is updated using a root-finder. For that purpose either the bisection method or the Newton-Raphson Method are suitable choices; however, other iteration schemes also can be used. In what follows, we assume all functions are real-valued functions of a real variable because that setting is all that is needed in Chapter 7.

In typical situations where iterative root-finding methods are used to find a root (or roots) of a function $f(x)$, an approximate location of a root of interest, say an interval I that contains the root, is often found first and an initial guess (or guesses) of the root is (are) chosen from that interval. Then the root-finding method generates a sequence of (hopefully) increasingly accurate approximations for the root. If $f(x)$ has more than one root, or more than one root of interest, the first step in applying the method is to determine an interval I so there is only one root, say r, of the function in the interval. For simplicity, we describe the bisection method and the Newton-Raphson method in this context.

2.9.1 Bisection Method

The bisection method is based on the intermediate value theorem: if a real-valued continuous function changes sign on an interval it must have a root in that interval. Suppose $f(x)$ is defined on an interval I that contains one root r of the function and that there are points a and b in I with $f(a)f(b) < 0$. The numbers $a = a_0$ and $b = b_0$ are a pair of initial guesses at the root r. The pair of initial guesses determines the first approximation $c_1 = (a_0 + b_0)/2$ of r and the error estimate $|c_1 - r| \le (b - a)/2$. If $f(c_1) = 0$, then $c_1 = r$ and the root is found. If $f(c_1) \ne 0$, then either $f(a_0)f(c_1) < 0$ or $f(b_0)f(c_1) < 0$. In the former case, set $a_1 = a_0$ and $b_1 = c_1$ and in the latter case set $a_1 = c_1$ and $b_1 = b_0$. This determines a new subinterval $[a_1, b_1]$ of $[a_0, b_0]$ that contains the root r and is half as long as the initial interval $[a_0, b_0]$. The new subinterval determines the second approximation $c_2 = (a_1 + b_1)/2$ and the error estimate

$$|c_2 - r| \le \frac{b_1 - a_1}{2} = \frac{b - a}{2^2}.$$

Continuing this successive halving procedure determines approximations

$$c_n = \frac{a_n + b_n}{2}$$

of r and error estimates

$$|c_n - r| \le \frac{b - a}{2^n}$$

for $n = 1, 2, 3, \ldots$. In practice the procedure is stopped at iterate c_N where N is the smallest positive integer with $(b - a)/2^N$ less than a prescribed acceptable error.

Example 1. Use the bisection method to approximate the real root of the cubic $x^3 + 2x - 1$ accurate to five decimal places. (This root was approximated by a contraction mapping iteration scheme in the previous section.) The function $f(x) = x^3 + 2x - 1$ is increasing on $(-\infty, \infty)$ because $f'(x) = 3x^2 + 2 > 0$. Since $f(0) = -1$ and $f(1) = 2$, the function has a unique root r in the interval $[0, 1]$. To find that root accurate to five decimal places with the bisection method, we must find c_n with $1/2^n < 5 \times 10^{-6}$; that is, c_n with $n \geq 18$. In this case, the choices $a = 0$ and $b = 1$ lead to the approximate root $c_{18} = 0.45340$, correctly rounded.

The most attractive feature of the bisection method is that it has easily computed error bounds. On the negative side, it converges rather slowly compared to most popular root-finding methods. This downside is less important than it once was due to the high speed of modern computers.

2.9.2 Newton-Raphson Method

The Newton-Raphson method, often just called Newton's method, is one of the most robust and effective root-finding methods known. The method applies to suitably differentiable functions. Although we only need the method for real-valued functions of a real variable, it applies to systems of equations and even to equations with infinitely many unknowns (equations in a Banach space, where it is known as the Newton-Raphson-Kantorovich method).

The Newton-Raphson method is suggested by the tangent line approximation in Figure 2.2, which illustrates the situation for a function that is increasing and concave up near the root r.

The figure suggests that if x_0 is an initial guess at a simple zero r of a suitably differentiable function $f(x)$, then the x-intercept

$$x_1 = x_0 - \frac{f(x_0)}{f'(x_0)}$$

of the tangent line drawn to the graph of $y = f(x)$ at $(x_0, f(x_0))$ is very often a much better approximation of r than is x_0. Repeating this process with each new estimate of the root regarded as a new initial guess leads to the Newton-Raphson method

$$x_{n+1} = x_n - \frac{f(x_n)}{f'(x_n)}$$

for $n = 1, 2, 3, \ldots$ and where x_0 is a given initial guess. The iteration formula is due to Raphson. The tangent line approximation was implicit in Newton's original use of the method but his formulation was not as simple as Raphson's.

Sketches such as Figure 2.2 strongly suggest that if the function $f(x)$ is either increasing or decreasing and is either concave up or concave down on an interval I that contains a root r of

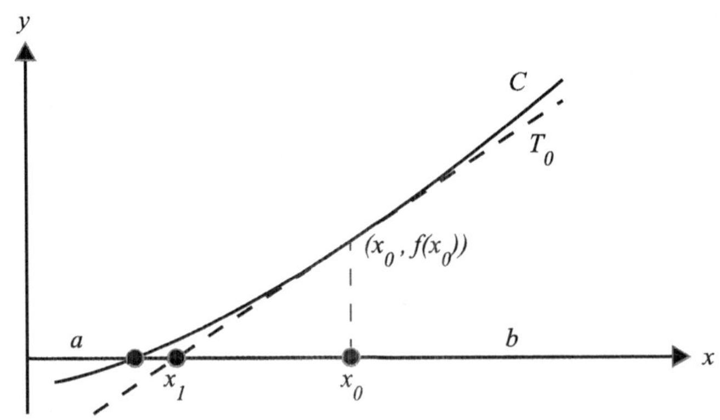

FIGURE 2.2: Newton-Raphson Method

$f(x)$, then it is easy to determine an initial guess x_0 in I such that all the Newton iterates x_n are defined and converge monotonically to r. See [1] for an elementary proof of this assertion.

When the Newton iterates converge to a simple root of $f(x)$, they do so very rapidly, at a quadratic rate. This means that

$$\lim_{n\to\infty} \frac{x_{n+1} - r}{(x_n - r)^2} = \frac{f''(r)}{2f'(r)}$$

holds if $f(x)$ is twice continuously differentiable near the simple root r. Complete statements and refinements of this result and others can be found in [21], [41], and [17]. In particular, the following result holds. It is formulated for real-valued functions of a real variable, the context in which it is used in Chapter 7 but it holds in a much more general setting.

Theorem 46 *If a real-valued function f is continuous on a closed bounded interval I, if f has a simple root r inside I, and if f' is continuous at r, then for all initial guesses x_0 sufficiently close to r, the Newton iterates x_n all exist and x_n converges to r as $n \to \infty$.*

Example 2. Use Newton's method to find the real root of $x^3 + 2x - 1$ correct to five decimal places. As in Example 1, the function $f(x) = x^3 + 2x - 1$ has a unique root r in the interval $[0, 1]$, is increasing there, and is concave up on $[0, 1]$ because $f''(x) = 6x$. In view of the remarks above, the initial guess $x_0 = 1$ will generate a sequence of Newton iterates

$$x_{n+1} = x_n - \frac{x_n^3 + 2x_n - 1}{3x_n^2 + 2}$$

that decrease to the root r. Here is a table of the first few Newton iterates

n	x_n
0	1
1	0.6
2	0.464935064935065
3	0.453467173827973
4	0.453397654028907
5	0.453397651516404
6	0.453397651516404

The table suggests that x_5 approximates r to five decimal places, indeed to several more decimal places. This can be confirmed by using the intermediate value theorem:

$$f(x_5 - 5 \times 10^{-6}) \approx -1.3084 \times 10^{-5} \quad \text{and} \quad f(x_5 + 5 \times 10^{-6}) \approx 1.3084 \times 10^{-5}.$$

It follows that $|x_5 - r| < 5 \times 10^{-6}$.

The most attractive features of Newton's method are its wide applicability and its rapid convergence in typical cases. The downside is that finding a suitable initial guess that generates the rapid convergence (or convergence at all) can be challenging, especially in higher dimensional situations. Also, not many practical (easily computable and relatively sharp) error bounds are available. The use of the intermediate value theorem at the end of Example 2 illustrates a practical way to estimate the error for real-valued functions of a real variable.

2.10 Maximum Principle

The Hopf maximum principle for elliptic partial differential equations and differential inequalities generalizes the classical maximum-minimum principle for harmonic functions. That principle, in turn, can be thought of as generalizing the following simple fact from calculus: if a continuous function y defined on a closed interval $[a, b]$ satisfies $y'' = 0$ on the open interval (a, b), then y cannot achieve a local maximum or local minimum value at a point in (a, b) unless y is identically constant. This follows because y is a linear function on $[a, b]$. Corresponding results with significant consequences hold for solutions y to certain ordinary differential equations and differential inequalities. Two of those results are presented here, in adequate generality for our purposes. The pioneering work of Eberhard Hopf is far deeper than what is suggested here.

We start with a simple observation: if a function y is continuous on $[a, b]$ and has a continuous second derivative on (a, b), then if y has a local maximum at a point c in (a, b),

$$y'(c) = 0 \quad \text{and} \quad y''(c) \leq 0.$$

Consequently, if $a(x)$ is a continuous function on $[a, b]$ and y satisfies the differential inequality

$$My = y'' + a(x)\ y' > 0 \text{ on } (a, b)$$

then y cannot have a local maximum at c. This is a *maximum principle* for solutions of the differential inequality $My > 0$. The following maximum principle for solutions of $My \geq 0$ is of more interest in part because it applies to solutions of certain differential equations. Note that any constant function will satisfy the differential inequality $My \geq 0$.

Theorem 47 *Let $a(x)$ be a continuous function on (a, b). If y is continuous on $[a, b]$, twice continuously differentiable on (a, b), and satisfies the differential inequality $My = y'' + a(x)y' \geq 0$ on (a, b), then y cannot achieve a global maximum at a point in (a, b) unless y is constant on $[a, b]$.*

Proof. The continuous function y assumes it global maximum at some point in $[a, b]$. Suppose the maximum is achieved at a point c in (a, b). We will show that y is constant on $[a, b]$. Indeed, the differential inequality $My \geq 0$ implies that

$$(A(x)y')' \geq 0$$

where

$$A(x) = \exp\left(\int_c^x a(t)\, dt\right) > 0$$

for $a < x < b$ is Euler's integrating factor used to solve first order linear differential equations. Integrate the inequality to find

$$A(x)y'(x) - A(c)y'(c) \geq 0 \quad \text{for } x > c \text{ in } (a, b)$$

and

$$A(c)y'(c) - A(x)y'(x) \geq 0 \quad \text{for } x < c \text{ in } (a, b).$$

Since $A(x) > 0$ and $y'(c) = 0$, it follows that

$$y'(x) \geq 0 \quad \text{for } x \text{ in } (c, b)$$

and

$$y'(x) \le 0 \quad \text{for } x \text{ in } (a,\, c).$$

This pair of inequalities shows that $y(c)$ is the global minimum of y on $[a,\, b]$. Since $y(c)$ is also the global maximum of y on $[a,\, b]$, y is constant on $[a,\, b]$. ■

Now let $Ly = My + b(x)y$ where $b(x)$ is continuous on $[a,\, b]$ and

$$Ly = y'' + a(x)y' + b(x)y.$$

Theorem 48 (Maximum Principle) *Assume $b(x) \le 0$ on $(a,\, b)$, that y has a continuous second derivative on $(a,\, b)$, and that $Ly \ge 0$ on $(a,\, b)$.*
(a) Then y cannot assume a positive *maximum in $(a,\, b)$ unless y is constant on $(a,\, b)$.*
(b) If, in addition, y is continuous on $[a,\, b]$, $y(a) \le 0$, and $y(b) \le 0$, then $y(x) \le 0$ on $[a,\, b]$.

Proof. (a) Since $My \ge -b(x)y$, if y achieves a positive maximum at c in $(a,\, b)$, then $My \ge 0$ on an interval $(a',\, b')$ that contains c and is contained in $[a,\, b]$. Thus $y = y(c)$ on $[a',\, b']$ by the previous theorem. Let b'' be the least upper bound of the endpoints b' of all open intervals containing c and contained in $[a,\, b]$ on which $y = y(c)$. Clearly $b'' \le b$. If $b'' < b$, then b'' belongs to $(a,\, b)$, y is continuous at b'', and

$$y(b'') = \lim_{b' \to b''} y(b') = y(c).$$

Now, just as we argued for c, b'' would be contained in an open interval in $[a,\, b]$ on which $y = y(c)$ and then b'' could not be the least upper bound of the right-hand endpoints of all such open intervals. This contradiction shows that $b'' = b$ and hence that $y = y(c)$ for $c \le x < b$. Likewise, $y = y(c)$ for $a < x \le c$; hence, $y = y(c)$ on $(a,\, b)$.

(b) Since y is continuous on $[a,\, b]$ it assumes its maximum value at some point, say c, in the interval. Suppose y could assume positive values. Then its maximum value is positive. Consequently, c cannot be a or b, y achieves its positive maximum at c in $(a,\, b)$, and y is nonconstant because $y(a) \le 0$. This contradicts (a). Thus y cannot assume any positive values. ■

The following direct consequence of the maximum principle implies that the Green's functions of many Dirichlet boundary value problems of practical importance maintain a fixed sign.

Theorem 49 *Let $a(x)$, $b(x)$, and $f(x)$ be continuous on $[a,\, b]$ and $Ly = y'' + a(x)y' + b(x)y$. If y is a solution to the Dirichlet problem*

$$\begin{cases} Ly = f, & a < x < b, \\ y(a) = 0, & y(b) = 0, \end{cases}$$

where $b(x) \le 0$ and $f(x) \ge 0$, then $y(x) \le 0$ on $[a,\, b]$.

Proof. A solution to the Dirichlet problem is a continuous function y on $[a,\, b]$ that satisfies the stated conditions. Since $f(x) \ge 0$, y satisfies the differential inequality $Ly \ge 0$ on $(a,\, b)$ as well as $y(a) \le 0$ and $y(b) \le 0$; hence, $y(x) \le 0$ on $[a,\, b]$ by the maximum principle. ■

Chapter 3

Integral Equations

The theory of integral equations was developed in part as a powerful tool for studying problems originally formulated in terms of ordinary or partial differential equations. It is natural that a problem formulated in terms of differential equations can be converted into an integral equation because differentiation and integration are inverse processes. One advantage of converting to an integral equation is that the integral operator that arises is better behaved than the differential operator in the original problem. Another advantage is that boundary conditions are incorporated directly into the integral equation and do not have to be treated separately.

In subsequent chapters, we will convert Sturm-Liouville eigenvalue problems into equivalent eigenvalue problems for an integral operator and use the theory of integral equations to establish the fundamental theoretical properties of such problems. The conversion uses the Green's function of the Sturm-Liouville problem and also leads to a convenient formula for the solution to Sturm-Liouville boundary value problems. In this chapter, we present those parts of the theory of integral equations that are needed for a unified study of Sturm-Liouville problems. But, first, we give an illustration of the conversion process.

We convert the eigenvalue problem

$$\begin{gathered} y'' + \lambda y = 0, \ \ 0 < x < l, \\ y(0) = 0, \ \ y(l) = 0, \end{gathered}$$

that we met earlier in Euler buckling and in the vibrations of a violin string to an eigenvalue problem in an integral equations setting. The eigenvalue problem at hand is for the differential operator $Ly = -y''$ together with the given boundary conditions because the differential equation can be expressed as $Ly = \lambda y$. We proceed along a path blazed by Lagrange, multiply the differential equation by a smooth function u and integrate by parts twice so as to reduce the order of derivatives of y,

$$\begin{aligned} \int_0^x (uy'' + \lambda uy)\, ds &= 0, \\ [uy']_0^x - \int_0^x u'y'\, ds + \int_0^x \lambda uy\, ds &= 0, \\ [uy' - u'y]_0^x + \int_0^x u''y\, ds + \int_0^x \lambda uy\, ds &= 0. \end{aligned}$$

Now restrict u by requiring that $Lu = -u'' = 0$ and $u(0) = 0$ to obtain

$$u(x)y'(x) - u'(x)y(x) + \int_0^x \lambda uy\, ds = 0.$$

In the same way, if v satisfies $Lv = -v'' = 0$ and $v(l) = 0$, then

$$-v(x)y'(x) + v'(x)y(x) + \int_x^l \lambda vy\, ds = 0.$$

Multiply the next to last equation by $v(x)$, the last by $u(x)$, and add to find

$$(u(x)v'(x) - u'(x)v(x))y(x) + v(x)\int_0^x \lambda uy\,ds + u(x)\int_x^l \lambda vy\,ds = 0.$$

Specific choices for u and v that satisfy the given requirements are $u = x$ and $v = l - x$. With these choices, the integral equation above becomes

$$-ly(x) + (l - x)\int_0^x \lambda sy(s)\,ds + x\int_x^l \lambda(l - s)y(s)\,ds = 0$$

or

$$y(x) = \lambda\int_0^l g(x, s)y(s)\,ds,$$

where

$$g(x, s) = \frac{1}{l}\begin{cases} (l - s)x & \text{for } 0 \le x \le s \le l \\ (l - x)x & \text{for } 0 \le s \le x \le l \end{cases}.$$

In this context, $g(x, s)$ is called the Green's function for the differential operator $Ly = -y''$ with the given boundary conditions. It is easy to confirm by direct differentiation that a continuous function y that is a solution of the integral equation with kernel $g(x, s)$ is a solution of the original eigenvalue problem; hence, the two eigenvalue problems are equivalent.

It is time to begin our discussion of the aspects of integral operators and equations that are essential for our treatment of Sturm-Liouville problems.

3.1 Integral Operators

Let $k(x, s)$ be a real or complex-valued function on $[a, b] \times [a, b]$. If $f(s)$ is a function defined on $[a, b]$ and $k(x, s)f(s)$ is integrable, say in the Riemann sense, then we regard $k(x, s)$, called a **kernel**, as transforming f into a new function on $[a, b]$ defined by $\int_a^b k(x, s)f(s)\,ds$. The new function is denoted by Kf and its value at x is

$$Kf(x) = \int_a^b k(x, s)f(s)\,ds.$$

We think of K as an **operator (transformation, mapping,** function) that has the function f as an input and outputs the function Kf. This notation is suggested by terminology for matrices. An $m \times n$ matrix A is often regarded as a transformation or operator that takes an n-vector x into the m-vector Ax gotten by matrix multiplication.

The kernels $k(x, s)$ that come up in the study of regular Sturm-Liouville problems are continuous on the closed square $[a, b] \times [a, b]$. For singular Sturm-Liouville problems, with singularity at $x = a$, the kernels have domains the closed square with (a, a) removed, $[a, b] \times [a, b] \backslash \{(a, a)\}$, and are continuous there. In the singular case, $Kf(x)$ is defined by an ordinary (proper) Riemann integral for $a < x \le b$ and $Kf(a)$ is defined by a convergent improper Riemann integral. These facts will be confirmed in later chapters. They are mentioned here to motivate the hypotheses in the theorems that follow.

Recall that a function space is a normed linear space of functions whose norm is chosen to fit the situation at hand. The functions f upon which K acts are elements of a function space $\mathcal{F}$

and, in most cases of interest, Kf is also a function in $\mathcal{F}$. In this situation, $\mathcal{F}$ is the domain of the operator K and its range, the collection of outputs Kf, is a subset of $\mathcal{F}$. We regard K as a mapping from $\mathcal{F}$ into $\mathcal{F}$ and write $K:\mathcal{F} \to \mathcal{F}$, using customary function notation. The same notation and terminology is used if $\int_a^b k(x, s)f(s)\,ds$ is a Lebesgue integral. There are situations in which Kf lies in a different function space, say $\mathcal{G}$, in which case we write $K:\mathcal{F} \to \mathcal{G}$ and otherwise use the same notation.

The only function spaces that are used in this book are $\mathcal{F} = C[a, b]$, $\mathcal{F} = \mathcal{L}_1[a, b]$, $\mathcal{F} = \mathcal{L}_2[a, b]$, and subspaces of these spaces. (See Section 2.5.2.)

In the setting just described, we call K an **integral operator** on $\mathcal{F}$ (or an integral operator from $\mathcal{F}$ to G). Integral operators are linear operators. This means that for all f and g in $\mathcal{F}$ and all scalars α the following properties hold

$$K(f+g) = Kf + Kg,$$
$$K(\alpha f) = \alpha Kf.$$

The properties hold because integration is a linear process. Set $\alpha = 0$ to see that any linear integral operator satisfies $K0 = 0$, where 0 is the zero function.

An integral operator $K:\mathcal{F} \to \mathcal{F}$ is **continuous at** g in $\mathcal{F}$ if given any $\varepsilon > 0$ there is a corresponding $\delta > 0$ such that $\|Kf - Kg\| < \varepsilon$ whenever $\|f - g\| < \delta$. K is **continuous** (on $\mathcal{F}$) if it is continuous at g for every g in $\mathcal{F}$. There is a convenient characterization of continuity for integral operators that holds because they are linear operators. An integral operator K is **bounded** if there is a constant M such that $\|Kf\| \le M\|f\|$ for all f in $\mathcal{F}$.

Lemma 50 *An integral operator $K:\mathcal{F} \to \mathcal{F}$ is continuous if and only if it is bounded.*

Proof. Assume K is continuous. In particular it is continuous at 0. So given $\varepsilon = 1$ there is a $\delta > 0$ such that $\|Kg - K0\| \le 1$ whenever $\|g - 0\| \le \delta$. That is $\|Kg\| \le 1$ whenever $\|g\| \le \delta$. If $f \ne 0$ is in $\mathcal{F}$, then $g = \delta f/\|f\|$ has norm δ; hence,

$$\left\| K\left(\frac{\delta f}{\|f\|}\right)\right\| = \frac{\delta}{\|f\|}\|Kf\| \le 1,$$

$$\|Kf\| \le \frac{1}{\delta}\|f\|.$$

Thus, K is bounded because this inequality also holds if $f = 0$.

If K is bounded, then for some constant M, $\|Kf\| \le M\|f\|$ for all f in $\mathcal{F}$ and

$$\|Kf - Kg\| = \|K(f-g)\| \le M\|f - g\|.$$

This inequality implies immediately that K is continuous. In fact, it establishes that K is uniformly continuous on $\mathcal{F}$. ■

Powers of integral operators are useful just as powers of matrices are. If $K{:}\,C[a, b] \to C[a, b]$ is an integral operator with continuous kernel $k(x, s)$ so that

$$Kf(x) = \int_a^b k(x, s)f(s)\,ds$$

for f in $C[a, b]$, then K^2 is also an integral operator with a continuous kernel. Indeed,

$$K^2 f(x) = K\big(Kf\big)(x) = \int_a^b k\big(x, s\big)Kf(s)\,ds$$
$$= \int_a^b k(x, s)\left(\int_a^b k(s, t)f(t)\,dt\right)ds = \int_a^b \left(\int_a^b k(x, s)k(s, t)\,ds\right)f(t)\,dt$$

so that

$$K^2 f(x) = \int_a^b k_2(x, t) f(t)\, dt$$

where

$$k_2(x, t) = \int_a^b k(x, s) k(s, t)\, ds.$$

Higher powers of K are also integral operators with continuous kernels. The kernel of K^n denoted $k_n(x, s)$ is called the nth **iterated kernel** of $k(x, s)$ and is given recursively by $k_1(x, s) = k(x, s)$ and

$$k_n(x, s) = \int_a^b k(x, t) k_{n-1}(t, s)\, dt,$$

for $n = 2, 3, \ldots$, a result established by repeated use of the reasoning above. Iterated kernels of a kernel that is not continuous are defined in the same way, provided the integrals above exist. In particular, this is the case for the singular kernels which are the Green's functions of the singular Sturm-Liouville problems in Chapters 5 and 6.

It turns out, although we will not pause to verify it, that the collection of all bounded linear operators on a normed space $\mathcal{F}$ is itself a normed linear space, often denoted by $\mathcal{L}(\mathcal{F})$. The norm on $\mathcal{L}(\mathcal{F})$ is defined as follows: if K is in $\mathcal{L}(\mathcal{F})$ there is a real number M so that $\|Kf\| \leq M\|f\|$ for all f in $\mathcal{F}$ and, consequently, there is a smallest number M with this property. The smallest M is by definition the norm of the operator K, denoted $\|K\|$. Thus, if K is a bounded operator

$$\|Kf\| \leq \|K\| \|f\| \text{ for all } f \text{ in } \mathcal{F}.$$

Useful formulas for $\|K\|$ are

$$\|K\| = \sup_{f \neq 0} \frac{\|Kf\|}{\|f\|} = \sup_{\|f\|=1} \|Kf\| = \sup_{\|f\| \leq 1} \|Kf\|.$$

We come next to a key property that many integral operators have. Its importance emerged when Hilbert and others began the systematic study of integral equations. The property was called complete continuity at first, but now is usually called compactness. An integral operator $K: \mathcal{F} \to \mathcal{F}$ is **compact** if for every bounded sequence $\{f_n\}$ in $\mathcal{F}$, the sequence $\{Kf_n\}$ has a convergent subsequence; that is, there is a function g in $\mathcal{F}$ and a subsequence $\{f_{n_p}\}_{p=1}^{\infty}$ of $\{f_n\}$ such that $Kf_{n_p} \to g$ as $p \to \infty$. Compactness has significant consequences. It takes time to fully appreciate its power and scope.

The next theorems establish compactness for the integral operators of importance to us; that is, integral operators whose kernels are Green's functions of regular or singular Sturm-Liouville problems. Theorems 51 and 53 establish compactness of the Green's functions for regular problems. Theorems 52 and 54 do the same for singular problems.

Theorem 51 *If $k(x, s)$ is a real or complex-valued continuous kernel defined on the square $[a, b] \times [a, b]$, then $K: C[a, b] \to C[a, b]$ and K is a bounded, linear, compact operator on $C[a, b]$ equipped with the maximum norm.*

Proof. Since $k(x, s)$ is continuous on $[a, b] \times [a, b]$, a closed bounded set, it is uniformly continuous there. That is, given any $\varepsilon > 0$ there is a $\delta > 0$ such that

$$|k(x', s') - k(x, s)| < \varepsilon \text{ whenever } |x' - x| < \delta,\ |s' - s| < \delta,$$

and (x', s') and (x, s) are points in $[a, b] \times [a, b]$. Consequently for any x and x_0 in $[a, b]$,

$$\left|k(x, s) - k(x_0, s)\right| < \varepsilon \text{ whenever } |x - x_0| < \delta \text{ and } s \text{ is in } [a, b].$$

From

$$Kf(x) - Kf(x_0) = \int_a^b (k(x, s) - k(x_0, s))f(s)\, ds$$

it follows that

$$|Kf(x) - Kf(x_0)| \leq \varepsilon(b - a)\|f\|_{\max} \text{ for } |x - x_0| < \delta.$$

Hence, Kf is uniformly continuous on $[a, b]$ and $K: C[a, b] \to C[a, b]$.

Since $|k(x, s)|$ is continuous on $[a, b] \times [a, b]$ it is bounded there, say by M, and

$$|Kf(x)| \leq \int_a^b |k(x, s)||f(s)|\, ds \leq M(b - a)\|f\|_{\max},$$

$$\|Kf\|_{\max} = \max_{a \leq x \leq b} |Kf(x)| \leq M(b - a)\|f\|_{\max},$$

which establishes that K is a bounded operator.

It remains to show that K is compact. If $\{f_n\}$ is a bounded sequence in $C[a, b]$, that is, for some constant M', $\|f_n\|_{\max} \leq M'$ for all n, then

$$\|Kf_n\|_{\max} \leq \|K\|\|f_n\|_{\max} \leq \|K\|M',$$

and, with $\varepsilon > 0$ and $\delta > 0$ chosen as above, for any x and x_0 in $[a, b]$,

$$|Kf_n(x) - Kf_n(x_0)| \leq \varepsilon(b - a)\|f_n\|_{\max} \leq \varepsilon(b - a)M' \text{ for } |x - x_0| < \delta.$$

Thus, $\{Kf_n\}$ is uniformly bounded and equicontinuous on $[a, b]$. By the Arzelà-Ascoli theorem it contains a subsequence $\{Kf_{n_p}\}$ that converges uniformly to a continuous function g on $[a, b]$. That is, $\|Kf_{n_p} - g\|_{\max} \to 0$ as $p \to \infty$, which establishes that K is a compact operator on $C[a, b]$ equipped with the maximum norm. ■

The Green's functions for singular Sturm-Liouville problems are only continuous on the square $[a, b] \times [a, b]$ with its lower left hand corner (a, a) removed and exhibit singular behavior near (a, a). The corresponding integral operator may be defined as a Lebesgue integral or as an improper Riemann integral. We choose the improper Riemann integral approach because it requires less mathematical background. The following theorem applies to such Green's functions. Not surprisingly, the proof is a variant on that for Theorem 51.

Theorem 52 *Let $k(x, s)$ be a continuous real or complex-valued kernel defined on $[a, b] \times [a, b] \backslash \{(a, a)\}$, the square with the point (a, a) removed. If*

(a) *for each f in $C[a, b]$, $Kf(a) = \int_a^b k(a, s)f(s)\, ds$ exists as a convergent improper Riemann integral,*

(b) *$\int_a^b |k(x, s)|\, ds \leq M$ for some constant M and all x in $[a, b]$,*

(c) *$\int_a^b |k(x, s) - k(a, s)|\, ds \to 0$ as $x \to a$,*

then $K: C[a, b] \to C[a, b]$ and K is a bounded, linear, compact operator on $C[a, b]$ equipped with the maximum norm.

Proof. Given f in $C[a, b]$, $Kf(a)$ is defined by (a) and for $x > a$ in $[a, b]$, $Kf(x)$ is given by a proper Riemann integral. So Kf is a well defined function on $[a, b]$. We claim that

$$\int_a^b |k(x, s) - k(x_0, s)|\, ds \to 0 \text{ as } x \to x_0$$

for each x_0 in $[a, b]$. If $x_0 = a$ the limit holds by (c). Fix $x_0 > a$ in $[a, b]$ and set $a' = (a + x_0)/2$. The kernel $k(x, s)$ is continuous on $[a', b] \times [a, b]$. Just as in the proof of Theorem 51, given $\varepsilon > 0$ there is a $\delta > 0$ such that

$$|k(x, s) - k(x_0, s)| < \varepsilon \quad \text{for } x \text{ in } [a', b] \quad \text{and} \quad s \text{ in } [a, b] \quad \text{when} \quad |x - x_0| < \delta.$$

Consequently for x in $[a', b]$,

$$\int_a^b |k(x, s) - k(x_0, s)|\, ds \leq \varepsilon(b - a) \quad \text{when} \quad |x - x_0| < \delta$$

and the claim is established for $x_0 > a$ in $[a, b]$. Thus, for f in $C[a, b]$,

$$|Kf(x) - Kf(x_0)| \leq \|f\|_{\max} \int_a^b |k(x, s) - k(x_0, s)|\, ds \to 0 \text{ as } x \to x_0,$$

the function Kf is continuous on $a \leq x \leq b$, and $K: C[a, b] \to C[a, b]$. By (b) the operator K is bounded because

$$|Kf(x)| \leq \|f\|_{\max} \int_a^b |k(x, s)|\, ds \leq M\|f\|_{\max},$$

$$\|Kf\|_{\max} \leq M\|f\|_{\max}.$$

It remains to show that K is a compact operator. If $\{f_n\}$ is a bounded sequence in $C[a, b]$, with $\|f_n\|_{\max} \leq M'$ for all n, then $\{Kf_n\}$ is uniformly bounded on $[a, b]$ because $\|Kf_n\|_{\max} \leq M\|f_n\|_{\max} \leq MM'$. Applying the inequality above for $|Kf(x) - Kf(x_0)|$ with $f = f_n$ yields

$$|Kf_n(x) - Kf_n(x_0)| \leq \|f_n\|_{\max} \int_a^b |k(x, s) - k(x_0, s)|\, ds$$

$$\leq M' \int_a^b |k(x, s) - k(x_0, s)|\, ds \to 0 \text{ as } x \to x_0.$$

Thus, $\{Kf_n\}$ is equicontinuous at x_0 for each x_0 in $[a, b]$ and $\{Kf_n\}$ is equicontinuous on $[a, b]$ by Proposition 42. The compactness of K follows from the Arzelà-Ascoli theorem by the same reasoning used in Theorem 51. ■

We will need the analogues of the previous two theorems when $C[a, b]$ is equipped with the 2-norm. The proofs require straightforward adjustments to previous arguments.

Theorem 53 *If $k(x, s)$ is a real or complex-valued continuous kernel defined on the square $[a, b] \times [a, b]$, then $K: C[a, b] \to C[a, b]$ and K is a bounded, linear, compact operator on $C[a, b]$ equipped with the 2-norm.*

Proof. Just as in the proof of Theorem 51, given any $\varepsilon > 0$ there is a $\delta > 0$ such that

$$|k(x', s') - k(x, s)| < \varepsilon \text{ whenever } |x' - x| < \delta,\ |s' - s| < \delta,$$

and (x', s') and (x, s) are points in $[a, b] \times [a, b]$. Consequently,

$$\int_a^b |k(x, s) - k(x_0, s)|\, ds < \varepsilon(b - a) \text{ whenever } |x - x_0| < \delta,$$

$$\int_a^b |k(x, s) - k(x_0, s)|^2\, ds < \varepsilon^2(b - a) \text{ whenever } |x - x_0| < \delta,$$

which establishes that

$$\int_a^b |k(x, s) - k(x_0, s)|\, ds \to \text{ as } x \to x_0$$

and

$$\int_a^b |k(x, s) - k(x_0, s)|^2\, ds \to 0 \text{ as } x \to x_0,$$

for any x_0 in $[a, b]$.

By the Schwarz inequality and the limits just established,

$$|Kf(x) - Kf(x_0)| \leq \left(\int_a^b |k(x, s) - k(x_0, s)|^2\, ds\right)^{1/2} \left(\int_a^b |f(s)|^2\, ds\right)^{1/2},$$

$$Kf(x) - Kf(x_0) \to 0 \text{ as } x \to x_0.$$

So Kf is a continuous function and $K{:}C[a, b] \to C[a, b]$.

In the same way,

$$|Kf(x)| \leq \left(\int_a^b |k(x, s)|^2\, ds\right)^{1/2} \left(\int_a^b |f(s)|^2\, ds\right)^{1/2},$$

$$\int_a^b |Kf(x)|^2\, dx \leq \int_a^b \left(\int_a^b |k(x, s)|^2\, ds\right) dx \|f\|_2^2,$$

$$\|Kf\|_2 \leq \max_{a \leq x \leq b} \left(\int_a^b \int_a^b |k(x, s)|^2\, ds\, dx\right)^{1/2} \|f\|_2.$$

Thus K is a bounded operator and

$$\|K\| \leq \max_{a \leq x \leq b} \left(\int_a^b \int_a^b |k(x, s)|^2\, ds\, dx\right)^{1/2}.$$

It remains to establish that $K{:}C[a, b] \to C[a, b]$ with the 2-norm is a compact operator. Let $\{f_n\}$ be a bounded sequence in the 2-norm. That is $\|f_n\|_2 \leq M$ for some M and all n. The first estimate of $|Kf(x)|$ above and the continuity of the kernel give

$$|Kf(x)| \leq \max_{a \leq x \leq b} \left(\int_a^b |k(x, s)|^2\, ds\right)^{1/2} \|f\|_2$$

for any f in $C[a, b]$ and for all x in $[a, b]$. In particular,

$$|Kf_n(x)| \leq \max_{a \leq x \leq b} \left(\int_a^b |k(x, s)|^2\, ds\right)^{1/2} M$$

for all n and $\{Kf_n\}$ is a uniformly bounded on $[a, b]$.

The estimate near the beginning of the proof applied to the functions f_n yields,

$$\begin{aligned} |Kf_n(x) - Kf_n(x_0)| &\leq \left(\int_a^b |k(x, s) - k(x_0, s)|^2\, ds\right)^{1/2} \left(\int_a^b |f_n(s)|^2\, ds\right)^{1/2}, \\ &\leq \left(\max_{a \leq x, s \leq b} |k(x, s) - k(x_0, s)|\right)(b - a)^{1/2} M \end{aligned}$$

because $\|f_n\|_2 \leq M$. It follows that $\{Kf_n\}$ is equicontinuous on $[a, b]$ because the kernel is uniformly continuous on $[a, b] \times [a, b]$. By the Arzelà-Ascoli theorem there is a g in $C[a, b]$ and a subsequence $\{Kf_{n_p}\}$ such that Kf_{n_p} converges uniformly on $[a, b]$ to g; that is, $\|Kf_{n_p} - g\|_{\max} \to 0$ as $p \to \infty$. Since

$$\|Kf_{n_p} - g\|_2 = \left(\int_a^b |Kf_{n_p}(s) - g(s)|^2 \, ds\right)^{1/2} \le \|Kf_{n_p} - g\|_{\max}(b-a)^{1/2},$$

it follows that $\|Kf_{n_p} - g\|_2 \to 0$ as $p \to \infty$ and compactness in the 2-norm is established. ■

Theorem 54 *Let $k(x, s)$ be a continuous real or complex-valued kernel defined on $[a, b] \times [a, b] \backslash \{(a, a)\}$, the square with the point (a, a) removed. If*

(a) *for each f in $C[a, b]$, $Kf(a) = \int_a^b k(a, s)f(s)\, ds$ exists as a convergent improper Riemann integral,*

(b) *$\int_a^b |k(x, s)|^2 \, ds \le M$ for some constant M and all x in $[a, b]$,*

(c) *$\int_a^b |k(x, s) - k(a, s)|^2 \, ds \to 0$ as $x \to a$, then $K{:}C[a, b] \to C[a, b]$ and K is a bounded, linear, compact operator on $C[a, b]$ equipped with the 2-norm.*

Proof. Given f in $C[a, b]$, $Kf(a)$ is defined by (a) and for $x > a$ in $[a, b]$, $Kf(x)$ is given by a proper Riemann integral. So Kf is a well defined function on $[a, b]$. We claim that

$$\int_a^b |k(x, s) - k(x_0, s)|^2 \, ds \to 0 \text{ as } x \to x_0$$

for each x_0 in $[a, b]$. If $x_0 = a$ the limit holds by (c). Fix $x_0 > a$ in $[a, b]$ and set $a' = (a + x_0)/2$. The kernel $k(x, s)$ is continuous on $[a', b] \times [a, b]$. Just as in the proof of Theorem 51, given $\varepsilon > 0$ there is a $\delta > 0$ such that

$$|k(x, s) - k(x_0, s)| < \varepsilon \quad \text{for } x \text{ in } [a', b] \quad \text{and} \quad s \text{ in } [a, b] \quad \text{when} \quad |x - x_0| < \delta.$$

Consequently for x in $[a', b]$,

$$\int_a^b |k(x, s) - k(x_0, s)|^2 \, ds \le \varepsilon^2 (b - a) \quad \text{when} \quad |x - x_0| < \delta$$

and the claim is established for $x_0 > a$ in $[a, b]$.

From the Schwarz inequality and the claim

$$\int_a^b |k(x, s) - k(x_0, s)| \, ds \le \left(\int_a^b |k(x, s) - k(x_0, s)|^2 \, ds\right)^{1/2} (b-a)^{1/2},$$

$$\int_a^b |k(x, s) - k(x_0, s)| \, ds \to 0 \text{ as } x \to x_0$$

for any x_0 in $[a, b]$. Thus, for f in $C[a, b]$,

$$|Kf(x) - Kf(x_0)| \le \left(\int_a^b |k(x, s) - k(x_0, s)|^2 \, ds\right)^{1/2} \|f\|_2 \to 0 \text{ as } x \to x_0,$$

the function Kf is continuous on $a \le x \le b$, and $K: C[a, b] \to C[a, b]$. By (b) the operator K is bounded because

$$|Kf(x)| \le \left(\int_a^b |k(x, s)|^2 \, ds \right)^{1/2} \|f\|_2 \le M^{1/2} \|f\|_2,$$

$$\int_a^b |Kf(x)|^2 \, dx \le M \|f\|_2^2 (b - a),$$

$$\|Kf\|_2 \le M^{1/2} (b - a)^{1/2} \|f\|_2.$$

It remains to show that K is a compact operator. If $\{f_n\}$ is a bounded sequence in $C[a, b]$, with $\|f_n\|_2 \le M'$ for all n, then $\{Kf_n\}$ is uniformly bounded on $[a, b]$ because by the inequality for $|Kf(x)|$ above applied to $f = f_n$,

$$|Kf_n(x)| \le M^{1/2} \|f_n\|_2 \le M^{1/2} M',$$

$$\|Kf_n\|_{\max} \le M^{1/2} M'.$$

Applying the inequality for $|Kf(x) - Kf(x_0)|$ with $f = f_n$ yields

$$\begin{aligned} |Kf_n(x) - Kf_n(x_0)| &\le \left(\int_a^b |k(x, s) - k(x_0, s)|^2 \, ds \right)^{1/2} \|f_n\|_2 \\ &\le M' \left(\int_a^b |k(x, s) - k(x_0, s)|^2 \, ds \right)^{1/2} \to 0 \text{ as } x \to x_0. \end{aligned}$$

Thus, $\{Kf_n\}$ is equicontinuous at x_0 for each x_0 in $[a, b]$ and $\{Kf_n\}$ is equicontinuous on $[a, b]$ by Proposition 42. The compactness of K follows from the Arzelà-Ascoli theorem by the same reasoning used in Theorem 53. ∎

3.2 More General Domains

All of the results of the last section and of this chapter extend by the same proofs to integral operators with continuous kernels defined on rather general domains D in $\mathbb{R}^n$. Only three types of domains will be important for us, balls, boxes, and simplices. In particular, $\Box_n$ is the n-dimensional box of points $x = (x_1, \ldots, x_n)$ with $a_i \le x_i \le b_i$ for $i = 1, 2, \ldots, n$ and Δ_n is the n-dimensional simplex of points $x = (x_1, \ldots, x_n)$ with $a \le x_1 \le \cdots \le x_n \le b$. The one dimensional simplex Δ_1 is the closed interval $[a, b]$, the two dimensional simplex Δ_2 is a solid triangle, and the three dimensional simplex Δ_3 is a solid tetrahedron. Just as in the last section, most results will be stated and proved only for the case $n = 1$ with additional discussion given in appendices, if warranted. Thus, the results established earlier in Chapter 3 for the integral operator K on $C[a, b]$ defined by

$$Kf(x) = \int_a^b k(x, s) f(s) \, ds$$

also hold for the integral operator K on $C(D)$ defined by

$$Kf(x) = \int_D k(x, s) f(s) \, ds$$

where $x = (x_1, \ldots, x_n)$, $s = (s_1, \ldots, s_n)$, $ds = ds_1 \cdots ds_n$, where D is a ball, a box, or a simplex in $\mathbb{R}^n$. The same is true for the mildly singular kernels and corresponding integral operators of the singular Sturm-Liouville problems in Chapters 5 and 6.

The Sturm-Liouville problems that are the primary focus of this book are defined on one dimensional intervals. So, for the most part, the integral operators of interest to us are one dimensional. Readers interested in higher dimensional problems and related eigenfunction expansions may wish to consult [2] and [3].

3.3 Eigenvalues of Operators and Kernels

A real or complex number μ is an **eigenvalue** of the integral operator K if $Kf = \mu f$ for some $f \neq 0$. (This terminology is standard for operators of any kind.) For historical reasons, a real or complex number λ is called an **eigenvalue of the kernel** $k(x, s)$ if $\lambda Kf = f$ for some $f \neq 0$. Consequently, eigenvalues of the kernel are necessarily nonzero and $\mu = 1/\lambda$ is an eigenvalue of the integral operator K if λ is an eigenvalue of the kernel $k(x, s)$, and conversely if $\mu \neq 0$. In either case, the nonzero function f is called an **eigenfunction** corresponding to the eigenvalue μ or λ.

The historical reason for this somewhat unsatisfactory terminology is the following: the matrix eigenvalue problem was expressed as $Ax = \mu x$, with eigenvalue μ and eigenvector $x \neq 0$. This usage was adopted in a general operator setting. On the other hand, applied problems involving differential equations led to Sturm-Liouville eigenvalue problems expressed as $Ly = \lambda y$ with L a differential operator and λ the eigenvalue parameter (for example, a separation constant). Solving the differential equation amounts to inverting L and leads to the equation $y = \lambda L^{-1} y$ which is usually expressed as $y = \lambda K y$ where K is an integral operator whose kernel is a Green's function.

There are several equivalent ways to define the multiplicities associated with an eigenvalue of an integral operator K. In a fundamental paper [13] Ivar Fredholm established in 1903 the following alternative, the **Fredholm alternative:** for a continuous kernel $k(x, s)$, either the integral equation, called a Fredholm integral equation of the second kind,

$$f(x) = \lambda \int_a^b k(x, s) f(s)\, ds + g(x) \tag{3.1}$$

in which $\lambda \neq 0$, has a unique solution $f(x)$ for every given continuous function $g(x)$, or the corresponding homogeneous equation

$$\phi(x) = \lambda \int_a^b k(x, s)\phi(s)\, ds$$

has a nontrivial solution $\phi(x)$. That is, either (3.1) has a unique solution for every $g(x)$ or λ is an eigenvalue of the kernel $k(x, s)$. Fredholm also found the following solution formula for (3.1),

$$f(x) = g(x) + \int_a^b \Gamma(x, s, \lambda) g(s)\, ds$$

where $\Gamma(x, s, \lambda)$, called the resolvent kernel of $k(x, s)$, can be expressed in a form analogous to Cramer's rule for solving systems of linear equations,

$$\Gamma(x, s, \lambda) = \frac{D(x, s, \lambda)}{D(\lambda)}.$$

Both functions on the right side are entire functions of the complex variable λ; that is, they are differentiable at every point in the complex plane. If $D(\lambda) \neq 0$, the solution to the integral equation is given by the foregoing formula for $f(x)$. Fredholm showed that λ is an eigenvalue of the kernel $k(x, s)$ if and only if $D(\lambda) = 0$. He went on to establish that if m is the multiplicity of λ as a root of $D(\lambda) = 0$, an integer now called the algebraic multiplicity of λ, then there is a largest integer n with $1 \leq n \leq m$ such that there are n linearly independent eigenfunctions corresponding to λ. The integer n is the geometric multiplicity of λ.

Fredholm's approach to multiplicity relies on non-elementary results of complex analysis. Subsequently Issai Schur observed that it would be desirable to give an elementary, but equivalent, formulation of Fredholm's multiplicities that is tied more closely to corresponding matrix results. He carried out that plan in his seminal paper [36] in which he established that every square complex matrix is unitarily equivalent to a lower triangular matrix. In the same paper, he showed that many matrix inequalities related to eigenvalues followed easily from the lower triangularization result. Among those inequalities was one due to Hadamard that was essential to Fredholm's proof that $D(x, s, \lambda)$ and $D(\lambda)$ are entire functions. Schur's approach to multiplicity, expressed in modern language, amounts to the following. Each eigenvalue μ of the integral operator K determines an **eigenspace**,

$$\mathcal{E}_1 = \mathcal{E}_1(\mu) = \{\phi{:}(\mu I - K)\phi = 0\},$$

the linear space of all eigenfunctions corresponding to μ and $\phi = 0$, and **generalized eigenspaces**

$$\mathcal{E}_p = \mathcal{E}_p(\mu) = \{\phi{:}(\mu I - K)^p\phi = 0\}$$

for a positive integer $p \geq 2$ whose elements, apart from $\phi = 0$ and the eigenfunctions corresponding to μ, are called **generalized eigenfunctions** corresponding to μ. By convention $(\mu I - K)^0 = I$ and $\mathcal{E}_0(\mu) = \{0\}$. Clearly $\mathcal{E}_p(\mu) \subset \mathcal{E}_q(\mu)$ for $p \leq q$. For a nonzero eigenvalue μ of the integral operator K, that is for $\lambda = 1/\mu$ an eigenvalue of the kernel $k(x, s)$, Schur proved that

$$\dim \mathcal{E}_p(\mu) \leq |\lambda|^2 \int_a^b \int_a^b |k(x, s)|^2 \, dx ds$$

for all p. It follows that all of the generalized eigenspaces corresponding to a nonzero eigenvalue μ are finite dimensional and that strict inclusion in $\mathcal{E}_p(\mu) \subset \mathcal{E}_{p+1}(\mu)$ can occur at most a finite number of times. It is easy to check that

$$\mathcal{E}_p(\mu) = \mathcal{E}_{p+1}(\mu) \Rightarrow \mathcal{E}_{p+1}(\mu) = \mathcal{E}_{p+2}(\mu) \Rightarrow \mathcal{E}_p(\mu) = \mathcal{E}_{p+1}(\mu) = \mathcal{E}_{p+2}(\mu) = \cdots.$$

Consequently, since $\mathcal{E}_0(\mu) \underset{\neq}{\subset} \mathcal{E}_1(\mu)$, if $\mu \neq 0$ there is a smallest positive integer m such that

$$\mathcal{E}_0(\mu) \underset{\neq}{\subset} \cdots \underset{\neq}{\subset} \mathcal{E}_m(\mu) = \mathcal{E}_{m+1}(\mu) = \mathcal{E}_{m+2}(\mu) \cdots.$$

By definition $\dim \mathcal{E}_m(\mu)$ is the **algebraic multiplicity** of μ as an eigenvalue of K or of $\lambda = 1/\mu$ as an eigenvalue of the kernel $k(x, s)$ and $\dim \mathcal{E}_1(\mu)$ is the **geometric multiplicity** of μ as an eigenvalue of K or of $\lambda = 1/\mu$ as an eigenvalue of the kernel $k(x, s)$. The geometric multiplicity of μ is the maximum number of linearly independent eigenfunctions that correspond to the eigenvalue. An eigenvalue μ is **simple** if its algebraic multiplicity is 1, in which case $\mathcal{E}_1(\mu) = \mathcal{E}_2(\mu)$, so there are no generalized eigenfunctions and, apart from nonzero constant multiples, there is only one eigenfunction corresponding to μ. We do not assign a multiplicity to $\mu = 0$ as an eigenvalue of K.

Evidently, a nonzero eigenvalue μ of K has no generalized eigenfunctions other than eigenfunctions if and only if its geometric and algebraic multiplicities are equal; that is, if and only if

$m = 1$ above. This is always the case for integral operators with self-adjoint kernels. See the next section.

If $k(x, s)$ is real-valued and λ is a real eigenvalue of $k(x, s)$, then it is usually convenient to work with corresponding real-valued eigenfunctions (and generalized eigenfunctions). The following lemma establishes that this is always possible for eigenfunctions. The corresponding result for generalized eigenfunctions can be established in the same way.

Lemma 55 *If $k(x, s)$ is real-valued and λ is a real eigenvalue of $k(x, s)$, then the eigenspace of λ has a basis consisting of real-valued orthonormal eigenfunctions.*

Proof. By Schur's results above, the eigenspace is finite dimensional and has a basis of complex-valued eigenfunctions, say $y_1, y_2, \ldots, y_m$, so that m is the dimension of the eigenspace. Express $y_j = u_j + iv_j$ with u_j and v_j real-valued. Since λ is real and $k(x, s)$ is real-valued, separating

$$y_j(x) = \lambda \int_a^b k(x, s) y_j(s)\, ds$$

into real and imaginary parts yields

$$u_j(x) = \lambda \int_a^b k(x, s) u_j(s)\, ds \quad \text{and} \quad v_j(x) = \lambda \int_a^b k(x, s) v_j(s)\, ds.$$

So either u_j is a real-valued eigenfunction corresponding to λ or $u_j = 0$ and the same holds for v_j. For any complex scalars c_j,

$$\sum c_j y_j = \sum c_j u_j + \sum i c_j v_j.$$

Consequently, the list $\{u_1, u_2, \ldots, u_m, v_1, \ldots, v_m\}$ must have at least m linearly independent functions; else the dimension of the eigenspace would be less than m. If there were more than m linearly independent functions in the list, the dimension of the eigenspace would be greater than m. Hence, there must be exactly m linearly independent functions in the list and they form a real-valued basis for the eigenspace.

Since any nontrivial finite linear combination of eigenfunctions belonging to λ is also an eigenfunction belonging to λ, we can apply the Gram-Schmidt process to the real-valued basis for the eigenspace of λ to obtain a real-valued orthonormal basis for the eigenspace of λ. ■

3.4 Self-Adjoint Operators and Kernels

Many eigenvalue problems that arise in applications are equivalent to eigenvalue problems for integral operators that are self-adjoint. This is fortunate because such eigenvalue problems behave in a way strictly analogous to eigenvalue problems for self-adjoint matrices. The eigenvalues for self-adjoint matrices are real and there is a corresponding set of orthonormal eigenvectors that are a basis for the underlying real or complex Euclidean space. Such a basis is strictly analogous to standard basis **i**, **j**, and **k** in three space and is equally useful for theoretical and computational purposes. Virtually all of the properties in the matrix setting carry over to the infinite dimensional, integral equations setting. (See Section 2.3 for further discussion of the matrix case.)

We make the following standing assumption throughout this section and its subsections:

> **Standing Assumption:** $k(x, s)$ is a real or complex-valued continuous kernel defined on $[a, b] \times [a, b]$ or on $[a, b] \times [a, b] \backslash \{(a, a)\}$ and the corresponding integral operator $K: C[a, b] \to C[a, b]$.

The choice of the domain of $k(x, s)$ in the standing assumption is dictated by the fact that we shall apply the results of this section to kernels that are Green's functions for regular or singular Sturm-Liouville problems. These Green's functions satisfy the standing assumption.

We equip $C[a, b]$ with the usual inner product

$$\langle f, g\rangle = \int_a^b f(s)\overline{g(s)}\,ds$$

and corresponding 2-norm $\|f\| = \sqrt{\langle f, f\rangle}$. We omit the subscript on the 2-norm in this section. Other norms will be indicated by appropriate subscripts. Sufficient conditions on $k(x, s)$ that guarantee that $K: C[a, b] \to C[a, b]$ and that K is a bounded linear compact operator on $C[a, b]$ equipped with the 2-norm are given in Theorems 53 and 54.

The following interchange of order of integration turns out to be unexpectedly important: If $K: C[a.b] \to C[a.b]$,

$$\begin{aligned}\langle Kf, g\rangle &= \int_a^b \left(\int_a^b k(s, t)f(t)\,dt\right)\overline{g(s)}\,ds = \int_a^b f(t)\left(\int_a^b k(s, t)\overline{g(s)}\,ds\right)dt \\ &= \int_a^b f(t)\left(\overline{\int_a^b \overline{k(s, t)}g(s)\,ds}\right)dt = \langle f, K^* g\rangle\end{aligned}$$

where $K^*: C[a.b] \to C[a.b]$, called the **adjoint (operator) of** K, is the integral operator with kernel $k^*(x, s) = \overline{k(s, x)}$. The kernel $k^*(x, s)$ is called the **adjoint kernel** to $k(x, s)$. The interchange of order is valid if $k(x, s)$ is continuous on $[a, b] \times [a, b]$ or is continuous on $[a, b] \times [a, b] \backslash \{(a, a)\}$ and mildly singular at (a, a), as is the case for the Green's functions of the singular Sturm-Liouville problems in Chapters 5 and 6. (See Section 3.7 for the definition of a mildly singular kernel.)

In this setting, much can be learned about the integral operator K through its adjoint K^*. An integral operator K is **self-adjoint** if its kernel satisfies

$$k(x, s) = k^*(x, s)$$

for all (x, s) in the domain of k, in which case the kernel $k(x, s)$ is called **self-adjoint**. The kernel $k(x, s)$ is called **symmetric** if it is real-valued and self-adjoint; that is, if $k(x, s)$ is real-valued and $k(x, s) = k(s, x)$ for all (x, s) in its domain.

The key relation between K and K^* that led to the operator K^* is

$$\langle Kf, g\rangle = \langle f, K^* g\rangle$$

for all f and g in $C[a, b]$.

The following useful properties of self-adjoint integral operators are well-known.

Lemma 56 *If K is a self-adjoint integral operator, then all eigenvalues of K are real and eigenfunctions corresponding to distinct eigenvalues are orthogonal.*

Proof. If μ is an eigenvalue of K and f a corresponding eigenfunction, then

$$\mu\langle f, f\rangle = \langle \mu f, f\rangle = \langle Kf, f\rangle = \langle f, Kf\rangle = \langle f, \mu f\rangle = \overline{\mu}\langle f, f\rangle.$$

Since $\langle f, f\rangle > 0$, $\mu = \overline{\mu}$ and μ is real. If $\nu \neq \mu$ is an eigenvalue of K with eigenfunction g, then

$$\mu\langle f, g\rangle = \langle \mu f, g\rangle = \langle Kf, g\rangle = \langle f, Kg\rangle = \langle f, \nu g\rangle = \nu\langle f, g\rangle$$

because ν is real. Since $\nu \neq \mu$ it follows that $\langle f, g\rangle = 0$. ■

Lemma 57 *The geometric and algebraic multiplicities of a nonzero eigenvalue of a self-adjoint integral operator K are equal. Hence, a self-adjoint integral operator has no generalized eigenfunctions.*

Proof. Suppose $\mu \neq 0$ and that ϕ belongs to $\mathcal{E}_2(\mu)$, that is, $(\mu I - K)^2\phi = 0$. Since K is self-adjoint

$$0 = \langle(\mu I - K)^2\phi, \phi\rangle = \langle(\mu I - K)\phi, (\mu I - K)\phi\rangle = \|(\mu I - K)\phi\|^2.$$

Thus $(\mu I - K)\phi = 0$. It follows that $\mathcal{E}_1(\mu) = \mathcal{E}_2(\mu)$ and the stated conclusion follows. ■

3.4.1 Hilbert-Schmidt Theorem

The following result is the key to extending the principal axis theorem for self-adjoint matrices to integral operators, and more generally to compact operators on inner product spaces. (The standing assumption remains in force.)

Theorem 58 *Let $\mathcal{F}$ be a nonzero subspace of $C[a, b]$. If $K:\mathcal{F} \to \mathcal{F}$ is self-adjoint, then*

$$\sup_{\|f\|=1} |\langle Kf, f\rangle| = \sup_{\|f\|=1} \|Kf\|,$$

or equivalently,

$$\sup_{f\neq 0}\frac{|\langle Kf, f\rangle|}{\langle f, f\rangle} = \sup_{f\neq 0}\frac{\|Kf\|}{\|f\|},$$

where the supremum on the right is $\|K\|$, the norm of the integral operator K. If the supremum on the left is achieved at f then both suprema are achieved at f and either $Kf = \|K\|f$ or $Kf = -\|K\|f$.

Proof. Let $s = \sup_{\|f\|=1} |\langle Kf, f\rangle|$ and $t = \sup_{\|f\|=1} \|Kf\| = \|K\|$, by the definition of the operator norm. If $\|f\| = 1$ then $|\langle Kf, f\rangle| \le \|Kf\|\|f\| \le \|K\|\|f\|^2 = \|K\| = t$. So $s \le t$. To establish the reverse inequality, expand the inner products on the left to obtain

$$\langle K(u+v), u+v\rangle - \langle K(u-v), u-v\rangle = 4\,\mathrm{Re}\,\langle Ku, v\rangle$$

for u and v in $\mathcal{F}$. Consequently,

$$4\mathrm{Re}\langle Ku, v\rangle \le s\|u+v\|^2 + s\|u-v\|^2 = 2s\big(\|u\|^2 + \|v\|^2\big).$$

If $\|u\| = 1$ and $Ku \neq 0$, set $v = Ku/\|Ku\|$ in the foregoing inequality to obtain

$$4\|Ku\| \le 2s(1+1) = 4s.$$

Since this inequality also holds if $Ku = 0$,

$$t = \sup_{\|u\|=1} \|Ku\| \le s.$$

Thus, $s = t$.

If the supremum s is achieved for some f with $\|f\| = 1$, then $\langle Kf, f\rangle = \mu$ for $\mu = \pm\|K\|$ and $\|Kf\| \leq \|K\| = |\mu|$. So $\|Kf\|^2 \leq \mu^2$ and

$$0 \leq \|Kf - \mu f\|^2 = \|Kf\|^2 - 2\mu\langle Kf, f\rangle + \mu^2 \leq 0,$$

$$Kf = \mu f,$$

and $\|Kf\| = |\mu| = \|K\|$. ■

The next theorem was established by David Hilbert ([10] p. 122) for integral operators with continuous or square summable kernels. It was generalized by Frigyes Riesz ([33] p. 232) to the setting of compact operators on a Hilbert space. The reasoning used here follows Riesz.

Theorem 59 *Let $\mathcal{F}$ be a nonzero subspace of $C[a, b]$. If $K:\mathcal{F} \to \mathcal{F}$ is a bounded, linear, self-adjoint, compact integral operator, then the extremal problem*

$$|\langle Kf, f\rangle| = \text{a maximum subject to the constraint } \|f\| = 1$$

has at least one solution. Moreover, every solution is an eigenfunction of K corresponding to an eigenvalue μ_1 with $|\mu_1| = \|K\|$.

Proof. If $\|K\| = 0$, then the maximum is achieved for all f with norm 1 and $Kf = 0 \cdot f$.

If $\|K\| \neq 0$, in view of the Theorem 58, all that remains to be proved is that $\sup_{\|f\|=1} |\langle Kf, f\rangle| = \|K\|$ is achieved for some f in $C[a, b]$. Choose a sequence f_n in $C[a, b]$ with $|\langle Kf_n, f_n\rangle| \to \|K\|$ and $\|f_n\| = 1$. A subsequence of $\langle Kf_n, f_n\rangle$ must converge to μ for μ either $\|K\|$ or $-\|K\|$. Replacing the original sequence by such a subsequence, we can assume without loss in generality that $\langle Kf_n, f_n\rangle \to \mu$. Since $\|f_n\| = 1$ and K is compact, Kf_n has a convergent subsequence. As above, we can assume without loss in generality that the full sequence $Kf_n \to g$. Since

$$\|Kf_n - \mu f_n\|^2 = \|Kf_n\|^2 - 2\mu\langle Kf_n, f_n\rangle + \mu^2 \leq 2\|K\|^2 - 2\mu\langle Kf_n, f_n\rangle \to 0$$

as $n \to \infty$, $Kf_n - \mu f_n \to 0$ and

$$\mu f_n = \mu f_n - Kf_n + Kf_n \to g, \ \|g\| = |\mu| \neq 0,$$

$$f_n \to \mu^{-1} g,$$

$$Kf_n \to \mu^{-1} Kg.$$

But $Kf_n \to g$ and, hence, $\mu^{-1}Kg = g$; thus, $Kg = \mu g$ with $\|g\| = |\mu| \neq 0$. Consequently, if $f = g/\|g\| = 1$, $\langle Kf, f\rangle = \langle Kg, g\rangle/\|g\|^2 = \mu$,

$$|\langle Kf, f\rangle| = \|K\|, \ \|f\| = 1,$$

and the supremum is achieved at f. ■

We shall establish the Hilbert-Schmidt theorem by repeated application of Theorem 59. First, observe that $C[a, b]$ is not finite dimensional because, for example, the powers $1, x, \ldots, x^n$ are linearly independent for all n. Consequently, no finite number of continuous functions can span $C[a, b]$.

Apply the theorem with $\mathcal{F} = \mathcal{F}_0 = C[a, b]$ to ascertain that K has an eigenfunction ϕ_1 with $\|\phi_1\| = 1$ that corresponds to an eigenvalue μ_1 with $|\mu_1| = \|K\|$. Define

$$\mathcal{F}_1 = \{f \in C[a, b]: \langle f, \phi_1\rangle = 0\}.$$

$\mathcal{F}_1$ is a nonzero subspace of $C[a, b]$ and is invariant under K: if f is in $\mathcal{F}_1$, then

$$\langle Kf, \phi_1\rangle = \langle f, K\phi_1\rangle = \mu_1\langle f, \phi_1\rangle = 0$$

and, hence, $K_1:\mathcal{F}_1 \to \mathcal{F}_1$ where $K_1 f = K_0 f$ and $K_0 = K$. By Theorem 59 applied to the self-adjoint compact operator K_1 there exist ϕ_2 in $\mathcal{F}_1$ with $\|\phi_2\| = 1$ such that ϕ_2 is an eigenfunction of K_1 with eigenvalue μ_2 satisfying $|\mu_2| = \|K_1\|$. Evidently ϕ_2 is an eigenfunction of K that is orthogonal to ϕ_1 and

$$|\mu_2| = \|K_1\| = \sup_{\substack{\|f\|=1,\\ f \text{ in } \mathcal{F}_1}} \|Kf\| \le \sup_{\substack{\|f\|=1,\\ f \text{ in } \mathcal{F}_0}} \|Kf\| = \|K\| = |\mu_1|.$$

Since $C[a, b]$ is not finite dimensional, proceed in this fashion to determine subspaces

$$\mathcal{F}_n = \{f \in C[a, b]:\langle f, \phi_1\rangle = 0, \ldots, \langle f, \phi_n\rangle = 0\}$$

and an infinite sequence of orthonormal eigenfunctions $\phi_1, \ldots, \phi_n, \ldots$ with corresponding eigenvalues $\mu_1, \ldots, \mu_n, \ldots$ that satisfy

$$|\mu_1| \ge \cdots \ge |\mu_n| \ge \cdots \text{ with } |\mu_{n+1}| = \|K_n\|.$$

It may happen that $\|K_N\| = 0$ for some N, in which case $\mu_{N+1} = \mu_{N+2} = \mu_{N+3} = \cdots = 0$. In any event, the sequence $\mu_n \to 0$: $|\mu_n|$ decreases to a positive limit or to 0. If the limit is positive, $\{\phi_n/\mu_n\}$ would be bounded and its image under K, $\{\phi_n\}$, would have a convergent subsequence because K is compact. This is impossible because $\|\phi_m - \phi_n\| = \sqrt{2}$ for $m \ne n$. Consequently, μ_n has limit 0 as asserted.

If $f \in C[a, b]$, then $f - \sum_{j=1}^{n}\langle f, \phi_j\rangle\phi_j \in \mathcal{F}_n$ and

$$\left\| f - \sum_{j=1}^{n}\langle f, \phi_j\rangle\phi_j \right\|^2 = \|f\|^2 - \sum_{j=1}^{n}|\langle f, \phi_j\rangle|^2 \le \|f\|^2.$$

Consequently,

$$\left\| K_n\left(f - \sum_{j=1}^{n}\langle f, \phi_j\rangle\phi_j \right) \right\| \le \|K_n\|\,\|f\|,$$

$$\left\| Kf - \sum_{j=1}^{n}\langle f, \phi_j\rangle K\phi_j \right\| \le |\mu_{n+1}|\,\|f\|,$$

or

$$\left\| Kf - \sum_{j=1}^{n}\langle Kf, \phi_j\rangle\phi_j \right\| \le |\mu_{n+1}|\,\|f\| \tag{3.2}$$

because μ_j is real and

$$\langle f, \phi_j\rangle K\phi_j = \langle f, \mu_j\phi_j\rangle\phi_j = \langle f, K\phi_j\rangle\phi_j = \langle Kf, \phi_j\rangle\phi_j.$$

Since $|\mu_{n+1}| \to 0$ it follows that

$$Kf = \sum_{j=1}^{\infty}\langle Kf, \phi_j\rangle\phi_j,$$

where the series converges in the 2-norm to Kf. If $\mu_{N+1} = 0$ for some N, the inequality (3.2) gives

$$Kf = \sum_{j=1}^{N} \langle Kf, \phi_j \rangle \phi_j.$$

Theorem 60 (Hilbert-Schmidt) *Let $K{:}C[a, b] \to C[a, b]$ be a bounded, linear, self-adjoint compact integral operator on $C[a, b]$ with the usual inner product. The construction above determines eigenvalues μ_n of K and corresponding eigenfunctions ϕ_n of K with the following properties:*

1. *The sequence of eigenvalues $\{\mu_n\}_{n=1}^{\infty}$ contains all the nonzero eigenvalues of K and satisfies*

$$|\mu_1| \geq |\mu_2| \geq \cdots \geq |\mu_n| \geq \cdots,$$

with $\mu_n \to 0$ as $n \to \infty$. Consequently, any nonzero *eigenvalue in the sequence $\{\mu_n\}_{n=1}^{\infty}$ is repeated at most a finite number of times.*

2. *The corresponding eigenfunctions are orthonormal*

$$\langle \phi_m, \phi_n \rangle = \delta_{mn}$$

where δ_{mn} is the Kronecker delta.

3. *Each nonzero eigenvalue μ_n is repeated to its (geometric) multiplicity in the sequence $\{\mu_n\}_{n=1}^{\infty}$. In other words, if μ is a nonzero eigenvalue of K, then the eigenspace of μ is*

$$\mathcal{E}_1(\mu) = \text{span}\{\phi_n{:}\lambda_n = \mu\}.$$

4. *If K has an infinite number of nonzero eigenvalues, the expansion*

$$Kf = \sum_{n=1}^{\infty} \langle Kf, \phi_n \rangle \phi_n = \sum_{n=1}^{\infty} \mu_n \langle f, \phi_n \rangle \phi_n$$

holds for each continuous function f on $[a, b]$ and convergence is in the 2-norm.

5. *If K has only N nonzero eigenvalues, then*

$$Kf = \sum_{n=1}^{N} \langle Kf, \phi_n \rangle \phi_n = \sum_{n=1}^{N} \mu_n \langle f, \phi_n \rangle \phi_n.$$

Proof. It only remains to establish that every nonzero eigenvalue of K appears in the sequence $\{\mu_n\}$ and the multiplicity assertion item 3. Suppose that the nonzero eigenvalue μ appears exactly m times in the sequence $\{\mu_n\}$. Then μ has at least m corresponding orthonormal eigenfunctions; hence, $\dim \mathcal{E}_1(\mu) \geq m$. Suppose that strict inequality holds. Then there is ϕ in $\mathcal{E}_1(\mu)$ that is linearly independent of the m eigenfunctions just mentioned. By subtracting from ϕ its projections along each of these m eigenfunctions as in the Gram-Schmidt process, we obtain a nonzero element ψ in $\mathcal{E}_1(\mu)$ that is orthogonal to those m eigenfunctions. Since K is self-adjoint, ψ also is orthogonal to the ϕ_n that correspond to eigenvalues $\mu_n \neq \mu$. Apply the eigenfunction expansion already established with $f = \psi$ to obtain

$$0 \neq \mu\psi = K\psi = \sum_{n=1}^{\infty} \langle K\psi, \phi_n \rangle \phi_n = \sum_{n=1}^{\infty} \mu_n \langle \psi, \phi_n \rangle \phi_n = 0$$

because $\langle \psi, \phi_n \rangle = 0$ for all n, a contradiction. Thus, $\dim \mathcal{E}_1(\mu) = m$ and each nonzero eigenvalue in $\{\mu_n\}$ is repeated to its geometric multiplicity. Now suppose that K has a nonzero eigenvalue $\mu \neq \mu_n$ for all n with $\mu_n \neq 0$ and let ψ be a corresponding eigenfunction. Then by self-adjointness ψ is orthogonal to all the ϕ_n and, just as above, the Hilbert-Schmidt expansion yields the contradiction $0 \neq \mu\psi = K\psi = 0$. ∎

In the language of inner product spaces, the Hilbert-Schmidt theorem says that the orthonormal set of eigenfunctions of the self-adjoint operator K is an orthonormal basis for the range of K.

The Hilbert-Schmidt theorem and $\mu_n \to 0$ proves that every nonzero eigenvalue of a self-adjoint integral operator K has finite (geometric) multiplicity. (This is also true in the non self-adjoint case as was noted earlier in Schur's algebraic approach to defining multiplicity.)

If K has infinitely many nonzero eigenvalues, relatively mild additional assumptions on the self-adjoint kernel $k(x, s)$ imply that the Hilbert-Schmidt expansion

$$Kf = \sum_{n=1}^{\infty} \langle Kf, \phi_n \rangle \phi_n$$

converges uniformly to Kf on $[a, b]$ in addition to the least squares (2-norm) convergence asserted in the theorem.

Corollary 61 (of the Hilbert-Schmidt Theorem) *If the self-adjoint kernel $k(x, s)$ has an infinite number of nonzero eigenvalues and satisfies the additional condition that*

$$\int_a^b |k(x, s)|^2 \, ds \leq M$$

for some constant M and all x in $[a, b]$, then for every f in $C[a, b]$

$$Kf = \sum_{n=1}^{\infty} \langle Kf, \phi_n \rangle \phi_n$$

and the series converges absolutely and uniformly on $[a, b]$ to Kf.

Note that the added condition is automatically satisfied if $k(x, s)$ is continuous on $[a, b] \times [a, b]$.

Proof. Since K is self-adjoint

$$\sum_{n=1}^{\infty} \langle Kf, \phi_n \rangle \phi_n = \sum_{n=1}^{\infty} \mu_n \langle f, \phi_n \rangle \phi_n$$

Fix x in $[a, b]$. The eigenvalue relation

$$\mu_n \phi_n(x) = \int_a^b k(x, s) \phi_n(s) \, ds$$

shows that $\mu_n \overline{\phi_n(x)}$ is the nth Fourier coefficient of the function of s, $k(x, s)$, with respect to the orthonormal set $\overline{\phi_1(s)}, \overline{\phi_2(s)}, \overline{\phi_3(s)}, \ldots$. Consequently, by Bessel's inequality,

$$\sum_{n=1}^{\infty} |\mu_n \phi_n(x)|^2 \leq \int_a^b |k(x, s)|^2 \, ds.$$

Also, by Bessel's inequality,

$$\sum_{n=1}^{\infty} |\langle f, \phi_n \rangle|^2 \leq \int_a^b |f(s)|^2 \, ds.$$

Hence, by the Schwarz inequality,

$$\sum_{n=N}^{\infty} |\mu_n \phi_n(x) \langle f, \phi_n \rangle| \leq \left(\sum_{n=N}^{\infty} |\mu_n \phi_n(x)|^2 \right)^{1/2} \left(\sum_{n=N}^{\infty} |\langle f, \phi_n \rangle|^2 \right)^{1/2},$$

$$\sum_{n=N}^{\infty} |\langle Kf, \phi_n \rangle \phi_n(x)| \leq M^{1/2} \left(\sum_{n=N}^{\infty} |\langle f, \phi_n \rangle|^2 \right)^{1/2}$$

for all x in $[a, b]$. Since the numerical series on the right converges to zero as $N \to \infty$, it follows, that the series $\sum_{n=1}^{\infty} \langle Kf, \phi_n \rangle \phi_n(x)$ on the left is absolutely and uniformly convergent on $[a, b]$. Now, by the Hilbert-Schmidt theorem

$$\lim_{N \to \infty} \int_a^b \left| Kf(x) - \sum_{n=1}^{N} \langle Kf, \phi_n \rangle \phi_n(x) \right|^2 dx = 0$$

and by the uniform convergence of the series the limit can be taken under the integral sign; hence,

$$\int_a^b \left| Kf(x) - \sum_{n=1}^{\infty} \langle Kf, \phi_n \rangle \phi_n(x) \right|^2 dx = 0.$$

Since the integrand is continuous,

$$\sum_{n=1}^{\infty} \langle Kf, \phi_n \rangle \phi_n(x) = Kf(x)$$

by Proposition 19 and the corollary is established. ■

Corollary 62 (of the Hilbert-Schmidt Theorem) *If the integral operator K in the theorem has a symmetric kernel $k(x, s)$, that is $k(x, s)$ is real-valued and $k(x, s) = k(s, x)$, then each eigenfunction ϕ_n in the orthonormal sequence $\{\phi_n\}_{n=1}^{\infty}$ can be chosen real-valued.*

Proof. By Part 2 of the theorem there is a complex-valued orthonormal sequence $\{\phi_n\}_{n=1}^{\infty}$ of eigenfunctions. Express $\phi_n = u_n + iv_n$ where u_n and v_n are real-valued. Since the eigenvalues of a self-adjoint operator are real,

$$K\phi_n = \lambda_n \phi_n$$

can be expressed as

$$Ku_n + iKv_n = \lambda_n u_n + i\lambda_n v_n.$$

Equate real and imaginary parts to obtain

$$Ku_n = \lambda_n u_n \text{ and } Kv_n = \lambda_n v_n.$$

Since one of u_n and v_n is nonzero, replace the complex-valued eigenfunction ϕ_n by $u_n/\|u_n\|$ if $u_n \neq 0$ and by $v_n/\|v_n\|$ if $u_n = 0$. If λ_n is an eigenvalue of multiplicity greater than 1, the real-valued eigenfunctions associated to the equal λ_n in this way can be replaced by the Gram-Schmidt process with real-valued orthonormal eigenfunctions, still called ϕ_n. These orthonormal eigenfunctions are orthogonal to all the other real-valued eigenfunctions constructed in this way because eigenfunctions belonging to distinct eigenvalues are orthogonal.

In this way, the possibly complex-valued orthonormal sequence of eigenfunctions can be replaced by a real-valued orthonormal sequence of eigenfunctions. ∎

If, in the construction leading to the Hilbert-Schmidt theorem, $\mu_{N+1}=0$ for some first N, the infinite series in the Hilbert-Schmidt theorem reduce to finite sum of N terms. In this case, the kernel $k(x, s)$ is called **degenerate** and can be expressed as

$$k(x, s) = \sum_{n=1}^{N} \mu_n \phi_n(x)\overline{\phi_n(s)}.$$

If $\mu_{N+1} \neq 0$ for all N the corresponding equality

$$k(x, s) = \sum_{n=1}^{\infty} \mu_n \phi_n(x)\overline{\phi_n(s)}$$

need not be true. However, it does hold for an important class of kernels, the positive definite symmetric kernels; see Mercer's theorem in the next section.

A system of real-valued orthogonal eigenfunctions $\phi_1(x)$, $\phi_2(x), \ldots.$ for a symmetric kernel $k(x, s)$ is called a **complete system** of orthogonal eigenfunctions for $k(x, s)$ if any eigenfunction of the kernel $k(x, s)$ is a finite linear combination of $\phi_1(x)$, $\phi_2(x), \ldots.$

Corollary 63 *(of the Hilbert-Schmidt theorem) The orthonormal eigenfunctions $\phi_1(x)$, $\phi_2(x), \ldots$ in the Hilbert-Schmidt theorem are a complete system of orthogonal eigenfunctions for the kernel $k(x, s)$.*

Proof. Let $\lambda_n = \mu_n^{-1}$ for each nonzero eigenvalue μ_n of the integral operator K in the Hilbert-Schmidt theorem. Let ψ be an eigenfunction of $k(x, s)$ and ρ its eigenvalue. Then $\mu = \rho^{-1}$ is a nonzero eigenvalue of K. By Item 1 in the Hilbert-Schmidt theorem $\rho = \lambda_{n_0}$ for some n_0 and by Item 3 ψ is a linear combination of the ϕ_n with $\lambda_n = \rho$. Thus, $\phi_1, \phi_2, \ldots$ is a complete orthogonal system for the kernel $k(x, s)$. ∎

The following result reveals the close connection between the eigenvalues and eigenfunctions of a self-adjoint kernel $k(x, s)$ and those of its iterated kernels $k_n(x, s)$. Recall that if K is the integral operator with kernel $k(x, s)$, then $k_n(x, s)$ is the kernel of the integral operator K^n.

Theorem 64 *Let $k(x, s)$ be a self-adjoint kernel and $k_n(x, s)$ be its nth iterated kernel. If the integral operator corresponding to K satisfies the hypotheses in the Hilbert-Schmidt theorem and $\phi_1(x)$, $\phi_2(x)$, ... is a complete system of orthogonal eigenfunctions for the kernel $k(x, s)$, then they are also a complete system of orthogonal eigenfunctions for the kernel $k_n(x, s)$.*

Proof. If λ_j is the eigenvalue of the kernel $k(x, s)$ with eigenfunction ϕ_j, then $\lambda_j K\phi_j = \phi_j$,

$$\lambda_j^2 K^2 \phi_j = \lambda_j K(\lambda_j K\phi_j) = \lambda_j K\phi_j = \phi_j,$$

and continuing in this way $\lambda_j^n K^n \phi_j = \phi_j$. Thus, λ_j^n is an eigenvalue of $k_n(x, s)$ with corresponding eigenfunction ϕ_j. Let ψ be an eigenfunction of $k_n(x, s)$ and ρ its eigenvalue. If $\rho \neq \lambda_j^n$ for all

j, then ψ is orthogonal to ϕ_j for all j and by the Hilbert-Schmidt expansion

$$\psi = \rho K^n \psi = \rho K\big(K^{n-1}\psi\big) = \rho \sum_{j=1}^{\infty} \langle K^{n-1}\psi,\, \phi_j\rangle \phi_j$$

$$= \rho \sum_{j=1}^{\infty} \langle \psi,\, K^{n-1}\phi_j\rangle \phi_j = \rho \sum_{j=1}^{\infty} \frac{\langle \psi,\, \phi_j\rangle}{\lambda_j^{n-1}} \phi_j = 0,$$

a contradiction. Therefore, $\rho = \lambda_{j_0}^n$ for some j_0. Let

$$\tilde{\psi} = \psi - \sum_{j \text{ with } \lambda_j = \lambda_{j_0}} \langle \psi,\, \phi_j\rangle \phi_j.$$

Then $\tilde{\psi} = \rho K^n \tilde{\psi}$ and $\tilde{\psi}$ is orthogonal to all ϕ_j with eigenvalues $\lambda_j = \lambda_{j_0}$. It is also orthogonal to all ϕ_j with eigenvalues $\lambda_j \neq \lambda_{j_0}$ because $k(x, s)$ is self-adjoint. Use of the Hilbert-Schmidt expansion, as above, leads to $\tilde{\psi} = 0$. Thus,

$$\psi = \sum_{j \text{ with } \lambda_j = \lambda_{j_0}} \langle \psi,\, \phi_j\rangle \phi_j$$

and the theorem is established. ■

3.4.2 Mercer's Theorem

An application of the Hilbert-Schmidt theorem will prepare the way for Mercer's theorem. Let $k(x, s)$ be a continuous symmetric kernel on $[a, b] \times [a, b]$ and $k_2(x, t)$ be its second iterated kernel. (See Section 3.1.) Since $f(x) = k(x, t)$, with t regarded as fixed, is continuous for x in $[a, b]$ and $Kf(x) = k_2(x, t)$ the Hilbert-Schmidt theorem gives

$$k_2(x, t) = \sum_{n=1}^{\infty} \langle Kf,\, \phi_n\rangle \phi_n(x)$$

where the series converges uniformly in x to $k_2(x, t)$ for each fixed t in $[a, b]$. Since

$$\begin{aligned}
\langle Kf,\, \phi_n\rangle &= \int_a^b Kf(x)\phi_n(x)\,dx \\
&\int_a^b \left(\int_a^b k(x, s)k(s, t)\,ds\right)\phi_n(x)\,dx \\
&= \int_a^b k(s, t)\left(\int_a^b k(x, s)\phi_n(x)\,dx\right)ds \\
&= \int_a^b k(s, t)\frac{\phi_n(s)}{\lambda_n}\,ds = \frac{\phi_n(t)}{\lambda_n^2},
\end{aligned}$$

we obtain

$$k_2(x, t) = \sum_{n=1}^{\infty} \frac{\phi_n(x)\phi_n(t)}{\lambda_n^2}$$

where, for each fixed t in $[a, b]$, the convergence is uniform for x in $[a, b]$. Set $x = t$ to obtain

$$k_2(t, t) = \sum_{n=1}^{\infty} \frac{\phi_n(t)^2}{\lambda_n^2}.$$

Now,

$$\lim_{N\to\infty}\int_a^b \Big[k(x,s)-\sum_{n=1}^{N}\frac{\phi_n(x)\phi_n(s)}{\lambda_n}\Big]^2 ds = \lim_{N\to\infty}\left[\int_a^b k(x,s)^2\,ds-\sum_{n=1}^{N}\frac{\phi_n(x)^2}{\lambda_n^2}\right]$$
$$= k_2(x,x)-\sum_{n=1}^{\infty}\frac{|\phi_n(x)|^2}{\lambda_n^2}=0 \tag{3.3}$$

from the expansion for $k_2(x, x)$ above. This establishes that for each x in $[a, b]$ the expansion

$$k(x,s)=\sum_{n=1}^{\infty}\frac{\phi_n(x)\phi_n(s)}{\lambda_n}$$

hold in the sense of least squares convergence (2-norm convergence).

Moreover, since

$$\sum_{n=1}^{m}\frac{\phi_n(t)^2}{\lambda_n^2}\le k_2(t,t)$$

for any m, integration yields

$$\sum_{n=1}^{m}\frac{1}{\lambda_n^2}\le\int_a^b k_2(t,t)\,dt$$

$$\sum_{n=1}^{\infty}\frac{1}{\lambda_n^2}\le\int_a^b k_2(t,t)\,dt<\infty$$

because $k_2(x, s)$ is continuous on $[a, b] \times [a, b]$. That is, for a symmetric kernel the series

$$\sum_{n=1}^{\infty}\frac{1}{\lambda_n^2}$$

converges.

A continuous symmetric kernel $k(x, s)$ is **positive definite** if all its eigenvalues λ_n are positive. If K is the corresponding self-adjoint integral operator, then the nonzero eigenvalues of K are $\mu_n = 1/\lambda_n$ and by the Hilbert-Schmidt theorem

$$\langle Kf,f\rangle=\left\langle\sum_{n=1}^{\infty}\langle Kf,\phi_n\rangle\phi_n,f\right\rangle=\sum_{n=1}^{\infty}\langle Kf,\phi_n\rangle\langle\phi_n,f\rangle=\sum_{n=1}^{\infty}\mu_n|\langle f,\phi_n\rangle|^2\ge 0$$

for every f in $C[a, b]$.

Theorem 65 (Mercer's Theorem) *If $k(x, s)$ is a continuous, symmetric, positive definite kernel on $[a, b] \times [a, b]$ and $K: C[a, b] \to C[a, b]$ is its corresponding integral operator with nonzero eigenvalues $\mu_n = 1/\lambda_n > 0$, where λ_n are eigenvalues of the kernel repeated to multiplicity and ϕ_n are corresponding real-valued orthonormal eigenfunctions, then*

$$k(x,s)=\sum_{n=1}^{\infty}\mu_n\phi_n(x)\phi_n(s)=\sum_{n=1}^{\infty}\frac{\phi_n(x)\phi_n(s)}{\lambda_n}$$

and the series is absolutely and uniform convergence on $[a, b] \times [a, b]$.

Proof. We show first that $k(x,x) \geq 0$ for $a \leq x \leq b$. To this end, assume the contrary so that $k(c,c) < 0$ for some c in (a,b). There is a $\delta > 0$ such that $k(x,s) < 0$ for (x,s) in $a \leq x, s \leq b$ with $|x-c| < \delta$ and $|s-c| < \delta$ because $k(x,s)$ is continuous. Fix any continuous function f such that $f \geq 0, f(c) = 1$, and $f(x) = 0$ for $|x-c| \geq \delta$. (A function with a piecewise linear graph can serve this purpose.) For such an f

$$\langle Kf, f\rangle = \int_a^b \left(\int_a^b k(x,s)f(s)\,ds\right) f(x)\,dx = \iint\limits_{\substack{|x-c|<\delta,\\ |s-c|<\delta}} k(x,s)f(s)f(x)\,dxds < 0,$$

which contradicts the positive definiteness of $k(x,s)$. Hence, $k(x,x) \geq 0$ for x in $[a,b]$ as asserted.

Now, the kernel

$$l(x,s) = k(x,s) - \sum_{j=1}^{n} \frac{\phi_j(x)\phi_j(s)}{\lambda_j}$$

satisfies the hypotheses of the theorem: positive definiteness of $l(x,s)$ follows from

$$\langle Lf, f\rangle = \langle Kf, f\rangle - \sum_{j=1}^{n} \frac{|\langle \phi_j, f\rangle|^2}{\lambda_j} = \sum_{j=n+1}^{\infty} \frac{|\langle \phi_j, f\rangle|^2}{\lambda_j} \geq 0$$

for any f in $C[a,b]$, by the expansion of $\langle Kf, f\rangle$ used above. Hence, if λ is an eigenvalue of the kernel $l(x,s)$ and ϕ a corresponding eigenfunction, then $\lambda^{-1}\langle \phi, \phi\rangle = \langle L\phi, \phi\rangle \geq 0$ and $\lambda > 0$ because 0 is not an eigenvalue of the kernel $l(x,s)$.

Consequently, $l(s,s) \geq 0$ on $[a,b]$ and

$$\sum_{j=1}^{n} \frac{|\phi_j(s)|^2}{\lambda_j} \leq k(s,s),$$

$$\sum_{j=1}^{\infty} \frac{|\phi_j(s)|^2}{\lambda_j} \leq k(s,s)$$

for s in $[a,b]$. By the Schwarz inequality

$$\sum_{j=n}^{n+p} \frac{|\phi_j(x)||\phi_j(s)|}{\lambda_j} = \sum_{j=n}^{n+p} \frac{|\phi_j(x)|}{\sqrt{\lambda_j}} \frac{|\phi_j(s)|}{\sqrt{\lambda_j}} \leq \sqrt{\sum_{j=n}^{n+p} \frac{|\phi_j(x)|^2}{\lambda_j}} \sqrt{\sum_{j=n}^{n+p} \frac{|\phi_j(s)|^2}{\lambda_j}},$$

$$\sum_{j=n}^{n+p} \frac{|\phi_j(x)||\phi_j(s)|}{\lambda_j} \leq \sqrt{k(x,x)k(s,s)}.$$

Since $k(t,t)$ is continuous on $[a,b]$, the right member is bounded and the series $\sum_{j=1}^{\infty} \frac{\phi_j(x)\phi_j(s)}{\lambda_j}$ is absolutely convergent for (x,s) in $[a,b] \times [a,b]$. Moreover, for each fixed s

$$\sum_{j=n}^{n+p} \frac{|\phi_j(x)||\phi_j(s)|}{\lambda_j} \leq \sqrt{k(x,x)} \sqrt{\sum_{j=n}^{n+p} \frac{|\phi_j(s)|^2}{\lambda_j}} \leq M \sqrt{\sum_{j=n}^{n+p} \frac{|\phi_j(s)|^2}{\lambda_j}}$$

for some constant M and for all x in $[a,b]$. Since the sum on the right can be made arbitrarily small for all p by choosing n suitably large, it follows that the series

$$\sum_{j=1}^{\infty} \frac{|\phi_j(x)||\phi_j(s)|}{\lambda_j}$$

is uniformly convergent in x for each s and conversely by symmetry. Thus,

$$\sum_{j=1}^{\infty} \frac{\phi_j(x)\phi_j(s)}{\lambda_j}$$

is absolutely convergent and uniformly convergent in x for each s and conversely. By the uniform convergence in s for each x we can pass to the limit under the integral in (3.3) to obtain

$$\int_a^b \left[k(x,s) - \sum_{n=1}^{\infty} \frac{\phi_n(x)\phi_n(s)}{\lambda_n} \right]^2 ds = 0$$

and the integrand is continuous in s for each x, again by the uniform convergence in s. It follows that

$$k(x,s) = \sum_{n=1}^{\infty} \frac{\phi_n(x)\phi_n(s)}{\lambda_n}$$

in the sense of pointwise convergence on $[a, b] \times [a, b]$. In fact, the convergence is uniform on $[a, b] \times [a, b]$. Indeed, by Dini's Theorem 25,

$$\sum_{n=1}^{\infty} \frac{|\phi_n(x)|^2}{\lambda_n}$$

converges uniformly on $[a, b]$ because

$$\sum_{n=1}^{\infty} \frac{|\phi_n(x)|^2}{\lambda_n} = k(x,x),$$

the series consists of nonnegative continuous terms, and its sum is continuous on $[a, b]$. Now the Schwarz inequality estimate above shows that

$$\sum_{j=1}^{\infty} \frac{|\phi_j(x)||\phi_j(s)|}{\lambda_j}$$

converges uniformly for (x, s) in $[a, b] \times [a, b]$ and, hence, the same is true for

$$\sum_{n=1}^{\infty} \frac{\phi_n(x)\phi_n(s)}{\lambda_n}.$$

■

An application of Mercer's theorem to Sturm-Liouville boundary value and eigenvalue problems is given in Theorem 122 and Theorem 126. Many of the most important Sturm-Liouville problems that occur in applied mathematics are covered by the theorems.

3.5 Nonnegative Kernels

Two main results from the theory of integral equations play a pivotal role in a unified study of Sturm-Liouville boundary value and eigenvalue problems. The first is the Hilbert-Schmidt theorem and the second is Jentzsch's theorem. The original theorem of Jentzsch asserts that an integral operator with a strictly positive continuous kernel has a positive eigenvalue which is simple and smallest in modulus among all the eigenvalues of the kernel and has a corresponding positive eigenfunction. Subsequent extensions of Jentzsch's theorem weaken the positivity assumptions on the kernel but maintain, in modified form, the essential conclusions of the original theorem. The results on such suitably positive kernels stand behind the rich oscillatory and approximation properties of the eigenfunctions of Sturm-Liouville eigenvalue problems and of corresponding results in other contexts. For the relevance of suitably positive kernels for Sturm-Liouville problems see Section 1.11.2.

The following holds throughout section:

> **Standing Assumptions:** $k(x, s) \geq 0$ *is a real-valued continuous kernel defined on* $[a, b] \times [a, b]$ *or on* $[a, b] \times [a, b] \backslash \{(a, a)\}$. *In the latter case, we also assume that the kernel satisfies (a), (b), and (c) in Theorem 52.*

Under the standing assumptions, the corresponding integral operator

$$Kf(x) = \int_a^b k(x, s) f(s)\, ds$$

maps the function space of real-valued continuous functions $C[a, b]$ into itself and is a compact, bounded linear operator when $C[a, b]$ is equipped with the maximum norm by Theorems 51 and 52.

The choice of the domain of $k(x, s)$ and the assumptions (a), (b), and (c) are dictated by the fact that we shall apply the results of this section to kernels that are Green's functions for regular or singular Sturm-Liouville problems. Those Green's function satisfy the standing assumptions.

The reasoning used here also applies, without essential change, when the interval $[a, b]$, a 1-dimensional simplex, is replaced by an n-dimensional simplex Δ_n. See the concluding remarks at the end of the section.

The results established below apply to kernels $k(x, s)$ that are nonnegative and subject to certain additional positivity requirements. The corresponding integral operators $K: C[a, b] \to C[a, b]$ map nonnegative functions into nonnegative functions. Thus, it is convenient to let

$$P = \{f \text{ in } C[a, b]: f \geq 0 \text{ on } [a, b]\}.$$

The set $P \backslash \{0\}$ is P with the zero function removed.

As usual,

$$\langle f, g \rangle = \int_a^b f(s)\overline{g(s)}\, ds.$$

3.5.1 Positive Kernels

In this section $k(x, s) > 0$ on its domain, which is $[a, b] \times [a, b]$, possibly with the point (a, a) removed. See the standing assumptions above. Kernels of this type that come up in applied mathematics include the Gauss kernel $\exp(-(x - s)^2)$, $\exp(-xs)$, and $\max(x, s)$. Define

$$r(K) = \sup\{\mu \geq 0\text{: there exists } p \text{ in } P \backslash \{0\} \text{ such that } \mu p \leq Kp\}.$$

If $r(K)p \leq Kp$ for some $p \in P \backslash \{0\}$, then p is called an **extremal function** for K.

Theorem 66 *Assume that* $k(x, s)$ *is strictly positive on its domain, in addition to the standing assumptions. The following hold. (1)* $r(K) > 0$. *(2a) Extremal functions exist and every extremal function is a positive eigenfunction of* K *corresponding to the eigenvalue* $r(K)$. *(2b)*

If ϕ is an eigenfunction corresponding to the eigenvalue $r(K)$, then $|\phi|$ is an extremal function corresponding to K and, hence, a positive eigenfunction corresponding to $r(K)$. Consequently if ϕ is a real-valued eigenfunction corresponding to $r(K)$, then $\phi > 0$ or $\phi < 0$ on $[a, b]$. (3) $r(K)$ has geometric multiplicity 1. (4) $r(K)$ has algebraic multiplicity 1. (5) Every eigenvalue μ of K different from $r(K)$ satisfies $|\mu| < r(K)$. Hence,

$$r(K) = \max\{|\mu| : \mu \text{ is an eigenvalue of } K\}.$$

Proof. Let $r = r(K)$.

(1) If $e(x) = 1$ for x in $[a, b]$, then Ke has a positive minimum m, $me \leq Ke$; hence, $r \geq m > 0$.

(2a) There exist $p_n \in P\backslash\{0\}$ with $\|p_n\|_{\max} = 1$ such that $\mu_n p_n \leq Kp_n$ and $\mu_n \to r$ as $n \to \infty$. Since K is compact we can assume without loss in generality that $Kp_n \to q$ in $C[a, b]$ as $n \to \infty$. Since $p_n(x_n) = 1$ for some x_n in $[a, b]$,

$$\begin{aligned} \|Kp_n\|_{\max} &\geq \mu_n, \\ \|q\|_{\max} &\geq r > 0, \\ \mu_n Kp_n &\leq K(Kp_n) \\ rq &\leq Kq \text{ and } q \in P\backslash\{0\}. \end{aligned}$$

Consequently, q is an extremal function for K. For any such extremal function q equality must hold in $rq \leq Kq$; otherwise, $Kq - rq \in P\backslash\{0\}$ and

$$K(Kq - rq) > 0 \text{ on } [a, b]$$

because $k(x, s) > 0$ on $[a, b] \times [a, b]$. Since $K(Kq - rq)$ assumes its minimum value which is positive,

$$K(Kq - rq) > \varepsilon Kq \text{ on } [a, b]$$

for some $\varepsilon > 0$. Then

$$K(Kq) > (r + \varepsilon)Kq$$

and since $Kq \in P\backslash\{0\}$ this contradicts the definition of r. Thus, $Kq = rq$ with $q \in P\backslash\{0\}$ for any extremal function q. That is, any extremal function of K is an eigenfunction of K corresponding to the eigenvalue r. Finally, $rq = Kq$, q in $P\backslash\{0\}$, and $k(x, s) > 0$ imply that $q > 0$ on $[a, b]$.

(2b) If $r\phi = K\phi$ with $\phi \neq 0$, then $r|\phi| \leq K|\phi|$. Hence $|\phi|$ is an extremal function of K. By (2a) it is a positive eigenfunction of K corresponding to the eigenvalue r. If ϕ is real-valued, then $\phi > 0$ or $\phi < 0$ on $[a, b]$ because $|\phi| > 0$ on $[a, b]$ implies ϕ never takes the value 0 in $[a, b]$.

(3) From (1) and (2a), $Kp = rp$ for some $p > 0$ on $[a, b]$. If $Ky = ry$ for some real-valued nonzero $y \in C[a, b]$, then

$$z = y - \frac{\langle y, p\rangle}{\langle p, p\rangle} p$$

is orthogonal to p. If $z \neq 0$, then it is an eigenfunction belonging to r and must maintain a fixed sign on $[a, b]$ by (2b). This contradicts the orthogonality of p and z on $[a, b]$. Thus, $z = 0$ and all real-valued eigenfunctions corresponding to the eigenvalue r are nonzero multiples of p. If y is a complex-valued eigenfunction of K corresponding to the eigenvalue r, then $y = u + iv$ where u and v are real-valued, $Ku = ru$ and $Kv = rv$. Either u is an eigenfunction of K corresponding to r or $u = 0$. In either case, $u = c_1 p$ for some real constant c_1. Likewise, $v = c_2 p$ for some real constant c_2 and $y = cp$ where $c = c_1 + ic_2$. This establishes that the eigenspace of r is one dimensional, consisting of all multiples of p. Thus, the geometric multiplicity of $r(K)$ is 1, the dimension of its eigenspace.

(4) Suppose $(K - rI)^2 w = 0$ for some real-valued w in $C[a, b]$ and that $(K - rI)w \neq 0$. Then $y = (K - rI)w$ is a real-valued eigenfunction of K corresponding to the eigenvalue r. So $r^{-1}Ky = y$ and, by replacing w by $-w$ if need be, we can assume without loss in generality that $y > 0$ on $[a, b]$. Apply the operator $r^{-1}K$ repeatedly to $Kw - rw = y$ to obtain

$$\begin{aligned} Kw - rw &= y, \\ r^{-1}K^2 w - Kw &= y, \\ r^{-2}K^3 w - r^{-1}K^2 w &= y, \\ &\cdots \\ r^{-n+1}K^n w - r^{-n+2}K^{n-1} w &= y \\ r^{-n}K^{n+1} w - r^{-n+1}K^n w &= y. \end{aligned}$$

Add the first n of these equations to get

$$r^{-n+1}K^n w - rw = ny,$$

$$K^n w = r^{n-1}(rw + ny) > 0$$

for n sufficiently large. For such n, the last equation in the chain above gives

$$Kp - rp = r^n y > 0 \quad \text{for } p = K^n w > 0.$$

Consequently, there is an $\varepsilon > 0$ such that $Kp - rp > \varepsilon p$, which contradicts the definition of r. Hence, $(K - rI)^2 w = 0$ for some real-valued w in $C[a, b]$ implies $(K - rI)w = 0$. Now suppose w is complex-valued and satisfies $(K - rI)^2 w = 0$ and $(K - rI)w \neq 0$. If $w = u + iv$ with u and v real-valued, then $(K - rI)^2 u = 0$ and if $(K - rI)u \neq 0$ we reach a contradiction as above. Likewise for v. Hence, $(K - rI)^2 w = 0$ implies $(K - rI)w = 0$. The reverse implication is evident. Thus, the generalized eigenspace $\mathcal{E}_2(r)$ and the eigenspace $\mathcal{E}_1(r)$ are equal. By (3), $\dim \mathcal{E}_2(r) = \dim \mathcal{E}_1(r) = 1$ and the algebraic multiplicity of $r(K)$ is 1.

(5) If $Ky = \mu y$ with $y \neq 0$, then

$$|\mu||y| = |Ky| \leq K|y| \quad \text{and} \quad |y| \in P\backslash\{0\},$$
$$|\mu| \leq r.$$

If $|\mu| = r$ then $|y|$ is an extremal function for K, $|y| > 0$ is an eigenfunction corresponding to r and

$$r|y| = |Ky| \leq K|y| = r|y|.$$

Thus equality holds throughout. In particular, $|Ky|(c) = K|y|(c)$ for $c = (a + b)/2$ and by the condition for equality in the triangle inequality for integrals (Proposition 22) the values of $k(c, s)y(s)$ for $a \leq s \leq b$ lie along a ray emanating from the origin in the complex plane; that is,

$$k(c, s)y(s) = e^{i\theta_c} u_c(s) \quad \text{for some real } \theta_c \quad \text{and} \quad u_c(s) \geq 0.$$

It follows that $u_c(s) > 0$ and that $y = e^{i\theta_c} p$ where $p(s) = u_c(s)/k(c, s) > 0$. Then $Ky = \mu y$ implies $Kp = \mu p$; hence, μ is real and positive and $\mu = |\mu| = r$. Thus, all eigenvalues of K different from r are less than r in modulus. ■

The basic conclusions in Theorem 66 are due to Jentzsch [22]. The original proofs were quite different and relied on some rather deep results in complex analysis and the Fredholm theory of integral equations. The proof given here is motivated by corresponding results about positive matrices and an inequality of Collatz [8].

The strict positivity assumed in Jentzsch's original theorem can be relaxed quite a lot and such variants of Jentzsch's theorem have important applications. The continuity and strict positivity condition cannot be relaxed too much: the Volterra kernel

$$k(x, s) = \begin{cases} 1 & \text{for } a \le x < s \le b \\ 0 & \text{for } a \le s \le x \le b \end{cases}$$

is known to have no eigenvalues. Note that $k(x, x) = 0$ for $a \le x \le b$.

3.5.2 Kernels Positive on the Open Diagonal

Many Green's functions determined by Sturm-Liouville problems are nonnegative but not strictly positive because they vanish at the endpoints of the interval of interest. For example, the Green's function $k(x, s)$ for a vibrating string of length l with ends pinned is

$$k(x, s) = \begin{cases} (l - x)s & \text{for } 0 \le s \le x \le l \\ (l - s)x & \text{for } 0 \le x \le s \le l \end{cases}.$$

Thus, we need to extend the results of the last section to nonnegative kernels that are suitably positive so as to embrace such Green's functions.

In addition to the standing assumptions, assume that $k(x, x) > 0$ for $a < x < b$; that is, $k(x, s)$ is positive on the diagonal of the square with its endpoints removed, a set we refer to as the open diagonal of the square. Let $K: C[a, b] \to C[a, b]$ be the corresponding integral operator and K_n be the integral operator on $C[a, b]$ with strictly positive kernel $k_n(x, s) = k(x, s) + n^{-1}$. Both K and K_n are compact linear operators on $C[a, b]$ with the maximum norm.

Lemma 67 *$r(K) > 0$, $r(K_n) \ge r(K)$ and the sequence $r(K_n)$ is decreasing.*

Proof. Since $k(x, x) > 0$ for $a < x < b$ there is a subinterval $[c, d]$ of $[a, b]$ with $a < c < d < b$ such that $k(x, s) > 0$ on $[c, d] \times [c, d]$. Theorem 66 applies to the integral operator $L: C[c, d] \to C[c, d]$ defined by $Lf(x) = \int_c^d k(x, s)f(s)\, ds$. Consequently, L has a positive eigenvector ψ on $[c, d]$ that corresponds to a positive eigenvalue $\mu = r(L)$,

$$\mu\psi(x) = \int_c^d k(x, s)\psi(s)\, ds, \quad c \le x \le d.$$

Since the integral in the right member of this equality is defined for all x in $[a, b]$, we extend ψ to a continuous, nonnegative function on $[a, b]$ by this formula. Then

$$\mu\psi(x) = \int_c^d k(x, s)\psi(s)\, ds \le \int_a^b k(x, s)\psi(s)\, ds = K\psi(x), \quad a \le x \le b,$$

$$\mu\psi \le K\psi,$$
$$0 < \mu \le r(K).$$

If $\mu p \le Kp$ for $p \ge 0$ and $p \ne 0$, then $\mu p \le K_n p$ and, hence, $r(K) \le r(K_n)$. Likewise, if $m \ge n$ and $\mu p \le K_m p$ for $p \ge 0$ and $p \ne 0$, then $\mu p \le K_n p$ and, hence, $r(K_m) \le r(K_n)$. ■

Theorem 68 *If $k(x, s) \ge 0$ on its domain and $k(x, x) > 0$ for $a < x < b$, in addition to the standing assumptions, then the corresponding integral operator K has $r(K) > 0$, $r(K)$ is a (positive) eigenvalue of K with a corresponding nonnegative eigenfunction p. Any such eigenfunction is positive on $a < x < b$. Moreover,*

$$r(K) = \max\{|\mu| : \mu \text{ is an eigenvalue of } K\}.$$

Proof. By the lemma $r(K) > 0$ and the sequence $r_n = r(K_n)$ decreases to a limit r' with $r' \geq r(K)$. Since $k_n(x, s) > 0$ on its domain, r_n is a positive eigenvalue of K_n with a corresponding positive continuous eigenfunction p_n on $[a, b]$, $r_n p_n = K_n p_n$ and $\|p_n\|_{\max} = 1$. The sequence $\{p_n\}$ is uniformly bounded (by 1) and it is easy to check that $\{p_n\}$ is equicontinuous on $[a, b]$: since $r_n p_n = K_n p_n$ and r_n decreases to r'

$$|p_n(x) - p_n(x_0)| \leq \frac{1}{r'} \int_a^b |k(x, s) - k(x_0, s)|\, ds$$

for x and x_0 in $[a, b]$. If $k(x, s)$ is continuous on $[a, b] \times [a, b]$, then the integral on the right tends uniformly to 0 as x tends to x_0 by the uniform continuity of the kernel. In the case when the kernel is defined and continuous on $[a, b] \times [a, b] \backslash \{(a, a)\}$ and satisfies the standing assumption (c), it was established in the proof of Theorem 52 that $\int_a^b |k(x, s) - k(x_0, s)|\, ds$ tends to zero as x tends to x_0 for every x_0 in $[a, b]$. Thus, $\{p_n\}$ is equicontinuous at x_0 for every x_0 in $[a, b]$. By Proposition 42, $\{p_n\}$ is equicontinuous on $[a, b]$. Consequently, in either case, by the Arzelà-Ascoli theorem $\{p_n\}$ has a subsequence that converges uniformly to a continuous function p on $[a, b]$. Without loss in generality we can assume that the full sequence converges to p. Let $n \to \infty$ in $r_n p_n = K_n p_n$ and $\|p_n\|_{\max} = 1$ to obtain

$$r' p = Kp \quad \text{with} \quad p \geq 0 \quad \text{and} \quad \|p\|_{\max} = 1.$$

So p is an eigenfunction of K corresponding to the eigenvalue r' and $r' \leq r(K)$. The reverse inequality follows from the lemma because $r_n \geq r(K)$ for all n. Thus, $r' = r(K)$.

Moreover, it follows that $p > 0$ on (a, b). Indeed, $p(c) > 0$ for some c in (a, b). If $\alpha = \inf\{x \in [a, c] : p > 0 \text{ on } (x, c]\}$ and $\beta = \sup\{x \in [c, b] : p > 0 \text{ on } [c, x)\}$, then $p(\beta) = 0$ and $p(\alpha) = 0$. If $\beta < b$ there is $\varepsilon > 0$ such that $k(\beta, s) > 0$ and $p(s) > 0$ for $0 < \beta - s < \varepsilon$. This leads to the contradiction

$$0 = r' p(\beta) = \int_a^b k(\beta, s) p(s)\, ds \geq \int_{\beta-\varepsilon}^{\beta} k(\beta, s) p(s)\, ds > 0.$$

Hence, $\beta = b$ and likewise $\alpha = a$. Thus $p > 0$ on (a, b).

If $\mu\phi = K\phi$ with $\phi \neq 0$, then $|\mu||\phi| \leq K|\phi|$ and, hence, $|\mu| \leq r(K)$ and the final assertion in the theorem is established. ∎

Now set $r = r(K)$ and suppose in addition to the positivity assumptions that $k(x, s) = k(s, x)$ so the kernel k is symmetric and K is self-adjoint. By the theorem $rp = Kp$ with $p > 0$ on (a, b). So p is an extremal function for K. If q is any extremal function for K, then $rq \leq Kq$ with $q \in P \backslash \{0\}$. If equality does not hold in $rq \leq Kq$, then

$$r\langle q, p\rangle < \langle Kq, p\rangle = \langle q, Kp\rangle = r\langle q, p\rangle,$$

which is a contradiction because $\langle q, p\rangle > 0$. Hence equality holds in $rq \leq Kq$; that is, $rq = Kq$ and q is a nonnegative eigenfunction corresponding to the eigenvalue $r(K)$. Thus every extremal function of K is an eigenfunction of K corresponding to the eigenvalue $r(K)$ and the extremal function is positive on (a, b). This establishes (1) and (2a) of the following theorem.

Theorem 69 *If, in addition to the standing assumptions, $k(x, x) > 0$ for $a < x < b$ and $k(x, s)$ is symmetric, then the following hold. (1) $r(K) > 0$. (2a) Extremal functions exist and every extremal function is positive on (a, b) and is an eigenfunction of K corresponding to the eigenvalue $r(K)$. (2b) If ϕ is an eigenfunction corresponding to the eigenvalue $r(K)$, then $|\phi|$ an extremal function corresponding to K and, hence, $|\phi|$ is an eigenfunction* *corresponding to* $r(K)$ *and* $|\phi| > 0$ *on* (a, b). *Consequently, if* ϕ *is real-valued, then* $\phi > 0$ *or* $\phi < 0$ *on* (a, b). *(3)* $r(K)$ *has geometric multiplicity* 1. (4) $r(K)$ *has algebraic multiplicity* 1. *(5) Every eigenvalue* μ *of* K *different from* $r(K)$ *satisfies* $|\mu| < r(K)$. *Hence,*

$$r(K) = \max\{|\mu| : \mu \text{ is an eigenvalue of } K\}.$$

Proof. Let $r = r(K)$.

(1) and (2a) have been established.

(2b) If $r(K)\phi = K\phi$ with $\phi \neq 0$, then $r(K)|\phi| \leq K|\phi|$ so $|\phi|$ is an extremal function for K. The remaining conclusions follow from (2a).

(3) The conclusion is established by reasoning as for (3) in Theorem 66.

(4) If K is self-adjoint and $(K - rI)^2 w = 0$, then

$$0 = \left\langle (K - rI)^2 w, w \right\rangle = \langle (K - rI)w, (K - \mu I)w \rangle = \|(K - rI)w\|^2$$

which implies that $(K - rI)w = 0$. Thus, $(K - rI)^2 w = 0$ implies $(K - rI)w = 0$. Since the reverse implication evidently is true, the subspace of functions w satisfying $(K - rI)^2 w = 0$ is one dimensional by (3); that is, the algebraic multiplicity of $r(K)$ is 1.

(5) The conclusion is established much as for (5) in Theorem 66 but with an adjustment because the kernel is no longer strictly positive on $[a, b] \times [a, b]$: the integral operator K^n has kernel $k_n(x, s)$, the nth iterated kernel of $k(x, s)$, and under the assumptions on $k(x, s)$, the iterated kernels satisfy $k_n(x, s) \geq 0$ on $[a, b] \times [a, b]$ and $k_n(x, x) > 0$ for $a < x < b$. Just as in the previous proof, if $Ky = \mu y$ with $y \neq 0$, then $|\mu||y| = |Ky| \leq K|y|$, $|y| \in P\backslash\{0\}$, and $|\mu| \leq r$. If $|\mu| = r$ then $|y|$ is an extremal function for K, $|y| > 0$ on (a, b), $|y|$ is an eigenfunction corresponding to r, $r|y| = K|y|$. Apply K repeatedly to the relations $\mu y = Ky$ and $r|y| = K|y|$ to obtain $\mu^n y = K^n y$ and $r^n|y| = K^n|y|$ for $n = 1, 2, \ldots$. Consequently,

$$r^n|y| = |\mu^n y| = |K^n y| \leq K^n|y| = r^n|y|$$

and equality holds throughout. In particular, $|K^n y|(c) = K^n|y|(c)$ for $c = (a + b)/2$ and by Proposition 22

$$k_n(c, s)y(s) = e^{i\theta_c} u_c(s) \quad \text{for some real } \theta_c \text{ and } u_c(s) \geq 0.$$

Take absolute values on both sides of the equality to see that $u_c(s)$ is continuous on $[a, b]$. Since $k(c, c) > 0$ there is a $\delta > 0$ such that $k(c, s) > 0$ for $|s - c| < \delta$ and s in $[a, b]$. It follows that $k_n(c, s) > 0$ for $|s - c| < n\delta$ and s in $[a, b]$. Assume this for the moment. Fix n so that $n\delta > (b - a)/2$. Then $k_n(c, s) > 0$ for all s in $[a, b]$ and the displayed equation implies that $u_c(s) > 0$ on (a, b) and that $y = e^{i\theta_c} p$ where $p(s) = u_c(s)/k_n(c, s) > 0$ on (a, b). Then $Ky = \mu y$ implies $Kp = \mu p$; hence, μ is real and positive and $\mu = |\mu| = r$. Thus, all eigenvalues of K different from r are less than r in modulus.

A simple inductive argument shows that $k_n(c, s) > 0$ for $|s - c| < n\delta$ and s in $[a, b]$ if $k(c, s) > 0$ for $|s - c| < \delta$ and s in $[a, b]$. Indeed, suppose $c < s$, then $k(c, t) > 0$ for $c < t < c + \delta$ and $k(t, s) > 0$ for $s - \delta < t < s$ and consequently $k_2(c, s) = \int_a^b k(c, t)k(t, s)\, dt > 0$ if the two open intervals overlap which is the case if $s - \delta < c + \delta$; that is, if $s - c < 2\delta$. Likewise, $k_2(c, s) > 0$ if $s < c$ and $c - s < 2\delta$. Thus, $k_2(c, s) > 0$ for $|s - c| < 2\delta$ and s in $[a, b]$. Similarly, if $c < s$, $k_3(c, s) = \int_a^b k(c, t)k_2(t, s)\, dt > 0$ if the open intervals $c < t < c + \delta$ and $s - 2\delta < t < s$ overlap which occurs if $s - 2\delta < c + \delta$; that is, $s - c < 3\delta$. Likewise, $k_3(c, s) > 0$ if $s < c$ and $c - s < 3\delta$. Thus, $k_3(c, s) > 0$ for $|s - c| < 3\delta$ and s in $[a, b]$. The general assertion follows by mathematical induction. ■

Theorem 69 remains true if the symmetry assumption on $k(x, s)$ is deleted. However, the proof requires further knowledge about integral equations. The following are standard results from the theory of integral operators. If $k(x, s)$ satisfies the standing assumptions in Section 3.4 and K and K^* are the corresponding integral operators with kernels $k(x, s)$ and $k^*(x, s) = \overline{k(s, x)}$, then

(a) $\mu \neq 0$ is an eigenvalue of K if and only if $\overline{\mu}$ is an eigenvalue of K^*.
(b) If $\mu \neq 0$, the null spaces of $(K - \mu I)^2$ and $(K^* - \overline{\mu} I)^2$ have the same dimension.

Proof. (Theorem 69 in the nonsymmetric case.) Let $r = r(K)$.

(1) and (2a): by Theorem 68 $r(K) > 0$ and $rp = Kp$ with $p > 0$ on (a, b). Thus p is an extremal function for K. If q is any extremal function for K, then $rq \leq Kq$ with $q \in P\backslash\{0\}$. Since $k^*(x, s)$ satisfies the hypothesis of Theorem 68 when $k(x, s)$ does and (a) implies that $r(K^*) = r(K) = r$, there exists $p^* > 0$ on (a, b) such that $rp^* = K^*p^*$. If equality does not hold in $rq \leq Kq$, then

$$r\langle q, p^*\rangle < \langle Kq, p^*\rangle = \langle q, K^*p^*\rangle = r\langle q, p^*\rangle,$$

which is a contradiction because $\langle q, p^*\rangle > 0$. Hence equality holds in $rq \leq Kq$; that is, $rq = Kq$ and q is a nonnegative eigenfunction corresponding to the eigenvalue $r(K)$. Thus every extremal function of K is an eigenfunction of K corresponding to the eigenvalue $r(K)$ and the extremal function is positive on (a, b).

(2b), (3), and (5) follow by the arguments used to prove (2b), (3), and (5) of Theorem 66.

(4) Suppose the dimension of the null space of $(K - rI)^2$ is greater than 1. By (3) there must be a function ψ such that $(K - rI)^2\psi = 0$ and $\phi = (K - rI)\psi \neq 0$. Suppose for the moment that ψ is real-valued. Since ϕ is an eigenfunction of K corresponding to r, by replacing ψ by $-\psi$ if need be, we can assume that $\phi > 0$ on (a, b). By the same reasoning and (b) there is real-valued function ψ^* such that $(K^* - rI)^2\psi^* = 0$ and $\phi^* = (K^* - rI)\psi^* > 0$ on (a, b). This leads to the contradiction

$$0 = \left\langle (K - rI)^2\psi, \psi^* \right\rangle = \langle (K - rI)\psi, (K^* - rI)\psi^*\rangle = \langle \phi, \phi^*\rangle > 0.$$

Now suppose $\psi = u + iv$ is complex-valued with u and v real-valued. Then $(K - rI)^2u = 0$, $(K - rI)^2v = 0$, and one of $(K - rI)u$ and $(K - rI)v$ is not zero. Without loss of generality assume $(K - rI)u \neq 0$. By replacing ψ by $-\psi$ if need be, we can assume that $w = (K - rI)u > 0$ on (a, b). Likewise, there is a real-valued function u^* such that $(K^* - rI)^2u^* = 0$ and $w^* = (K - rI)u^* > 0$ on (a, b). This leads to the contradiction

$$0 = \left\langle (K - rI)^2u, u^* \right\rangle = \langle (K - rI)u, (K^* - rI)u^*\rangle = \langle w, w^*\rangle > 0$$

and completes the proof of (4). ■

Under the standing hypotheses of this section, the same line of reasoning establishes the results of this section for a suitably positive kernel $k(x, s)$ defined on $\Delta_n \times \Delta_n$. The same is true for singular kernels in higher dimensions once the theorems are properly formulated. See Appendix A.

3.5.3 Summary of Results

Under suitable assumptions on a nonnegative kernel $k(x, s)$ the big picture is as follows:

1. If it is only known that $k(x, s) \geq 0$, then the kernel may have no eigenvalues but, given enough positivity, will have a positive eigenvalue with a corresponding nonnegative eigenfunction.

2. If $k(x, s) > 0$ on its domain, then the kernel has a simple positive eigenvalue that is smallest in modulus of all the eigenvalues of the kernel and the corresponding eigenfunction is positive.

3. If $k(x, s) \geq 0$ and $k(x, x) > 0$ on the open diagonal, then the kernel has a simple positive eigenvalue that is smallest in modulus of all the eigenvalues of the kernel and the corresponding eigenfunction is positive except possibly at the endpoints of the underlying interval.

4. If $k(x, s)$ is a Kellogg kernel, a generalization of Item 3, then the kernel has an infinite sequence of simple positive eigenvalues and the corresponding eigenfunctions exhibit a rich oscillation structure.

3.6 Kellogg Kernels and Total Positivity

As we mentioned in Section 1.11.2, Kellogg discovered a property of a symmetric kernel that implies all the familiar oscillatory and approximation properties possessed by common orthogonal systems, including certain trigonometric functions, Legendre polynomials, and Bessel's functions which are all eigenfunctions of particular Sturm-Liouville eigenvalue problems. Kellogg assumed eigenvalues and eigenfunctions existed because he considered only problems with symmetric kernels, where that existence had already been established. Later Gantmacher and Krein used results of Jentzsch and Schur to simultaneously prove existence of eigenvalues and eigenfunctions for a class of not necessarily self-adjoint kernels and to show that such kernels have all the oscillatory and approximation properties established by Kellogg in the symmetric case. In this section, which follows the approach of Gantmacher and Krein in [16], we establish for symmetric kernels the existence of an infinite sequence of eigenvalues, their simplicity, and the oscillatory and approximation properties of the eigenfunctions. The restriction to self-adjoint kernels will cover all the applications to Sturm-Liouville eigenvalue problems that come in later chapters. The arguments of this section complement rather than replace the existence results of eigenvalues and eigenfunctions leading to the Hilbert-Schmidt theorem because, in particular, they do not lead to the critically important Hilbert-Schmidt expansion theorem and its corollaries. Although our primary focus is on the symmetric case, we point out what adjustments are needed to establish the same results in the nonsymmetric case.

Throughout this section and its subsections $k(x, s)$ is a real-valued continuous kernel defined on $[a, b] \times [a, b]$ and K1 and K2 denote the conditions:

$$\text{K1. } \det\left[k(x_i, x_j)\right]_{n\times n} > 0 \quad \text{for } a < x_1 < \cdots < x_n < b,$$

$$\text{K2. } \det\left[k(x_i, s_j)\right]_{n\times n} \geq 0 \quad \text{for } \begin{array}{l} a \leq x_1 < \cdots < x_n \leq b, \\ a \leq s_1 < \cdots < s_n \leq b, \end{array}$$

and for $n = 1, 2, 3, \ldots$ and all choices of $x_1, x_2, \ldots, x_n$ and $s_1, s_2, \ldots, s_n$ that satisfy the given conditions. As noted in Section 1.11.2 the importance of K1 and K2 for oscillatory and approximation properties was discovered by Kellogg.

A kernel $k(x, s)$ defined on $I \times J$, where I and J are intervals of real numbers of positive length, is **totally positive** if $\det [k(x_i, s_j)]_{n \times n} \geq 0$ for all $x_1 < x_2 < \cdots < x_n$ with x_i in I, for all $s_1 < s_2 < \cdots < s_n$ with s_j in J, and for all $n = 1, 2, \ldots$. Consequently, a kernel $k(x, s)$ that satisfies K2 is totally positive on $[a, b] \times [a, b]$. A symmetric kernel $k(x, s)$ that satisfies K1 is positive definite.

A kernel $k(x, s)$ is **strictly totally positive** on $I \times J$ if $\det [k(x_i, s_j)]_{n \times n} > 0$ for all $x_1 < \cdots < x_n$ with x_i in I, for all $s_1 < \cdots < s_n$ with s_j in J, and for all $n = 1, 2, \ldots$. We have already confirmed in Section 2.4 that the kernel

$$k(x, s) = e^{-(x-s)^2/\sigma}, \; \sigma > 0,$$

is strictly totally positive on $(-\infty, \infty) \times (-\infty, \infty)$.

3.6.1 Compound Kernels

It is useful to express determinants such as those in conditions K1 and K2 by the notation

$$k_{[n]}(x, s) = k\begin{pmatrix} x_1, x_2, \ldots, x_n \\ s_1, s_2, \ldots, s_n \end{pmatrix} = \det\big[k(x_i, s_j)\big]_{n \times n}$$

where $a \leq x_1 \leq \cdots \leq x_n \leq b$, $a \leq s_1 \leq \cdots \leq s_n \leq b$. Recall that

$$\Delta_n = \{x \in \mathbb{R}^n : a \leq x_1 \leq \cdots \leq x_n \leq b\}$$

is a simplex in $\mathbb{R}^n$ which we sometimes call the standard simplex (based on the interval $[a, b]$) to distinguish it for other simplices coming later. The kernel $k_{[n]}(x, s)$ is defined on $\Delta_n \times \Delta_n$ and is called the nth **compound kernel** of $k(x, s)$. As usual, the integral operator on $C[a, b]$ with kernel $k(x, s)$ is denoted by K. The integral operator on $C(\Delta_n)$ with kernel $k_{[n]}(x, s)$ is denoted by $K_{[n]}$. In this paragraph and in what follows we use the following convention: it will be clear from the context whether x and s are real variables or elements of $\mathbb{R}^n$. For example, x and s are real variables in $k(x, s)$ and elements of $\mathbb{R}^n$ in $k_{[n]}(x, s)$.

If the kernel $k(x, s)$ is symmetric on $[a, b] \times [a, b]$, then its compound kernel $k_{[n]}(x, s)$ is symmetric on $\Delta_n \times \Delta_n$ because a matrix and its transpose have the same determinant.

Our interest in compound kernels stems from work of Schur that establishes a fundamental connection between the eigenvalues, eigenfunctions, and generalized eigenfunctions of a kernel $k(x, s)$ and those of its compound kernels $k_{[n]}(x, s)$. If $k(x, s)$ is a symmetric kernel there are no generalized eigenfunctions; see Lemma 57. This will be the case of primary interest for us.

Several preliminary observations prepare the way for the result of Schur just mentioned. Let $f(t)$ be a continuous real-valued function on the n-dimensional box

$$\square_n = \{t = (t_1, \ldots, t_n) \in \mathbb{R}^n : a \leq t_1, t_2, \ldots, t_n \leq b\}$$

and

$$\Delta_n^\sigma = \{t = (t_1, \ldots, t_n) \in \mathbb{R}^n : a \leq t_{\sigma(1)} \leq \cdots \leq t_{\sigma(n)} \leq b\}$$

be the simplex in $\square_n$ determined by the permutation σ. (For permutations see Section 2.3.1.) If $n = 2$, $\square_2$ is a square in the plane and Δ_2^σ is the subtriangle of $\square_2$ with $t_1 \leq t_2$ when $\sigma = \text{id} = (1)(2)$, the identity permutation, and is the subtriangle $t_2 \leq t_1$ when $\sigma = (2, 1)$. The linear change of variables $u_i = t_{\sigma(i)}$, which simply amounts to relabeling the coordinates, maps the simplex Δ_n^σ onto the standard simplex

$$\Delta_n = \Delta_n^{\text{id}} = \{u = (u_1, \ldots, u_n) \in \mathbb{R}^n : a \leq u_1 \leq \cdots \leq u_n \leq b\}$$

and gives

$$\int_{\Delta_n^\sigma} f(t)\,dt = \int_{\Delta_n} f(u)\,du = \int_{\Delta_n} f(t)\,dt,$$

where dt is short for $dt_1 \cdots dt_n$ and du is short for $du_1 \cdots du_n$. Hence,

$$\int_{\square_n} f(t)\,dt = \sum_\sigma \int_{\Delta_n^\sigma} f(t)\,dt = n! \int_{\Delta_n} f(t)\,dt$$

because $\square_n$ is the union of the $n!$ nonoverlapping simplices Δ_n^σ.

Next we establish the **basic composition formula** of total positivity theory: if the kernels $k(x,s)$, $l(x,s)$, and $m(x,s)$ are related by

$$m(x,s) = \int_a^b k(x,t)l(t,s)\,dt,$$

then their compound kernels are related by

$$m\begin{pmatrix} x_1, x_2, \ldots, x_n \\ s_1, s_2, \ldots, s_n \end{pmatrix} = \int_{\Delta_n} k\begin{pmatrix} x_1, \ldots, x_n \\ t_1, \ldots, t_n \end{pmatrix} l\begin{pmatrix} t_1, \ldots, t_n \\ s_1, \ldots, s_n \end{pmatrix} dt, \tag{3.4}$$

where $dt = dt_1 \cdots dt_n$, or, more briefly, by

$$m_{[n]}(x,s) = \int_{\Delta_n} k_{[n]}(x,t) l_{[n]}(t,s)\,dt. \tag{3.5}$$

For our purposes the kernels can be assumed to be continuous and the integral over the simplex is an ordinary n-fold Riemann integral.

The basic composition formula follows from the following identity, a lemma of Schur:

Lemma 70 *If $\phi_i(t)$ and $\psi_j(t)$ are continuous functions on $[a,b]$ for $i, j = 1, 2, \ldots, n$, then*

$$\begin{aligned}\det\left[\int_a^b \phi_i(t)\psi_j(t)\,dt\right]_{n\times n} &= \frac{1}{n!}\int_{\square_n} \det[\phi_i(t_j)]\det[\psi_i(t_j)]\,dt_1\cdots dt_n \\ &= \int_{\Delta_n} \det[\phi_i(t_j)]\det[\psi_i(t_j)]\,dt_1\cdots dt_n.\end{aligned}$$

Proof. If D is the determinant on the left, then

$$\begin{aligned} D &= \begin{vmatrix} \int_a^b \phi_1(t_1)\psi_1(t_1)\,dt_1 & \cdots & \int_a^b \phi_1(t_1)\psi_n(t_1)\,dt_1 \\ \int_a^b \phi_2(t_2)\psi_1(t_2)\,dt_2 & \cdots & \int_a^b \phi_2(t_2)\psi_n(t_2)\,dt_2 \\ & \cdots & \\ \int_a^b \phi_n(t_n)\psi_1(t_n)\,dt_n & \cdots & \int_a^b \phi_n(t_n)\psi_n(t_n)\,dt_n \end{vmatrix} \\ &= \int_a^b \cdots \int_a^b \phi_1(t_1)\phi_2(t_2)\cdots\phi_n(t_n)\det[\psi_i(t_j)]\,dt_1\,dt_2\cdots dt_n \end{aligned}$$

because a determinant is a linear function of each of its rows. Relabel the variables $t_1, \ldots, t_n$ in this result by $t_{\sigma(1)}, \ldots, t_{\sigma(n)}$ where σ is any permutation of $\{1, 2, \ldots, n\}$ to get

$$D = \int_a^b \cdots \int_a^b \phi_1(t_{\sigma(1)})\phi_2(t_{\sigma(2)})\cdots\phi_n(t_{\sigma(n)})\det[\psi_i(t_{\sigma(j)})]\,dt_1\,dt_2\cdots dt_n.$$

If m interchanges of the columns in $\det[\psi_i(t_{\sigma(j)})]$ put the columns in the order $t_1, \ldots, t_n$, then $\operatorname{sgn}\sigma = (-1)^m$ and

$$D = \int_a^b \cdots \int_a^b (\operatorname{sgn}\sigma)\phi_1(t_{\sigma(1)})\phi_2(t_{\sigma(2)}) \cdots \phi_n(t_{\sigma(n)}) \det[\psi_i(t_j)]\, dt_1\, dt_2 \cdots dt_n.$$

Summing over the $n!$ permutations gives

$$n!D = \int_a^b \cdots \int_a^b \det[\phi_i(t_j)] \det[\psi_i(t_j)]\, dt_1\, dt_2 \cdots dt_n$$

and the lemma is established. ■

Apply the lemma with

$$\phi_i(t) = k(x_i, t) \quad \text{and} \quad \psi_j(t) = l(t, s_j)$$

to find that

$$\det\left[\int_a^b k(x_i, t) l(t, s_j)\, dt\right] = \int_{\Delta_n} \det[k(x_i, t_r)] \det[l(t_r, s_j)]\, dt_1\, dt_2 \cdots dt_n;$$

that is

$$m\begin{pmatrix} x_1, x_2, \ldots, x_n \\ s_1, s_2, \ldots, s_n \end{pmatrix} = \int_{\Delta_n} k\begin{pmatrix} x_1, \ldots, x_n \\ t_1, \ldots, t_n \end{pmatrix} l\begin{pmatrix} t_1, \ldots, t_n \\ s_1, \ldots, s_n \end{pmatrix} dt_1\, dt_2 \cdots dt_n,$$

which is the basic composition formula.

It is useful to regard determinants such as $\det[\phi_i(t_j)]$ in the lemma as the values of a function on the simplex Δ_n. Define the function $\phi_1 \wedge \phi_2 \wedge \cdots \wedge \phi_n$ with domain the simplex Δ_n by

$$\phi_1 \wedge \phi_2 \wedge \cdots \wedge \phi_n(x) = \det[\phi_i(x_j)]$$

for $x = (x_1, \ldots, x_n)$ in Δ_n. This function is called the **wedge product** of $\phi_1, \phi_2, \ldots, \phi_n$. The wedge product $\phi_1 \wedge \phi_2 \wedge \cdots \wedge \phi_n$ is a linear function of any one of its factors because a determinant is a linear function of any one of its rows.

Lemma 71 *The functions $\phi_1, \phi_2, \ldots, \phi_n$ are linearly independent on $[a, b]$ if and only if $\phi_1 \wedge \phi_2 \wedge \cdots \wedge \phi_n$ is not the zero function on Δ_n.*

Proof. $\Rightarrow$: We use an inductive argument. If $n = 1$ the forward implication in the lemma is true because ϕ_1 is not the zero function and the wedge product is ϕ_1. Assume the forward implication is true for any n linearly independent functions on $[a, b]$. Let $\phi_1, \phi_2, \ldots, \phi_{n+1}$ be linearly independent on $[a, b]$. Suppose, contrary to what we want to prove, that $\det[\phi_i(x_j)]_{(n+1)\times(n+1)} = 0$ for all choices of $x_1, \ldots, x_{n+1}$ in $[a, b]$ with $x_1 < \cdots < x_{n+1}$. We will show this is impossible, a contradiction that advances the induction step and proves the forward implication. Since $\phi_1, \phi_2, \ldots, \phi_n$ are linearly independent on $[a, b]$, there are points $x_1, \ldots, x_n$ in $[a, b]$ with $x_1 < \cdots < x_n$ such that $\det[\phi_i(x_j)] \neq 0$. Consider the determinant

$$\begin{vmatrix} \phi_1(x_1) & \cdots & \phi_1(x_n) & \phi_1(x) \\ \vdots & & & \vdots \\ \phi_n(x_1) & \cdots & \phi_n(x_n) & \phi_n(x) \\ \phi_{n+1}(x_1) & \cdots & \phi_{n+1}(x_n) & \phi_{n+1}(x) \end{vmatrix}$$

for x in $[a, b]$. The determinant is zero for $x > x_n$ by our supposition; hence, it is zero for all x in $[a, b]$ by elementary properties of determinants. Expand the determinant by its last column to find

$$\sum_{j=1}^{n+1} c_j \phi_j(x) = 0$$

for all x in $[a, b]$ and with $c_{n+1} = \det\left[\phi_i(x_j)\right]_{n\times n} \neq 0$, which contradicts the linear independence of $\phi_1, \phi_2, \ldots, \phi_{n+1}$ and completes the proof.

$\Leftarrow$: Assume, to the contrary of what we want to prove, that $\phi_1, \phi_2, \ldots, \phi_n$ were linearly dependent. Then there are constant c_i not all zeros such that

$$\sum_{i=1}^{n} c_i \phi_i(x) = 0 \quad \text{for all } x \text{ in } \quad [a, b]$$

and given any point $x = (x_1, \ldots, x_n)$ in Δ_n

$$\sum_{i=1}^{n} c_i \phi_i(x_j) = 0 \quad \text{for } j = 1, \ldots, n.$$

Since this homogeneous system has a nontrivial solution

$$\phi_1 \wedge \phi_2 \wedge \cdots \wedge \phi_n(x) = \det\left[\phi_i(x_j)\right] = 0. \qquad \blacksquare$$

Clearly the lemma holds with $[a, b]$ replaced by any interval I of positive length and with Δ_n being the simplex based on the interval I.

Apply the Lemma 70 with $\phi_i(t) = k(x_i, t)$ to obtain

$$\det\left[\int_a^b k(x_i, t)\psi_j(t)\, dt\right]_{n\times n} = \int_{\Delta_n} \det\left[k(x_i, t_j)\right] \det\left[\psi_i(t_j)\right] dt_1 \cdots dt_n$$

or, in wedge product notation,

$$K\psi_1 \wedge K\psi_2 \wedge \cdots \wedge K\psi_n(x) = K_{[n]}\left(\psi_1 \wedge \psi_2 \wedge \cdots \wedge \psi_n\right)(x),$$

where K is the integral operator on $C[a, b]$ with kernel k and $K_{[n]}$ is the integral operator on $C(\Delta_n)$ with kernel $k_{[n]}$. Thus,

$$K_{[n]}\left(\psi_1 \wedge \psi_2 \cdots \wedge \psi_n\right) = K\psi_1 \wedge K\psi_2 \cdots \wedge K\psi_n \tag{3.6}$$

for any functions $\psi_1, \psi_2, \ldots, \psi_n$ in $C[a, b]$.

There is an important connection between the iterated kernels $k_n(x, s)$ of a kernel $k(x, s) = k_1(x, s)$ defined by

$$k_{n+1}(x, s) = \int_a^b k(x, t) k_n(t, s)\, dt$$

for $n = 1, 2, \ldots$ and iterated kernels of the compound kernels of $k(x, s)$. Recall that $k_n(x, s)$ is the kernel of the integral operator K^n.

The iterated kernels of the compound kernels $k_{[m]}(x, s)$, which are kernels of the integral operators $\left(K_{[m]}\right)^n$, are defined by the recursion formula

$$\left(k_{[m]}\right)_{n+1}(x, s) = \int_{\Delta_n} k_{[m]}(x, t)\left(k_{[m]}\right)_n(t, s)\, dt$$

for $n=1, 2, \dots$ and $\left(k_{[m]}\right)_1 = k_{[m]}$. On the other hand, since

$$k_{n+1}(x, s) = \int_a^b k(x, t) k_n(t, s)\, dt,$$

the basic composition formula gives

$$(k_{n+1})_{[m]}(x, s) = \int_{\Delta_n} k_{[m]}(x, t)(k_n)_{[m]}(t, s)\, dt.$$

That is, the kernels $(k_n)_{[m]}(x, s)$ satisfy the recursion formula for the iterated kernels of the compound kernel $k_{[m]}(x, s)$ and have the same initial kernel $(k_1)_{[m]} = k_{[m]}$. It follows that

$$\left(k_{[m]}\right)_n = (k_n)_{[m]} \tag{3.7}$$

for $n, m = 1, 2, \dots$. In words, the nth iterated kernel of the mth compound kernel of k is the mth compound kernel of the nth iterated kernel of k. The displayed equality means that

$$\left(K_{[m]}\right)_n = (K^n)_{[m]}$$

when expressed in terms of the corresponding integral operators.

3.6.2 Spectral Properties of Compound Kernels

In [37] Schur gave a complete description of the relationship between the eigenvalues, eigenfunctions, and generalized eigenfunctions of a kernel and those of its compound kernels. Schur's work in this area, which applies to both symmetric and nonsymmetric kernels, is not as well known as it should be. In this section, we establish for *symmetric* kernels one of the key conclusions of Schur in [37]. The result for symmetric kernels will cover all the applications to Sturm-Liouville boundary value problems and eigenvalue problems presented later in the book.

The third corollary to the Hilbert-Schmidt theorem established that a symmetric kernel $k(x, s)$ has a complete system of real-valued orthogonal eigenfunctions. Denote such a system by $\phi_0, \phi_1, \phi_2, \dots$ with corresponding eigenvalues $\lambda_0, \lambda_1, \lambda_2, \dots$. Recall that complete means that any eigenfunction of $k(x, s)$ is a finite linear combination of $\phi_0, \phi_1, \phi_2, \dots$.

As usual K is the integral operator on $C[a, b]$ with kernel $k(x, s)$ and $K_{[n]}$ is the integral operator on $C(\Delta_n)$ with kernel $k_{[n]}(x, s)$. We maintain the convention that the context determines the dimension of the variables x and s. That is, x and s are real variables in $k(x, s)$ and are points in $\mathbb{R}^n$ in $k_{[n]}(x, s)$, and so on.

Theorem 72 (Schur) *Let $k(x, s)$ be a continuous symmetric kernel on $[a, b] \times [a, b]$ that is not identically zero and let $k_{[n]}(x, s)$ be its nth compound kernel. If $\phi_0(x), \phi_1(x), \phi_2(x), \dots$ is a complete system of orthogonal eigenfunctions for a symmetric kernel $k(x, s)$ defined on $[a, b] \times [a, b]$, then $\phi_{i_1} \wedge \phi_{i_2} \wedge \cdots \wedge \phi_{i_n}(x)$ forms a complete system of orthogonal eigenfunctions for the (symmetric) compound kernel $k_{[n]}(x, s)$ when the indices $i_1, i_2, \dots, i_n$ with $0 \le i_1 < i_2 < \cdots < i_n$ vary over all subsets of indices appearing in $\phi_0(x), \phi_1(x), \phi_2(x), \dots$.*

The theorem is interpreted to mean that if $k(x, s)$ has only a finite number of eigenvalues repeated to multiplicity, say $\lambda_0, \dots, \lambda_N$, then only the compound kernels for $n=1, \dots, N+1$ have the given eigenfunctions. (In fact, the higher order compound kernels are identically zero and have no eigenvalues.)

Proof. Since

$$
\begin{aligned}
K_{[n]}\phi_{i_1} \wedge \phi_{i_2} \wedge \cdots \wedge \phi_{i_n} &= K\phi_{i_1} \wedge K\phi_{i_2} \wedge \cdots \wedge K\phi_{i_n} \\
&= \lambda_{i_1}^{-1}\phi_{i_1} \wedge \lambda_{i_2}^{-1}\phi_{i_2} \wedge \cdots \wedge \lambda_{i_n}^{-1}\phi_{i_n} \\
&= \left(\lambda_{i_1}\lambda_{i_2}\cdots\lambda_{i_n}\right)^{-1}\phi_{i_1} \wedge \phi_{i_2} \wedge \cdots \wedge \phi_{i_n},
\end{aligned}
$$

$\lambda_{i_1}\lambda_{i_2}\cdots\lambda_{i_n}$ is an eigenvalue of $k_{[n]}(x, s)$ with corresponding eigenfunction $\phi_{i_1} \wedge \phi_{i_2} \wedge \cdots \wedge \phi_{i_n}$. Furthermore, it follows directly from Lemma 70 that the wedge products $\phi_{i_1} \wedge \phi_{i_2} \wedge \cdots \wedge \phi_{i_n}$ are mutually orthogonal because $\phi_0(x), \phi_1(x), \phi_2(x), \ldots$ are.

It remains to show that $\phi_{i_1} \wedge \phi_{i_2} \wedge \cdots \wedge \phi_{i_n}$ forms a complete system of eigenfunctions for the kernel $k_{[n]}(x, s)$. The proof proceeds in two steps. First, we establish the completeness when the kernel is positive definite, which is often the case in applications. Second, we show that the general case follows from the positive definite case.

Step 1. If $k(x, s)$ is positive definite, then

$$
k(x, s) = \sum_{n=1}^{\infty} \frac{\phi_n(x)\phi_n(s)}{\lambda_n}
$$

and the series converges absolutely and uniformly on $[a, b] \times [a, b]$ by Mercer's theorem. It follows that

$$
k_{[n]}(x, s) = \sum_{0 \le i_1 < \cdots < i_n} \frac{\phi_{i_1} \wedge \phi_{i_2} \wedge \cdots \wedge \phi_{i_n}(x)\phi_{i_1} \wedge \phi_{i_2} \wedge \cdots \wedge \phi_{i_n}(s)}{\lambda_{i_1}\lambda_{i_2}\cdots\lambda_{i_n}}, \tag{3.8}
$$

with absolute and uniform convergence for x and s in Δ_n. Indeed,

$$
\begin{aligned}
k_{[n]}(x, s) &= \begin{vmatrix} \sum_{i_1=1}^{\infty} \frac{\phi_{i_1}(x_1)\phi_{i_1}(s_1)}{\lambda_{i_1}} & \cdots & \sum_{i_1=1}^{\infty} \frac{\phi_{i_1}(x_1)\phi_{i_1}(s_n)}{\lambda_{i_1}} \\ \vdots & & \vdots \\ \sum_{i_n=1}^{\infty} \frac{\phi_{i_n}(x_n)\phi_{i_n}(s_1)}{\lambda_{i_n}} & \cdots & \sum_{i_n=1}^{\infty} \frac{\phi_{i_n}(x_n)\phi_{i_n}(s_n)}{\lambda_{i_n}} \end{vmatrix} \\
&= \sum_{i_1=1}^{\infty} \cdots \sum_{i_n=1}^{\infty} \frac{\phi_{i_1}(x_1)\cdots\phi_{i_n}(x_n)}{\lambda_{i_1}\cdots\lambda_{i_n}} \begin{vmatrix} \phi_{i_1}(s_1) & \cdots & \phi_{i_1}(s_n) \\ \vdots & & \vdots \\ \phi_{i_n}(s_1) & \cdots & \phi_{i_n}(s_n) \end{vmatrix}.
\end{aligned}
$$

The determinant is zero if any pair of indices have the same value; hence,

$$
k_{[n]}(x, s) = \sum_{\substack{0 \le i_1, \ldots, i_n \\ i_r \ne i_s \text{ if } r \ne s}} \frac{\phi_{i_1}(x_1)\cdots\phi_{i_n}(x_n)}{\lambda_{i_1}\cdots\lambda_{i_n}} \begin{vmatrix} \phi_{i_1}(s_1) & \cdots & \phi_{i_1}(s_n) \\ \vdots & & \vdots \\ \phi_{i_n}(s_1) & \cdots & \phi_{i_n}(s_n) \end{vmatrix}.
$$

Fix a set of n distinct indices $0 \le i_1 < \cdots < i_n$. This set of indices and all its permutations occur exactly once in the sum above. Thus

$$
\begin{aligned}
k_{[n]}(x, s) = &\sum_{0 \le i_1 < \cdots < i_n} \sum_{\sigma} \frac{\phi_{\sigma(i_1)}(x_1)\cdots\phi_{\sigma(i_n)}(x_n)}{\lambda_{\sigma(i_1)}\cdots\lambda_{\sigma(i_n)}} \\
&\times \begin{vmatrix} \phi_{\sigma(i_1)}(s_1) & \phi_{\sigma(i_1)}(s_2) & \cdots & \phi_{\sigma(i_1)}(s_n) \\ \vdots & & & \vdots \\ \phi_{\sigma(i_n)}(s_1) & \phi_{\sigma(i_n)}(s_2) & \cdots & \phi_{\sigma(i_n)}(s_n) \end{vmatrix}.
\end{aligned}
$$

Since $\lambda_{\sigma(i_1)} \cdots \lambda_{\sigma(i_n)} = \lambda_{i_1} \cdots \lambda_{i_n}$ and $\operatorname{sgn} \sigma = (-1)^m$ if m row interchanges will put the row indices of the determinant in the order $i_1, \ldots, i_n$,

$$\begin{aligned} k_{[n]}(x, s) &= \sum_{0 \le i_1 < \cdots < i_n} \frac{1}{\lambda_{i_1} \cdots \lambda_{i_n}} \\ &\quad \times \left(\sum_{\sigma} (\operatorname{sgn} \sigma) \phi_{\sigma(i_1)}(x_1) \cdots \phi_{\sigma(i_n)}(x_n) \right) \phi_{i_1} \wedge \phi_{i_2} \wedge \cdots \wedge \phi_{i_n}(s) \\ &= \sum_{0 \le i_1 < \cdots < i_n} \frac{\phi_{i_1} \wedge \phi_{i_2} \wedge \cdots \wedge \phi_{i_n}(x) \phi_{i_1} \wedge \phi_{i_2} \wedge \cdots \wedge \phi_{i_n}(s)}{\lambda_{i_1} \cdots \lambda_{i_n}}, \end{aligned}$$

with absolute and uniform convergence inherited from that of $\sum_{n=1}^{\infty} \phi_n(x)\phi_n(s)/\lambda_n$.

We use the expansion (3.8) to show that the orthonormal eigenfunctions $\phi_{i_1} \wedge \phi_{i_2} \wedge \cdots \wedge \phi_{i_n}$ are a complete system for $k_{[n]}(x, s)$. Let ψ be an eigenfunction of the kernel $k_{[n]}(x, s)$ with eigenvalue ρ. If $\rho \neq \lambda_{i_1} \cdots \lambda_{i_n}$ for all $0 \le i_1 < \cdots < i_n$, then ψ is orthogonal to all the eigenfunctions $\phi_{i_1} \wedge \phi_{i_2} \wedge \cdots \wedge \phi_{i_n}$ because the kernel $k_{[n]}(x, s)$ is symmetric and

$$\begin{aligned} \psi &= \rho K_{[n]} \psi \\ &= \rho \sum_{0 \le i_1 < \cdots < i_n} \frac{\phi_{i_1} \wedge \phi_{i_2} \wedge \cdots \wedge \phi_{i_n}(x)}{\lambda_{i_1} \cdots \lambda_{i_n}} \int_{\Delta_n} \phi_{i_1} \wedge \phi_{i_2} \wedge \cdots \wedge \phi_{i_n}(s) \psi(s)\, ds \\ &= 0, \end{aligned}$$

with the interchange of order of summation and integration justified by the uniform convergence of the series. This contradiction shows that $\rho = \lambda_{i_1} \cdots \lambda_{i_n}$ for some $i_1 < \cdots < i_n$. Let

$$\tilde{\psi} = \psi - \sum{}' \langle \psi, \phi_{i_1} \wedge \phi_{i_2} \wedge \cdots \wedge \phi_{i_n} \rangle \phi_{i_1} \wedge \phi_{i_2} \wedge \cdots \wedge \phi_{i_n}$$

where the prime means the sum if over all $0 \le i_1 < \cdots < i_n$ with $\lambda_{i_1} \lambda_{i_2} \cdots \lambda_{i_n} = \rho$. Clearly $\tilde{\psi}$ is orthogonal to all the $\phi_{i_1} \wedge \phi_{i_2} \wedge \cdots \wedge \phi_{i_n}$ with $\lambda_{i_1} \cdots \lambda_{i_n} = \rho$, $\rho K_{[n]} \tilde{\psi} = \tilde{\psi}$, and $\tilde{\psi}$ is orthogonal to all the $\phi_{i_1} \wedge \phi_{i_2} \wedge \cdots \wedge \phi_{i_n}$ with $\lambda_{i_1} \cdots \lambda_{i_n} \neq \rho$ because $k_{[n]}(x, s)$ is symmetric. Consequently, ψ is orthogonal to all the $\phi_{i_1} \wedge \phi_{i_2} \wedge \cdots \wedge \phi_{i_n}$ and, just as above, this implies $\tilde{\psi} = 0$. That is, ψ is a linear combination of some of the eigenfunctions $\phi_{i_1} \wedge \phi_{i_2} \wedge \cdots \wedge \phi_{i_n}$ and the system is complete. This establishes the theorem in the case of a positive definite kernel.

Step 2. The iterated kernel $k_2(x, s)$ has eigenvalues λ_n^2 where λ_n are the eigenvalues of $k(x, s)$; hence, $k_2(x, s)$ is positive definite. By Theorem 64 the eigenfunctions $\phi_0(x), \phi_1(x), \phi_2(x), \ldots$ which are a complete orthogonal system for $k(x, s)$, also form a complete orthogonal system of eigenfunctions for the iterated kernel $k_2(x, s)$. Consequently, by Step 1, $\phi_{i_1} \wedge \phi_{i_2} \wedge \cdots \wedge \phi_{i_n}$ form a complete system of eigenfunctions for the compound kernel $(k_2)_{[n]}(x, s) = (k_{[n]})_2(x, s)$ and

$$(k_{[n]})_2(x, s) = \sum_{0 \le i_1 < \cdots < i_n} \frac{\phi_{i_1} \wedge \phi_{i_2} \wedge \cdots \wedge \phi_{i_n}(x) \phi_{i_1} \wedge \phi_{i_2} \wedge \cdots \wedge \phi_{i_n}(s)}{\left(\lambda_{i_1} \lambda_{i_2} \cdots \lambda_{i_n}\right)^2}$$

with absolute and uniform convergence.

Now we reason in the same fashion as at the end of Step 1 to show that $\{\phi_{i_1} \wedge \phi_{i_2} \wedge \cdots \wedge \phi_{i_n}\}_{0 \le i_1 < \cdots < i_n}$ is a complete orthogonal system for the kernel $k_{[n]}$. Let ψ be an eigenfunction of the kernel $k_{[n]}$ and ρ its eigenvalue so that $\rho K_{[n]} \psi = \psi$. If $\rho \neq \lambda_{i_1} \lambda_{i_2} \cdots \lambda_{i_n}$ for all choices $0 \le i_1 < \cdots < i_n$, then ψ is orthogonal to $\phi_{i_1} \wedge \phi_{i_2} \wedge \cdots \wedge \phi_{i_n}$ for all choices because the kernel is symmetric, and

$$\psi = \rho^2 \left(K_{[n]}\right)^2 \psi = 0,$$

where the last equality uses term-by-term integration in the series expansion of $\left(k_{[n]}\right)_2(x, s)$. This contradiction implies that $\rho = \lambda_{i_1}\lambda_{i_2}\cdots\lambda_{i_n}$ for some $i_1, \ldots, i_n$. Let

$$\tilde{\psi} = \psi - \sum{}' \langle \psi, \phi_{i_1} \wedge \phi_{i_2} \wedge \cdots \wedge \phi_{i_n} \rangle \phi_{i_1} \wedge \phi_{i_2} \wedge \cdots \wedge \phi_{i_n}$$

where the prime means the sum if over all $0 \leq i_1 < \cdots < i_n$ with $\lambda_{i_1}\lambda_{i_2}\cdots\lambda_{i_n} = \rho$. Consequently, $\tilde{\psi}$ is orthogonal to all $\phi_{i_1} \wedge \phi_{i_2} \wedge \cdots \wedge \phi_{i_n}$ with $\lambda_{i_1}\lambda_{i_2}\cdots\lambda_{i_n} = \rho$ and is orthogonal to all the other eigenfunctions $\phi_{i_1} \wedge \phi_{i_2} \wedge \cdots \wedge \phi_{i_n}$ belonging to eigenvalues different from ρ. Hence,

$$\tilde{\psi} = \rho^2 \left(K_{[n]}\right)^2 \tilde{\psi} = 0,$$

$$\psi = \sum{}' \langle \psi, \phi_{i_1} \wedge \phi_{i_2} \wedge \cdots \wedge \phi_{i_n} \rangle \, \phi_{i_1} \wedge \phi_{i_2} \wedge \cdots \wedge \phi_{i_n},$$

and the system $\left\{\phi_{i_1} \wedge \phi_{i_2} \wedge \cdots \wedge \phi_{i_n}\right\}$ is complete for the kernel $k_n(x, s)$. ■

Schur's general version of Theorem 72 in [37] asserts that if $\phi_0(x), \phi_1(x), \phi_2(x), \ldots$ is a complete system of eigenfunction and generalized eigenfunctions for a not necessarily symmetric kernel $k(x, s)$, then $\phi_{i_1} \wedge \phi_{i_2} \wedge \cdots \wedge \phi_{i_n}$ for $0 \leq i_1 < \cdots < i_n$ is a complete system of eigenfunctions and generalized eigenfunctions for the compound kernel $k_{[n]}(x, s)$. The general result also establishes that eigenvalues of $k_{[n]}(x, s)$ can only arise as n-fold products of eigenvalues of the kernel $k(x, s)$. If the general Schur's theorem is cited in the proofs in the next section and the complete systems of orthogonal eigenfunctions are replaced by complete systems of eigenfunctions and generalized eigenfunctions for the kernel, then small adjustments to the arguments given there establish the results obtained there for nonsymmetric and symmetric kernels at the same time. Nevertheless, we will present the reasoning in the context of a symmetric kernel because we only use the symmetric case in later chapters.

3.6.3 Spectral Properties of Kellogg Kernels

A kernel $k(x, s)$ is called a **Kellogg kernel** if $k(x, s)$ is continuous and symmetric on $[a, b] \times [a, b]$ and for $n = 1, 2, 3, \ldots$ satisfies:

K1. $\det\left[k(x_i, x_j)\right]_{n\times n} > 0, \quad a < x_1 < \cdots < x_n < b,$

K2. $\det\left[k(x_i, s_j)\right]_{n\times n} \geq 0, \quad a \leq x_1 \leq \cdots \leq x_n \leq b, \; a \leq s_1 \leq \cdots \leq s_n \leq b.$

In this section, we establish the principal properties of the eigenvalues and eigenfunctions of a Kellogg kernel.

We know from the Hilbert-Schmidt theorem that a Kellogg kernel has an infinite sequence of eigenvalues $\lambda_0, \lambda_1, \ldots$ and a corresponding complete orthogonal system of eigenfunctions $\phi_0, \phi_1, \ldots$. Moreover, the notation can be chosen so that the eigenvalues are listed by increasing absolute values and repeated eigenvalues occur in the list according to their geometric multiplicity as

$$|\lambda_0| \leq |\lambda_1| \cdots.$$

Furthermore, $|\lambda_n| \to \infty$ as $n \to \infty$ by the Hilbert-Schmidt theorem.

By K1 and K2 with $n = 1$, the kernel $k(x, s)$ satisfies the hypothesis in Jentzsch's theorem (Theorem 69). Thus, the kernel $k(x, s)$ has a positive, simple eigenvalue that is strictly smaller in modulus than any other eigenvalue of $k(x, s)$ and it has a corresponding eigenfunction that is positive on (a, b). This eigenvalue is λ_0 and since it is simple, ϕ_0 is a constant multiple of the eigenfunction that is positive on (a, b). Consequently, ϕ_0 maintains a strict sign (always positive or always negative) on (a, b). Since λ_0 has the smallest modulus of any eigenvalue, it follows from Jentzsch's theorem that $|\lambda_0| = \lambda_0 > 0$ and

$$0 < \lambda_0 < |\lambda_1| \cdots.$$

By K1 and K2 with $n = 2$, the kernel $k_{[2]}(x, s)$ satisfies the hypothesis in Jentzsch's theorem and hence has a positive, simple eigenvalue that is smaller in modulus than any other eigenvalue of $k_{[2]}(x, s)$. It follow from Schur's theorem (Theorem 72) and the ordering $|\lambda_0| \leq |\lambda_1| \cdots$ that the eigenvalue of $k_{[2]}(x, s)$ of minimum modulus is $\lambda_0\lambda_1$ and $\phi_0 \wedge \phi_1$ is a corresponding eigenfunction. By Jentzsch's theorem, $\lambda_0\lambda_1 > 0$, hence $\lambda_1 > 0$, and $\phi_0 \wedge \phi_1$ maintains a strict sign (always positive or always negative) on the interior of Δ_2. Proceeding step-by-step in this manner it follows that the eigenvalues of the kernel $k(x, s)$ are all simple and positive,

$$0 < \lambda_0 < \lambda_1 < \lambda_2 < \cdots,$$

and the corresponding eigenfunctions $\phi_0, \phi_1, \phi_2, \ldots$ have the property that

$$\phi_0 \wedge \cdots \wedge \phi_n(x) = \det\left[\phi_i(x_j)\right]$$

maintains a strict sign (always positive or always negative) for all $x = (x_1, x_2, \ldots, x_{n+1})$ with $a < x_1 < x_2 < \cdots < x_{n+1} < b$. Consequently,

$$\phi_0 \wedge \cdots \wedge \phi_{n-1} \wedge \pm\phi_n(x) > 0$$

for a specific choice of sign ± 1 and $a < x_1 < x_2 < \cdots < x_{n+1} < b$. Consequently, $\phi_0, \phi_1, \ldots, \phi_{n-1}, \phi_n$ or $\phi_0, \phi_1, \ldots, \phi_{n-1}, -\phi_n$ is a Tchebycheff system on (a, b) for each $n = 0, 1, 2\ldots$ and we have established the following theorem.

Theorem 73 *All the eigenvalues of a Kellogg kernel $k(x, s)$ on $[a, b] \times [a, b]$ are positive and simple. If*

$$\lambda_0 < \lambda_1 < \lambda_2 < \cdots$$

are the eigenvalues, then $\lambda_n \to \infty$ as $n \to \infty$. If $\phi_0, \phi_1, \phi_2, \ldots$ is the corresponding complete set of (orthogonal) eigenfunctions for $k(x, s)$, then for each $n = 0, 1, 2\ldots$ either $\phi_0, \phi_1, \ldots, \phi_{n-1}, \phi_n$ or $\phi_0, \phi_1, \ldots, \phi_{n-1}, -\phi_n$ is a Tchebycheff system on (a, b).

The fact that $\phi_0, \phi_1, \ldots, \phi_n$ or $\phi_0, \phi_1, \ldots, \phi_{n-1}, -\phi_n$ is a Tchebycheff system on (a, b) and the orthogonality of the eigenfunctions leads to

Theorem 74 *If $k(x, s)$ is a Kellogg kernel on $[a, b] \times [a, b]$, $\lambda_0 < \lambda_1 < \lambda_2 < \cdots$ are all its eigenvalues, and $\phi_0, \phi_1, \phi_2 \ldots$ are corresponding (orthogonal) eigenfunctions, then for any n, the eigenfunctions $\phi_0, \phi_1, \phi_2, \ldots, \phi_n$ have the following oscillatory and approximation properties*:

1. *Given any $n + 1$ points in (a, b) and any $n + 1$ values $b_0, \ldots, b_n$, there is a unique ϕ-polynomial $\phi(x) = \sum_{i=0}^{n} a_i\phi_i(x)$ that take on the prescribed values at the given points.*
2. *A nontrivial ϕ-polynomial has at most n zeros in (a, b) where nonnodal zeros are counted twice and nodal zeros once.*
3. *A nontrivial ϕ-polynomial $\phi(x) = \sum_{i=m}^{n} a_i\phi_i(x)$ has at least m nodal zeros in (a, b) and has at most n zeros there, counting zeros as in Property 2.*
4. *ϕ_n has n nodal zeros in (a, b) and no other zeros there.*
5. *The zeros of ϕ_{n-1} and ϕ_n strictly interlace on (a, b).*

Proof. Since $\phi_0, \phi_1, \phi_2 \ldots$ is a complete set of eigenfunctions for the kernel $k(x, s)$, either $\phi_0, \phi_1, \ldots, \phi_n$ or $\phi_0, \phi_1, \ldots, \phi_{n-1}, -\phi_n$ is a Tchebycheff system on (a, b). For definiteness and without loss in generality assume $\phi_0, \phi_1, \ldots, \phi_n$ is a Tchebycheff system on (a, b).

Properties 1 and 2, that hold for any Tchebycheff system, were established in Section 2.4.

To prove Property 3, first recall that the eigenfunctions $\phi_0, \ldots, \phi_n, \ldots$ are mutually orthogonal because the kernel is symmetric. Assume that ϕ has exactly $p < m$ nodal zeros in (a, b), say $a < x_1 < \cdots < x_p < b$ and form the function

$$\psi(x) = \begin{vmatrix} \phi_0(x_1) & \cdots & \phi_0(x_p) & \phi_0(x) \\ \phi_1(x_1) & \cdots & \phi_1(x_p) & \phi_1(x) \\ \cdots & \cdots & \cdots & \cdots \\ \phi_p(x_1) & \cdots & \phi_p(x_p) & \phi_p(x) \end{vmatrix}$$

for x in (a, b). Expand by the last column to see that $\psi(x)$ is a linear combination of $\phi_0, \ldots, \phi_p$. Let $a = x_0$ and $b = x_{p+1}$. For x in $x_j < x < x_{j+1}$ with $j = 0, \ldots, p$

$$\begin{vmatrix} \phi_0(x_1) & \cdots & \phi_0(x_j) & \phi_0(x) & \phi_0(x_{j+1}) & \cdots & \phi_0(x_p) \\ \vdots & \vdots & \vdots & \vdots & \vdots & \vdots & \vdots \\ \phi_p(x_1) & \cdots & \phi_p(x_j) & \phi_p(x) & \phi_p(x_{j+1}) & \cdots & \phi_p(x_p) \end{vmatrix} > 0$$

because $\phi_0, \ldots, \phi_p$ is a Tchebycheff system on (a, b). Since $p - j$ interchanges of adjacent columns move the $(j+1)$-st column to the last column and gives the determinant defining ψ while each such interchange changes the sign of the determinant, it follows that $\psi(x)$ is nonzero on $x_j < x < x_{j+1}$ and has sign $(-1)^{p-j}$ there. Consequently, the only zeros of $\psi(x)$ are nodal zeros at $x_1 < \cdots < x_p$. By assumption these are also the nodal zeros of the $\phi(x) = \sum_{i=m}^{n} a_i \phi_i(x)$. Consequently, $\langle \psi, \phi \rangle \neq 0$. But ϕ is orthogonal to $\phi_0, \ldots, \phi_p$ because we have assumed $p < m$. Thus, $\langle \psi, \phi \rangle = 0$, a contradiction. The original assumption that $p < m$ must be false. So $p \geq m$ and Property 3 is established.

Property 4 follows directly from Property 3 by setting $m = n$.

It remains to prove Property 5. Consider the function

$$f(x) = \frac{\phi_{n-1}(x)}{\phi_n(x)} \quad \text{for } x_i < x < x_{i+1}$$

where $x_1 < \cdots < x_n$ are the n nodal zeros of $\phi_n(x)$, $x_0 = a$, and $x_{n+1} = b$. The continuous function $f(x)$ must be strictly increasing or decreasing on $x_i < x < x_{i+1}$ for $i = 0, \ldots, n$. If this assertion were false for some (fixed) i, then $f(x)$ has either a local maximum or a local minimum at some point, say ξ_i, with $x_i < \xi_i < x_{i+1}$. (See Theorem 9.) Let $y_i = f(\xi_i)$ and form the ϕ-polynomial

$$\phi(x) = \phi_n(x)(f(x) - y_i) = \phi_{n-1}(x) - y_i \phi_n(x).$$

Since y_i is a local maximum or minimum value of $f(x)$, the function $f(x) - y_i$ has ξ_i as a zero and maintains a fixed sign (≥ 0 or ≤ 0) in some interval containing ξ_i. The same is true for $\phi(x)$. So $\phi(x)$ has a nonnodal zero at ξ_i and also has the nodal zeros $x_1 < \cdots < x_n$. So $\phi(x)$ has at least $n + 2$ zeros, counting zeros as in Property 2. This contradicts Property 2 and establishes that $f(x)$ is either strictly increasing or decreasing on $x_i < x < x_{i+1}$ for $i = 0, \ldots, n$.

Since $f(x)$ is strictly monotone on $x_i < x < x_{i+1}$ for $i = 0, \ldots, n$, the following limits exist, finite or infinite ($\pm\infty$):

$$\lim_{x \to x_i+} f(x) = l_i^+ \quad \text{for } i = 0, \ldots, n,$$

$$\lim_{x \to x_i-} f(x) = l_i^- \quad \text{for } i = 1, \ldots, n + 1.$$

We show next that none of the one-sided limits at $x_1, \dots, x_n$ is finite. The proof is by contradiction. Consider the case where for some interior node x_i of $\phi_n(x)$ the limit l_i^+ is finite. (The case l_i^- finite is treated in the same way.) Since $f(x) = \phi_{n-1}(x)/\phi_n(x)$, l_i^+ finite can happen only if x_i is also a zero of $\phi_{n-1}(x)$. So x_i is a nodal zero of both $\phi_{n-1}(x)$ and of $\phi_n(x)$; consequently, $f(x)$ does not change its sign as x increases through x_i and l_i^- has the same sign as l_i^+. There are four possibilities that might occur:

(1) l_i^- is infinite.

(2) l_i^- is finite and $l_i^- \neq l_i^+$.

(3) $l_i^- = l_i^+$ and as x increases through x_i the function $f(x)$ maintains its monotonicity.

(4) $l_i^- = l_i^+$ and as x increases through x_i the function $f(x)$ reverses its monotonicity; hence has a local extreme value at x_i.

It may be helpful to sketch graphs of $f(x)$ for x near x_i that illustrate the four possibilities. In cases (1) and (2) there is a value say y_i strictly between l_i^- and l_i^+. In case (3), we set $y_i = l_i^- = l_i^+$. It follows that the ϕ-polynomial

$$\phi(x) = \phi_n(x)\big(f(x) - y_i\big) = \phi_{n-1}(x) - y_i\phi_n(x)$$

is zero at x_i but does not change sign as x increases through x_i because both $\phi_n(x)$ and $f(x) - y_i$ change sign at x_i. Consequently, x_i is a nonnodal zero of $\phi(x)$, $\phi(x)$ also has the $n-1$ other zeros of $\phi_n(x)$. Thus, $\phi(x)$ has at least $n+1$ zeros counted as in Property 2, a contradiction. So none of cases (1), (2), or (3) can occur. Suppose case (4) occurs and let $y_i = l_i^- \pm \varepsilon$ where $\varepsilon > 0$ and the plus sign is used if $l_i^- = l_i^+$ is a local minimum and the minus sign for a local maximum. For $\varepsilon > 0$ chosen sufficiently small $f(x) - y_i$ has two nodal zeros in (x_{i-1}, x_{i+1}), one slightly less than x_i and the other slightly greater than x_i. Hence, $\phi(x)$ has the same two nodal zeros as well as the n nodal zeros of $\phi_n(x)$. Thus, $\phi(x)$ has at least $n+2$ zeros, contradicting Property 2. Thus, none of cases (1)-(4) can occur. This contradiction establishes that none of the limits l_i^- or l_i^+ at $x_1, \dots, x_n$ can be finite.

Since $f(x)$ is strictly monotone, continuous, and varies from $-\infty$ to ∞ or vice versa on the $n-1$ intervals (x_i, x_{i+1}) for $i = 1, \dots, n-1$, $f(x) = \phi_{n-1}(x)/\phi_n(x)$ must have $n-1$ zeros, say $\xi_1, \dots, \xi_{n-1}$, with $x_i < \xi_i < x_{i+1}$. The $n-1$ zeros ξ_i are also zeros of $\phi_{n-1}(x)$. By Property 4, they are all nodal zeros and $\phi_{n-1}(x)$ has no other zeros in (a, b). This establishes Property 5. ∎

In applications to vibrating mechanical systems, if $k(x, s)$ is a Green's function, then $k(a, a) = 0$ means that a unit force applied at $s = a$ causes no displacement at $x = a$. This means the point a is an immovable point and it is expected there cannot be a nonzero displacement at any other point of the system; that is $k(x, a) = 0$ for all x in $[a, b]$. The following corollary confirms this behavior and an implication for the eigenfunctions of the kernel.

Corollary 75 *If a Kellogg kernel satisfies $k(a, a) = 0$, then $k(x, a) = 0$ for all x in $[a, b]$ and all the eigenfunctions of the kernel vanish at $x = a$. Likewise, $k(x, b) = 0$ for all x in $[a, b]$ and all the eigenfunctions vanish at $x = b$ if $k(b, b) = 0$.*

Proof. Since $k(x, s)$ is a Kellogg kernel

$$\begin{vmatrix} k(x_1, s_1) & k(x_1, s) \\ k(x, s_1) & k(x, s) \end{vmatrix} \geq 0$$

for $a < x_1 < x < b$ and $a < s_1 < s < b$. Let $s_1 \to a$ and $x_1 \to a$ to obtain

$$\begin{vmatrix} k(a, a) & k(a, s) \\ k(x, a) & k(x, s) \end{vmatrix} \geq 0$$

for $a < x, s < b$. Set $x = s$ and use $k(a, a) = 0$ to find that $k(a, s) = 0$ for $a \leq s \leq b$. By symmetry of the kernel, $k(x, a) = 0$ for $a \leq x \leq b$ and

$$\phi_n(a) = \int_a^b k(a, s)\phi_n(s)\, ds = 0.$$

■

3.7 Singular Kellogg Kernels

Readers not interested in singular problems can skip this section. Analogs of the results for Kellogg kernels are established in this section for kernels $k(x, s)$ that are continuous on $[a, b] \times [a, b] \backslash \{(a, a)\}$ and are mildly singular at (a, a). The basic conclusions are the same as for Kellogg kernels. We outline the changes needed to establish them in the singular case. Some proofs will be given in appendices in order to concentrate on the forest and not the trees.

A real-valued kernel $k(x, s)$ with domain $[a, b] \times [a, b] \backslash \{(a, a)\}$ is **mildly singular** if either

$$\text{(i)} \quad k(x, s) = h(x, s) \ln(\max(x, s) - a)$$

for all (x, s) in its domain and where $h(x, s)$ is a continuous function on $[a, b] \times [a, b]$; or

$$\text{(ii)} \quad k(x, s) \text{ is bounded and continuous}$$

for all (x, s) in its domain and the kernel does not have a continuous extension to $[a, b] \times [a, b]$. The Green's functions of the singular Sturm-Liouville problems in Chapter 5 are mildly singular of type (i) and the Green's functions of the singular Sturm-Liouville problems in Chapter 6 are mildly singular of type (ii).

It is established in Appendices A and B that a mildly singular kernel $k(x, s)$ has the following properties that are assumed to hold throughout this section:

> **Standing Assumptions:** *$k(x, s)$ is a real-valued continuous kernel defined on $[a, b] \times [a, b] \backslash \{(a, a)\}$ that satisfies (a), (b) and (c) in Theorem 52 and has compound kernels that satisfy $(a)_n$, $(b)_n$, and $(c)_n$ of Theorem 76.*

Under the standing assumptions, the integral operator

$$Kf(x) = \int_a^b k(x, s)f(s)\, ds$$

maps the function space of real-valued continuous functions $C[a, b]$ into itself and is a compact, bounded, linear operator when $C[a, b]$ is equipped with the maximum norm by Theorem 52.

3.7.1 Compound Kernels

The n-th compound kernel $k_{[n]}(x, s)$ of a kernel $k(x, s)$ that is defined and continuous on $[a, b] \times [a, b] \backslash (a, a)$ is

$$k_{[n]}(x, s) = k\begin{pmatrix} x_1, x_2, \dots, x_n \\ s_1, s_2, \dots, s_n \end{pmatrix} = \det\left[k(x_i, s_j)\right]_{n \times n}$$

for all $x = (x_1, \ldots, x_n)$ and $s = (s_1, \ldots, s_n)$ in Δ_n for which the determinant makes sense; that is, each entry $k(x_i, s_j)$ of the determinant is defined. Since $k(x, s)$ is continuous in a neighborhood of each point (x_i, s_j) in its domain, the compound kernel $k_{[n]}(x, s)$ is continuous in a neighborhood of each point (x, s) in its domain. We continue to use the convention that the context determines the dimension of the variables x and s. Thus, in $k(x, s)$ the variables x and s are real numbers while in $k_{[n]}(x, s)$ they are elements of $\mathbb{R}^n$.

It takes a little care to determine the domain of $k_{[n]}(x, s)$. To this end, let

$$\Delta_n = \{u = (u_1, \ldots, u_n) : a \le u_1 \le \cdots \le u_n \le b\},$$

$$\tilde{\Delta}_n = \{u \in \Delta_n : u_1 > a\},$$

$$F_1 = \{u \in \Delta_n : u_1 = a\}.$$

In geometric terms, F_1 is the face of the simplex Δ_n that lies in the hyperplane perpendicular to the u_1-axis at $u_1 = a$ and $\tilde{\Delta}_n$ is the simplex Δ_n with its face F_1 removed. When $n = 2$ and the u_1u_2-plane is given its usual orientation, Δ_2 is a solid triangle, F_1 is the vertical side of the solid triangle, and $\tilde{\Delta}_2$ is the solid triangle with its vertical side removed. Now $k_{[n]}(x, s) = \det\big[k(x_i, s_j)\big]_{n\times n}$ is not defined at (x, s) in $\Delta_n \times \Delta_n$ if and only if $(x_i, s_j) = (a, a)$ for some i and j; which holds if and only if

$$a = x_1 = \cdots = x_i \quad \text{and} \quad a = s_1 = \cdots = s_j$$

for some i and j; which holds if and only if $x_1 = a$ and $s_1 = a$. Thus, (x, s) in $\Delta_n \times \Delta_n$ is in the domain of $k_{[n]}$ if and only if $s_1 > a$ when $x_1 = a$ or $x_1 > a$ when $s_1 = a$; that is,

$$\text{domain of } k_{[n]} = (\Delta_n \times \tilde{\Delta}_n) \cup (\tilde{\Delta}_n \times \Delta_n).$$

The compound kernel $k_{[n]}(x, s)$ is continuous on its domain, as we noted above, and may exhibit singular behavior, reflecting that of $k(x, s)$, as (x, s) approaches a point (x^0, s^0) in $\Delta_n \times \Delta_n$ with $x_1^0 = a$ and/or $s_1^0 = a$.

The analogue of Theorem 52 for the singular compound kernels of $k(x, s)$ is

Theorem 76 *Let $k_{[n]}(x, s)$ be a continuous real or complex-valued kernel defined on $(\Delta_n \times \tilde{\Delta}_n) \cup (\tilde{\Delta}_n \times \Delta_n)$. If*

$(a)_n$ *for each f in $C(\Delta_n)$ and x^0 in F_1, $K_{[n]}f(x^0) = \int_{\Delta_n} k_{[n]}(x^0, s)f(s)\, ds$ exists as a convergent improper Riemann integral,*

$(b)_n$ *$\int_{\Delta_n} |k_{[n]}(x, s)|\, ds \le M$ for some constant M and all x in Δ_n,*

$(c)_n$ *$\int_{\Delta_n} |k_{[n]}(x, s) - k_{[n]}(x^0, s)|\, ds \to 0$ as $x \to x^0$ for each x^0 in F_1,*

then $K_{[n]} : C(\Delta_n) \to C(\Delta_n)$ and $K_{[n]}$ is a bounded, linear, compact operator on $C(\Delta_n)$ equipped with the maximum norm.

The proof is essentially the same as for Theorem 52. It is given in Appendix A as is the easy check that the theorem for the compound kernels reduces to Theorem 52 when $n = 1$.

Since the integral operator $K_{[n]}$ is a compact, bounded, linear operator on $C(\Delta_n)$ and the proof of Theorem 76 establishes that

$$\int_{\Delta_n} |k_{[n]}(x, s) - k_{[n]}(x^0, s)|\, ds \to 0 \text{ as } x \to x^0$$

for each x^0 in Δ_n, the reasoning used in Section 3.5 when $n=1$ extends directly to any positive integer n and establishes the following version of Jentzsch's theorem:

Theorem 77 *If $k(x, s)$ is a mildly singular kernel on $[a, b] \times [a, b]\backslash\{(a, a)\}$, $k_{[n]}(x, s) \geq 0$ on its domain and $k_{[n]}(x, x) > 0$ for x in Δ_n with $a < x_1 < \cdots < x_n < b$, then the following hold. (1) $r(K_{[n]}) > 0$. (2a) Extremal functions exist and every extremal function is positive on $a < x_1 < \cdots < x_n < b$ and is an eigenfunction of $K_{[n]}$ corresponding to the eigenvalue $r(K_{[n]})$. (2b) If $\phi(x)$ is an eigenfunction corresponding to the eigenvalue $r(K_{[n]})$, then $|\phi|$ an extremal function corresponding to $K_{[n]}$ and, hence, $|\phi|$ is an eigenfunction corresponding to $r(K)$ and $|\phi(x)| > 0$ for x in Δ_n with $a < x_1 < \cdots < x_n < b$. Consequently, if ϕ is real-valued, then $\phi(x) > 0$ or $\phi(x) < 0$ for x in Δ_n with $a < x_1 < \cdots < x_n < b$. (3) $r(K_{[n]})$ has geometric multiplicity 1. (4) $r(K_{[n]})$ has algebraic multiplicity 1. (5) Every eigenvalue μ of $K_{[n]}$ different from $r(K_{[n]})$ satisfies $|\mu| < r(K_{[n]})$. Hence,*

$$r(K_{[n]}) = \max\{|\mu| : \mu \text{ is an eigenvalue of } K_{[n]}\}.$$

Next we extend to singular kernels $k(x, s)$ two key results established for continuous kernels:

$$K_{[n]}(\psi_1 \wedge \psi_2 \cdots \wedge \psi_n) = K\psi_1 \wedge K\psi_2 \cdots \wedge K\psi_n$$

for any functions $\psi_1, \psi_2, \ldots, \psi_n$ in $C[a, b]$ and the basic composition formula. To establish the first result let u be a point in Δ_n with $u_1 > a$ and $\psi_1, \psi_2, \ldots, \psi_n$ be continuous functions on $[a, b]$. Then $\phi_i(t) = k(u_i, t)$ is continuous on $[a, b]$ and by Lemma 70

$$\begin{aligned}\det[K\psi_j(u_i)]_{n\times n} &= \det\left[\int_a^b k(u_i, t)\psi_j(t)\, dt\right]_{n\times n} \\ &= \int_{\Delta_n} \det[k(u_i, t_r)]_{n\times n} \det[\psi_j(t_r)]_{n\times n} dt_1 \cdots dt_n \\ &= \int_{\Delta_n} k_{[n]}(u, t)\psi_1 \wedge \psi_2 \cdots \wedge \psi_n(t)\, dt,\end{aligned}$$

where $dt = dt_1 \cdots dt_n$. That is,

$$K\psi_1 \wedge K\psi_2 \cdots \wedge K\psi_n(u) = K_{[n]}(\psi_1 \wedge \psi_2 \cdots \wedge \psi_n)(u)$$

for any u in Δ_n with $u_1 > a$. Given any x in Δ_n there are points u in Δ_n with $u_1 > a$ and $u \to x$. Since K maps $C[a, b]$ into itself and $K_{[n]}$ maps $C(\Delta_n)$ into itself, both sides of the last equation are continuous functions on Δ_n. Thus, letting $u \to x$ in that equation gives

$$K\psi_1 \wedge K\psi_2 \cdots \wedge K\psi_n(x) = K_{[n]}(\psi_1 \wedge \psi_2 \cdots \wedge \psi_n)(x)$$

for all x in Δ_n. That is, $K\psi_1 \wedge K\psi_2 \cdots \wedge K\psi_n = K_{[n]}(\psi_1 \wedge \psi_2 \cdots \wedge \psi_n)$ as claimed.

The basic composition formula asserts: if the kernels $k(x, s)$, $l(x, s)$, and $m(x, s)$ are related by

$$m(x, s) = \int_a^b k(x, t)l(t, s)\, dt,$$

then their compound kernels are related by

$$m\begin{pmatrix} x_1, x_2, \ldots, x_n \\ s_1, s_2, \ldots, s_n \end{pmatrix} = \int_{\Delta_n} k\begin{pmatrix} x_1, \ldots, x_n \\ t_1, \ldots, t_n \end{pmatrix} l\begin{pmatrix} t_1, \ldots, t_n \\ s_1, \ldots, s_n \end{pmatrix} dt,$$

where $dt = dt_1 \cdots dt_n$, or, more briefly, by

$$m_{[n]}(x, s) = \int_{\Delta_n} k_{[n]}(x, t) l_{[n]}(t, s)\, dt.$$

We established this formula in Section 3.6 when the kernels $k(x, t)$ and $l(t, s)$ were continuous on $[a, b] \times [a, b]$. The same proof establishes the formula when $k(x, t)$ is continuous on $[a, b] \times [c, d]$ and $l(t, s)$ is continuous on $[c, d] \times [a, b]$ and Δ_n is the simplex based on the interval $[c, d]$. For our purposes, it is enough to establish that the basic composition formula holds for mildly singular kernels $k(x, t)$ and $l(t, s)$ with the same mildly singular behavior. For such mildly singular kernels

$$m(x, s) = \int_a^b k(x, t) l(t, s)\, dt$$

exists as an improper Riemann integral,

$$m(x, s) = \lim_{a' \to a} \int_{a'}^b k(x, t) l(t, s)\, dt,$$

and $m(x, s)$ is continuous on $[a, b] \times [a, b]$. (See Appendix B.) Fix a' with $a < a' < b$. The kernel $k(x, t)$ is continuous on $[a, b] \times [a', b]$ and the kernel $l(t, s)$ is continuous on $[a', b] \times [a, b]$. Consequently, if

$$m'(x, s) = \int_{a'}^b k(x, t) l(t, s)\, dt$$

the basic composition formula for continuous kernels gives

$$m'_{[n]}(x, s) = \int_{\Delta'_n} k_{[n]}(x, t) l_{[n]}(t, s)\, dt.$$

for x and s in Δ_n and where $\Delta'_n = \{t \in \Delta_n : t_1 \geq a'\}$ is a subsimplex of Δ_n. Since the Riemann integrals $m'(x_i, s_j)$ that are the entries of the determinant $m'_{[n]}(x, s)$ converge to $m(x_i, s_j)$ as $a' \to a$,

$$m_{[n]}(x, s) = \lim_{a' \to a} \int_{\Delta'_n} k_{[n]}(x, t) l_{[n]}(t, s)\, dt.$$

The existence of the limit on the right means that the improper Riemann integral of $k_{[n]}(x, t) l_{[n]}(t, s)$ over Δ_n exists and equals $m_{[n]}(x, s)$; that is,

$$m_{[n]}(x, s) = \int_{\Delta_n} k_{[n]}(x, t) l_{[n]}(t, s)\, dt,$$

and the basic composition formula holds for mildly singular kernels $k(x, t)$ and $l(t, s)$.

It follows from Appendix B that the iterated kernels $k_n(x, s)$ of a mildly singular kernel $k(x, s)$ exist and are continuous on $[a, b] \times [a, b]$ for $n \geq 2$. Use of the basic composition formula just as at the end of Section 3.6.1 gives

$$(k_{[m]})_n = (k_n)_{[m]}$$

for $n, m = 1, 2, \ldots$. The displayed equality means that

$$\left(K_{[m]}\right)_n = \left(K^n\right)_{[m]}$$

when expressed in terms of corresponding integral operators.

3.7.2 Spectral Properties of Compound Kernels

If the mildly singular kernel $k(x, s)$ is self-adjoint, in particular if it is symmetric, the Hilbert-Schmidt theorem implies that the kernel has a *complete system* of orthogonal eigenfunctions $\phi_0, \phi_1, \phi_2, \ldots$ with corresponding eigenvalues $\lambda_0, \lambda_1, \lambda_2, \ldots$, listed to multiplicity, that can be labeled so that

$$|\lambda_0| \leq |\lambda_1| \leq |\lambda_2| \leq \cdots .$$

We use this notation and ordering throughout this subsection.

Schur's theorem (Theorem 72) holds for the symmetric, mildly singular kernels: the wedge products $\phi_{i_1} \wedge \phi_{i_2} \wedge \cdots \wedge \phi_{i_n}(x)$ form a complete system of orthogonal eigenfunctions for the (symmetric) compound kernel $k_{[n]}(x, s)$ when the indices $i_1, i_2, \ldots, i_n$ with $0 \leq i_1 < i_2 < \cdots < i_n$ vary over all subsets of indices appearing in $\phi_0(x), \phi_1(x), \phi_2(x), \ldots$.

The proof of Schur's theorem in the mildly singular case is essentially given in Step 2 of the argument for the case when $k(x, s)$ is continuous. We reprise that reasoning here. Just as in Section 3.6.2,

$$K_{[n]}(\phi_{i_1} \wedge \phi_{i_2} \wedge \cdots \wedge \phi_{i_n}) = \left(\lambda_{i_1}\lambda_{i_2}\cdots\lambda_{i_n}\right)^{-1}\phi_{i_1} \wedge \phi_{i_2} \wedge \cdots \wedge \phi_{i_n}.$$

Thus $\phi_{i_1} \wedge \phi_{i_2} \wedge \cdots \wedge \phi_{i_n}$ is an eigenfunction of $k_{[n]}(x, s)$ and $\lambda_{i_1}\lambda_{i_2}\cdots\lambda_{i_n}$ is its corresponding eigenvalue.

It remains to show that the system is complete. Under our standing assumptions, the iterated kernel

$$k_2(x, s) = \int_a^b k(x, t)k(t, s)\, dt$$

is continuous on $[a, b] \times [a, b]$ and by Theorem 64 $\phi_0, \phi_1, \phi_2, \ldots$ form a complete system of eigenfunctions for $k_2(x, s)$ with corresponding eigenvalues $\lambda_0^2, \lambda_1^2, \lambda_2^2, \ldots$. By Mercer's theorem

$$k_2(x, s) = \sum_{i=0}^{\infty} \frac{\phi_i(x)\phi_i(s)}{\lambda_i^2}$$

and the series converges absolutely and uniformly on $[a, b] \times [a, b]$. Use this expansion and the reasoning at the beginning of Step 1 of the proof of Schur's theorem in the continuous case to obtain

$$(k_2)_{[n]}(x, s) = \sum_{0 \leq i_1 < \cdots < i_n} \frac{\phi_{i_1} \wedge \phi_{i_2} \wedge \cdots \wedge \phi_{i_n}(x)\phi_{i_1} \wedge \phi_{i_2} \wedge \cdots \wedge \phi_{i_n}(s)}{\left(\lambda_{i_1}\lambda_{i_2}\cdots\lambda_{i_n}\right)^2}$$

for x and s in Δ_n, with absolute and uniform convergence inherited from the expansion for $k_2(x, s)$. Since $(k_2)_{[n]}(x, s) = (k_{[n]})_2(x, s)$,

$$(k_{[n]})_2(x, s) = \sum_{0 \leq i_1 < \cdots < i_n} \frac{\phi_{i_1} \wedge \phi_{i_2} \wedge \cdots \wedge \phi_{i_n}(x)\phi_{i_1} \wedge \phi_{i_2} \wedge \cdots \wedge \phi_{i_n}(s)}{\left(\lambda_{i_1}\lambda_{i_2}\cdots\lambda_{i_n}\right)^2}$$

with absolute and uniform convergence.

This expansion implies that $\{\phi_{i_1} \wedge \phi_{i_2} \wedge \cdots \wedge \phi_{i_n}\}_{0 \le i_1 < \cdots < i_n}$ is a complete orthogonal system for the kernel $k_{[n]}$: let ψ be an eigenfunction of the kernel $k_{[n]}$ and ρ its eigenvalue so that $\rho K_{[n]}\psi = \psi$. If $\rho \neq \lambda_{i_1}\lambda_{i_2}\cdots\lambda_{i_n}$ for all choices $0 \le i_1 < \cdots < i_n$, then ψ is orthogonal to $\phi_{i_1} \wedge \phi_{i_2} \wedge \cdots \wedge \phi_{i_n}$ for all choices because the kernel is symmetric, and

$$\psi = \rho^2 \left(K_{[n]}\right)^2 \psi = 0,$$

where the last equality uses term-by-term integration in the series expansion of $(k_{[n]})_2(x, s)$. This contradiction implies that $\rho = \lambda_{i_1}\lambda_{i_2}\cdots\lambda_{i_n}$ for some $i_1, \ldots, i_n$. Let

$$\tilde{\psi} = \psi - \sum{}' \langle \psi, \phi_{i_1} \wedge \phi_{i_2} \wedge \cdots \wedge \phi_{i_n} \rangle \phi_{i_1} \wedge \phi_{i_2} \wedge \cdots \wedge \phi_{i_n}$$

where the prime means the sum if over all $0 \le i_1 < \cdots < i_n$ with $\lambda_{i_1}\lambda_{i_2} \cdots \lambda_{i_n} = \rho$. Consequently, $\tilde{\psi}$ is orthogonal to all $\phi_{i_1} \wedge \phi_{i_2} \wedge \cdots \wedge \phi_{i_n}$ with $\lambda_{i_1}\lambda_{i_2}\cdots\lambda_{i_n} = \rho$ and is orthogonal to all the other eigenfunctions $\phi_{i_1} \wedge \phi_{i_2} \wedge \cdots \wedge \phi_{i_n}$ belonging to eigenvalues different from ρ because $\rho K_{[n]}\tilde{\psi} = \tilde{\psi}$ and $K_{[n]}$ is self-adjoint. Hence, using term-by-term integration as above,

$$\tilde{\psi} = \rho^2 \left(K_{[n]}\right)^2 \tilde{\psi} = 0,$$

$$\psi = \sum{}' \langle \psi, \phi_{i_1} \wedge \phi_{i_2} \wedge \cdots \wedge \phi_{i_n} \rangle \phi_{i_1} \wedge \phi_{i_2} \wedge \cdots \wedge \phi_{i_n},$$

and the system $\left\{\phi_{i_1} \wedge \phi_{i_2} \wedge \cdots \wedge \phi_{i_n}\right\}$ is complete for the kernel $k_{[n]}(x, s)$.

3.7.3 Spectral Properties of Kellogg Kernels

A symmetric, mildly singular kernel $k(x, s)$ with domain $[a, b] \times [a, b] \backslash \{(a, a)\}$ that satisfies

K1. $\det\left[k(x_i, x_j)\right]_{n\times n} > 0, \quad a < x_1 < \cdots < x_n < b,$

K2. $\det\left[k(x_i, s_j)\right]_{n\times n} \ge 0, \quad \text{for } (x, s) \text{ in } \left(\Delta_n \times \tilde{\Delta}_n\right) \cup \left(\tilde{\Delta}_n \times \Delta_n\right),$

is called a **mildly singular Kellogg kernel.** A mildly singular Kellogg kernel $k(x, s)$ and its compound kernels $k_{[n]}(x, s)$ determine integral operators $K : C[a, b] \to C[a, b]$ and $K_{[n]} : C(\Delta_n) \to C(\Delta_n)$ that are self-adjoint, compact, bounded, linear operators. The arguments given in Section 3.6.3 apply without change to establish that the results in Theorems 73 and 74 hold for mildly singular Kellogg kernels. In particular, they hold for the Green's functions of the singular Sturm-Liouville problems in Chapters 5 and 6.

Chapter 4

Regular Sturm-Liouville Problems

The last section of the chapter on eigenvalues and eigenfunctions of regular Sturm-Liouville problems will be of primary interest to many readers. It contains results of great practical importance. There are two equally important parts of the discussion in that section. The first part establishes the basic properties of the eigenvalues and eigenfunctions related to their existence, multiplicity, orthogonality, and eigenfunction expansions. These results follow from the Hilbert-Schmidt theorem and can be found in many books on applied mathematics. The second part develops the oscillatory and approximation properties of the eigenfunctions from a unified perspective that has been largely overlooked in the English literature and slipped into obscurity in the Russian and German literature where it once appeared. This is the approach based on Jentzsch's theorem, Schur's theorem, and the Kellogg conditions; see Section 1.11.2 and Section 3.6.2. The reader primarily interested in the spectral results can skim the necessary background results in Chapter 3 and the properties of Green's functions established in this chapter and concentrate on the material on eigenvalue problems in Section 4.4 and Section 4.8. Readers seeking a fuller account of Sturm-Liouville initial value problems, boundary value problems, their adjoint problems, and Green's functions will find a readable account in the intervening sections.

Results in the chapter often are established for Sturm-Liouville problems involving complex-valued data and therefore admit complex-valued solutions. When solutions must be real-valued, theorems to that effect are established. As noted in Section 1.13 most problems of applied interest involve only real-valued data and the physically relevant solutions are real-valued. Readers of the chapter interested only in such problems can assume all data is real-valued and solutions are real-valued without any essential loss.

4.1 Sturm-Liouville Form

The general second order linear inhomogeneous differential equation on the interval (a, b) is

$$a(x)y'' + b(x)y' + c(x)y = g(x), \quad a < x < b, \tag{4.1}$$

where $a(x)$, $b(x)$, $c(x)$, and $g(x)$ are given real or complex-valued functions for $a < x < b$. It is sufficient for our purposes to assume that $a(x)$, $b(x)$, $c(x)$, and $g(x)$ are continuous on $a < x < b$ that $a(x) \neq 0$ there.

It is often useful to express (4.1) in **formally self-adjoint** form by applying Euler's method for solving first order linear equations to the first two terms on the left of (4.1): express the equation as

$$y'' + \frac{b(x)}{a(x)}y' + \frac{c(x)}{a(x)}y = \frac{g(x)}{a(x)},$$

and multiply through by

$$p(x) = \exp\left(\int^x \frac{b(s)}{a(s)}\,ds\right),$$

where the integral notation stands for any particular antiderivative of the integrand, to obtain

$$(p(x)y')' + p(x)\frac{c(x)}{a(x)}y = p(x)\frac{g(x)}{a(x)}.$$

So (4.1) can be expressed as

$$-(p(x)y'(x))' + q(x)y(x) = f(x), \quad a < x < b,$$

where

$$p(x) = \exp\left(\int^x \frac{b(s)}{a(s)}\,ds\right) \neq 0,$$

$$q(x) = -p(x)\frac{c(x)}{a(x)} \quad \text{and} \quad f(x) = -p(x)\frac{g(x)}{a(x)}.$$

Thus, the inhomogeneous linear second order differential equation

$$a(x)y'' + b(x)y' + c(x)y = g(x), \quad a < x < b,$$

can be put in the form

$$-(p(x)y'(x))' + q(x)y(x) = f(x), \quad a < x < b,$$

where $p(x) \neq 0$ is continuous on $a < x < b$ and $q(x)$ and $f(x)$ are real or complex-valued continuous functions on $a < x < b$. It is useful to observe that, in the reduction above, $p(x)$ is continuously differentiable, and $p(x) > 0$ if $a(x)$ and $b(x)$ are real-valued. It is common to call this form of (4.1) its formally self-adjoint form. The word formally means that a related boundary value or eigenvalue problem will be self-adjoint when appropriate boundary conditions are chosen but will not be self-adjoint with other boundary conditions. We prefer to avoid this somewhat ambiguous terminology.

Linear second order ordinary differential equations can always be put into formally self-adjoint form, as we have just seen. This is not always possible for higher order equations. In the second order case, self-adjointness is determined by the boundary conditions attached to the differential equation. In the higher order problems, it is determined by both the differential equation and boundary conditions.

A differential equation of the form $-(p(x)y'(x))' + q(x)y(x) = f(x)$ for $a < x < b$ often is derived directly from physical laws, where, in certain physical contexts, it is natural to assume only that $p(x) \neq 0$ is continuous on (a, b). It is for this reason that we do not assume further smoothness on $p(x)$.

4.2 Sturm-Liouville Differential Equations

A differential equation of the form

$$-(p(x)y'(x))' + q(x)y(x) = f(x), \quad a < x < b, \tag{4.2}$$

is a **Sturm-Liouville (differential) equation**. We always assume that $p(x) \neq 0$ on $a < x < b$ and $p(x)$, $q(x)$ and $f(x)$ are real or complex-valued continuous functions on

$a < x < b$. We use the same terminology if the differential equation is defined on any of the other three intervals with endpoints a and b.

A little care and discussion are needed about a suitable definition of a solution y to a Sturm-Liouville differential equation. By a **solution** to (4.2), we mean a real or complex-valued function y such that $(p(x)y'(x))'$ exists for each x in (a, b) and (4.2) holds for each x in (a, b). (The meaning of a solution is defined in the same way if the differential equation is defined on any of the other three intervals with endpoints a and b.) Several comments about this definition are in order. The definition implies that $y'(x)$ exists on (a, b) so that $y(x)$ is continuous on (a, b), that $p(x)y'(x)$ is continuous on (a, b), and, hence, that $y'(x)$ is continuous on (a, b) because $p(x) \neq 0$ there. Since $y(x)$ is continuous on (a, b), the differential equation implies that $(p(x)y'(x))'$ is continuous on (a, b). We summarize these observations as

Lemma 78 *A solution $y(x)$ to the Sturm-Liouville differential (4.2) is continuously differentiable on (a, b) and $(p(x)y'(x))'$ is continuous on (a, b).*

The solution to a Sturm-Liouville differential equation can be defined in an alternative integrated form: let $y(x)$ be a solution to (4.2) and c and x be any two points in (a, b). Apply the fundamental theorem of calculus (Theorem 14) on the closed interval with endpoints c and x to obtain

$$p(x)y'(x) - p(c)y'(c) = \int_c^b (p(s)y'(s))' \, ds.$$

Thus, if y is a solution to (4.2), the differential equation can be integrated to obtain

$$-(p(x)y'(x) - p(c)y'(c)) + \int_c^x q(s)y(s) \, ds = \int_c^x f(s) \, ds \tag{4.3}$$

for any x and c in (a, b). Conversely, if $y(x)$ is a solution to this integrated equation, by which we mean that $y'(x)$ exists for all x in (a, b) and the integrated equation is satisfied, then $y(x)$ is continuous on (a, b) and

$$\frac{p(x)y'(x) - p(c)y'(c)}{x - c} = \frac{1}{x - c} \int_c^x (q(s)y(s) - f(s)) \, ds.$$

By Theorem 13, another form of the fundamental theorem of calculus, the limit on the right exists and, hence, there exists

$$(p(x)y'(x))'|_{x=c} = -(q(c)y(c) - f(c))$$

and the differential equation (4.2) is satisfied at any c in (a, b). Thus, if $y(x)$ is a solution to (4.3), then it is a solution to (4.2). In summary, a solution y to (4.2) can be defined directly as we did initially or by means of the integrated form (4.3), according to the convenience of the moment.

The following comments shed further light on the definition of a solution of (4.2). The comments are based on the relation

$$\begin{aligned}\frac{p(x+h)y'(x+h) - p(x)y'(x)}{h} &= \frac{p(x+h) - p(x)}{h} y'(x+h) \\ &\quad + \frac{y'(x+h) - y'(x)}{h} p(x).\end{aligned} \tag{4.4}$$

1. If $p(x)$ is differentiable, which is the case when a general second order linear differential equation with nonzero leading coefficient is put in Sturm-Liouville form and in many

applications that lead directly to the Sturm-Liouville form, then (4.4) shows that py' is differentiable at x if and only if y' is differentiable at x, in which case the usual product rule

$$(py')' = p'y' + py''$$

holds. Consequently, under our definition of a solution y to (4.2), $y''(x)$ exists at any x in (a, b) where $p'(x)$ exists. If $p(x)$ is differentiable on (a, b), then y' is differentiable on (a, b).

2. The definition of a solution has at least one unexpected consequence when $p(x)$ is merely continuous. It turns out that most continuous functions are not differentiable at any point in their domain in a sense that is made precise in analysis courses. If $p(x) \neq 0$ is chosen as a continuous function that is not differentiable at any point, then under our definition of a solution y, the first difference quotient in (4.4) has a finite limit, the second never has a finite limit, and, hence, the third difference quotient cannot have a finite limit at any x in (a, b). That is, $y''(x)$ does not exist for any x in (a, b). We are left in the awkward situation in which a solution y to a second order differential equation does not have to have an ordinary second derivative at a single point in (a, b).

3. For those who prefer it, an alternative definition of a solution to (4.2) is a function y defined on (a, b) such that $p(x)y'(x)$ is absolutely continuous on (a, b), and the differential equation holds at each x for which $p(x)y'(x))'$ exists. Under this definition, a solution y satisfies (4.3). In general, an absolutely continuous function is differentiable for almost all x. However, reasoning from (4.3) as above, the derivative of $p(x)y'(x)$ exists for all x in (a, b) and (4.2) holds for all x in (a, b).

Let I be one of the four intervals with endpoints a and b. The Sturm-Liouville equation $-(p(x)y'(x))' + q(x)y(x) = f(x)$ for x in I is **regular** if $p(x) \neq 0$, $q(x)$, and $f(x)$ are continuous on the *closed* interval $a \leq x \leq b$.

Although the coefficients of a regular Sturm-Liouville differential equation are defined on the closed interval $[a, b]$, the interval I on which the differential equation is known to hold may, depending on the context, exclude one or both of the endpoints a and b of I. For example, in a physical system modeled as a one-dimensional continuum, the interval $a \leq x \leq b$, an equation of state typically is derived at each interior point of the interval while the coefficients that occur in that equation are often defined and continuous throughout the full continuum.

If a solution $y(x)$ to a *regular* Sturm-Liouville differential equation defined on $a < x < b$ has a continuous extension to the close interval $a \leq x \leq b$, then the extended function, which we still denoted by $y(x)$, has additional smoothness properties that will be useful when we study initial value problems, boundary value problems, and eigenvalue problems. We will show later that such a continuous extension always exists for a regular Sturm-Liouville differential equation; see Theorem 85.

Lemma 79 *Assume $y(x)$ for $a < x < b$ is a solution of the regular Sturm-Liouville differential equation*

$$-(p(x)y'(x))' + q(x)y(x) = f(x), \quad a < x < b.$$

(a) If $y(x)$ extends to a continuous function on $a \leq x < b$, then $y(x)$ is continuously differentiable on $a \leq x < b$ and satisfies the Sturm-Liouville differential equation there.
(b) If $y(x)$ extends to a continuous function on $a < x \leq b$, then $y(x)$ is continuously differentiable on $a < x \leq b$ and satisfies the Sturm-Liouville differential equation there.
(c) If $y(x)$ extends to a continuous function on $[a, b]$, then $y(x)$ is continuously differentiable on $[a, b]$ and satisfies the Sturm-Liouville differential equation at every point in $[a, b]$.

Proof. We know that any solution y to a Sturm-Liouville differential equation is continuously differentiable on the open interval $a < x < b$. Since the differential equation is regular, $p(x) \neq 0$, $q(x)$, and $f(x)$ are continuous on $[a, b]$.

(a) For x and c in (a, b) integrate the differential equation to get

$$p(x)y'(x) = p(c)y'(c) + \int_c^x (q(s)y(s) - f(s))\, ds$$

and

$$y'(x) = \frac{1}{p(x)}\left(p(c)y'(c) + \int_c^x (q(s)y(s) - f(s))\, ds\right).$$

Since the integrand is continuous on $a \leq x < b$, there exists

$$\lim_{x \to a} y'(x) = \frac{1}{p(a)}\left(p(c)y'(c) + \int_c^a (q(s)y(s) - f(s))\, ds\right).$$

Since $y(x)$ is continuous on $a \leq x < b$, by Lemma 11, $y(x)$ is differentiable at a,

$$y'(a) = \frac{1}{p(a)}\left(p(c)y'(c) - \int_a^c (q(s)y(s) - f(s))\, ds\right),$$

and $y'(x)$ is continuous at $x = a$. Consequently,

$$\frac{p(c)y'(c) - p(a)y'(a)}{c - a} = \frac{1}{c - a}\int_a^c (q(s)y(s) - f(s))\, ds.$$

Let $c \to a$ and use fundamental theorem of calculus to find that there exists

$$(p(x)y'(x))'|_{x=a} = q(a)y(a) - f(a);$$

that is, the Sturm-Liouville differential equation holds at $x = a$.

Part (b) is established by the same line of reasoning and (c) follows from (a) and (b). ■

It is convenient to introduce the **Sturm-Liouville differential operator**

$$Ly = -(py')' + qy.$$

We call the operator Ly **regular** (on $[a, b]$) if $p(x) \neq 0$ on $[a, b]$ and $p(x)$ and $q(x)$ are continuous on $[a, b]$. Later we shall need to determine a natural domain for L. The foregoing discussion will help determine that domain because the y's of interest will be those that satisfy an equation of the form $Ly = f$ or $Ly = \lambda ry$ on (a, b) together with appropriate initial or boundary conditions at $x = a$ and $x = b$.

Lemma 80 (Lagrange Identity) *Let $Ly = -(py')' + qy$ where $p \neq 0$ and q are real or complex-valued continuous functions on an interval I of any type. If y and z are real or complex-valued functions such that $(py')'$ and $(pz')'$ exist on I, then*

$$yLz - zLy = (p(zy' - yz'))'.$$

Consequently, if $(py')'$ and $(pz')'$ are continuous on I, then

$$\int_c^d (yLz - zLy)\, ds = p(zy' - yz')\Big|_c^d$$

for any c, d in I.

Proof. Lagrange's identity follows from an elementary calculation,

$$yLz - zLy = y(-pz')' - z(-py')' = (zpy')' - (ypz')' = (p(zy' - yz'))'.$$

If $(py')'$ and $(pz')'$ are continuous on I, then $yLz - zLy$ is continuous on I, hence, integrable on any bounded subinterval of I, and the final conclusion of the lemma follows from the fundamental theorem of calculus. ■

Both results of the lemma are referred to as Lagrange's identity. The stronger hypotheses in the integrated form of the identity are satisfied whenever y and z are solutions to a regular Sturm-Liouville boundary value problem or eigenvalue problem on $[a, b]$.

In the typical case when $p(x) \neq 0$ is real-valued in (4.2), it is sometimes useful to know that (4.2) can be expressed in the more common form (4.1) by a change of variables. Suppose first that $p(x) > 0$. The change of variable

$$\xi = \int_c^x \frac{1}{p(s)}\, ds \text{ with } c \text{ fixed in } (a, b)$$

is increasing, differentiable with $d\xi/dx = 1/p(x)$, and maps the interval (a, b) onto the interval (A, B) where

$$A = \int_c^a \frac{1}{p(s)}\, ds \quad \text{and} \quad B = \int_c^b \frac{1}{p(s)}\, ds.$$

If $P(\xi) = p(x)$, $Q(\xi) = q(x)$, $Y(\xi) = y(x)$ and $F(\xi) = f(x)$ where ξ and x are corresponding points under the change of variable and a prime denotes $d/d\xi$ for functions of ξ and d/dx for functions of x, then

$$(py')' = \frac{d}{dx}\left(p(x)\frac{dY}{d\xi}\frac{d\xi}{dx}\right) = \frac{d\xi}{dx}\frac{d}{d\xi}\left(p(x)\,Y'(\xi)\frac{1}{p(x)}\right) = \frac{1}{P(\xi)}\,Y''(\xi)$$

and (4.2) transforms to

$$-(1/P(\xi))\,Y'' + Q(\xi)\,Y = F(\xi), \quad A < \xi < B,$$

where $P(\xi) \neq 0$ and $P(\xi)$, $Q(\xi)$, and $F(\xi)$ are continuous on $[A, B]$. If $p(x) < 0$ the change of variables is decreasing and the same conclusion is reached with the endpoints A and B interchanged. In particular, this transformation can be used to transfer many results established for an equation given in the standard form (4.1) to equations expressed in Sturm-Liouville form (4.2).

4.3 Initial Value Problems

Although our primary focus is on boundary value problems and eigenvalue problems, initial value problems play an important background role and are an essential component of an effective numerical method used to determine eigenvalues and eigenfunctions. Throughout this section, we deal with linear second order differential equations and always assume that they are expressed in Sturm-Liouville form

$$-(p(x)y'(x))' + q(x)y(x) = f(x).$$

The existence, uniqueness, and continuous dependence results that follow are established in a more general setting than is usual because no smoothness beyond continuity is assumed on the coefficient $p(x)$. Two situations arise frequently: the Sturm-Liouville differential equation holds on a closed interval $[a, b]$ or the differential equation holds on an open interval (a, b). The latter case occurs when physical assumptions leading to the differential equation of state only hold on (a, b). Even in this case, the coefficients in the differential equation and right member are usually defined and continuous on the closed interval $[a, b]$, which models the underlying physical continuum. These observations lead to the three forms of the basic existence and uniqueness theorem for initial value problems that follow. Slight adjustments to the proof of the first theorem establish the other two.

Theorem 81 (Basic Existence and Uniqueness Theorem) *Fix c in $[a, b]$ and real or complex constants c_0 and c_1. If $p(x) \neq 0$ on $[a, b]$ and $p(x)$, $q(x)$ and $f(x)$ are real or complex-valued continuous functions on $[a, b]$, then the initial value problem*

$$\begin{aligned} &-(p(x)y')' + q(x)y = f(x), \quad a \leq x \leq b, \\ &y(c) = c_0, \quad y'(c) = c_1, \end{aligned}$$

has a unique solution y.

Proof. Of course, by a solution to the initial value problem we mean a function $y(x)$ that satisfies the differential equation on $[a, b]$ and the given initial conditions at $x = c$. If y is a solution of the initial value problem, then y is continuous on the interval $[a, b]$ and for x in $[a, b]$

$$\begin{aligned} p(x)y'(x) - p(c)y'(c) &= \int_c^x (q(u)y(u) - f(u))\, du, \\ y(x) - y(c) &= p(c)y'(c) \int_c^x \frac{1}{p(u)}\, du \\ &\quad + \int_c^x \left(\frac{1}{p(u)} \int_c^u (q(t)y(t) - f(t))\, dt \right) du, \end{aligned}$$

and

$$\begin{aligned} y(x) &= c_0 + p(c)c_1 \int_c^x \frac{1}{p(u)}\, du \\ &\quad + \int_c^x \left(\frac{1}{p(u)} \int_c^u (q(t)y(t) - f(t))\, dt \right) du. \end{aligned}$$

If $T : C[a, b] \to C[a, b]$ is defined by

$$Ty(x) = c_0 + p(c)c_1 \int_c^x \frac{1}{p(u)}\, du + \int_c^x \left(\frac{1}{p(u)} \int_c^u (q(t)y(t) - f(t))\, dt \right) du$$

for $a \leq x \leq b$, we have shown: if y is a solution to the initial value problem in the theorem, then y is continuous on $[a, b]$ and $y(x) = Ty(x)$ for all x in $[a, b]$. Conversely, if y is continuous on $[a, b]$ and $y(x) = Ty(x)$ for all x in $[a, b]$, then two differentiations of $y(x) = Ty(x)$ for x in $[a, b]$ shows that y is a solution of the initial value problem in the theorem.

Thus, y is a solution of the initial value problem in the theorem if and only if y is continuous on $[a, b]$ and $y(x) = Ty(x)$ for all x in $[a, b]$.

We use the contraction mapping theorem to establish that there exists a unique continuous function y on $[a, b]$ that satisfies $y = Ty$. This will show that the initial value problem in the theorem has a unique solution.

To this end, let $T: C[a, b] \to C[a, b]$ be the transformation defined above and equip $C[a, b]$, the space of complex-valued continuous functions on $[a, b]$, with the norm $\|y\|_L = \max_{a \le x \le b} e^{-L(x-a)}|y(x)|$ where $L > 0$ is a constant to be determined shortly. This norm is equivalent (see Section 2.5.2) to the maximum norm for every choice of $L > 0$; hence, $C[a, b]$ is a Banach space with the L-norm. We claim that T is a contraction on $C[a, b]$ when a suitable choice for L is made: for y and z in $C[a, b]$

$$Ty(x) - Tz(x) = \int_c^x \left(\frac{1}{p(u)} \int_c^u q(t)(y(t) - z(t))\, dt \right) du$$

and, consequently,

$$\begin{aligned} |Ty(x) - Tz(x)| &\le \int_a^b \frac{1}{|p(u)|} \left| \int_c^x |q(t)||y(t) - z(t)|\, dt \right| du, \\ &\le \frac{(b-a)\|q\|_{\max}}{\min_{a \le u \le b} |p(u)|} \int_a^x |y(t) - z(t)|\, dt. \end{aligned}$$

Since

$$\begin{aligned} \int_a^x |y(t) - z(t)|\, dt &= \int_a^x e^{L(t-a)} e^{-L(t-a)} |y(t) - z(t)|\, dt \\ &\le \|y - z\|_L \int_a^x e^{L(t-a)}\, dt = \frac{e^{L(x-a)} - 1}{L} \|y - z\|_L, \end{aligned}$$

$$\begin{aligned} |Ty(x) - Tz(x)| &\le \frac{(b-a)\|q\|_{\max}}{\min_{a \le u \le b} |p(u)|} \frac{e^{L(x-a)} - 1}{L} \|y - z\|_L, \\ e^{-L(x-a)} |Ty(x) - Tz(x)| &\le \frac{(b-a)\|q\|_{\max}}{\min_{a \le u \le b} |p(u)|} \frac{1 - e^{-L(x-a)}}{L} \|y - z\|_L, \\ \|Ty - Tz\|_L &\le \frac{(b-a)\|q\|_{\max}}{\min_{a \le u \le b} |p(u)|} \frac{1}{L} \|y - z\|_L. \end{aligned}$$

Fix L so that

$$\frac{(b-a)\|q\|_{\max}}{\min_{a \le u \le b} |p(u)|} \frac{1}{L} < \frac{1}{2}.$$

Then

$$\|Ty - Tz\|_L \le \frac{1}{2} \|y - z\|_L$$

and $T: C[a, b] \to C[a, b]$ is a contraction. Thus, T has a unique fixed point y_0 in $C[a, b]$. As noted above, this is equivalent to the assertion that the initial value problem in the theorem has a unique solution, namely y_0. ■

Theorem 82 (Basic Existence and Uniqueness Theorem) *Fix c in $[a, b]$ and real or complex constants c_0 and c_1. If $p(x) \neq 0$ on $[a, b]$ and $p(x)$, $q(x)$ and $f(x)$ are real or complex-valued continuous functions on $[a, b]$, then the initial value problem*

$$\begin{aligned} &-(p(x)y')' + q(x)y = f(x), \quad a < x < b, \\ &y(c) = c_0, \quad y'(c) = c_1, \end{aligned}$$

has a unique solution y; moreover, y extends to a continuously differentiable function on $[a, b]$ that satisfies the differential equation at $x = a$ and $x = b$.

Proof. Let y_0 for $a \leq x \leq b$ be the unique solution to the initial value problem in Theorem 81. It is convenient to present the proof in two cases: (a) The point c satisfies $a < c < b$ and (b) either $c = a$ or $c = b$.

(a) If $a < c < b$, then

$$y(x) = y_0(x)$$

for $a < x < b$ is a solution to the initial value problem in the current theorem and $y_0(x)$ extends $y(x)$ to a continuous function on $[a, b]$. Suppose $z(x)$ is also a solution to the initial value problem in the current theorem. Let a' and b' satisfy $a < a' < c < b' < b$ but otherwise be arbitrary. Then y and z are solutions to the initial value problem

$$\begin{aligned} &-(p(x)w')' + q(x)w = f(x), \quad a' \leq x \leq b', \\ &w(c) = c_0, \quad w'(c) = c_1. \end{aligned}$$

By Theorem 81 this initial value problem has a unique solution; hence, $y(x) = z(x)$ for $a' \leq x \leq b'$. Since a' and b' can be chosen arbitrarily subject to the constraint above, it follows that $y(x) = z(x)$ for $a < x < b$ and uniqueness is established.

Thus, the initial value problem in Theorem 82 has the unique solution $y(x) = y_0(x)$ for x in (a, b) and $y_0(x)$ extends $y(x)$ to a continuous function on $[a, b]$. Since the solution $y(x)$ has a continuous extension to the closed interval $[a, b]$ it follows from Lemma 79 that y is continuously differentiable on $[a, b]$ and satisfies the differential equation there. This completes the proof of the theorem in case (a).

(b) Assume $c = a$. As in case (a),

$$y(x) = y_0(x)$$

for $a \leq x < b$ solves the initial value problem in the current theorem and $y_0(x)$ extends $y(x)$ to a continuous function on $[a, b]$. Suppose $z(x)$ is also a solution to the initial value problem in the current theorem. Then y and z satisfy the initial value problem

$$\begin{aligned} &-(p(x)w')' + q(x)w = f(x), \quad a \leq x \leq b', \\ &w(a) = c_0, \quad w'(a) = c_1, \end{aligned}$$

for any b' with $a < b' < b$. By Theorem 81 this initial value problem has a unique solution; hence, $y(x) = z(x)$ for $a \leq x \leq b'$. Since b' can be chosen arbitrarily, it follows that $y(x) = z(x)$ for $a \leq x < b$ and uniqueness is established.

Thus, the initial value problem when $c = a$ has a unique solution $y(x) = y_0(x)$ for $a \leq x < b$ and $y_0(x)$ extends $y(x)$ to a continuous function on $[a, b]$. Since the solution $y(x)$ has a continuous extension to the closed interval $[a, b]$ it follows from Lemma 79 that y is continuously differentiable on $[a, b]$ and satisfies the differential there. This completes the proof of the theorem in case (b) when $c = a$. The proof is similar when $c = b$. ■

An initial value problem is called **regular** if the differential equation is regular. So Theorem 82 applies to regular initial value problems. If the coefficients $p(x)$, $q(x)$, and $f(x)$ are only continuous on the open interval $a < x < b$, the following theorem follows easily from the regular case.

Theorem 83 (Basic Existence and Uniqueness Theorem) *Fix c in (a, b) and real or complex constants c_0 and c_1. If $p(x) \neq 0$ on (a, b) and $p(x)$, $q(x)$ and $f(x)$ are real or complex-valued continuous functions on (a, b), then the initial value problem*

$$\begin{aligned} &-(p(x)y')' + q(x)y = f(x), \quad a < x < b, \\ &y(c) = c_0, \quad y'(c) = c_1, \end{aligned}$$

has a unique solution y defined on $a < x < b$.

Proof. Let $a_n = a + 1/n$ and $b_n = b - 1/n$ for positive integers n such that $a_n < c < b_n$. By Theorem 81 the initial value problem

$$-(p(x)y')' + q(x)y = f(x), \quad a_n \leq x \leq b_n,$$
$$y(c) = c_0, \quad y'(c) = c_1,$$

has a unique solution, say y_n. Define a function $y(x)$ on $a < x < b$ by

$$y(x) = y_n(x) \quad \text{if } x \text{ is in } [a_n, b_n].$$

The function y is well-defined: if x is in the domain of y_n and also in the domain of y_m we can choose the labeling so that $m > n$ in which case both y_n and y_m solve the same regular initial value problem on $a_n \leq x \leq b_n$ and by uniqueness of the solution $y_m = y_n$ on $[a_n, b_n]$. Since x belongs to $[a_n, b_n]$, it follows that $y_m(x) = y_n(x)$. This establishes that y is well-defined and that $y(x) = y_n(x)$ on $[a_n, b_n]$ for every n. Consequently, y satisfies the given initial conditions and the differential equation on (a, b).

If z also solves the initial value problem in the theorem, then y and z are both solutions to

$$-(p(x)y')' + q(x)y = f(x), \quad a_n \leq x \leq b_n,$$
$$y(c) = c_0, \quad y'(c) = c_1.$$

Since this problem has a unique solution, $z = y$ on $[a_n, b_n]$ for every n; hence, $z = y$ on (a, b). ■

It is natural to expect that solutions to initial value problems whose data is all real-valued will be real-valued. This is confirmed in

Theorem 84 *If the coefficients $p(x)$ and $q(x)$ are real-valued, $f(x)$ is real-valued, and c_0 and c_1 are real numbers in any of the initial value problems above, then the solution y to the problem is real-valued.*

Proof. If $y = y_1 + iy_2$ with y_1 and y_2 real-valued, then separating the initial value problem into real and imaginary parts reveals that y_2 satisfies the corresponding homogeneous initial value problem. The unique solution to that problem is clearly $y_2 = 0$ and $y = y_1$ is real-valued. ■

Since any solution to a regular Sturm-Liouville differential equation on a bounded open interval, solves an initial value problem, we have the following result.

Theorem 85 *If y is a solution to the regular Sturm-Liouville differential equation*

$$-(p(x)y'(x)) + q(x)y(x) = f(x)$$

for $a < x < b$, then y extends to a continuously differentiable function on the closed interval $a \leq x \leq b$ and satisfies the Sturm-Liouville equation there.

Proof. Let $c = (a + b)/2$. The solution y to the differential equation is a solution to the regular initial value problem

$$-(p(x)z')' + q(x)z = f(x), \quad a < x < b,$$
$$z(c) = y(c), \quad z'(c) = y'(c).$$

By Theorem 82 this initial value problem has a unique solution $z_0(x)$ that extends to a continuously differentiable function on $[a, b]$, still called $z_0(x)$, and that satisfies the differential equation there. By uniqueness, $y(x) = z_0(x)$ for $a < x < b$. So z_0 is the desired continuously differentiable extension of y that satisfies the differential equation at $x = a$ and $x = b$. ■

4.3.1 Basis of Solutions

Fix c in (a, b) and let $p(x) \neq 0$ and $q(x)$ be continuous on (a, b). The initial value problem

$$-(p(x)y')' + q(x)y = 0, \quad a < x < b,$$
$$y(c) = c_0, \quad y'(c) = c_1.$$

has a unique solution u determined by the choices $c_0 = 1$ and $c_1 = 0$ and a unique solution v determined by the choices $c_0 = 0$ and $c_1 = 1$. If y is any solution to a regular Sturm-Liouville differential equation $-(p(x)y')' + q(x)y = 0,\ a < x < b$, then y satisfies the initial value problem above when $c_0 = y(c)$ and $c_1 = y'(c)$. The function $z = y(c)u(x) + y'(c)v(x)$ satisfies the same initial value problem. By the uniqueness assertion in Theorem 83, $y = y(c)u(x) + y'(c)v(x)$. Thus, all solutions to the homogeneous Sturm-Liouville equation

$$-(p(x)y')' + q(x)y = 0, \quad a < x < b,$$

are expressible as linear combinations of u and v.

The solutions u and v are linearly independent on (a, b): if

$$d_0 u(x) + d_1 v(x) = 0 \quad \text{for } a < x < b,$$

set $x = c$ to obtain $d_0 = 0$ and $d_1 v(x) = 0$ for $a < x < b$. Since $v'(c) = 1$, $d_1 = 0$, and u and v are linearly independent. Therefore, the solution space of the homogeneous equation

$$-(p(x)y')' + q(x)y = 0, \quad a < x < b,$$

is two dimensional and u and v are a basis for it. Consequently, any two linearly independent solutions to the differential equation are a basis for the solution space.

The **Wronskian** of any two solutions, u and v, to the homogeneous Sturm-Liouville equation is

$$W_{u,v}(x) = \begin{vmatrix} u(x) & v(x) \\ u'(x) & v'(x) \end{vmatrix}.$$

Lemma 86 *$p(x)\, W_{u,v}(x)$ is constant for $a < x < b$.*

Proof. We check this standard result for completeness: if u and v are solutions to $-(py')' + qy = 0$, then

$$(p\, W_{u,v})'(x) = (u(pv') - (pu')v)' = u(pv')' - (pu')'v$$
$$= uqv - quv = 0$$

for $a < x < b$ and the desired conclusion follows. ■

If u and v are linearly dependent solutions of a homogeneous Sturm-Liouville equation, then there are constants d_0 and d_1, not both zero, such that

$$d_0 u(x) + d_1 v(x) = 0$$

for x in (a, b). Consequently, the linear system for d_0 and d_1

$$d_0 u(x) + d_1 v(x) = 0,$$
$$d_0 u'(x) + d_1 v'(x) = 0,$$

has a nontrivial solution for each x in (a, b); hence, its determinant $W_{u,v}(x) = 0$ for each x in (a, b). Consequently, if $W_{u,v}(x) \neq 0$ for some x in (a, b) (and, hence, for all x in (a, b)), then u and v are linearly independent on (a, b). Thus, we arrive at the familiar result that solutions u and v to a homogeneous Sturm-Liouville equation are linearly independent if and only if $W_{u,v}(x) \neq 0$ for some x in (a, b).

Suppose now that the homogeneous Sturm-Liouville differential equation

$$-(p(x)y')' + q(x)y = 0, \quad a < x < b,$$

is regular; that is, that $p(x) \neq 0$ and $q(x)$ are continuous on the closed interval $[a, b]$. Since any solution to a regular Sturm-Liouville equation extends to a continuously differentiable function on $[a, b]$ and satisfies the differential there, all the assertions established earlier in this section hold on the closed interval $[a, b]$ for regular equations.

4.3.2 Variation of Parameters

The general inhomogeneous initial value problem can be solved using any basis of solutions to the corresponding homogeneous Sturm-Liouville equation and the method of variation of parameters.

Theorem 87 (Variation of Parameters) *Fix c in $[a, b]$. The initial value problem*

$$\begin{aligned} &-(p(x)y')' + q(x)y = f(x), \quad a < x < b, \\ &y(c) = 0, \quad y'(c) = 0, \end{aligned}$$

has the unique solution

$$y(x) = A(x)u(x) + B(x)v(x),$$

where u and v are any two linearly independent solutions of the corresponding homogeneous differential equation,

$$A(x) = \int_c^x \frac{v(s)f(s)}{p(s)\,W_{u,v}(s)}\,ds,$$

$$B(x) = -\int_c^x \frac{u(s)f(s)}{p(s)\,W_{u,v}(s)}\,ds,$$

$W_{u,\,v}$ is the Wronskian of u and v, and $p(s)\,W_{u,v}(s)$ is constant.

The theorem is confirmed by the standard variation of parameters technique: substitution of $y = A(x)u(x) + B(x)v(x)$ into the differential equation $-(p(x)y')' + q(x)y = f(x)$ shows that y will be a solution of the differential equation if A' and B' satisfy the system of equations

$$\begin{aligned} uA' + vB' &= 0, \\ (pu')A' + (pv')B' &= -f. \end{aligned}$$

Solving for A' and B' and using the antiderivatives A and B in the theorem give the stated result. Since $uA' + vB' = 0$, the variation of parameters solution satisfies

$$y'(x) = A(x)u'(x) + B(x)v'(x),$$

a result that will be useful later and also makes it easy to confirm that $y'(c) = 0$.

If $u(x)$ and $v(x)$ are linearly independent solutions of the homogeneous differential equation and $y_p(x)$ is the solution to the initial value problem in the theorem, then the inhomogeneous differential equation has general solution

$$y = Du(x) + Ev(x) + y_p(x)$$

where D and E are arbitrary constants.

4.3.3 Continuous Dependence

Finally, we will need a special case of the following result on continuous dependence of solutions to initial value problems. That special case is that only the datum $q(x)$ varies. However, the strategy of the proof is the same with a few more triangle inequality estimates as more of the data is allowed to vary. Even more general results are developed in [9] for first order systems; see especially, Chapter 1, Section 7, Theorem 7.4 and its proof as it applies to a linear second order system.

Theorem 88 *Fix c in $[a, b]$ and denote the solution to the regular initial value problem*

$$-(\tilde{p}y')' + \tilde{q}y = \tilde{f}, \quad a < x < b,$$
$$y(c) = \tilde{c}_0, \quad y'(c) = \tilde{c}_1,$$

by $\tilde{y} = \tilde{y}(x)$. If $y = y(x)$ is the solution to the regular initial value problem

$$-(py')' + qy = f, \quad a < x < b,$$
$$y(c) = c_0, \quad y'(c) = c_1,$$

then given $\varepsilon > 0$ there is a $\delta > 0$ such that if

$$\|p - \tilde{p}\|_{\max}, \ \|q - \tilde{q}\|_{\max}, \ \left\|f - \tilde{f}\right\|_{\max}, \ |c_0 - \tilde{c}_0|, \ |c_1 - \tilde{c}_1| < \delta,$$

then

$$\left|y(x) - \tilde{y}(x)\right| < \varepsilon \quad \text{and} \quad \left|y'(x) - \tilde{y}'(x)\right| < \varepsilon$$

for $a \leq x \leq b$.

Proof. The solutions y and $\tilde{y}$ are in $C^1[a, b]$ as we have already established. The proof of continuous dependence on the data follows from the corresponding continuous dependence result for fixed points of a family of contraction mappings, Theorem 45. Let $\mathcal{M}$ be the linear space of points $m = (p, q, f, c_0, c_1)$ with p, q, and f in $C[a, b]$ and c_0 and c_1 in $\mathbb{C}$ with componentwise addition and scalar multiplication as the vector space operations. Equip $\mathcal{M}$ with the norm

$$\|m\|_{\mathcal{M}} = \max\left\{\|p\|_{\max}, \|q\|_{\max}, \left\|f\right\|_{\max}, |c_0|, |c_1|\right\}.$$

Convergence in this norm is uniform convergence on $[a, b]$ for the functions and the usual convergence in $\mathbb{C}$. Let S be the set of points in $\mathcal{M}$ such that

$$\|p - \tilde{p}\|_{\max} < \tilde{m}, \ \|q - \tilde{q}\|_{\max} < 1, \ |c_0 - \tilde{c}_0| < 1, \ |c_1 - \tilde{c}_1| < 1,$$

where

$$\tilde{m} = \frac{1}{2} \min_{a \leq x \leq b} \left|\tilde{p}(x)\right|.$$

Define $F\colon C[a,b]\times S\to C[a,b]$ to be the transformation that takes the pair (y,s) into the continuous function $F(y,s)$ whose value at x in $[a,b]$ is

$$F(y,s)(x)=c_0+p(c)c_1\int_c^x\frac{1}{p(u)}\,du+\int_c^x\left(\frac{1}{p(u)}\int_c^u(q(t)y(t)-f(t))\,dt\right)du,$$

where $s=(p,q,f,c_0,c_1)$ is in S. For fixed s in S, let $T_sy=F(y,s)$. That is, T_s is the integral operator corresponding to the initial value problem with data s that was used in the proof of Theorem 81. Just as in the proof of that theorem,

$$\|T_sy-T_sz\|\le\frac{(b-a)\|q\|_{\max}}{\min_{a\le x\le b}|p(x)|}\frac{1}{L}\|y-z\|_L,$$

where $\|y\|_L=\max_{a\le x\le b}e^{-L(x-a)}|y(x)|$ is a norm on $C[a,b]$ that is equivalent to the maximum norm for any choice of $L>0$. By the triangle inequality, for s in S,

$$|p(x)|\ge|\tilde{p}(x)|-|p(x)-\tilde{p}(x)|\ge\tilde{m}-\frac{\tilde{m}}{2}=\frac{\tilde{m}}{2},$$

$$\min_{a\le x\le b}|p(x)|\ge\frac{\tilde{m}}{2},$$

and, hence,

$$\|T_sy-T_sz\|_L\le\frac{2(b-a)(\|\tilde{q}\|_{\max}+1)}{\tilde{m}}\frac{1}{L}\|y-z\|_L.$$

Fix L such that

$$\frac{2(b-a)(\|\tilde{q}\|_{\max}+1)}{\tilde{m}}\frac{1}{L}<\frac{1}{2}$$

to find that

$$\|T_sy-T_sz\|_L\le\frac{1}{2}\|y-z\|_L$$

for all s in S. That is, $1/2$ is a (uniform) contraction constant for the family of contractions $\{T_s\}$ for s in S. If y_s is the unique fixed point of T_s for $s=(p,q,f,c_0,c_1)$, then y_s is the unique solution to the Sturm-Liouville initial value problem with data s. We show next that y_s varies continuously with s in S. Fix y in $C[a,b]$. Then $F(y,s)$ is a continuous function on $[a,b]$ for each s in S. For $s=(p,q,f,c_0,c_1)$ and $s_n=(p_n,q_n,f_n,c_{0,n},c_{1,n})$ a sequence in S with $s_n\to s$, the uniform convergence of p_n,q_n,f_n to p,q,f on $[a,b]$ justifies taking the limit under the integrals in the following evaluation and the uniform convergence to the limit:

$$\begin{aligned}\lim_{n\to\infty}F(y,s_n)(x)&=\lim_{n\to\infty}\Bigg(c_{0,n}+p_n(c)c_{1,n}\int_c^x\left(\frac{1}{p_n(u)}\right)du\\&\quad+\int_c^x\left(\frac{1}{p_n(u)}\int_c^u(q_n(t)y(t)-f_n(t))\,dt\right)du\Bigg)\\&=c_0+p(c)c_1\int_c^x\left(\frac{1}{p(u)}\right)du\\&\quad+\int_c^x\left(\frac{1}{p(u)}\int_c^u(q(t)y(t)-f(t))\,dt\right)du\\&=F(y,s)(x)\end{aligned}$$

uniformly for x in $[a,b]$. (See Theorem 16.) That is, for each fixed y,

$$\|F(y,s_n)-F(y,s)\|_{\max}\to0\text{ as }n\to\infty.$$

Thus, $F(y, s)$ is continuous in s for each fixed y. By Theorem 45 the unique fixed point y_s varies continuously with s in S. Thus, given $\varepsilon > 0$ there is a $\delta_0 > 0$ such that

$$\|p - \tilde{p}\|_{\max}, \ \|q - \tilde{q}\|_{\max}, \ \left\|f - \tilde{f}\right\|_{\max}, \ |c_0 - \tilde{c}_0|, \ |c_1 - \tilde{c}_1| < \delta_0,$$

implies $\left\|y - \tilde{y}\right\|_{\max} < \varepsilon$ where y is the solution to the initial value problem with data s and $\tilde{y}$ is the solution to the problem with data $\tilde{s}$.

It remains to establish that a corresponding δ_1 exists so that $\left\|y' - \tilde{y}'\right\|_{\max} < \varepsilon$ as well. To this end, integrate the respective initial value problems with respective data s and $\tilde{s}$ in S to obtain

$$p(x)y'(x) = p(c)c_1 + \int_c^x \big(q(u)y(u) - f(u)\big)\, du,$$

$$\tilde{p}(x)\tilde{y}'(x) = \tilde{p}(c)\tilde{c}_1 + \int_c^x \big(\tilde{q}(u)\tilde{y}(u) - \tilde{f}(u)\big)\, du,$$

and

$$\begin{aligned} p(x)y'(x) - p(x)\tilde{y}'(x) &= \tilde{p}(x)\tilde{y}'(x) - p(x)\tilde{y}'(x) + p(x)y'(x) - \tilde{p}(x)\tilde{y}'(x) \\ &= \tilde{p}(x)\tilde{y}'(x) - p(x)\tilde{y}'(x) \\ &\quad + p(c)c_1 - \tilde{p}(c)\tilde{c}_1 \\ &\quad + \int_c^x \big(q(u)y(u) - f(u)\big)\, du \\ &\quad - \int_c^x \big(\tilde{q}(u)\tilde{y}(u) - \tilde{f}(u)\big)\, du. \end{aligned}$$

Now

$$\left|\tilde{p}(x)\tilde{y}'(x) - p(x)\tilde{y}'(x)\right| \le \left\|\tilde{y}'\right\|_{\max} \|p - \tilde{p}\|_{\max},$$

and

$$\begin{aligned} \left|p(c)c_1 - \tilde{p}(c)\tilde{c}_1\right| &\le \left|p(c)c_1 - \tilde{p}(c)\tilde{c}_1\right| + \left|\tilde{p}(c)\tilde{c} - \tilde{p}(c)\tilde{c}_1\right| \\ &\le \|p - \tilde{p}\|_{\max}(|\tilde{c}_1| + 1) + \tilde{p}(c)|c_1 - \tilde{c}_1|, \end{aligned}$$

and

$$\begin{aligned} &\left|\int_c^x (q(u)y(u) - f(u))\, du - \int_c^x \big(\tilde{q}(u)\tilde{y}(u) - \tilde{f}(u)\big)\, du\right| \\ &\quad \le \left|\int_c^x \left|q(u)y(u) - q(u)\tilde{y}(u) + q(u)\tilde{y}(u) - \tilde{q}(u)\tilde{y}(u)\right|\, du\right| \\ &\quad\quad + \left|\int_c^x \left|f(u) - \tilde{f}(u)\right|\, du\right| \\ &\quad \le (b - a)\Big((\|\tilde{q}\| + 1)\left\|y - \tilde{y}\right\|_{\max} + \left\|\tilde{y}\right\|_{\max}\|q - \tilde{q}\|_{\max} + \|f - \tilde{f}\|_{\max}\Big). \end{aligned}$$

Combining estimates there is a constant $\tilde{M}$ depending on the initial value problem with data $\tilde{s}$ such that

$$\begin{aligned} \left|p(x)\right|\left|y'(x) - \tilde{y}'(x)\right| \le \tilde{M} \max \{&\left\|y - \tilde{y}\right\|_{\max}, \|p - \tilde{p}\|_{\max}, \|q - \tilde{q}\|_{\max}, \\ &\|f - \tilde{f}\|_{\max}, |c_1 - \tilde{c}_1|\} \end{aligned}$$

for all x in $[a, b]$. Since s is in S, $|p(x)| \geq \tilde{m}/2$, and

$$|y'(x) - \tilde{y}'(x)| \leq (2\tilde{M}/\tilde{m}) \max\{\|y - \tilde{y}\|_{\max}, \|p - \tilde{p}\|_{\max}, \|q - \tilde{q}\|_{\max}, \\ \|f - \tilde{f}\|_{\max}, |c_1 - \tilde{c}_1|\}$$

for all x in $[a, b]$. In view of the continuous dependence result already established for $\|y - \tilde{y}\|_{\max}$, given $\varepsilon > 0$ there is a $\delta > 0$ so that $\|y - \tilde{y}\|_{\max} < \varepsilon$ and the right member of the foregoing inequality is less than ε if

$$\|p - \tilde{p}\|_{\max}, \|q - \tilde{q}\|_{\max}, \|f - \tilde{f}\|_{\max}, |c_0 - \tilde{c}_0|, |c_1 - \tilde{c}_1| < \delta.$$

Consequently,

$$\|p - \tilde{p}\|_{\max}, \|q - \tilde{q}\|_{\max}, \|f - \tilde{f}\|_{\max}, |c_0 - \tilde{c}_0|, |c_1 - \tilde{c}_1| < \delta$$

implies

$$|y(x) - \tilde{y}(x)| < \varepsilon \quad \text{and} \quad |y'(x) - \tilde{y}'(x)| < \varepsilon$$

for all x in $[a, b]$ and the continuous dependence proof is complete. ■

4.4 BVPs and EVPs - Examples

In Chapter 1 we made a survey of regular and singular Sturm-Liouville boundary value problems and eigenvalue problems and how they arise from science and engineering problems. In the sections that follow, we develop the basic theory of regular Sturm-Liouville boundary value problems and eigenvalue problems, including a careful development of Green's functions and their characteristic properties. Most problems of interest cannot be solved explicitly. Consequently, in Chapter 7 we present effective numerical methods for calculating eigenvalues and eigenfunctions.

A principal tool in our study will be the Green's function, when it exists, of a Sturm-Liouville differential operator and its accompanying boundary conditions. The lead example in Chapter 3 showed how to convert a particular Sturm-Liouville eigenvalue problem to an equivalent eigenvalue problem for an integral operator with a symmetric kernel. That kernel is the Green's function for the Sturm-Liouville differential operator and the given boundary conditions.

It is natural to expect that a Sturm-Liouville boundary value problem or eigenvalue problem can be solved by integration of the differential equation, either explicitly or in principle. This is indeed the case. In the examples that follow, the integrations can be carried out explicitly and lead naturally to an integral representation of the solution in terms of a Green's function. Subsequent sections of the chapter develop corresponding results for general Sturm-Liouville problems and cover the case where explicit solutions are not available.

In the examples that follow and are revisited throughout the chapter, we point out some important properties of Green's functions and typical behavior shared by many Sturm-Liouville eigenvalue problems that come up in applications. See Section 1.10 for a physical motivation for Green's functions. The general definition of a Green's function is given in the next section. The examples also show that initial value problems play an important background role and point to the shooting method in Chapter 7 used to obtain accurate numerical approximations for eigenvalues and eigenfunctions.

Example 1a. Let $a > 0$ and $l > 0$ be fixed and $f(x)$ be continuous on $[0, l]$. Solve the Sturm-Liouville boundary value problem

$$-y'' + ay = f(x), \quad 0 < x < l,$$
$$y(0) = 0, \ y(l) = 0.$$

$Ly = -y'' + ay$ is the Sturm-Liouville differential operator and $y(0) = 0, y(l) = 0,$ called Dirichlet boundary conditions, are the boundary conditions associated with L.

The homogeneous equation $-y'' + ay = 0$ has linearly independent exponential solutions $e^{\sqrt{a}x}$ and $e^{-\sqrt{a}x}$ or, alternatively, the linearly independent solutions

$$\frac{e^{\sqrt{a}x} + e^{-\sqrt{a}x}}{2} = \cosh(\sqrt{a}x) \quad \text{and} \quad \frac{e^{\sqrt{a}x} - e^{-\sqrt{a}x}}{2} = \sinh(\sqrt{a}x).$$

The boundary value problem can be solved by variation of parameters; see Theorem 87: by that theorem applied with $u = \cosh(\sqrt{a}x)$ and $v = \sinh(\sqrt{a}x)$, the inhomogeneous differential equation has the particular solution

$$y_p(x) = -\frac{1}{\sqrt{a}} \int_0^x \sinh\sqrt{a}(x - s) f(s)\, ds.$$

The general solution of $-y'' + ay = f$ is

$$y = A\cosh(\sqrt{a}x) + B\sinh(\sqrt{a}x) - \frac{1}{\sqrt{a}} \int_0^x \sinh\sqrt{a}(x - s) f(s)\, ds,$$

where A and B are arbitrary constants, and it will satisfy the Dirichlet boundary conditions if and only if $y(0) = A = 0$ and

$$y(l) = B\sinh(\sqrt{a}l) - \frac{1}{\sqrt{a}} \int_0^l \sinh\sqrt{a}(l - s) f(s)\, ds = 0,$$

$$B = \frac{1}{\sqrt{a}\sinh(\sqrt{a}l)} \int_0^l \sinh\sqrt{a}(l - s) f(s)\, ds.$$

Thus, the boundary problem has the unique solution

$$y = \frac{\sinh(\sqrt{a}x)}{\sqrt{a}\sinh(\sqrt{a}l)} \int_0^l \sinh\sqrt{a}(l - s) f(s)\, ds - \frac{1}{\sqrt{a}} \int_0^x \sinh\sqrt{a}(x - s) f(s)\, ds.$$

For later purposes, we express the solution in the following way:

$$y(x) = \int_x^l \frac{\sinh(\sqrt{a}x)}{\sqrt{a}\sinh(\sqrt{a}l)} \sinh(\sqrt{a}(l - s)) f(s)\, ds$$
$$+ \int_0^x \left(\frac{\sinh(\sqrt{a}x)\sinh(\sqrt{a}(l - s))}{\sqrt{a}\sinh(\sqrt{a}l)} - \frac{1}{\sqrt{a}} \sinh\sqrt{a}(x - s) \right) f(s)\, ds.$$

Use of the hyperbolic identity

$$\sinh(\alpha - \beta) = \sinh\alpha\cosh\beta - \cosh\alpha\sinh\beta$$

in the factor multiplying $f(s)$ in second integrand expresses the solution to the boundary value problem as

$$y(x) = \int_x^l \frac{\sinh(\sqrt{a}x)\sinh(\sqrt{a}(l-s))}{\sqrt{a}\sinh(\sqrt{a}l)} f(s)\,ds + \int_0^x \frac{\sinh\sqrt{a}s\sinh\sqrt{a}(l-x)}{\sqrt{a}\sinh(\sqrt{a}l)} f(s)\,ds$$

or

$$y(x) = \int_0^l g(x, s) f(s)\,ds$$

where

$$g(x, s) = \frac{1}{\sqrt{a}\sinh\sqrt{a}l}\begin{cases}\sinh\sqrt{a}x\sinh\sqrt{a}(l-s), & 0 \le x \le s \le l\\ \sinh\sqrt{a}s\sinh\sqrt{a}(l-x), & 0 \le s \le x \le l\end{cases}.$$

The function $g(x, s)$ is the Green's function for the differential operator $Ly = -y'' + ay$ and the boundary conditions $y(0) = 0$ and $y(l) = 0$.

Notice the following important features of the Green's function. The Green's function is continuous on $[0, l] \times [0, l]$,

$$g(0, s) = 0, \ \ g(l, s) = 0, \ \ g(x, s) = g(s, x),$$

and, on each interval $s \le x$ and $x \le s$, it is a product of two factors, each a solution to the homogeneous equation $Ly = 0$. Thus, g regarded as a function of x for fixed s satisfies $Lg = 0$ for $x \neq s$, and satisfies $Lg = 0$ as a function of $s \neq x$ for fixed x. We will see that this is typical behavior for Green's functions of Strum-Liouville boundary value problems when the boundary conditions are separated (each boundary condition involves only one endpoint of the underlying interval). In this example, the Green's function is positive on $0 < x, s < l$. This is typical of many problems with homogeneous Dirichlet boundary conditions.

Example 1b. Let $a > 0$ and $l > 0$ be fixed. Solve the Sturm-Liouville eigenvalue problem

$$-y'' + ay = \lambda y, \ \ 0 < x < l,$$
$$y(0) = 0, \ \ y(l) = 0.$$

As in Example 1a, $Ly = -y'' + ay$ is the Sturm-Liouville differential operator and $y(0) = 0$, $y(l) = 0$ are the associated boundary conditions.

We will give two solutions. First we solve the eigenvalue problem by straightforward analytic means. Second we use a shooting method that illustrates the theoretical underpinnings used to accurately estimate eigenvalues and eigenvectors in the typical situation where exact solutions are not available.

Express the eigenvalue problem as

$$y'' + (\lambda - a)y = 0, \ \ 0 < x < l,$$
$$y(0) = 0, \ \ y(l) = 0.$$

It turns out that all the eigenvalues of this problem are real because the Green's function is symmetric. Although this can be established directly by elementary means, we prefer just to use this fact. We will do the same in Examples 2b, 3b, and 4b.

If $\lambda - a \le 0$, then any solution y to the problem above is $y = 0$ by the maximum principle (Theorem 48(b) applied to y and $-y$). Thus, any eigenvalue satisfies $\lambda > a$. For such λ the differential equation $y'' + (\lambda - a)y = 0$ has general solution

$$y = A\cos\sqrt{\lambda - a}x + B\sin\sqrt{\lambda - a}x$$

and this solution will satisfy the boundary conditions if and only if $y(0) = A = 0$ and

$$y(l) = B \sin \sqrt{\lambda - a}\,l = 0.$$

Since $A = 0$, y will be a nontrivial solution and λ and eigenvalue if and only if $B \neq 0$ and $\sin \sqrt{\lambda - a}\,l = 0$; that is, the eigenvalues are

$$\lambda = \lambda_n = \left(\frac{n\pi}{l}\right)^2 + a$$

and the corresponding eigenfunctions are the nonzero multiples of

$$y_n(x) = \sin \frac{n\pi x}{l}$$

for $n = 1, 2, \ldots$. This is typical of Sturm-Liouville eigenvalue problems with separated boundary conditions. Each eigenvalue has only one corresponding eigenfunction up to a constant multiple.

The second approach to solving the eigenvalue problem is a theoretical shooting method. (Note that $\lambda = a$ is not an eigenvalue.) We start with the simpler initial value problem

$$-y'' + ay = \lambda y, \quad 0 < x < l,$$
$$y(0) = 0, \quad y'(0) = 1,$$

and try to determine λ, the shooting parameter, so that the solution to the initial value problem also solves the eigenvalue problem. The general solution to the differential equation,

$$y = A \cos \sqrt{\lambda - a}\,x + B \sin \sqrt{\lambda - a}\,x,$$

will satisfy the initial conditions if and only if $y(0) = A = 0$ and

$$y'(0) = B\sqrt{\lambda - a} = 1.$$

Thus, the initial value problem has solution

$$y(x) = \frac{\sin \sqrt{\lambda - a}\,x}{\sqrt{\lambda - a}}.$$

This solution to the initial value problem will be an eigenfunction and λ a corresponding eigenvalue if and only if $y(l) = 0$. Consequently, the eigenvalues are the roots of the equation $\sin \sqrt{\lambda - a}\,l = 0$; that is, the eigenvalues are $\lambda = \lambda_n = a + (n\pi/l)^2$ and the corresponding eigenfunctions are the nonzero multiples of $\sin(n\pi x/l)$.

Example 2a. Let $a < 0$ and $l > 0$ be fixed and $f(x)$ be continuous on $[0, l]$. Solve the Sturm-Liouville boundary value problem

$$-y'' + ay = f(x), \quad 0 < x < l,$$
$$y(0) = 0, \quad y(l) = 0.$$

$Ly = -y'' + ay$ is the Sturm-Liouville differential operator and $y(0) = 0$, $y(l) = 0$ are the associated boundary conditions.

The solution to the boundary value problem proceeds as in Example 1a with one notable exception, a Green's function does not always exist. The homogeneous equation $-y'' + ay = 0$

has linearly independent exponential solutions $e^{i\sqrt{-a}x}$ and $e^{-i\sqrt{-a}x}$ or, more conveniently, has the pair of linearly independent solutions

$$\frac{e^{i\sqrt{-a}x}+e^{-i\sqrt{-a}x}}{2}=\cos\big(\sqrt{-a}x\big) \quad \text{and} \quad \frac{e^{i(\sqrt{-a})x}-e^{-i\sqrt{-a}x}}{2i}=\sin\big(\sqrt{-a}x\big).$$

Use of variation of parameters as in Example 1a with $u=\cos(\sqrt{-a}x)$ and $v=\sin(\sqrt{-a}x)$ leads to

$$y=A\cos(\sqrt{-a}x)+B\sin(\sqrt{-a}x)-\frac{1}{\sqrt{-a}}\int_0^x \sin(\sqrt{-a}(x-s))f(s)\,ds$$

as the general solution to $-y''+ay=f(x)$. This solution will satisfy the boundary conditions if and only if $y(0)=A=0$ and

$$B\sin(\sqrt{-a}l)-\frac{1}{\sqrt{-a}}\int_0^l \sin(\sqrt{-a}(l-s))f(s)\,ds=0,$$

$$B=\frac{1}{\sqrt{-a}\sin(\sqrt{-a}l)}\int_0^l \sin(\sqrt{-a}(l-s))f(s)\,ds,$$

provided $\sin(\sqrt{-a}l)\neq 0$. If this inequality holds, then the boundary value problem has the unique solution

$$y(x)=\frac{\sin(\sqrt{-a}x)}{\sqrt{-a}\,\sin(\sqrt{-a}l)}\int_0^l \sin\sqrt{-a}(l-s)f(s)\,ds-\frac{1}{\sqrt{-a}}\int_0^x \sin\sqrt{-a}(x-s)f(s)\,ds.$$

$$y(x)=\int_x^l \frac{\sin(\sqrt{-a}x)\,\sin(\sqrt{-a}(l-s))}{\sqrt{-a}\,\sin(\sqrt{-a}l)}f(s)\,ds$$
$$+\int_0^x\left(\frac{\sin(\sqrt{-a}x)\,\sin(\sqrt{-a}(l-s))}{\sqrt{-a}\,\sin(\sqrt{-a}l)}-\frac{1}{\sqrt{-a}}\sin(\sqrt{-a}(x-s))\right)f(s)\,ds$$

Manipulating this solution much as we did for the solution to Example 1a leads to

$$y(x)=\int_x^l \frac{\sin(\sqrt{-a}x)\,\sin(\sqrt{-a}(l-s))}{\sqrt{-a}\,\sin(\sqrt{-a}l)}f(s)ds$$
$$+\int_0^x \frac{\sin(\sqrt{-a}s)\,\sin(\sqrt{-a}(l-x))}{\sqrt{-a}\,\sin(\sqrt{-a}l)}f(s)ds$$

or

$$y(x)=\int_0^l g(x,s)f(s)ds$$

where

$$g(x,s)=\frac{1}{\sqrt{-a}\sin\big(\sqrt{-a}l\big)}\begin{cases}\sin\big(\sqrt{-a}x\big)\sin\big(\sqrt{-a}(l-s)\big), & 0\le x\le s\le l\\ \sin\big(\sqrt{-a}s\big)\sin\sqrt{-a}(l-x), & 0\le s\le x\le l\end{cases}.$$

The function $g(x,s)$ is the Green's function for the differential operator $Ly=-y''+ay$ and the boundary conditions $y(0)=0$ and $y(l)=0$.

Just as in Example 1a, the Green's function is continuous on $[0,l]\times[0,l]$,

$$g(0,s)=0,\ g(l,s)=0,\ g(x,s)=g(s,x),$$

and, on each interval $s \le x$ and $x \le s$, it is a product of two factors, each a solution to the homogeneous equation $Ly = 0$. Thus, g regarded as a function of x for fixed s satisfies $Lg = 0$ for $x \ne s$, and satisfies $Lg = 0$ as a function of $s \ne x$ for fixed x. In this example, the Green's function is positive on $0 < x,\ s < l$ only when $l \le \pi/\sqrt{-a}$.

The discussion that led to the Green's function assumed that $\sin(\sqrt{-a}l) \ne 0$. If this is not the case, the equation

$$B\sin(\sqrt{-a}l) - \frac{1}{\sqrt{-a}}\int_0^l \sin(\sqrt{-a}(l-s))f(s)\ ds = 0$$

used to determine B needs a closer look. If $\sin(\sqrt{-a}l) = 0$, that is, if $l = n\pi/\sqrt{-a}$ or $l\sqrt{-a} = n\pi$ for some $n = 1, 2, \ldots$, then the equation above reduces to

$$B \cdot 0 + \frac{(-1)^n}{\sqrt{-a}}\int_0^l \sin(\sqrt{-a}s)f(s)\ ds = 0$$

and either has no solution if

$$\int_0^l \sin(\sqrt{-a}s)f(s)\ ds \ne 0$$

or is satisfied for any B if

$$\int_0^l \sin(\sqrt{-a}s)f(s)\ ds = 0.$$

In the first case the boundary value problem has no solution and in the second case it has infinitely many solutions, namely,

$$y = B\sin(\sqrt{-a}x) - \frac{1}{\sqrt{-a}}\int_0^x \sin(\sqrt{-a}(x-s))f(s)\ ds$$

for any choice of B and with $l\sqrt{-a} = n\pi$ for some $n = 1, 2, \ldots$. Both of these possibilities preclude the possibility that there is a function $g(x, s)$ for which

$$y(x) = \int_0^l g(x, s)f(s)\ ds$$

is the only solution to the boundary value problem.

Consequently, in Example 2a there is a Green's function if and only if $l\sqrt{-a} \ne n\pi$ for any $n = 1, 2, 3, \ldots$.

Example 2b. Let $a < 0$ and $l > 0$ be fixed. Solve the Sturm-Liouville eigenvalue problem

$$\begin{gathered} -y'' + ay = \lambda y,\ \ 0 < x < l, \\ y(0) = 0,\ \ y(l) = 0. \end{gathered}$$

As in Example 2a, $Ly = -y'' + ay$ is the Sturm-Liouville differential operator and $y(0) = 0$, $y(l) = 0$ are the associated boundary conditions.

The solution is the same as in Example 1b. Express the eigenvalue problem as

$$\begin{gathered} y'' + (\lambda - a)y = 0,\ \ 0 < x < l, \\ y(0) = 0,\ \ y(l) = 0. \end{gathered}$$

As in the previous solution we assume that all the eigenvalues are real and use the maximum principle to find that any eigenvalue λ satisfies $\lambda > a$. For such λ the differential equation $y'' + (\lambda - a)y = 0$ has general solution

$$y = A\cos\sqrt{\lambda - a}x + B\sin\sqrt{\lambda - a}x$$

and this solution will satisfy the boundary conditions if and only if $y(0) = A = 0$ and

$$y(l) = B\sin\sqrt{\lambda - a}l = 0.$$

Since $A = 0$, y will be a nontrivial solution and λ and eigenvalue if and only if $B \neq 0$ and $\sin\sqrt{\lambda - a}l = 0$; that is, the eigenvalues are

$$\lambda = \lambda_n = a + \left(\frac{n\pi}{l}\right)^2$$

and the corresponding eigenfunctions are the nonzero multiples of

$$y_n(x) = \sin\frac{n\pi x}{l}$$

for $n = 0, 1, 2, \ldots$.

In contrast to the case $a > 0$, when $a < 0$ a finite number of the eigenvalues $\lambda_n = (n\pi/l)^2 + a$ may be negative, depending on the choice of l. If $l > \pi/\sqrt{-a}$, equivalently $a + (\pi/l)^2 > 0$, then all the eigenvalues are positive and by inspection $g(x, s) \geq 0$ while if $a + (\pi/l)^2 < 0$ a finite number of the eigenvalues are negative and by inspection $g(x, s)$ is not nonnegative. In either case, $a > 0$ or $a < 0$, $\lambda_n \to \infty$ as $n \to \infty$. It is typical of Sturm-Liouville eigenvalue problems that at most a finite number of the eigenvalues are negative.

In Example 2a, the Green's function for the boundary value problem exists if and only if $l\sqrt{-a} \neq n\pi$, equivalently $a + (n\pi/l)^2 \neq 0$. So the Green's function exists if and only if 0 is not an eigenvalue of the corresponding eigenvalue problem. We will show later that a Sturm-Liouville boundary value problem has a Green's function if and only if $\lambda = 0$ is not an eigenvalue of the corresponding eigenvalue problem.

Example 3a. Let $l > 0$ be fixed and $f(x)$ be continuous on $[0, l]$. Solve the Sturm-Liouville boundary value problem

$$-y'' = f(x), \quad 0 < x < l,$$
$$y(0) = 0, \quad y(l) = 0.$$

Here $Ly = -y''$ is the Sturm-Liouville differential operator and $y(0) = 0$, $y(l) = 0$ are the associated boundary conditions. This is the case $a = 0$ in the context of Examples 1a and 2a.

The general solution to the differential equation $-y'' = f$ is

$$y = A + Bx + \int_0^x (s - x)f(s)\, ds$$

via variation of parameters with $u = 1$ and $v = x$ as in Examples 1a and 2a. This solution satisfies the boundary conditions if and only if $y(0) = A = 0$ and

$$y(l) = Bl + \int_0^l (s - l)f(s)\, ds = 0,$$

$$B = \frac{1}{l}\int_0^l (l - s)f(s)\, ds,$$

and the boundary value problem has the unique solution

$$\begin{aligned} y(x) &= \int_0^l \frac{x(l-s)}{l} f(s)\, ds + \int_0^x (s-x) f(s)\, ds \\ &= \int_x^l \frac{x(l-s)}{l} f(s)\, ds + \int_0^x \left(\frac{x(l-s)}{l} + (s-x) \right) f(s)\, ds \\ &= \int_x^l \frac{x(l-s)}{l} f(s)\, ds + \int_0^x \frac{s(l-x)}{l} f(s)\, ds \\ &= \int_0^l g(x,\, s) f(s)\, ds, \end{aligned}$$

where

$$g(x,\, s) = \frac{1}{l} \begin{cases} x(l-s), & 0 \le x \le s \le l \\ s(l-x), & 0 \le s \le x \le l \end{cases}$$

is the Green's function for the differential operator $Ly = -y''$ and the boundary conditions $y(0) = 0,\ y(l) = 0$. Notice that the Green's function for Example 3a has all the general properties of the Green's function in Example 1a.

Example 3b. Let $l > 0$ be fixed. Solve the Sturm-Liouville eigenvalue problem

$$\begin{aligned} -y'' &= \lambda y, \quad 0 < x < l, \\ y(0) &= 0, \quad y(l) = 0, \end{aligned}$$

for the differential operator $Ly = -y''$ with the boundary conditions $y(0) = 0,\ y(l) = 0$.

As usual, we assume that all the eigenvalues are real. Just as in Examples 1b and 2b, express the eigenvalue problem as

$$\begin{aligned} y'' + \lambda y = 0, \quad 0 < x < l, \\ y(0) = 0, \quad y(l) = 0, \end{aligned}$$

and use the maximum principle to see that any eigenvalue $\lambda > 0$. The general solution to the differential equation $y'' + \lambda y = 0$,

$$y = A \cos(\sqrt{\lambda} x) + B \sin(\sqrt{\lambda} x),$$

satisfies the boundary conditions if and only if $y(0) = A = 0$ and

$$B \sin(\sqrt{\lambda} l) = 0.$$

Since $A = 0$, λ will be an eigenvalue if and only if $\sin(\sqrt{\lambda} l) = 0$; that is $\lambda = (n\pi/l)^2$ and $\sin(n\pi x/l)$ is not identically zero. Hence, the eigenvalues are

$$\lambda = \lambda_n = \left(\frac{n\pi}{l} \right)^2$$

for $n = 1, 2, 3, \ldots$ and the corresponding eigenfunctions are nonzero multiples of

$$y_n(x) = \sin\left(\frac{n\pi x}{l} \right).$$

Example 4a. Fix $l > 0$ and let $f(x)$ be continuous on $[0, l]$. Solve the Sturm-Liouville boundary value problem

$$-y'' = f(x), \quad 0 < x < l,$$
$$y(0) - y'(0) = 0, \quad y(l) + y'(l) = 0,$$

for the Sturm-Liouville differential operator $Ly = -y''$ with the separated boundary conditions $y(0) - y'(0) = 0$, $y(l) + y'(l) = 0$.

We start with the simpler initial value problem

$$-y'' = f(x), \quad 0 < x < l,$$
$$y(0) = C, \quad y'(0) = C,$$

which satisfies the boundary condition $y(0) - y'(0) = 0$ for any choice of C. We seek to determine C so that the solution to the initial value problem also solves the boundary value problem; that is, it satisfies the boundary condition at $x = l$. The initial value problem has solution

$$y(x) = C(1 + x) - \int_0^x (x - s)f(s)\, ds$$

by use of variation of parameters with $u = 1$ and $v = x$ as in Examples 1a, 2a, and 3a. The boundary condition at $x = l$ will be satisfied if and only if

$$y(l) + y'(l) = C(1 + l) - \int_0^l (l - s)f(s)\, ds + C - \int_0^l f(s)\, ds = 0,$$

$$C = \frac{1}{l+2}\int_0^l (l + 1 - s)f(s)\, ds,$$

and the solution to the boundary value problem is

$$y(x) = \int_0^l \frac{(1 + x)(l + 1 - s)}{l + 2} f(s)\, ds - \int_0^x (x - s)f(s)\, ds.$$

Since

$$\frac{(1 + x)(l + 1 - s)}{l + 2} - (x - s) = \frac{(1 + s)(l + 1 - x)}{l + 2},$$

the solution can be expressed as

$$y(x) = \int_x^l \frac{(1 + x)(l + 1 - s)}{l + 2} f(s)\, ds + \int_0^x \frac{(1 + s)(l + 1 - x)}{l + 2} f(s)\, ds$$

or

$$y(x) = \int_0^l g(x, s)f(s)\, ds$$

where

$$g(x, s) = \frac{1}{l+2}\begin{cases} (1 + x)(l + 1 - s), & 0 \le x \le s \le l, \\ (1 + s)(l + 1 - x), & 0 \le s \le x \le l, \end{cases}$$

is the Green's function for $Ly = -y''$ and the boundary conditions $y(0) - y'(0) = 0$, $y(l) + y'(l) = 0$.

The Green's function has all the attributes pointed out at the end of the solution to Example 1a and, in this case, $g(x, s) > 0$ for all x and s in $0 \le x, s \le l$.

Example 4b. Fix $l > 0$. Solve the eigenvalue problem

$$-y'' = \lambda y, \quad 0 < x < l,$$
$$y(0) - y'(0) = 0, \quad y(l) + y'(l) = 0,$$

for the Sturm-Liouville differential operator $Ly = -y''$ with the separated boundary conditions $y(0) - y'(0) = 0$, $y(l) + y'(l) = 0$.

As in Example 1b the eigenvalues of this problem are known to be real and we will use this fact. They are also positive, a fact we will establish shortly, but will use in the meanwhile. Express the eigenvalue problem as

$$y'' + \lambda y = 0,$$
$$y(0) - y'(0) = 0, \quad y(l) + y'(l) = 0.$$

The differential equation $y'' + \lambda y = 0$ has general solution

$$y(x) = A\cos\sqrt{\lambda}x + B\sin\sqrt{\lambda}x$$

and

$$y'(x) = -A\sqrt{\lambda}\sin\sqrt{\lambda}x + B\sqrt{\lambda}\cos\sqrt{\lambda}x.$$

The general solution will satisfy the boundary conditions if and only if

$$\begin{cases} A - \sqrt{\lambda}B = 0, \\ \left(\cos\sqrt{\lambda}l - \sqrt{\lambda}\sin\sqrt{\lambda}l\right)A + \left(\sin\sqrt{\lambda}l + \sqrt{\lambda}\cos\sqrt{\lambda}l\right)B = 0. \end{cases}$$

Nontrivial solutions for A and B (and hence for $y(x)$) exist if and only if the determinant of the system is zero,

$$(1 - \lambda)\sin\sqrt{\lambda}l + 2\sqrt{\lambda}\cos\sqrt{\lambda}l = 0.$$

If $\cos\sqrt{\lambda}l = 0$ for some eigenvalue λ, then $\sin\sqrt{\lambda}l \neq 0$ and the eigenvalue is $\lambda = 1$. So all eigenvalues different from 1 satisfy $\cos\sqrt{\lambda}l \neq 0$ and are the roots of the equation

$$\tan\sqrt{\lambda}l = \frac{2\sqrt{\lambda}}{\lambda - 1}.$$

If $\lambda = 1$ is an eigenvalue, then $\cos l = 0$ and $l = (2n+1)\pi/2$ for some integer $n \geq 0$. Conversely, if l has this form, then $\lambda = 1$ is an eigenvalue. In summary, the eigenvalues are the roots of the equation

$$\tan\sqrt{\lambda}l = \frac{2\sqrt{\lambda}}{\lambda - 1}$$

and, in case $l = (2n+1)\pi/2$ for some integer $n \geq 0$, the additional eigenvalue $\lambda = 1$.

The fact that the eigenvalues λ are all positive follows from the maximum principle. Suppose $\lambda \leq 0$ and that y satisfies $y'' + \lambda y = 0$ and $y(0) - y'(0) = 0$, $y(l) + y'(l) = 0$. Assume $y(0) > 0$, then $y'(0) = y(0) > 0$ and $y(0)$ is not the positive maximum of y on $[0, l]$. Since y is continuous on $[0, l]$ it has a positive maximum in $0 < x \leq l$. The positive maximum cannot occur at $x = l$ because $y(l) > 0$ implies $y'(l) = -y(l) < 0$ and $y(l)$ cannot be the positive maximum of y on $[0, l]$. So, if $y(0) > 0$, then y achieves its positive maximum at an interior point of $[0, l]$. By the maximum principle, Theorem 48(a), y must be a constant on $[0, l]$. But then $y(0) = y'(0)$ implies that $y(0) = 0$ and $y = 0$ on $[0, l]$. So any nontrivial solution y satisfies $y(0) \leq 0$. Likewise, any nontrivial solution satisfies $y(l) \leq 0$. But then y satisfies $y'' + \lambda y = 0$, $y(0) \leq 0$, $y(l) \leq 0$ and again by the maximum principle

$y \leq 0$ on $[0, l]$. Now $z = -y$ satisfies $z'' + \lambda z = 0$, $z(0) - z'(0) = 0$, $z(l) + z'(l) = 0$ and, hence, $z = -y \leq 0$ on $[0, l]$. Consequently, $y = 0$ on $[0, l]$ and $\lambda \leq 0$ is not an eigenvalue of the eigenvalue problem.

So all the eigenvalues are positive. Plot the graphs of $\tan \sqrt{\lambda} l$ and $2\sqrt{\lambda}/(\lambda - 1)$ on the same axes to see that the eigenvalues satisfy

$$0 < \lambda_1 < \lambda_2 < \cdots < \lambda_n \to \infty$$

as $n \to \infty$. In fact, the plot reveals that $\lambda_n \approx n\pi$ with the accuracy increasing as n increases. The relation $A - \sqrt{\lambda_n} B = 0$ with λ_n an eigenvalue shows that the corresponding eigenfunctions are the nonzero multiples of

$$y_n(x) = \sqrt{\lambda_n} \cos \sqrt{\lambda_n} x + \sin \sqrt{\lambda_n} x.$$

Once again we see that each eigenvalue has a unique eigenfunction up to a constant multiple and that the eigenvalues $\lambda_n \to \infty$ as $n \to \infty$.

A final observation is in order. It provides the key to the systematic study of Sturm-Liouville eigenvalue problems in the typical case in which explicit solutions are not available. To be concrete, consider the Green's function $g(x, s)$ determined by the differential operator $Ly = -y'' + ay$ and the boundary conditions $y(0) = 0$ and $y(l) = 0$ in Example 1a. The solution to $Ly = f$ and $y(0) = 0$, $y(l) = 0$ is

$$y(x) = \int_0^l g(x, s) f(s)\, ds.$$

The corresponding eigenvalue problem is $Ly = \lambda y$ and $y(0) = 0$, $y(l) = 0$. So if λ, y is an eigenvalue, eigenfunction pair of this eigenvalue problem, then y satisfies the integral equation

$$y(x) = \lambda \int_0^l g(x, s) y(s)\, ds.$$

To see this, just set $f = \lambda y$ in the previous formula. The converse is also true, although we will not pause to verify it now. That is, if λ, y is an eigenvalue, eigenfunction pair for the integral operator determined by the Green's function, then λ, y is an eigenvalue, eigenfunction pair for the Sturm-Liouville eigenvalue problem $Ly = \lambda y$ and $y(0) = 0$, $y(l) = 0$. It turns out, as we discussed at the start of Chapter 3, that replacing the differential equation eigenvalue problem by the equivalent eigenvalue problem for the integral operator has several advantages.

The reader may find it useful to revisit the four examples and the observations made about them while reading the rest of the chapter. We mention in passing that Examples 1 and 2 can be handled as a single example (and even Example 3 can be included as a limiting case) and the constant a can be any complex number. However, no new insights are gained by the added generality.

4.5 BVPs and EVPs - Notation

We use the following notation in the rest of the chapter

$$\begin{aligned}
Ly &= -(p(x)y')' + q(x)y, \quad a < x < b, \\
B_a y &= \alpha y(a) + \beta y'(a), \\
B_b y &= \gamma y(b) + \delta y'(b), \\
B_i y &= a_{i1} y(a) + a_{i2} y'(a) + b_{i1} y(b) + b_{i2} y'(b), \quad i = 1, 2,
\end{aligned}$$

where $p(x)$, $q(x)$, and $f(x)$ are real or complex-valued continuous functions on (a, b) and $p(x) \neq 0$ there, α, β, γ, δ, a_{ij}, and b_{ij} are given real or complex numbers.

Ly is called a Sturm-Liouville differential operator. Ly is called **regular** because $p(x) \neq 0$ on $[a, b]$ and $p(x)$ and $q(x)$ are continuous there. The boundary forms B_a, B_b, B_1, and B_2 are used to define the boundary conditions we shall consider, either separated boundary conditions or mixed boundary conditions.

Separated boundary conditions, boundary conditions that involve data at only one endpoint $x = a$ or $x = b$, are specified by

$$B_a y = c_a \quad \text{and} \quad B_b y = c_b,$$

where c_a and c_b are real or complex numbers. **Mixed boundary conditions**, boundary conditions that may involve data at both endpoints, are specified by

$$B_1 y = c_1 \quad \text{and} \quad B_2 y = c_2$$

where c_1 and c_2 are real or complex numbers. Of course, mixed boundary conditions include separated boundary conditions as a special case. However, it is advantageous to consider separated boundary conditions independently because most of the Sturm-Liouville boundary value problems that arise in applications have separated boundary conditions and certain theoretical simplifications occur. Our main interest in mixed boundary conditions is the case of periodic boundary conditions and, to a lesser extent, antiperiodic boundary conditions. Consequently, we treat problems with separated boundary conditions in depth and then give a briefer account of problems with mixed boundary conditions.

The general Sturm-Liouville boundary value problem with separated boundary conditions is

$$\begin{cases} -(p(x)y')' + q(x)y = f(x), & a < x < b, \\ \alpha y(a) + \beta y'(a) = c_a, & |\alpha| + |\beta| \neq 0, \\ \gamma y(b) + \delta y'(b) = c_b, & |\gamma| + |\delta| \neq 0, \end{cases} \tag{4.5}$$

and the general mixed boundary value problem is

$$\begin{cases} -\big(p(x)y'\big)' + q(x)y = f(x), \quad a < x < b, \\ a_{11}y(a) + a_{12}y'(a) + b_{11}y(b) + b_{12}y'(b) = c_1, \\ a_{21}y(a) + a_{22}y'(a) + b_{21}y(b) + b_{22}y'(b) = c_2. \end{cases} \tag{4.6}$$

The Sturm-Liouville boundary value problem with separated boundary conditions can be expressed compactly by

$$Ly = f,\ B_a y = c_a,\ B_b y = c_b,$$

and the problem with mixed boundary conditions by

$$Ly = f,\ B_1 y = c_1,\ B_2 y = c_2.$$

The **corresponding homogeneous problem** to $Ly = f$, $B_a y = c_a$, $B_b y = c_b$ is

$$Ly = 0,\ B_a y = 0,\ B_b y = 0,$$

and the corresponding homogeneous problem to $Ly = f$, $B_1 y = c_1$, $B_2 y = c_2$ is

$$Ly = 0,\ B_1 y = 0,\ B_2 y = 0.$$

Likewise, the **corresponding eigenvalue problem** when the boundary conditions are separated is

$$Ly = \lambda y,\ B_a y = 0,\ B_b y = 0$$

and is

$$Ly = \lambda y, \;\; B_1 y = 0, \;\; B_2 y = 0$$

when they are mixed.

Example 1a. (continued, with $f(x) = 1$) The boundary value problem

$$\begin{aligned} -y'' + ay &= 1, \;\; 0 < x < l, \\ y(0) &= 1 \;\; y(l) = 2, \end{aligned}$$

with $a > 0$ is a model for the steady-state temperature in the cross-sections of a rod of length l with heat loss permitted through the lateral surface of the rod, constant thermal coefficients, and constant heat generation along the rod.

We use this example to motivate and clarify the formal definition of a solution to a boundary problem. The three equations in the boundary value problem are satisfied by the doubly infinite family of the functions

$$y(x) = \begin{cases} A\cosh\sqrt{a}x + B\sinh\sqrt{a}x + a^{-1} & \text{for } x < x < 1 \\ 1 & \text{for } x = 0 \\ 2 & \text{for } x = l \end{cases},$$

where A and B can be any constants. The top line is the general solution to the differential equation. Which, if any, of these functions should be called a solution to the boundary value problem? The temperature must vary continuously on physical grounds. Since the expression for the temperature $y(x)$ is clearly continuous where it satisfies the differential equation, that is on $0 < x < l$, the temperature will vary continuously throughout the rod if it satisfies

$$\lim_{x\to 0} y(x) = y(0) = 1 \quad \text{and} \quad \lim_{x\to l} y(x) = y(l) = 2.$$

That is the temperature inside the rod tends to the temperature imposed on the boundary as x approaches either end of the rod. This pair of limit relations is also natural on mathematical grounds. There must be some relation among the three equations that comprise the boundary value problem; otherwise, why group them together? Continuity ties the three conditions in the boundary value problem together.

In this example, the limit relations yield the pair of equations

$$\begin{aligned} &\lim_{x\to 0} y(x) = A + a^{-1} = 1, \\ &\lim_{x\to l} y(x) = A\cosh\sqrt{a}l + B\sinh\sqrt{a}l + a^{-1} = 2, \end{aligned}$$

whose solution is

$$A = 1 - a^{-1}, \; B = \frac{2 - (1 - a^{-1})\cosh\sqrt{a}l}{\sinh\sqrt{a}l}.$$

These choices for A and B single out, from the doubly infinite collection of functions above, a unique continuous function

$$y(x) = A\cosh\sqrt{a}x + B\sinh\sqrt{a}x \quad \text{for } 0 \le x \le l$$

that satisfies the three conditions in the boundary value problem, is both physically and mathematically realistic, and should be called the solution of the boundary value problem.

These considerations lead us to define a **solution** to a Sturm-Liouville boundary value problem or eigenvalue problem to be a real or complex-valued function $y(x)$ that satisfies the given differential equation on $a < x < b$, satisfies the given boundary conditions, and is continuous on the closed interval $a \leq x \leq b$. The corresponding homogeneous problem always has the so-called **trivial solution** y identically zero; any other solution to the homogeneous problem is called **nontrivial**. Notice that $\lambda = 0$ is not an eigenvalue of the corresponding Sturm-Liouville eigenvalue problem if and only if the corresponding homogeneous problem has only the trivial solution.

We restrict our attention to regular Sturm-Liouville problems, unless explicitly stated to the contrary. A Sturm-Liouville boundary value problem or eigenvalue problem is **regular** if $p(x)$, $q(x)$, and $f(x)$ are real or complex-valued continuous functions on $[a, b]$ and $p(x) \neq 0$ there. That is, the problem is regular if the Sturm-Liouville equation in the problem is regular. A solution to a regular Sturm-Liouville problem has added smoothness.

Theorem 89 *If $y(x)$ is a solution to a regular Sturm-Liouville boundary value or eigenvalue problem, then $y(x)$ is continuously differentiable function on $[a, b]$ and satisfies the differential equation at every point in $[a, b]$.*

Proof. Since $y(x)$ is continuous on the closed interval $[a.b]$ and satisfies a regular Sturm-Liouville differential equation on the corresponding open interval, by Lemma 79 it is continuously differentiable on $[a, b]$ and satisfies the differential equation at the endpoints. ■

A convenient result that guarantees solutions are real-valued in expected cases follows.

Theorem 90 *If the Sturm-Liouville boundary value problem (4.5) or (4.6) has only real-valued data and its corresponding homogeneous boundary value problem has only the trivial solution, then any solution to (4.5) or (4.6) is real-valued.*

Proof. Let $y = y_1 + iy_2$ be a solution to (4.5) or (4.6), where y_1 and y_2 are the real and imaginary parts of y. Substitute $y = y_1 + iy_2$ into the equations in (4.5) or (4.6) and separate the equations into real and imaginary parts to find that y_2 is a solution of the corresponding homogeneous equation. Consequently, $y_2 = 0$ and $y = y_1$ is real-valued. ■

4.6 Green's Functions

Green's functions were introduced and motivated in Section 1.10. Examples of particular Green's functions where given in Section 4.4. In this section, we give a systematic development of Green's functions and their properties for regular Sturm-Liouville problems.

> We assume throughout the discussion that the boundary value problem is regular and has homogeneous boundary conditions. That is, $p(x)$, $q(x)$, and $f(x)$ are continuous functions on the closed interval $[a, b]$ and $p(x) \neq 0$ there.
>
> By Theorem 89 any solution y to such a problem is continuously differentiable on $[a, b]$ and satisfies the Sturm-Liouville differential equation there.

A regular Sturm-Liouville boundary value problem, with either separated or mixed homogeneous boundary conditions, has a **Green's function**, denoted by $g(x, s)$, if $g(x, s)$ is continuous on $a \leq x, s \leq b$ and for every continuous right member $f(x)$ of the Sturm-Liouville differential equation, the boundary value problem has a unique solution y given by

$$y(x) = \int_a^b g(x, s) f(s)\, ds$$

for $a \leq x \leq b$.

A physical motivation for the existence of Green's functions is given in Section 1.10. The superposition reasoning used there relied on the fact that the boundary conditions were homogeneous. In the sections that follow, we establish the existence of Green's functions by mathematical means and provide effective means for finding them. We will also see that the Green's function $g(x, s)$ determines a solution operator G such that the differential equation $Ly = f$ together with its boundary conditions is equivalent to the equation $y = Gf$, where G is the operation of integration of the Green's function against f; that is,

$$Gf(x) = \int_a^b g(x, s)f(s)\, ds.$$

A few preliminary observations are in order, before embarking on this program. The boundary value problem in Example 1a in Section 4.4 always has a Green's function, regardless of the choice of a. However, this is not the case in Example 2a. We can only expect to find a solution formula in terms of a Green's function when the boundary value problem has a unique solution for all right-hand sides. If this is so, the corresponding homogeneous problem must have the unique solution $y = 0$, the trivial solution. Equivalently, $\lambda = 0$ cannot be an eigenvalue of the corresponding eigenvalue problem. This was confirmed explicitly in Example 2a.

A Green's function is uniquely determined when it exists.

Theorem 91 *If a Sturm-Liouville boundary value problem with homogeneous boundary conditions has a Green's function $g(x, s)$, then the Green's function is unique.*

Proof. If $g(x, s)$ is a Green's function for the boundary value problem $Ly = f$, $B_1 y = 0$, $B_2 y = 0$, then $g(x, s)$ is continuous on $[a, b] \times [a, b]$ and the unique solution y to the problem is

$$y(x) = \int_a^b g(x, s)f(s)\, ds$$

for each right member $f(x)$ that is continuous on $[a, b]$. Suppose $h(x, s)$ also has the same property: $h(x, s)$ is continuous on $[a, b] \times [a, b]$ and

$$y(x) = \int_a^b h(x, s)f(s)\, ds$$

is the unique solution to the boundary value problem $Ly = f$, $B_1 y = 0$, $B_2 y = 0$ for each right member $f(x)$ that is continuous on $[a, b]$. Then, for each continuous function $f(x)$ on $[a, b]$,

$$\int_a^b h(x, s)f(s)\, ds = y(x) = \int_a^b g(x, s)f(s)\, ds,$$

where y is the unique solution to $Ly = f$, $B_a y = 0$, $B_b y = 0$. Hence,

$$\int_a^b (h(x, s) - g(x, s))f(s)\, ds = 0,$$

for all continuous functions $f(s)$ on $[a, b]$. By Corollary 20 it follows that for each x in $[a, b]$, $h(x, s) = g(x, s)$ for all s in $[a, b]$ and uniqueness of the Green's function is established. ■

We will consider two cases in the following sections. (1) The boundary conditions are separated, in which case, if the data is all real, then the Green's function will be real-valued and symmetric. (2) The boundary conditions are mixed, in which case, our emphasis will be on the special cases of periodic and antiperiodic boundary conditions.

4.6.1 Separated Boundary Conditions

The following lemma is a key ingredient in the construction of Green's functions when the boundary conditions are separated. It is followed by other lemmas that aid in establishing when (4.5) has a unique solution and how to construct the solution.

Lemma 92 *There is a continuously differentiable function $u(x)$ on $[a, b]$ that satisfies*

$$Lu = 0, \ B_a u = 0, \ u \neq 0$$

there, and there is a continuously differentiable function $v(x)$ on $[a, b]$ that satisfies

$$Lv = 0, \ B_b v = 0, \ v \neq 0$$

there. Moreover, if all the data in the problem is real-valued, u and v may be chosen to be real-valued.

Proof. Let $c = (a + b)/2$ and let y_1 and y_2 be the unique solutions to the initial value problems $Ly = 0$ with initial conditions $y_1(c) = 1$, $y_1'(c) = 0$ and $y_2(c) = 0$, $y_2'(c) = 1$. The solutions y_1 and y_2 are linearly independent on $[a, b]$. We established earlier that these solutions are (more properly extend to) continuously differentiable functions on $[a, b]$ and satisfy the differential equation there. (See Theorem 82.)

The function $u = c_1 y_1 + c_2 y_2$ satisfies $Lu = 0$ for any choice of constants c_1 and c_2. It will satisfy $B_a u = 0$ if and only if

$$c_1 B_a y_1 + c_2 B_a y_2 = 0.$$

If $B_a y_1 = 0$ choose $c_1 = 1$ and $c_2 = 0$; otherwise, choose $c_2 = -1$ and $c_1 = B_a y_2 / B_a y_1$ to find a nontrivial solution to $Lu = 0$ with $B_a u = 0$. Similar reasoning establishes the second assertion in the lemma.

If all the data in L, B_a, and B_b is real-valued, then the solutions y_1 and y_2 to the initial value problems at the beginning of the proof are real-valued by Theorem 84. The subsequent argument shows that u and v are real valued. ∎

Lemma 93 *There exist continuously differentiable functions u and v on $[a, b]$ that satisfy*

$$Lu = 0, \ B_a u = 0, \ u \neq 0,$$
$$Lv = 0, \ B_a v = 0, \ v \neq 0$$

there. For any such pair u and v,

$$Ly = 0, \ B_a y = 0, \ B_b y = 0$$

has only the trivial solution if and only if u and v are linearly independent.

Proof. Functions u and v with the stated properties exist by the previous lemma.

⇒: We use proof by contradiction. If u and v are linearly dependent on $[a, b]$, then $B_b u = 0$ because u is a nonzero multiple of v. Consequently, u is a nontrivial solution of $Ly = 0$, $B_a y = 0$, $B_b y = 0$, a contradiction. Hence, u and v are linearly independent on $[a, b]$.

⇐: We use a proof by contradiction again. Suppose $Ly = 0$, $B_a y = 0$, $B_b y = 0$ has a nontrivial solution y. Then $B_a y = 0$ and $B_a u = 0$; hence,

$$\begin{cases} \alpha y(a) + \beta y'(a) = 0 \\ \alpha u(a) + \beta u'(a) = 0 \end{cases} \quad \text{with } |\alpha| + |\beta| > 0.$$

Since the 2×2 system has a nontrivial solution its determinant which is $W_{y,u}(a)$ must be zero. Thus, u and y are linearly dependent on $[a, b]$. Likewise, v and y are linearly dependent on $[a, b]$. Since all three functions are nonzero, u and v are nonzero multiples of y. Consequently, $u = cv$ for some $c \neq 0$ and u and v are linearly dependent, a contradiction. Hence, $Ly = 0$, $B_a y = 0$, $B_v y = 0$ has only the trivial solution. ■

Lemma 94 *$Ly = 0$, $B_a y = 0$, $B_b y = 0$ has only the trivial solution if and only if*

$$Ly = 0,\ B_a y = c_a,\ B_v y = c_b$$

has a unique solution for each choice of data c_a and c_b.

Proof. $\Rightarrow$: By the previous lemma there are linearly independent, continuously differentiable functions u and v on $[a, b]$ such that

$$\begin{aligned} Lu &= 0,\ \ B_a u = 0, \\ Lv &= 0,\ \ B_a v = 0, \end{aligned}$$

and the general solution to $Ly = 0$ is $y = c_1 u + c_2 v$. Such a y satisfies $B_a y = c_a$ and $B_b y = c_b$ if and only if c_1 and c_2 satisfy

$$\begin{cases} c_1 B_a u + c_2 B_a v = c_a \\ c_1 B_b u + c_2 B_b v = c_b \end{cases}.$$

By assumption when $c_a = 0$ and $c_b = 0$ the only solution to the system is $c_1 = 0$ and $c_2 = 0$; hence, the determinant of the system must be nonzero

$$\begin{vmatrix} B_a u & B_a v \\ B_b u & B_b v \end{vmatrix} \neq 0$$

and the linear system for c_1 and c_2 has a unique solution for any choice for c_a and c_b. Thus, $y = c_1 u + c_2 v$, with these choices for c_1 and c_2, is the one and only solution to $Ly = 0$, $B_a y = c_a$, $B_b y = c_b$.

$\Leftarrow$: In particular, $Ly = 0$, $B_a y = 0$, $B_b y = 0$ has a unique solution. One solution is the trivial solution. So it must be the only solution to the homogeneous problem. ■

Suppose $Ly = 0$, $B_a y = 0$, $B_b y = 0$ has only the trivial solution. Let $\tilde{y}$ be the unique solution to

$$L\tilde{y} = 0,\ B_a \tilde{y} = c_a,\ B_b \tilde{y} = c_b.$$

Since L, B_a, and B_b are all linear operators, the equations

$$Ly = f,\ \ B_a y = c_a,\ \ B_b y = c_b$$

in (4.5) are satisfied if and only if

$$L(y - \tilde{y}) = f,\ \ B_a(y - \tilde{y}) = 0,\ \ B_b(y - \tilde{y}) = 0.$$

It follows that y is a solution of (4.5) and is its only solution if and only if (4.5) has a unique solution when $c_a = 0$ and $c_b = 0$ and $f(x)$ is an arbitrary continuous function on $[a, b]$.

Theorem 95 *The regular Sturm-Liouville boundary value problem (4.5) with $c_a = 0$ and $c_b = 0$*

$$Ly = f,\ \ B_a y = 0,\ \ B_b y = 0$$

has a unique solution for each function $f(x)$ that is continuous on $[a, b]$ if and only if the corresponding homogeneous problem

$$Ly = 0,\ B_a y = 0,\ B_b y = 0$$

has only the trivial solution.

Proof. $\Rightarrow$: If $f = 0$, then $y = 0$ is a solution and it is the only one by hypothesis. So the corresponding homogeneous problem has only the trivial solution.

$\Leftarrow$: First, if $Ly = f$, $B_a y = 0$, $B_b y = 0$ has a solution there can be only one because if y and z are solutions, then

$$L(y - z) = 0,\ B_a(y - z) = 0,\ B_b(y - z) = 0$$

and, hence $y = z$. Second, we provisionally assume that $Ly = f$, $B_a y = 0$, $B_b y = 0$ does have a (unique) solution and proceed to construct a formula for it. Once this formula is obtained we will check directly that it does in fact solve the problem.

So assume that y is a solution of $Ly = f$, $B_a y = 0$, $B_b y = 0$. By Lemma 92 there are continuously differentiable functions u and v on $[a, b]$ such that

$$Lu = 0,\ B_a u = 0,\ u \neq 0,$$
$$Lv = 0,\ B_b v = 0,\ v \neq 0.$$

Since $Ly = 0$, $B_a y = 0$, $B_b y = 0$ has only the trivial solution

$$u \text{ and } v \text{ are linearly independent on } [a, b]$$

by Lemma 93. Apply Lemma 80 (Lagrange's identity) with $z = u$ and y the solution to $Ly = f$, $B_a y = 0$, $B_b y = 0$ to obtain

$$\int_a^x -uf\, ds = p(uy' - yu')\big|_a^x.$$

Since $B_a u = \alpha u(a) + \beta u'(a) = 0$ and $B_a y = \alpha y(a) + \beta y'(a) = 0$, if $\beta \neq 0$, then

$$(uy' - yu')\big|_a = u(a)(-\alpha/\beta)y(a) - y(a)(-\alpha/\beta)u(a) = 0$$

and a similar calculation yields the same conclusion if $\alpha \neq 0$. Thus,

$$-\int_a^x uf\, ds = p(x)(u(x)y'(x) - y(x)u'(x)).$$

In the same way, replace z by v in Lagrange's identity to get

$$\int_x^b -vf\, ds = p(vy' - yv')\big|_x^b.$$

The evaluation at the upper limit gives 0 and

$$-\int_x^b vf\, ds = -p(x)(v(x)y'(x) - y(x)v'(x)).$$

Thus,

$$\int_a^x uf\, ds = p(x)(-u(x)y'(x) + y(x)u'(x))$$

and

$$\int_x^b vf\ ds = p(x)(v(x)y'(x) - y(x)v'(x)).$$

Multiply the last equation by $u(x)$, the equation above it by $v(x)$, and add to eliminate $y'(x)$ and obtain

$$v(x)\int_a^x uf\ ds + u(x)\int_x^b vf\ ds = y(x)p(x)(v(x)u'(x) - u(x)v'(x)).$$

The difference in parenthesis on the right is $-W_{u,v}(x)$. Since $p(x)W_{u,v}(x) = -C$ for x in $[a, b]$ by Lemma 86 and $C \neq 0$ because u and v are linearly independent,

$$y(x) = \frac{1}{C}\left(v(x)\int_a^x u(s)f(s)\ ds + u(x)\int_x^b v(s)f(s)\ ds\right)$$

where $p(x)W_{u,v}(x) = -C$. This formula was obtained under the assumption that a solution to (4.5) with $c_a = c_b = 0$ did exist. It is easy to check that this formula does in fact solve that problem: from the fundamental theorem of calculus

$$y'(x) = \frac{1}{C}\left(v'(x)\int_a^x u(s)f(s)\ ds + u'(x)\int_x^b v(s)f(s)\ ds\right)$$

and

$$\begin{aligned}(p(x)y'(x))' = \frac{1}{C}[&(p(x)v'(x))'\int_a^x u(s)f(s)\ ds + (p(x)u'(x))'\int_x^b v(s)f(s)\ ds \\ &+ p(x)v'(x)u(x)f(x) - p(x)u'(x)v(x)f(x)]\end{aligned}$$

for all x in $[a, b]$. Hence,

$$\begin{aligned}Ly &= -(p(x)y')' + q(x)y \\ &= \frac{1}{C}\left((Lv)\int_a^x u(s)f(s)\ ds + (Lu)\int_x^b v(s)f(s)\ ds - p(x)W_{u,v}(x)f(x)\right) \\ &= f(x)\end{aligned}$$

because $Lu = 0$, $Lv = 0$, and $p(x)W_{u,v}(x) = -C$. Thus, $Ly = f$ holds for all x in $[a, b]$. Since

$$y(a) = \frac{1}{C}\left(u(a)\int_a^b v(s)f(s)\ ds\right) \quad \text{and} \quad y'(a) = \frac{1}{C}\left(u'(a)\int_a^b v(s)f(s)\ ds\right),$$

and B_a is linear,

$$B_a y = \left(\frac{1}{C}\int_a^b v(s)f(s)\ ds\right)B_a u = 0$$

and likewise, $B_b y = 0$. Thus, under the assumption that $Ly = 0$, $B_a y = 0$, $B_b y = 0$ has only the trivial solution, we have established that

$$y(x) = \frac{1}{C}\left(v(x)\int_a^x u(s)f(s)\ ds + u(x)\int_x^b v(s)f(s)\ ds\right),$$

where $p(x)W_{u,v}(x) = -C$, is the unique solution to (4.5) when $c_a = c_b = 0$. ■

As in Examples 1a and 2a of Section 4.4, the explicit solution formula developed in the proof of the previous theorem leads us to the Green's function for the boundary value problem and a more convenient expression for the solution. In the proof, u and v can be chosen as any linearly independent, continuously differentiable functions on $[a, b]$ that satisfy

$$\begin{aligned} Lu &= 0, \ B_a u = 0, \\ Lv &= 0, \ B_b v = 0. \end{aligned}$$

Consequently, their Wronskian $W_{u,v}$ satisfies $p(x)\,W_{u,v}(x) = -C$ for x in $[a, b]$ and some $C \neq 0$. The replacement of v by v/C gives a new pair of functions u and v satisfying the first pair of conditions above and $p(x)\,W_{u,v}(x) = -1$. With this choice for u and v, the solution to (4.5) when $c_a = c_b = 0$ is

$$\begin{aligned} y(x) &= v(x) \int_a^x u(s)f(s)\ ds + u(x) \int_x^b v(s)f(s)\ ds \\ &= \int_a^b g(x, s)f(s)\ ds \end{aligned}$$

where

$$g(x, s) \begin{cases} u(x)v(s) & \text{for } a \le x \le s \le b \\ u(s)v(x) & \text{for } a \le s \le x \le b \end{cases}.$$

If all the data in L, B_a, and B_b is real-valued, then the functions u and v above can be chosen real-valued by Lemma 92. We summarize this discussion as

Theorem 96 *If the corresponding homogeneous problem $Ly = 0$, $B_a y = 0$, $B_b y = 0$ has only the trivial solution, then the regular Sturm-Liouville boundary value problem $Ly = f$, $B_a y = 0$, $B_b y = 0$, where f is a given continuous function on $[a, b]$, has a unique solution y. Moreover, there are continuously differentiable functions u and v on $[a, b]$ that satisfy*

$$\begin{aligned} Lu &= 0, \ B_a u = 0, \\ Lv &= 0, \ B_b v = 0, \end{aligned}$$

$p(x)\,W_{u,v}(x) = -1$ *for all x in $[a, b]$, and the unique solution y is given by*

$$y(x) = \int_a^b g(x, s)f(s)\ ds$$

where

$$g(x, s) \begin{cases} u(x)v(s) & \text{for } a \le x \le s \le b \\ u(s)v(x) & \text{for } a \le s \le x \le b \end{cases}.$$

That is, $g(x, s)$ is the Green's function for the boundary value problem $Ly = f$, $B_a y = 0$, $B_b y = 0$. Moreover, if all the data in L, B_a, and B_b is real-valued, then u and v can be chosen real-valued and the Green's function $g(x, s)$ is real-valued and $g(x, s) = g(s, x)$; that is, $g(x, s)$ is a symmetric kernel whose corresponding integral operator is self-adjoint.

We have established that a Green's function can exist only if the corresponding homogeneous problem has only the trivial solution, and under that assumption, we have established in Theorem 96 that there is a Green's function and have found a formula for it. This establishes

Theorem 97 *The regular boundary value problem $Ly = f$, $B_a y = 0$, $B_b y = 0$ has a Green's function if and only if the corresponding homogeneous problem has only the trivial solution, in which case the Green's function is given by the expression in Theorem 96.*

The Green's function $g(x, s)$ for $Ly = f$, $B_a y = 0$, $B_b y = 0$ has the following properties (when it exists):

1. $g(x, s)$ is continuous on the square $[a, b] \times [a, b]$ and has continuous partial derivatives on the upper triangle ($x \leq s$) of the square and on the lower triangle ($s \leq x$) of the square.
2. $g(x, s)$, regarded as a function of x for fixed s in $[a, b]$, satisfies the differential equation $Ly = 0$ for $x \neq s$ in $[a, b]$.
3. $g(x, s)$, regarded as a function of x for fixed s in (a, b), satisfies the homogeneous boundary conditions of the problem.
4. $g(x, s)$, regarded as a function of x for fixed s in (a, b), has a jump in its derivative with respect to x at $x = s$ given by

$$\frac{\partial g}{\partial x}(s+, s) - \frac{\partial g}{\partial x}(s-, s) = -\frac{1}{p(s)}.$$

The four properties can be verified directly from the formula for the Green's function in Theorem 96. The formula for the Green's function reveals that $g(x, s) = g(s, x)$. Consequently, Properties 1-4 hold with the roles of x and s interchanged.

Properties 1-4 characterize the Green's function:

Theorem 98 *If a function $g(x, s)$ exists with Properties 1-4, then the regular Sturm-Liouville boundary value problem $Ly = 0$, $B_a y = 0$, $B_b y = 0$ has only the trivial solution, $g(x, s)$ is the Green's function for the differential operator $Ly = -(py')' + qy$ and boundary conditions $B_a y = 0$, $B_b y = 0$, and $g(x, s) = g(s, x)$.*

Proof. Let $g(x, s)$ be a function with Properties 1-4. Fix s with $a < s < b$ and define functions z_1 and z_2 by

$$z_1(x) = g(x, s) \quad \text{for } a \leq x \leq s \quad \text{and} \quad z_2(x) = g(x, s) \quad \text{for } s \leq x \leq b.$$

By Properties 2 and 3, $z_1(x)$ satisfies $Lz_1 = 0$ on $a \leq x < s$, $B_a z_1 = 0$ and $z_2(x)$ satisfies $Lz_2 = 0$ on $s < x \leq b$, $B_b z_2 = 0$. Since these problems are regular, z_1 and z_2 are continuously differentiable on $[a, s]$ and $[s, b]$ respectively and the differential equation holds at $x = s$ in both cases. We show first that $Ly = 0$, $B_a y = 0$, $B_b y = 0$ has only the trivial solution. Assume the contrary and let $z(x)$ be a nontrivial solution. Then

$$Lz = 0 \quad \text{for } a \leq x \leq s, \; B_a z = 0,$$

and

$$Lz_1 = 0 \quad \text{for } a \leq x \leq s, \; B_a z_1 = 0.$$

Consequently,

$$\begin{cases} \alpha z(a) + \beta z'(a) = 0 \\ \alpha z_1(a) + \beta z_1'(a) = 0 \end{cases} \quad \text{with} \quad |\alpha| + |\beta| \neq 0;$$

hence, $W_{z,z_1}(a) = 0$ and z and z_1 are linearly dependent on $a \le x \le s$. So $d(s)z(x) + d_1(s)z_1(x) = 0$ for some scalars $d(s)$ and $d_1(s)$, dependent on the fixed value s, with $|d(s)| + |d_1(s)| \ne 0$. If $d_1(s) = 0$ then $z(x) = 0$ on $[a, s]$, $z(s) = z'(s) = 0$, and z solves the initial value problem $Lz = 0$ on (a, b), $z(s) = z'(s) = 0$. Consequently, $z = 0$ on $[a, b]$, a contradiction, which implies that $d_1(s) \ne 0$. Thus, $z_1(x) = c_1(s)z(x)$ on $a \le x < s$ for $c_1(s) = -d(s)/d_1(s)$. Likewise, $z_2(x) = c_2(s)z(x)$ on $s \le x \le b$ for some scalar $c_2(s)$. Since $g(x, s)$ is continuous at $x = s$ by Property 1,

$$g(s+, s) - g(s-, s) = c_2(s)z(s) - c_1(s)z(s) = 0.$$

Since z is nontrivial, there exist s_0 in (a, b) where $z(s_0) \ne 0$ and, hence, $c_1(s_0) = c_2(s_0)$ and

$$g_x(s_0+, s_0) - g_x(s_0-, s_0) = c_2(s_0)z'(s_0) - c_1(s_0)z'(s_0) = 0,$$

which contradicts the jump condition in Property 4. Hence, $Ly = 0$, $B_a y = 0$, $B_b y = 0$ has only the trivial solution and $Ly = f$, $B_a y = 0$, $B_b y = 0$ has a Green's function.

Finally we establish that a function $g(x, s)$ with Properties 1-4 is the Green's function. To this end, for any continuous function f, let y be the unique solution to $Ly = f$, $B_a y = 0$, $B_b y = 0$. Fix s in (a, b), regard $g(x, s)$ as a function of x in $[a, b]$ and let $a < r < s < t < b$. By Property 2

$$0 = \int_a^r yLg\, dx = \int_a^r y(-pg')'\, dx + \int_a^r yqg\, dx.$$

Integration by parts gives

$$\begin{aligned} 0 &= -ypg'\big|_a^r + \int_a^r pg'y'\, dx + \int_a^r qyg\, dx \\ &= -ypg'\big|_a^r + py'g\big|_a^r - \int_a^r g(py')'\, dx + \int_a^r qyg\, dx \\ &= \left(py'g - ypg'\right)\big|_a^r + \int_a^r gLy\, dx \\ &= \left(py'g - ypg'\right)\big|_a^r + \int_a^r gf\, dx. \end{aligned}$$

Thus,

$$-p(y'g - yg')\big|_a^r = \int_a^r gf\, dx.$$

Since

$$\begin{cases} \alpha y(a) + \beta y'(a) = 0 \\ \alpha g(a) + \beta g'(a) = 0 \end{cases}$$

with $|\alpha| + |\beta| > 0$, the determinant of the 2×2 system is 0 and the contribution to the evaluated term above at $x = a$ is 0. Thus,

$$-p\left(y'g - yg'\right)\big|_{x=r} = \int_a^r gf\, dx.$$

Let r tend to s with $r < s$ and use Property 1 and the fact that y is continuously differentiable on (a, b) to obtain

$$-p\left(y'g - yg'\right)\big|_{x=s-} = \int_a^s gf\, dx.$$

In the same way,

$$-p(y'g - yg')\big|_t^b = \int_t^b gf\,dx,$$

$$p(y'g - yg')\big|_{x=t} = \int_t^b gf\,dx,$$

and

$$p(y'g - yg')\big|_{x=s+} = \int_s^b gf\,dx.$$

Combining evaluations gives

$$p(y'g - yg')\big|_{x=s-}^{x=s+} = \int_a^b gf\,dx.$$

Since p, y', and g are continuous on (a, b),

$$py'g\big|_{x=s-}^{x=s+} = 0$$

and by the jump condition (Property 4)

$$-pyg'\big|_{x=s-}^{x=s+} = -p(s)y(s)\left(-\frac{1}{p(s)}\right) = y(s),$$

it follows that

$$y(s) = \int_a^b g(x, s)f(x)\,dx$$

for $a < s < b$. Since both members of this equality are continuous functions on $a \le s \le b$, the equality holds for all s in $[a, b]$. By definition $g(s, x)$ is the Green's function for the differential operator L and the boundary conditions $B_a y = 0$ and $B_b y = 0$. By uniqueness, $g(s, x)$ must be given by the formula in Theorem 96 which shows that $g(s, x) = g(x, s)$. ■

We reprise parts of Examples 1a and 4a of Section 4.4 to illustrate these results.

Example 1a. (reprise) Fix $a > 0$ and $l > 0$ and let $f(x)$ be continuous on $[0, l]$. Find the Green's function for

$$-y'' + ay = f(x), \quad 0 < x < l,$$
$$y(0) = 0, \; y(l) = 0.$$

Here $Ly = -y'' + ay$ so that $p(x) = 1$ and $B_0 y = y(0)$, $B_l y = y(l)$.

The homogeneous equation $Ly = -y'' + ay = 0$ has linearly independent solutions $\cosh\sqrt{a}x$ and $\sinh\sqrt{a}x$ and also $\cosh\sqrt{a}(l - x)$ and $\sinh\sqrt{a}(l - x)$. We will find a continuous function $g(x, s)$ for $0 \le x, s \le l$ that has Properties 1-4. This function must be the Green's function. Fix s and regard g as a function of x. For $x < s$, g must satisfy the homogeneous equation $Lg = 0$ and the boundary condition $g(0, s) = 0$. Hence,

$$g(x, s) = c_1(s)\cosh\sqrt{a}x + c_2(s)\sinh\sqrt{a}x$$

and c_1 and c_2 must satisfy $g(0, s) = c_1(s) = 0$. Thus,

$$g(x, s) = c_2(s)\sinh\sqrt{a}x \quad \text{for } x < s.$$

Likewise for $x > s$ the Green's function must satisfy $Lg = 0$ and $g(l, s) = 0$; that is,

$$g(x, s) = d_1(s)\cosh\sqrt{a}(l-x) + d_2(s)\sinh\sqrt{a}(l-x) \quad \text{and} \quad g(l, s) = d_1(s) = 0.$$

So

$$g(x, s) = d_2(s)\sinh\sqrt{a}(l-x) \quad \text{for } s < x.$$

Since the Green's function must be continuous we must have $g(s-, s) = g(s+, s)$; that is,

$$c_2(s)\sinh\sqrt{a}s = d_2(s)\sinh\sqrt{a}(l-s).$$

To satisfy the jump condition

$$\frac{\partial g}{\partial x}(s+, s) - \frac{\partial g}{\partial x}(s-, s) = -\frac{1}{p(s)},$$

$c_2(s)$ and $d_2(s)$ must satisfy

$$-d_2(s)\sqrt{a}\cosh\sqrt{a}(l-s) - c_2(s)\sqrt{a}\cosh\sqrt{a}s = -1.$$

So $c_2(s)$ and $d_2(s)$ must satisfy

$$\begin{cases} c_2(s)\cosh\sqrt{a}s + d_2(s)\cosh\sqrt{a}(l-s) = 1/\sqrt{a} \\ c_2(s)\sinh\sqrt{a}s - d_2(s)\sinh\sqrt{a}(l-s) = 0 \end{cases}.$$

Solve to find $c_2(s) = \sinh(\sqrt{a}(l-s))/\sqrt{a}\sinh\sqrt{a}l$ and $d_2(s) = \sinh(\sqrt{a}s)/\sqrt{a}\sinh\sqrt{a}l$. Hence, the Green's function is

$$g(x, s) = \frac{1}{\sqrt{a}\sinh\sqrt{a}l}\begin{cases} \sinh\sqrt{a}x\,\sinh\sqrt{a}(l-s) & \text{for } 0 \le x \le s \le l \\ \sinh\sqrt{a}s\,\sinh\sqrt{a}(l-x) & \text{for } 0 \le s \le x \le l \end{cases}.$$

We mention in passing that the foregoing reasoning is valid for a any nonzero complex number. Here if a is not negative, $\sqrt{a}$ may be chosen as the square root of a with positive real part and if $a < 0$, then $\sqrt{a} = i\sqrt{|a|}$. In this generality, the Green's function assumes non-real values, except when $a < 0$.

Example 4a. (reprise) Fix $l > 0$ and let $f(x)$ be continuous on $[0, l]$. Find the Green's function for

$$\begin{gathered} -y'' = f(x), \quad 0 < x < l, \\ y(0) - y'(0) = 0, \ y(l) + y'(l) = 0. \end{gathered}$$

Here $Ly = -y''$ so that $p(x) = 1$ and $B_a y = y(0) - y'(0)$, $B_l y = y(l) + y'(l)$.

This time we will find the Green's function using Theorem 96 rather than seeking a function $g(x, s)$ that has properties 1-4. The general solution $u = c_1 + c_2 x$ to $-y'' = 0$ satisfies

$$y(0) - y'(0) = c_1 - c_2 = 0$$

if and only if $c_1 = c_2$. In particular $u(x) = 1 + x$ satisfies

$$Lu = 0, \ u(0) - u'(0) = 0, \ u \neq 0.$$

The general solution $v = d_1 + d_2 x$ to $-y'' = 0$ satisfies

$$y(l) + y'(l) = d_1 + d_2 l + d_2 = 0$$

if and only if $d_1 = -d_2(l+1)$. In particular $v = -d_2(l+1) + d_2 x$ with $d_2 \neq 0$ satisfies

$$Lv = 0, \; v(l) + v'(l) = 0, \; v \neq 0.$$

The solutions u and v will satisfy the jump condition $p(x)\,W_{u,v}(x) = -1$ if and only if

$$d_2 \begin{vmatrix} 1+x & -(l+1)+x \\ 1 & 1 \end{vmatrix} = -1$$

$$d_2 = -\frac{1}{l+2.}$$

Hence, $v(x) = (l+1-x)/(l+2)$ and the Green's function is

$$g(x,\, s) = \frac{1}{l+2}\begin{cases} (1+x)(l+1-s) & \text{for } 0 \le x \le s \le l \\ (1+s)(l+1-x) & \text{for } 0 \le s \le x \le l \end{cases}.$$

If the fully inhomogeneous problem (4.5) has a unique solution, it can be found by adding the solution $\tilde{y}$ to $Ly = 0$, $B_a y = c_a$, $B_b y = c_b$ to the Green's function solution of $Ly = f$, $B_a y = 0$, $B_b y = 0$. Alternatively, the solution to the fully inhomogeneous problem can be expressed directly in terms of the Green's function for $Ly = f$, $B_a y = 0$, $B_b y = 0$, as we show next. The reasoning that follows is a slight variant on that used to prove Theorem 98 so we compress some of the details. Suppose that $Ly = 0$, $B_a y = 0$, $B_b y = 0$ has only the trivial solution so that $Ly = f$, $B_a y = c_a$, $B_b y = c_b$ has a unique solution that we will denote by y and let $g(x,\, s)$ be the Green's function for $Ly = f$, $B_a y = 0$, $B_b y = 0$. Fix x in $(a,\, b)$, regard $g(x,\, s)$ as a function of s, denote derivatives with respect to s by primes, and use Properties 1-4 with the roles of x and s interchanged to obtain

$$0 = \int_a^b yLg\; ds = \int_a^b y(-pg')'\; ds + \int_a^b yqg\; ds$$

and

$$\begin{aligned} \int_a^b y(-pg')'\; ds &= \int_a^x y(-pg')'\; ds + \int_x^b y(-pg')'\; ds \\ &= -ypg'\big|_a^x + \int_a^x pg'y'\; ds + -ypg'\big|_x^b + \int_x^b pg'y'\; ds \\ &= -ypg'\big|_a^b - ypg'\big|_{x+}^{x-} + py'g\big|_a^x - \int_a^x g(py')'\; ds \\ &\quad + py'g\big|_x^b - \int_x^b g(py')'\; ds \\ &= \left(py'g - ypg'\right)\big|_a^b - ypg'\big|_{x+}^{x-} - \int_a^b g(py')'\; ds \end{aligned}$$

because $py'g$ is continuous for s in $[a,\, b]$. Now,

$$-ypg'\big|_{x+}^{x-} = y(x)p(x)\left(\frac{\partial g}{\partial s}(x,\, x+) - \frac{\partial g}{\partial s}(x,\, x-)\right) = -y(x)$$

by the jump condition. Combine these results to find

$$0 = \left(py'g - ypg'\right)\big|_a^b - y(x) + \int_a^b gLy\; ds$$

$$y(x) = p\left(y'g - yg'\right)\big|_a^b + \int_a^b gLy\; ds$$

and, since $Ly = f$,

$$y(x) = p(s)\Delta(x, s)\big|_a^b + \int_a^b gf\, ds,$$

where $\Delta(x, s) = y'(s)g(x, s) - y(s)g'(x, s)$, primes indicates derivatives with respect to s, and x is fixed in (a, b). The left and right members of the displayed equation for $y(x)$ are continuous at $x = a$ and $x = b$. Hence, that equation holds on the closed interval $[a, b]$. The boundary term can be expressed in terms of the Green's function and the data as follows. Since $B_b y = c_b$ and $B_b g = 0$, we have

$$\begin{cases} \gamma y(b) + \delta y'(b) = c_b \\ \gamma g(b) + \delta g'(b) = 0 \end{cases}$$

and, by the standard elimination process,

$$\gamma\Delta(x, b) = -g'(x, b)c_b \quad \text{and} \quad \delta\Delta(x, b) = c_b g(x, b).$$

Recall x is fixed in the foregoing argument and derivatives are with respect to s. Likewise,

$$\alpha\Delta(x, a) = -g'(x, a)c_a \quad \text{and} \quad \beta\Delta(x, a) = c_a g(x, a).$$

Using these results in the formula for $y(x)$ above yields

Theorem 99 *If $g(x, s)$ is the Green's function determined by the regular Sturm-Liouville differential operator $Ly = -(py')' + qy$ and the separated boundary conditions $B_a y = 0$, $B_b y = 0$, then the Sturm-Liouville boundary value problem (4.5) has the unique solution*

$$y(x) = p(s)\Delta(x, s)\big|_{s=a}^b + \int_a^b g(x, s)f(s)\, ds,$$

where

$$\Delta(x, a) = \begin{cases} -c_a g_s(x, a)/\alpha & \text{if } \alpha \neq 0 \\ c_a g(x, a)/\beta & \text{if } \alpha = 0 \end{cases}$$

and

$$\Delta(x, b) = \begin{cases} -c_b g_s(x, b)/\gamma & \text{if } \gamma \neq 0 \\ c_b g(x, b)/\delta & \text{if } \gamma = 0 \end{cases}$$

for x in $[a, b]$.

Green's functions for Sturm-Liouville boundary value problems with mixed boundary conditions have representations similar to that in Theorem 96 for problems with separated conditions. However, the condition $g(x, s) = g(s, x)$ does not always hold for mixed boundary conditions. A brief treatment of such Green's functions is given in the next section.

4.6.2 Mixed Boundary Conditions

Consider the regular Sturm-Liouville boundary value problem

$$\begin{cases} -(p(x)y')' + q(x)y = f(x) & a < x < b, \\ B_1 y = 0, \ B_2 y = 0, \end{cases} \tag{4.7}$$

with the boundary conditions determined by the linear forms

$$B_i y = a_{i1} y(a) + a_{i2} y'(a) + b_{i1} y(b) + b_{i2} y'(b)$$

for $i = 1, 2$ and given real or complex constants a_{ij} and b_{ij}. Since the problem is regular, $p(x) \neq 0$ on $[a, b]$, and $p(x)$, $q(x)$, and $f(x)$ are continuous on $[a, b]$. The boundary value problem (4.7) is expressed concisely as

$$Ly = f,\ B_1 y = 0,\ B_2 y = 0,$$

where $Ly = -(py')' + qy$. We inquire about the existence of a Green's function for this problem. A necessary condition for the existence of a Green's function is that the corresponding homogeneous problem $Ly = 0$, $B_1 y = 0$, $B_2 y = 0$ has only the trivial solution, just as for separated boundary conditions. Assume this condition holds.

A natural way to construct the Green's function in the case of mixed boundary data is through the variation of parameters formula for solving inhomogeneous initial value problems. The variations of parameters solution to the initial value problem

$$Ly = f,\ y(a) = 0,\ y'(a) = 0$$

is

$$y(x) = \int_a^x (v(x)u(s) - u(x)v(s))f(s)\, ds$$

where u and v satisfy $Lu = 0$ on $[a, b]$, $Lv = 0$ on $[a, b]$, and the Wronskian condition $p(x)\, W_{u,v}(x) = -1$ there. See Theorem 87. The functions u and v can be chosen real-valued when L has all real-valued coefficients and all the coefficients in the boundary conditions are real numbers. Define

$$\tilde{g}(x, s) = \begin{cases} 0 & \text{for } a \leq x \leq s \leq b \\ v(x)u(s) - u(x)v(s) & \text{for } a \leq s \leq x \leq b \end{cases}$$

and observe that $\tilde{g}(x, s)$ is continuous on the square $[a, b] \times [a, b]$ and

$$\frac{\partial g}{\partial x}(s+, s) - \frac{\partial g}{\partial x}(s-, s) = -\frac{1}{p(s)}$$

for s in (a, b). Then

$$y(x) = \int_a^b \tilde{g}(x, s)f(s)\, ds$$

satisfies $Ly = f$ but probably does not satisfy the boundary conditions $B_1 y = 0$ and $B_2 y = 0$. We modify $\tilde{g}(x, s)$ so the modified function satisfies both $Ly = f$ and the boundary conditions: set

$$g(x, s) = c_1 u(x) + c_2 v(x) + \tilde{g}(x, s)$$

where $c_1 = c_1(s)$ and $c_2 = c_2(s)$ are to be determined. The function $g(x, s)$, regarded as a function of x for each fixed s in $[a, b]$ will satisfy the boundary conditions $B_1 g = 0$ and $B_2 g = 0$ if and only if

$$\begin{cases} c_1 B_1 u + c_2 B_1 v = -B_1 \tilde{g} \\ c_1 B_2 u + c_2 B_2 v = -B_2 \tilde{g} \end{cases}$$

where $c_1 = c_1(s)$ and $c_2 = c_2(s)$ are scalars that depend on the fixed value of s. The determinant of the system

$$\begin{vmatrix} B_1 u & B_1 v \\ B_2 u & B_2 v \end{vmatrix} \neq 0;$$

otherwise, the corresponding homogeneous problem $Ly = 0$, $B_1 y = 0$, $B_2 y = 0$ would have a nontrivial solution. Thus, c_1 and c_2 are uniquely determined by the 2×2 system above.

Cramer's rule or explicit solution of the system reveals that $c_1(s)$ and $c_2(s)$ are continuously differentiable on $[a, b]$. For these choices,

$$y(x) = \int_a^b g(x, s)f(s)\, ds = \int_a^b \big(c_1(s)u(x) + c_2(s)v(x) + \tilde{g}(x, s)\big)f(s)\, ds$$

satisfies

$$\begin{aligned} Ly(x) &= Lu(x)\int_a^b c_1(s)f(s)\, ds + Lv(x)\int_a^b c_2(s)f(s)\, ds \\ &\quad + L\bigg(\int_a^b \tilde{g}(x, s)f(s)\, ds\bigg) \\ &= 0 + 0 + f(x) = f(x) \end{aligned}$$

and

$$B_j y = \int_a^b \big(B_j g(x, s)\big)f(s)\, ds = 0$$

for $j = 1, 2$ by choice of the scalars $c_1(s)$ and $c_2(s)$. Thus,

$$y(x) = \int_a^b g(x, s)f(s)\, ds$$

is the unique solution to (4.7) and $g(x, s)$ is the Green's function.

Theorem 100 *The regular mixed boundary value problem (4.7) has a Green's function $g(x, s)$ if and only if the corresponding homogeneous problem $Ly = 0$, $B_1 y = 0$, $B_2 y = 0$ has only the trivial solution, in which case the Green's function can be constructed as follows: let u and v satisfy $Lu = 0$ on $[a, b]$, $Lv = 0$ on $[a, b]$, and the Wronskian condition $p(x)W_{u,v}(x) = -1$ and let*

$$\tilde{g}(x, s) = \begin{cases} 0 & \text{for } a \le x \le s \le b \\ v(x)u(s) - u(x)v(s) & \text{for } a \le s \le x \le b \end{cases}.$$

For each fixed s in $[a, b]$ let $c_1 = c_1(s)$ and $c_2 = c_2(s)$ be the unique solution to

$$\begin{cases} c_1 B_1 u + c_2 B_1 v = -B_1 \tilde{g} \\ c_1 B_2 u + c_2 B_2 v = -B_2 \tilde{g} \end{cases}$$

where B_1 and B_2 act on $\tilde{g}$ regarded as a function of x for fixed s. Then

$$g(x, s) = c_1(s)u(x) + c_2(s)v(x) + \tilde{g}(x, s)$$

for (x, s) in $[a, b] \times [a, b]$. Moreover, if L has real-valued coefficients and all the coefficients in the boundary data are real, then u and v can be chosen real-valued and $g(x, s)$ is real-valued.

A review of the derivation leading to Theorem 100 confirms that the Green's function $g(x, s)$ for $Ly = f$, $B_1 y = 0$, $B_2 y = 0$ (when it exists) has the following properties:

1. $g(x, s)$ is continuous on the square $[a, b] \times [a, b]$ and has continuous partial derivatives on the upper triangle ($x \le s$) of the square and on the lower triangle ($s \le x$) of the square.

2. $g(x, s)$, regarded as a function of x for fixed s in $[a, b]$, satisfies the differential equation $Ly = 0$ for $x \ne s$ in $[a, b]$.

3. $g(x, s)$, regarded as a function of x for fixed s in $[a, b]$, satisfies the boundary conditions $B_1 y = 0$ and $B_2 y = 0$.

4. $g(x, s)$, regarded as a function of x for fixed s in (a, b), has a jump in its derivative with respect to x at $x = s$ given by

$$\frac{\partial g}{\partial x}(s+, s) - \frac{\partial g}{\partial x}(s-, s) = -\frac{1}{p(s)}.$$

If a Green's function exists these four properties characterize it.

Theorem 101 *Let L be a regular Sturm-Liouville differential operator on $[a, b]$. If $Ly = 0$, $B_1 y = 0$, $B_2 y = 0$ has only the trivial solution and a function $g(x, s)$ exists with Properties 1-4, then it is the Green's function for the mixed Sturm-Liouville boundary value problem $Ly = f$, $B_1 y = 0$, $B_2 y = 0$.*

Proof. Since $Ly = 0$, $B_1 y = 0$, $B_2 y = 0$ has only the trivial solution, there is a Green's function $g(x, s)$ that has Properties 1-4. We must show that no other function $h(x, s)$ defined on $[a, b] \times [a, b]$ has Properties 1-4. Suppose $h(x, s)$ were such a function. Then $z(x) = h(x, s) - g(x, s)$ regarded as a function of x for each fixed s, is continuous and satisfies $B_1 z = 0$ and $B_2 z = 0$. Since $Lz = 0$ for $x \neq s$, z' exists and is continuous for $x \neq s$. By Property 1, z is continuously differentiable on $[a, s]$ and on $[s, b]$ and by the jump condition

$$z'(s+) - z'(s-) = 0.$$

It follows that z is continuously differentiable on $[a, b]$. Now integrate $Lz = 0$ from c in $[a, s)$ to x in $[a, s)$ and let c tend to s to get

$$p(x)z'(x) - p(s)z'(s) = \int_s^x q(t)z(t)\, dt.$$

Similar reasoning on $(s, b]$ establishes the same result for x in $(s, b]$. Consequently, for $x \neq s$ in $[a, b]$,

$$\frac{p(x)z'(x) - p(s)z'(s)}{x - s} = \frac{1}{x - s}\int_s^x q(t)z(t)\, dt$$

and by the fundamental theorem of calculus there exists

$$\left(p(x)z'(x)\right)'\big|_{x=s} = q(s)z(s)$$

and z satisfies $Lz = 0$ at $x = s$. Thus z is a solution to the corresponding homogeneous problem $Ly = 0$, $B_1 y = 0$, $B_2 y = 0$ and by assumption $z(x) = 0$ for all x in $[a, b]$ and each fixed s in $[a, b]$. That is, $h(x, s) = g(x, s)$ for all (x, s) in $[a, b] \times [a, b]$. ■

Example 5. Fix $l > 0$ and let $f(x)$ be continuous on $[0, l]$. Find the Green's function for

$$\begin{cases} -y'' + y = f(x), & 0 < x < l, \\ y'(0) = 0, & y(0) - y(l) = 0. \end{cases}$$

Here $Ly = -y'' + y$, $p(x) = 1$, $B_1 y = y'(0)$, and $B_2 y = y(0) - y(l)$.

It is easy to check that the corresponding homogeneous problem has only the trivial solution so the Green's function $g(x, s)$ exists. We will use Theorem 100 to find it. The functions

$$u = \cosh x \quad \text{and} \quad v = -\sinh x$$

satisfy $Lu = 0$, $Lv = 0$, and $p(x)\,W_{u,v}(x) = -1$. So

$$v(x)u(s) - u(x)v(s) = -\sinh x \cosh s + \cosh x \sinh s = \sinh(s - x),$$

and

$$\tilde{g}(x, s) = \begin{cases} 0 & \text{for } 0 \le x \le s \le l \\ -\sinh(x - s) & \text{for } 0 \le s \le x \le l \end{cases}.$$

The 2×2 system for c_1 and c_2 in Theorem 100 is

$$\begin{cases} c_1(0) + c_2(-1) = 0 \\ c_1(1 - \cosh l) + c_2(\sinh l) = -\sinh(l - s) \end{cases},$$

$$c_1 = -\frac{\sinh(l - s)}{1 - \cosh l}, \quad \text{and} \quad c_2 = 0.$$

Thus, the Green's function is

$$g(x, s) = \frac{\sinh(l - s)}{\cosh l - 1}\cosh x - \begin{cases} 0 & \text{for } 0 \le x \le s \le l \\ \sinh(x - s) & \text{for } 0 \le s \le x \le l \end{cases}.$$

Example 6. Let $f(x)$ be continuous on $[0, l]$. Find the Green's function for

$$\begin{cases} -y'' + ay = f(x), & 0 < x < l, \\ y(0) = y(l), & y'(0) = y'(l), \end{cases}$$

where $a > 0$. Here $Ly = -y'' + ay$, $p(x) = 1$, $B_1 y = y(0) - y(l)$, and $B_2 y = y'(0) - y'(l)$.

It is easy to check that the corresponding homogeneous problem has only the trivial solution so the Green's function exists. We will use Theorem 100 to find it. The real-valued functions

$$u = \sinh\sqrt{a}x \quad \text{and} \quad v = \left(\sqrt{a}\right)^{-1}\cosh\sqrt{a}x$$

satisfy $Lu = 0$, $Lv = 0$, $p(x)\,W_{u,v}(x) = -1$,

$$v(x)u(s) - u(x)v(s) = -\left(\sqrt{a}\right)^{-1}\sinh\sqrt{a}(x - s).$$

Hence,

$$\tilde{g}(x, s) = \begin{cases} 0 & \text{for } a \le x \le s \le b \\ -\left(\sqrt{a}\right)^{-1}\sinh\sqrt{a}(x - s) & \text{for } a \le s \le x \le b \end{cases}.$$

The 2×2 system for c_1 and c_2 in Theorem 100 is

$$\begin{cases} c_1(-\sinh\sqrt{a}l) + c_2\left(\left(\sqrt{a}\right)^{-1}(1 - \cosh\sqrt{a}l)\right) = -\left(\sqrt{a}\right)^{-1}\sinh\sqrt{a}(l - s) \\ c_1\left(\sqrt{a}(1 - \cosh\sqrt{a}l)\right) + c_2(-\sinh\sqrt{a}l) = -\cosh\sqrt{a}(l - s) \end{cases}.$$

The system has determinant $\Delta = 2\left(\cosh\sqrt{a}l - 1\right)$. Solving the system, say by Cramer's rule, and using hyperbolic identities gives

$$\begin{aligned} c_1 &= \frac{1}{\Delta\sqrt{a}}\left(\sinh\sqrt{a}l\sinh\sqrt{a}(l - s) + \cosh\sqrt{a}(l - s) - \cosh\sqrt{a}l\cosh\sqrt{a}(l - s)\right) \\ &= \frac{1}{\Delta\sqrt{a}}\left(\cosh\sqrt{a}(l - s) - \cosh\sqrt{a}s\right) \end{aligned}$$

and

$$\begin{aligned} c_2 &= \frac{1}{\Delta}\left(\sinh\sqrt{a}l\cosh\sqrt{a}(l-s)+\sinh\sqrt{a}(l-s)-\cosh\sqrt{a}l\sinh\sqrt{a}(l-s)\right) \\ &= \frac{1}{\Delta}\left(\sinh\sqrt{a}(l-s)+\sinh\sqrt{a}s\right) \end{aligned}$$

So

$$\begin{aligned} c_1u(x)+c_2v(x) &= \frac{1}{\Delta\sqrt{a}}\left(\cosh\sqrt{a}(l-s)-\cosh\sqrt{a}s\right)\sinh\sqrt{a}x \\ &\quad+\frac{1}{\Delta}\left(\sinh\sqrt{a}(l-s)+\sinh\sqrt{a}s\right)\frac{1}{\sqrt{a}}\cosh\sqrt{a}x \\ &= \frac{1}{\Delta\sqrt{a}}\left(\cosh\sqrt{a}(l-s)\sinh\sqrt{a}x+\sinh\sqrt{a}(l-s)\cosh\sqrt{a}x\right) \\ &\quad+\frac{1}{\Delta\sqrt{a}}\left(\sinh\sqrt{a}s\cosh\sqrt{a}x-\cosh\sqrt{a}s\sinh\sqrt{a}x\right) \\ &= \frac{1}{\Delta\sqrt{a}}\left(\sinh\sqrt{a}(l-s+x)+\sinh\sqrt{a}(s-x)\right) \end{aligned}$$

and the Green's function is

$$g(x,s)=\frac{1}{\Delta\sqrt{a}}\begin{cases} \sinh\sqrt{a}(l+x-s)-\sinh\sqrt{a}(x-s) & \text{for } 0\le x\le s\le l \\ \sinh\sqrt{a}(l+x-s)-(1+\Delta)\sinh\sqrt{a}(x-s) & \text{for } 0\le s\le x\le l \end{cases}$$

where

$$\Delta = 2\left(\cosh\sqrt{a}l-1\right).$$

An alternative convenient expression for the Green's function follows from another use a hyperbolic identity. Since

$$\sinh\sqrt{a}(l+x-s)=\sinh\sqrt{a}l\cosh\sqrt{a}(x-s)+\cosh\sqrt{a}l\sinh\sqrt{a}(x-s),$$

$$\begin{aligned} \sinh\sqrt{a}(l+x-s)-\sinh\sqrt{a}(x-s) &= \sinh\sqrt{a}l\cosh\sqrt{a}(x-s) \\ &\quad+\left(\cosh\sqrt{a}l-1\right)\sinh\sqrt{a}(x-s) \\ &= \sinh\sqrt{a}l\cosh\sqrt{a}(x-s) \\ &\quad+\frac{\Delta}{2}\sinh\sqrt{a}(x-s) \end{aligned}$$

and

$$\begin{aligned} &\sinh\sqrt{a}(l+x-s)-(1+\Delta)\sinh\sqrt{a}(x-s)= \\ &\sinh\sqrt{a}l\cosh\sqrt{a}(x-s)-\frac{\Delta}{2}\sinh\sqrt{a}(x-s) \end{aligned}.$$

Hence

$$g(x,s)=\frac{1}{\sqrt{a}}\begin{cases} 2^{-1}\sinh\sqrt{a}(x-s)+\Delta^{-1}\sinh\sqrt{a}l\cosh\sqrt{a}(x-s), & x\le s \\ -2^{-1}\sinh\sqrt{a}(x-s)+\Delta^{-1}\sinh\sqrt{a}l\cosh\sqrt{a}(x-s), & s\le x \end{cases}.$$

This representation makes it easy to confirm directly that $g(x,s)=g(s,x)$. So the Green's function is a real-valued, symmetric kernel.

We assumed in Example 6 that $a > 0$. However, the solution is valid for any real or complex constant $a \neq 0$ for which the Green's function exists; that is, for which $\Delta = 2(\cosh\sqrt{a}l - 1) \neq 0$, equivalently $a \neq -(2\pi n/l)^2$ for some positive integer n.

Example 6. (continued) The most important choices for a are $a > 0$ and $a < 0$. In the latter case, $a = -|a|$ and $\sqrt{a} = i\sqrt{|a|}$. Since

$$\cosh it = \cos t \quad \text{and} \quad \sinh it = i \sin t,$$

the formulas for the Green's function can be expressed in terms of trigonometric functions as

$$g(x, s) = \frac{1}{\Delta\sqrt{|a|}}\begin{cases} \sin\sqrt{|a|}(l + x - s) - \sin\sqrt{|a|}(x - s), & 0 \leq x \leq s \leq l \\ \sin\sqrt{|a|}(l + x - s) - (1 + \Delta)\sin\sqrt{|a|}(x - s), & 0 \leq s \leq x \leq l \end{cases}$$

or as

$$g(x, s) = \frac{1}{\sqrt{|a|}}\begin{cases} 2^{-1}\sin\sqrt{|a|}(x - s) + \Delta^{-1}\sin\sqrt{|a|}l\cos\sqrt{|a|}(x - s), & x \leq s \\ -2^{-1}\sin\sqrt{|a|}(x - s) + \Delta^{-1}\sin\sqrt{|a|}l\cos\sqrt{|a|}(x - s), & s \leq x \end{cases}.$$

The same results can be obtained directly for Theorem 100 using the real-valued functions

$$u = \sin\sqrt{|a|}x \quad \text{and} \quad v = \left(\sqrt{|a|}\right)^{-1}\cos\sqrt{|a|}x$$

that satisfy $Lu = 0$, $Lv = 0$, $p(x)\,W_{u,v}(x) = -1$.

4.7 Adjoint Operators and Problems

In this section, we take a closer look at the Sturm-Liouville differential operator L and introduce its adjoint operator, adjoint boundary conditions, and adjoint boundary value problem. Throughout this section, we assume that the Sturm-Liouville differential operator $L = -(py')' + qy$ is **regular** on $[a, b]$, that is, $p(x) \neq 0$ on $[a, b]$ and $p(x)$ and $q(x)$ are continuous on $[a, b]$ and that $C\,[a, b]$ is the inner product space of real or complex-valued continuous functions with the usual inner product

$$\langle y, z\rangle = \int_a^b y(s)\overline{z(s)}\,ds.$$

The **domain** of the Sturm-Liouville differential operator

$$Ly(x) = -(p(x)y'(x))' + q(x)y(x)$$

is

$$D = \{y \in C\,[a, b] : Ly \in C\,[a, b]\}.$$

Equivalently,

$$D = \{y \in C\,[a, b] : (py')' \in C\,[a, b]\}.$$

The motivation for this choice for the domain of L is that we are interested in functions y that are solutions to $Ly = f$ and $Ly = \lambda ry$ where all coefficients and function data are continuous on $[a, b]$. Thus Ly is continuous there. Note that y in D implies y' is continuous on $[a, b]$ because py' is continuous and $p \neq 0$ there. If the coefficient p is continuously differentiable, then y in D implies y'' is continuous on $[a, b]$. Conversely, if p is continuously differentiable and y'' is continuous on $[a, b]$, then $(py')' = py'' + p'y$ is continuous on $[a, b]$; see Section 4.2. In summary, if p is continuously differentiable on $[a, b]$, then

$$D = C^2\,[a, b].$$

For y in D and z suitably smooth, integration by parts yields

$$\begin{aligned}\int_a^b -\left(py'\right)'\bar{z}\,ds &= -py'\bar{z}\big|_a^b + \int_a^b py'\bar{z}'\,ds \\ &= \left(py\bar{z}' - py'\bar{z}\right)\big|_a^b - \int_a^b \left(p\bar{z}'\right)'y\,ds \\ &= p\left(y\bar{z}' - y'\bar{z}\right)\big|_a^b + \int_a^b y\overline{\left(-\bar{p}z'\right)'}\,ds.\end{aligned}$$

Hence,

$$\begin{aligned}\langle Ly, z\rangle &= \int_a^b Ly\bar{z}\,ds = p\left(y\bar{z}' - y'\bar{z}\right)\big|_a^b + \int_a^b y\overline{L^*z}\,ds \\ &= p\left(y\bar{z}' - y'\bar{z}\right)\big|_a^b + \langle y, L^*z\rangle \end{aligned} \tag{4.8}$$

where

$$L^*z = -\left(\bar{p}z'\right)' + \bar{q}z.$$

L^* is the **adjoint differential operator** of L and has domain

$$D^* = \{z \in C\,[a, b] : L^*z \in C\,[a, b]\}.$$

It follows that $D^* = \overline{D}$. Consequently, z in D^* implies z' exists and is continuous on $[a, b]$, $(\bar{p}z')'$ is continuous there, and (4.8) holds. If p is continuously differentiable on $[a, b]$, then $D^* = C^2\,[a, b]$.

We split our treatment of adjoint boundary conditions into two cases: the case of separated boundary conditions and the case of mixed boundary conditions. In the latter case, we restrict our attention mainly to periodic boundary conditions. They are the problems with mixed boundary conditions that arise in practice, for example, when separating variables in the Laplace equation on a circular domain.

4.7.1 Separated Adjoint Boundary Conditions

Let

$$B_a^*z = \alpha^*z(a) + \beta^*z'(a) \quad \text{where} \quad |\alpha^*| + |\beta^*| > 0$$

and

$$B_b^*z = \gamma^*z(b) + \delta^*z'(b) \quad \text{where} \quad |\gamma^*| + |\delta^*| > 0.$$

The separated boundary conditions $B_a^*z = 0$ and $B_b^*z = 0$ are called **adjoint boundary conditions** to $B_ay = 0$ and $B_by = 0$ if

$$B(y, z) = p\left(y\bar{z}' - y'\bar{z}\right)\big|_a^b = 0$$

for all continuously differentiable functions y and z that satisfy $B_ay = 0$, $B_by = 0$ and $B_a^*z = 0$ and $B_b^*z = 0$. For any set of boundary conditions and adjoint boundary conditions, we have

$$\langle Ly, z\rangle = \langle y, L^*z\rangle$$

for all y in the domain of L and all z in the domain of L^* that satisfy the respective boundary conditions.

Assume that $B_a y = 0$, $B_b y = 0$ and $B_a^* z = 0$ and $B_b^* z = 0$ are adjoint boundary conditions. Among the functions y and z that satisfy the boundary conditions are those with $y(b) = y'(b) = 0$. For such y and z,

$$B(y, z) = -p(a)(y(a)\bar{z}'(a) - y'(a)\bar{z}(a)).$$

If $\alpha\alpha^* \neq 0$, then

$$\begin{aligned} B(y, z) &= -p(a)((-\beta/\alpha)y'(a)\bar{z}'(a) - y'(a)(-\overline{\beta^*}/\overline{\alpha^*})\bar{z}'(a)) \\ &= -p(a)y'(a)\bar{z}'(a)(\alpha\overline{\beta^*} - \beta\overline{\alpha^*})/\alpha\overline{\alpha^*} \end{aligned}$$

and functions y and z can be chosen that satisfy the boundary conditions and assume arbitrarily prescribed values for $y'(a)$ and $z'(a)$. It follows that

$$\alpha\overline{\beta^*} - \beta\overline{\alpha^*} = 0$$

because the boundary conditions are adjoint to each other. If $\alpha = 0$, then $y(a)$ can be chosen arbitrarily, $\beta \neq 0$ so y must be chosen with $y'(a) = 0$ to satisfy $B_a y = 0$ and

$$B(y, z) = -p(a)y(a)\bar{z}'(a) = 0$$

because the boundary conditions are adjoint to each other. This requires $\alpha^* = 0$; otherwise, $z'(a)$ can be chosen arbitrarily in determining a z with $B_a^* z = 0$ and $B(y, z) = 0$ cannot hold for all admissible choices of y and z. Thus, $\alpha = 0$ implies $\alpha^* = 0$ for adjoint boundary conditions. Likewise, $\alpha^* = 0$ implies $\alpha = 0$ for adjoint boundary conditions. Consequently,

$$\alpha\overline{\beta^*} - \beta\overline{\alpha^*} = 0$$

is a necessary condition for the boundary conditions to be adjoint to each other. Likewise,

$$\gamma\overline{\delta^*} - \delta\overline{\gamma^*} = 0$$

is a necessary condition for the boundary conditions to be adjoint to each other. Retracing the reasoning above with small adjustments confirms that these necessary conditions are also sufficient conditions. We have established

Lemma 102 *The separated boundary conditions $B_a y = 0$, $B_b y = 0$ and $B_a^* z = 0$ and $B_b^* z = 0$ are adjoint to each other if and only if $\alpha\overline{\beta^*} - \beta\overline{\alpha^*} = 0$ and $\gamma\overline{\delta^*} - \delta\overline{\gamma^*} = 0$.*

An important special case of the lemma is: if α, β, γ, and δ are real, then the boundary conditions are adjoint to themselves because the conditions in the lemma are satisfied by the choices $\alpha^* = \alpha$, $\beta^* = \beta$, $\gamma^* = \gamma$, and $\delta^* = \delta$. Boundary conditions that are adjoint to themselves are called **self-adjoint (boundary conditions).**

We call the boundary value problems

$$Ly = f, \; B_a y = 0, \; B_b y = 0$$

and

$$L^* z = h, \; B_a^* z = 0, \; B_b^* z = 0$$

where h is a give continuous function on $[a, b]$ **adjoint boundary value problems** if $B_a y = 0$, $B_b y = 0$ and $B_a^* z = 0$, $B_b^* z = 0$ are adjoint boundary conditions. There is a close relation between the Green's function $g^*(x, s)$ for the latter problem and the Green's function $g(x, s)$ of the former problem. The key to this relationship is

Lemma 103 *If* $\alpha\overline{\beta^*} - \beta\overline{\alpha^*} = 0$ *then* c *and* d *satisfy* $\alpha c + \beta d = 0$ *if and only if* c *and* d *satisfy* $\overline{\alpha^*} c + \overline{\beta^*} d = 0$. *If* $\gamma\overline{\delta^*} - \delta\overline{\gamma^*} = 0$ *then* c *and* d *satisfy* $\gamma c + \delta d = 0$ *if and only if* c *and* d *satisfy* $\overline{\gamma^*} c + \overline{\delta^*} d = 0$.

Proof. Assume $\alpha\overline{\beta^*} - \beta\overline{\alpha^*} = 0$. If $\alpha = 0$, then $\beta \neq 0$ and hence $\alpha^* = 0$ and $\beta^* \neq 0$. In this case, the common solution set of the two equations is c arbitrary and $d = 0$. The same conclusion is reached if $\alpha^* = 0$. Now, assume $\alpha\alpha^* \neq 0$. Then

$$\overline{\alpha^*} c + \overline{\beta^*} d = \overline{\alpha^*}\left(c + \frac{\overline{\beta^*}}{\overline{\alpha^*}} d \right) = \overline{\alpha^*}\left(c + \frac{\beta}{\alpha} d \right) = \frac{\overline{\alpha^*}}{\alpha}(\alpha c + \beta d)$$

and the first assertion in the lemma is established. The second is established in the same way. ■

It follows by taking complex conjugates throughout, that the equations

$$L^* z = 0, \ B_a^* z = 0, \ B_b^* z = 0 \quad \text{and} \quad y = \bar{z}$$

hold if and only if the equations

$$Ly = 0, \ \overline{B_a^*} y = 0, \ \overline{B_b^*} y = 0, \quad \text{and} \quad z = \bar{y}$$

hold, where

$$\overline{B_a^*} y = \overline{\alpha^*} y(a) + \overline{\beta^*} y'(a) \quad \text{and} \quad \overline{B_b^*} y = \overline{\gamma^*} y(a) + \overline{\delta^*} y'(a).$$

By the lemma the homogeneous boundary conditions $\overline{B_a^*} y = 0$ and $\overline{B_b^*} y = 0$ hold if and only if $B_a y = 0$ and $B_b y = 0$ hold. So the equations

$$L^* z = 0, \ B_a^* z = 0, \ B_b^* z = 0 \quad \text{and} \quad y = \bar{z}$$

hold if and only if the equations

$$Ly = 0, \ B_a y = 0, \ B_b y = 0, \quad \text{and} \quad z = \bar{y}$$

hold. Consequently, the adjoint boundary value problem has a Green's function $g^*(x, s)$ if and only if the original boundary value problem has a Green's function $g(x, s)$, in which case, by Theorem 96, there are functions u and v such that

$$\begin{aligned} Lu &= 0, \ B_a u = 0, \\ Lv &= 0, \ B_b v = 0, \\ pW_{u,v} &= -1, \end{aligned}$$

and

$$g(x, s) = \begin{cases} u(x)v(s) & \text{for } a \le x \le s \le b \\ u(s)v(x) & \text{for } a \le s \le x \le b \end{cases}.$$

Take complex conjugates to obtain

$$\begin{aligned} L^* \bar{u} &= 0, \ \overline{B_a} \bar{u} = 0, \\ L\bar{v} &= 0, \ \overline{B_b} \bar{v} = 0, \\ \bar{p} W_{\bar{u},\bar{v}} &= -1. \end{aligned}$$

By the lemma the boundary conditions $\overline{B_a}\bar{u} = 0$ and $\overline{B_b}\bar{v}$ hold if and only if $B_a^*\bar{u} = 0$ and $B_b^*\bar{v} = 0$ hold. By Theorem 96 the Green's function for the adjoint boundary value problem is

$$g^*(x, s) = \begin{cases} \bar{u}(x)\bar{v}(s) & \text{for } a \le x \le s \le b \\ \bar{u}(s)\bar{v}(x) & \text{for } a \le s \le x \le b \end{cases}$$

and, hence, $g^*(x, s) = \overline{g(x, s)}$. Since $g(x, s) = g(s, x)$, the Green's function for L with boundary conditions $B_a y = 0$ and $B_b y = 0$ and for L^* with boundary conditions $B_a^* z = 0$ and $B_b^* y = 0$ are related by

$$g^*(x, s) = \overline{g(s, x)}$$

for x and s in $[a, b]$. That is, $g^*(x, s)$ is the adjoint kernel of $g(x, s)$ as defined in Section 3.4. In summary,

Theorem 104 *If $Ly = f$, $B_a y = 0$, $B_b y = 0$ and $L^* z = h$, $B_a^* z = 0$, $B_b^* z = 0$ are adjoint boundary value problems, then the first problem has a Green's function $g(x, s)$ if and only if the second problem has a Greens's function $g^*(x, s)$, in which case*

$$g^*(x, s) = \overline{g(s, x)}.$$

If $G : C[a, b] \to C[a, b]$ and $G^* : C[a, b] \to C[a, b]$ are the integral operators with kernels $g(x, s)$ and $g^*(x, s)$, respectively, then

$$\langle Gf, h\rangle = \langle f, G^* h\rangle$$

for all continuous functions f and h in $C[a, b]$. This follows from the results in Section 3.4 or directly from the interplay between Sturm-Liouville operators and their Green's functions: given f and h in $C[a, b]$, let y and z be the solutions of $Ly = f$, $B_a y = 0$, $B_b y = 0$ and $L^* z = h$, $B_a^* z = 0$, $B_b^* z = 0$ respectively, then

$$\langle Gf, h\rangle = \langle y, L^* z\rangle = \langle Ly, z\rangle = \langle f, G^* h\rangle.$$

The differential operator L with boundary conditions $B_a y = 0$ and $B_b y = 0$ is called **self-adjoint** if $L^* = L$ and the boundary conditions $B_a y = 0$ and $B_b y = 0$ are adjoint to themselves; that is, $p = \bar{p}$ and $q = \bar{q}$ and the choices $\alpha^* = \alpha$, $\beta^* = \beta$, $\gamma^* = \gamma$, $\delta^* = \delta$ satisfy

$$\alpha\overline{\beta^*} - \beta\overline{\alpha^*} = 0 \quad \text{and} \quad \gamma\overline{\delta^*} - \delta\overline{\gamma^*} = 0.$$

These conditions for self-adjointness hold if and only if

$$p \text{ and } q \text{ are real-valued and } \alpha\bar{\beta} \text{ and } \gamma\bar{\delta} \text{ are real.}$$

Consequently,

Theorem 105 *The regular Sturm-Liouville differential operator $Ly = -(py')' + qy$ with separated boundary conditions $B_a y = 0$ and $B_b y = 0$ is self-adjoint if p and q are real-valued and α, β, γ,* and δ are real numbers.

In the self-adjoint case, the boundary condition at $x = a$ can be expressed with all real coefficients because

$$\alpha y(a) + \beta y'(a) = \begin{cases} \bar{\beta}^{-1}(\alpha\bar{\beta}y(a) + |\beta|^2 y'(a)) & \text{if } \beta \ne 0 \\ \bar{\alpha}^{-1}(|\alpha|^2 y(a) + \bar{\alpha}\beta y'(a)) & \text{if } \alpha \ne 0 \end{cases}$$

Likewise, the boundary condition at $x = b$ can be expressed with all real coefficients. Since p and q are real-valued and the boundary conditions can be expressed with real coefficients in the self-adjoint case, the Green's function $g(x, s)$ (when it exists) is a symmetric kernel (that is, $g(x, s)$ is real-valued and $g(x, s) = g(s, x)$) by Theorem 96. Hence,

$$g^*(x, s) = \overline{g(s, x)} = g(s, x) = g(x, s)$$

and the Green's function is a self-adjoint kernel.

Theorem 106 *If a self-adjoint regular Sturm-Liouville differential operator with separated boundary conditions has a Green's function $g(x, s)$, then $g(x, s)$ is a real-valued, symmetric, self-adjoint kernel.*

Consequently, if f is real-valued, the solution y to a self-adjoint Sturm-Liouville boundary value problem $Ly = f$, $B_a y = 0$, and $B_b y = 0$ is real-valued.

4.7.2 Mixed Adjoint Boundary Conditions

We continue with the notation from the previous sections. The Sturm-Liouville differential operator L and its adjoint operator L^* satisfy

$$Ly = -(py')' + qy,$$
$$L^* z = -(\bar{p}y')' + \bar{q}y,$$
$$\langle Ly, z\rangle = p(y\bar{z}' - y'\bar{z})\big|_a^b + \langle y, L^* z\rangle$$

for all y in the domain of L and all z in the domain of L^*. The linear forms

$$B_i y = a_{i1} y(a) + a_{i2} y'(a) + b_{i1} y(b) + b_{i2} y'(b)$$

for $i = 1, 2$ and for real or complex constants a_{ij} and b_{ij} define the linear homogeneous boundary conditions $B_i y = 0$ for $i = 1, 2$. Let

$$B_i^* y = a_{i1}^* y(a) + a_{i2}^* y'(a) + b_{i1}^* y(b) + b_{i2}^* y'(b)$$

be linear forms that determine the boundary conditions $B_i^* y = 0$ for $i = 1, 2$. The mixed boundary conditions $B_1^* z = 0$ and $B_2^* z = 0$ are called **adjoint boundary conditions** to $B_a y = 0$ and $B_b y = 0$ if

$$B(y, z) = p(y\bar{z}' - y'\bar{z})\big|_a^b = 0$$

for all continuously differentiable functions y and z that satisfy $B_1 y = 0$, $B_2 y = 0$ and $B_1^* z = 0$ and $B_2^* z = 0$. For any set of boundary conditions and adjoint boundary conditions, we have

$$\langle Ly, z\rangle = \langle y, L^* z\rangle$$

for all y in the domain of L and all z in the domain of L^* that satisfy the respective boundary conditions.

We call the boundary value problems

$$Ly = f, \ B_1 y = 0, \ B_2 y = 0$$

and

$$L^* z = h, \ B_1^* z = 0, \ B_2^* z = 0$$

where h is a give continuous function on $[a, b]$ **adjoint boundary value problems** if $B_1 y = 0$, $B_2 y = 0$ and $B_1^* z = 0$, $B_2^* z = 0$ are adjoint boundary conditions. There is a close relation between the Green's function $g^*(x, s)$ for the adjoint problem and the Green's function $g(x, s)$, that we present next.

Lemma 107 *For adjoint boundary value problems, $Ly = 0$, $B_1 y = 0$, $B_2 y = 0$ has only the trivial solution $y = 0$ if and only if $L^* z = 0$, $B_1^* z = 0$, $B_2^* z = 0$ has only the trivial solution $z = 0$.*

Proof. If $Ly = 0$, $B_1y = 0$, $B_2y = 0$ has only the trivial solution, the Green's function $g(x, s)$ exists. If z is a solution of $L^*z = 0$, $B_1^*z = 0$, $B_2^*z = 0$, then

$$\langle Ly, z\rangle = \langle y, L^*z\rangle = 0$$

for all y in the domain of L that satisfy $B_1y = 0$ and $B_2y = 0$. Since the Green's function $g(x, s)$ exists and z is continuous on $[a, b]$, the problem $Ly = z$, $B_1y = 0$, $B_2y = 0$ has a unique solution y. This choice for y in the displayed equation above gives $\langle z, z\rangle = 0$ and $z = 0$. The converse assertion is proven in the same way. ■

By the lemma, if $Ly = f$, $B_1y = 0$, $B_2y = 0$ has a Green's function $g(x, s)$, then $L^*z = h$, $B_1^*z = 0$, $B_2^*z = 0$ has a Green's function $g^*(x, s)$, and conversely. In this case, given any two continuous functions f and h on $[a, b]$, the solution y to $Ly = f$, $B_1y = 0$, $B_2y = 0$ is $y = Gf$ and the solution z to $L^*z = h$, $B_1^*z = 0$, $B_2^*z = 0$ is $z = G^*h$ where G and G^* are the integral operators with kernels $g(x, s)$ and $g^*(x, s)$, respectively. Substitution into $\langle Ly, z\rangle = \langle y, L^*z\rangle$ yields

$$\langle f, G^*h\rangle = \langle Gf, h\rangle$$

for any continuous functions f and g on $[a, b]$.

Theorem 108 *If $Ly = f$, $B_1y = 0$, $B_2y = 0$ and $L^*z = h$, $B_1^*z = 0$, $B_2^*z = 0$ are adjoint boundary value problems, then the first problem has a Green's function $g(x, s)$ if and only if the second problem has a Greens's function $g^*(x, s)$, in which case*

$$g^*(x, s) = \overline{g(s, x)}.$$

Proof. The first conclusion has already been established. Since

$$Gf(x) = \int_a^b g(x, s)f(s)\, ds \quad \text{and} \quad G^*h(x) = \int_a^b g^*(x, s)h(s)\, ds,$$

the relation $\langle f, G^*h\rangle = \langle Gf, h\rangle$ can be expressed as

$$\int_a^b f(x)\overline{\int_a^b g^*(x, s)h(s)\, ds}\, dx = \int_a^b \left(\int_a^b g(x, s)f(s)\, ds\right)\overline{h(x)}\, dx,$$

$$\int_a^b \int_a^b \overline{g^*(x, s)}f(x)\overline{h(s)}\, dsdx = \int_a^b \int_a^b g(s, x)f(x)\overline{h(s)}\, dsdx.$$

Thus

$$\int_a^b \left(\int_a^b \left(\overline{g^*(x, s)} - g(s, x)\right)\overline{h(s)}\, ds\right)f(x)\, dx = 0$$

for all continuous functions f and h on $[a, b]$. Apply Corollary 20 twice to obtain $\overline{g^*(x, s)} - g(s, x) = 0$ for all (x, s) in $[a, b] \times [a, b]$ and the theorem is established. ■

The differential operator L with boundary conditions $B_1y = 0$ and $B_2y = 0$ is called **self-adjoint** if $L^* = L$, that is, $p = \bar{p}$ and $q = \bar{q}$ and the boundary conditions $B_1y = 0$ and $B_2y = 0$ are adjoint to themselves.

Theorem 109 *If a self-adjoint regular Sturm-Liouville differential operator with mixed boundary conditions whose coefficients are real numbers has a Green's function $g(x, s)$, then $g(x, s)$ is a real-valued, symmetric, self-adjoint kernel.*

Proof. By Theorem 100, $g(x, s)$ is real-valued because all coefficients in the differential equation and boundary conditions are real-valued. Since the problem $Ly = f$, $B_1y = 0$, $B_2y = 0$ is adjoint to itself and a Green's function is unique when it exists, $g^*(x, s) = g(x, s)$. On the other hand, by the previous theorem $g^*(x, s) = \overline{g(s, x)} = g(s, x)$ because g is real-valued. Consequently, $g(x, s) = g(s, x)$ and the Green's function is real-valued and symmetric. ■

The mixed boundary conditions of primary interest to us are periodic boundary conditions, and, to a lesser extent, antiperiodic boundary conditions and some close relatives. Specifically we consider the mixed boundary conditions determined by the linear forms

$$B_1 y = y(a) - \sigma_0 y(b), \quad \text{and} \quad B_2 y = y'(a) - \sigma_1 y'(b)$$

where σ_0 and σ_1 are given real or complex numbers. These boundary conditions have adjoint boundary conditions of the form

$$B_1^* z = z(a) - \tau_0 z(b) \quad \text{and} \quad B_2^* z = z'(a) - \tau_1 z'(b),$$

where τ_0 and τ_1 are given real or complex numbers and σ_0, σ_1, τ_0, and τ_1 are suitably related.

Lemma 110 *Assume σ_0, σ_1, τ_0, and τ_1 are all nonzero real or complex numbers. The boundary conditions $y(a) = \sigma_0 y(b)$, $y'(a) = \sigma_1 y'(b)$, and $z(a) = \tau_0 z(b)$, $z'(a) = \tau_1 z'(b)$ are adjoint to each other if and only if*

$$p(b) = p(a)\sigma_0\overline{\tau_1} \quad \text{and} \quad p(b) = p(a)\sigma_1\overline{\tau_0}.$$

Proof. The bilinear form

$$\begin{aligned} B(y, z) &= p\left(y\bar{z}' - y'\bar{z}\right)\Big|_a^b \\ &= p(b)(y(b)\bar{z}'(b) - y'(b)\bar{z}(b)) \\ &\quad - p(a)(\sigma_0 y(b)\overline{\tau_1}\bar{z}'(b) - \sigma_1 y'(b)\overline{\tau_0}\bar{z}(b)) \\ &= (p(b) - p(a)\sigma_0\overline{\tau_1})y(b)\bar{z}'(b) + (-p(b) + p(a)\sigma_1\overline{\tau_0})y'(b)\bar{z}(b). \end{aligned}$$

Functions y and z can be chosen that satisfy the respective boundary conditions and for which $y(b)$, $\bar{z}'(b)$, $y'(b)$, and $\bar{z}(b)$ can assume arbitrary values. Consequently, $B(y, z) = 0$ for all such y and z and the boundary conditions are adjoint to each other if and only if

$$p(b) - p(a)\sigma_0\overline{\tau_1} = 0 \quad \text{and} \quad -p(b) + p(a)\sigma_1\overline{\tau_0} = 0,$$

which establishes the lemma. ■

Theorem 111 *If $p(b) = p(a)$, then the periodic boundary conditions*

$$y(a) = y(b), \; y'(a) = y'(b)$$

are self-adjoint (adjoint to themselves) and the antiperiodic boundary conditions

$$y(a) = -y(b), \; y'(a) = -y'(b)$$

are self-adjoint. If $Ly = -(py')' + qy$ has real-valued coefficients, then, when they exist, the Green's function for Ly with periodic boundary conditions and the Green's function for Ly with antiperiodic boundary conditions are real-valued and symmetric.

Proof. The choices $\sigma_0 = 1$, $\sigma_1 = 1$, $\tau_0 = 1$, and $\tau_1 = 1$ give periodic boundary conditions and the lemma show that periodic boundary conditions are adjoint to themselves. Likewise, the choices $\sigma_0 = -1$, $\sigma_1 = -1$, $\tau_0 = -1$, and $\tau_1 = -1$ give antiperiodic boundary conditions and the lemma show that antiperiodic boundary conditions are adjoint to themselves. The last pair of assertions follow from the preceding theorem. ■

4.8 Eigenvalue Value Problems

We continue to use the notation of the previous sections:

$$\begin{aligned} Ly &= -(p(x)y')' + q(x)y, \\ B_a y &= \alpha y(a) + \beta y'(a), \\ B_b y &= \gamma y(b) + \delta y'(b), \\ B_i y &= a_{i1} y(a) + a_{i2} y'(a) + b_{i1} y(b) + b_{i2} y'(b) \end{aligned}$$

for $i = 1, 2$ and where Ly is a regular Sturm-Liouville differential operator on $[a, b]$.

We surveyed several regular Sturm-Liouville eigenvalue problems in Chapter 1. Those problems are representative of the vast majority of regular eigenvalue problems that come up in applications: they involve homogeneous boundary conditions with all real data and a regular Sturm-Liouville differential equation of the form

$$-(p(x)y')' + q(x)y = \lambda r(x)y$$

for $a < x < b$ and where $p(x) > 0$, $q(x)$ is real-valued, $r(x) > 0$, and λ is the eigenvalue parameter, which may be real or complex. Consequently, we always assume the following in our treatment of regular eigenvalue problems in this chapter.

Standing Assumptions:

(1) *The Sturm-Liouville* differential operator is regular on $[a, b]$.

(2) $p(x) > 0$ on $[a, b]$ and $q(x)$ is real-valued.

(3) $r(x) > 0$ on $[a,b]$.

(4) The coefficients in B_a, B_b, B_1, and B_2 are real numbers.

Singular problems will be treated in the next two chapters.

Except for an occasional appearance of periodic boundary conditions and antiperiodic boundary conditions, the vast majority of applied problems involve separated boundary conditions. For this reason, we first treat eigenvalue problems with separated boundary conditions and after that consider some problems with mixed boundary conditions. Since the coefficients in the boundary conditions are real, Sturm-Liouville eigenvalue problems with separated boundary conditions are self-adjoint. We only develop general results for Sturm-Liouville eigenvalue problems with mixed boundary conditions in the self-adjoint case, which includes the cases of periodic and antiperiodic boundary conditions. We study Sturm-Liouville eigenvalue problems by converting them to equivalent eigenvalue problems for self-adjoint integral operators whose kernels are Green's functions. We set the stage for this conversion next and explain how to handle the case when there is no Green's function; that is, when zero is an eigenvalue. In subsequent sections, we develop the general theoretical properties of Sturm-Liouville eigenvalue problems. In Chapter 7 we present effective numerical means for calculating eigenvalues and eigenfunctions for the typical situation in which exact evaluations are not possible.

Occasionally we reprise, from a differential equations perspective, proofs of results that were established earlier in the context of integral equations.

We call the eigenvalue problem with separated boundary conditions

$$Ly = \lambda ry, \;\; B_a y = 0, \;\; B_b y = 0 \tag{4.9}$$

and the eigenvalue problem

$$Ly = \lambda ry, \ B_1y = 0, \ B_2y = 0 \tag{4.10}$$

with mixed boundary conditions a **regular Sturm-Liouville eigenvalue problem** when the standing assumptions (1)-(4) are in force. In this context, $r(x) > 0$ is called a **weight function**. Often $r(x) = 1$, as in the case of the eigenvalue problem corresponding to a Sturm-Liouville boundary value problem. Each weight function r determines an inner product on $C[a, b]$ by

$$\langle y, z \rangle_r = \int_a^b y(x)\overline{z(x)}r(x)\, dx$$

and we say y and z are **orthogonal with respect to the weight function** r if $\langle y, z\rangle_r = 0$. The weighted inner product determines the norm $\|y\|_r = \sqrt{\langle y, y\rangle_r}$.

Since the eigenvalue problem with separated boundary conditions is a special case of the problem with mixed boundary conditions, the definitions and observations that follow apply to both problems.

A real or complex number λ is an **eigenvalue** of a Sturm-Liouville eigenvalue problem and a real or complex-valued function $y \neq 0$ is a corresponding **eigenfunction** if y is continuous on $[a, b]$ and (4.10) is satisfied for the pair λ and y. We also say the eigenfunction y belongs to the eigenvalue λ. When we say y satisfies (4.10), we mean that y satisfies the differential equation on (a, b), and satisfies the given boundary conditions. Just as for boundary value problems, this definition implies further smoothness for y. See Theorem 89; a partial restatement of the theorem is given here for convenient reference.

Theorem 112 *If $y(x)$ is an eigenfunction of the regular eigenvalue problem (4.9) or (4.10), then $y(x)$ is continuously differentiable on $[a, b]$ and satisfies the Sturm-Liouville differential equation at every point in $[a, b]$.*

The eigenvalue problem $Ly = \lambda ry$, $B_1y = 0$, $B_2y = 0$ is **self-adjoint** if $L = L^*$ and the boundary conditions are self-adjoint. Consequently,

$$\langle Ly, z\rangle = \langle y, Lz\rangle$$

for all y and z in the domain of L that satisfy the given boundary conditions.

Theorem 113 *If $L = -(py')' + qy$ and $B_ay = 0$, $B_by = 0$ are the differential operator and separated boundary conditions of a regular eigenvalue problem, then the eigenvalue problem is self-adjoint. Moreover, if $\lambda = 0$ is not an eigenvalue of the problem, then the Green's function $g(x, s)$ determined by the differential operator and boundary conditions is real-valued and symmetric.*

Proof. By our standing assumptions, the problem is regular and all the coefficients in the differential operator and separated boundary conditions are real-valued. By Theorem 105 the eigenvalue problem is self-adjoint and by Theorem 106 the Green's function is real-valued and symmetric. ■

Theorem 114 *If the differential operator $L = -(py')' + qy$ and mixed boundary conditions $B_1y = 0$, $B_2y = 0$ determines a self-adjoint eigenvalue problem and if $\lambda = 0$ is not an eigenvalue of the problem, then the corresponding Green's function $g(x, s)$ is real-valued and symmetric.*

Proof. By our standing assumptions, a self-adjoint eigenvalue problem is regular and all the coefficients and data are real-valued. The desired conclusion follows at once from Theorem 109. ■

Lemma 115 *Any eigenvalue of a self-adjoint regular Sturm-Liouville eigenvalue problem is real and eigenfunctions belonging to distinct eigenvalues are orthogonal with respect to the weight function r.*

Proof. If $Ly = \lambda ry$ with $y \neq 0$, then

$$\lambda\langle y, y\rangle_r = \langle \lambda ry, y\rangle = \langle Ly, y\rangle = \langle y, Ly\rangle = \langle y, \lambda ry\rangle = \bar{\lambda}\langle y, y\rangle_r.$$

Since $\langle y, y\rangle_r > 0$, it follows that $\lambda = \bar{\lambda}$ and λ is real. If $Lz = \mu z$ with $z \neq 0$, then

$$\lambda\langle y, z\rangle_r = \langle \lambda ry, y\rangle = \langle Ly, z\rangle = \langle y, Lz\rangle = \langle y, \mu rz\rangle = \mu\langle y, z\rangle_r$$

because μ is real. If $\lambda \neq \mu$ then $\langle y, z\rangle_r = 0$. ■

4.8.1 Recasting the Problem

Assume that $\lambda = 0$ is not an eigenvalue of (4.10) so that the regular Sturm-Liouville differential operator $Ly = -(py')' + qy$ and boundary conditions $B_1 y = 0$, $B_2 y = 0$ determine a Green's function for the boundary value problem

$$Ly = f, \; B_1 y = 0, \; B_2 y = 0.$$

The Green's function can be used to express the Strum-Liouville eigenvalue problem

$$Ly = \lambda ry, \; B_1 y = 0, \; B_2 = 0,$$

as an equivalent integral equation. Simply let $f = \lambda ry$ in the Green's function representation for the solution of the boundary value problem to find the equivalent integral equation eigenvalue problem

$$y(x) = \lambda \int_a^b g(x, s)r(s)y(s)\; ds. \tag{4.11}$$

A few comments about the equivalence are in order. A pair λ, y is a solution to the Sturm-Liouville eigenvalue problem (4.10) if y satisfies $Ly = \lambda ry$ on (a, b) and satisfies the boundary conditions $B_1 y = 0$ and $B_2 y = 0$, in which case y is continuous on $[a, b]$ by Theorem 112. A pair λ, y is a solution to the integral equation (4.11) if y is continuous on $[a, b]$ and the integral equation is satisfied there. The substitution $f = \lambda ry$ used to obtain (4.11) and the fact that a solution y to (4.10) is continuous on $[a, b]$ shows at once that a solution to (4.10) is a solution to (4.11). That the converse holds follows from the four characteristic properties of the Green's function, Properties 1-4 in Section 4.6.2. Simply express the integral equation as

$$y(x) = \lambda \int_a^x g(x, s)r(s)y(s)\; ds + \lambda \int_x^b g(x, s)r(s)y(s)\; ds$$

and differentiate twice using the fundamental theorem of calculus and properties of the Green's function to confirm that the pair λ, y is a solution to (4.10). In summary, λ is an eigenvalue and y is a corresponding eigenfunction of the Sturm-Liouville eigenvalue problem (4.10) if and only if λ is an eigenvalue and y is a corresponding eigenfunction of the kernel $g(x, s)r(s)$.

In the case where the Green's function is real-valued and λ is a real eigenvalue it is useful to know that a corresponding eigenfunction can be chosen real-valued. This is true even if the Green's function is not symmetric. This assertion follows from Lemma 55.

We shall study the eigenvalue problem (4.10) through its equivalent integral equation eigenvalue problem (4.11). This approach requires us to assume that $\lambda = 0$ is not an eigenvalue of the eigenvalue problem (4.10). This is not a serious restriction for the self-adjoint eigenvalue problems considered here for the following reasons: for any constant q_0, the pair λ, y is an eigenvalue, eigenfunction pair for the eigenvalue problem

$$-(py')' + qy = \lambda ry, \ B_1 y = 0, \ B_2 y = 0,$$

if and only if $\tilde{\lambda}$, y is an eigenvalue, eigenfunction pair for the eigenvalue problem

$$-(py')' + (q + q_0 r)y = \tilde{\lambda} ry, \ B_1 y = 0, \ B_2 y = 0,$$

where $\tilde{\lambda} = \lambda + q_0$. We establish in the next theorem that for self-adjoint problems a real constant q_0 can be chosen so that the modified eigenvalue problem does not have zero as an eigenvalue and, hence, there is an equivalent integral equation formulation of the modified eigenvalue problem. Once properties of the eigenvalues and eigenfunctions, $\lambda + q_0$ and y, of modified problem are established, the corresponding properties of the eigenvalues and eigenfunctions, λ and y, of original problem follow at once. In addition, q_0 can be chosen so that $q + q_0 r > 0$ on $[a, b]$, which means that, when it is advantageous to do so, we can assume $q > 0$ on $[a, b]$ when establishing properties of eigenvalues and eigenfunctions of Sturm-Liouville eigenvalue problems.

The assertions about q_0 in the previous paragraph are a consequence of the following theorem.

Theorem 116 *Either every complex number λ is an eigenvalue of the eigenvalue problem $Ly = \lambda ry$, $B_1 y = 0$, $B_2 y = 0$ or the eigenvalue problem has at most a finite number of eigenvalues in any bounded region of the complex plane. The second alternative always holds for a self-adjoint eigenvalue problem.*

Proof. Since the second order differential equation $Ly = \lambda ry$ for $a < x < b$ is expressible as the first order linear system $Z' = (A(x) + \lambda B(x))Z$ for $a < x < b$ where

$$Z = \begin{bmatrix} y \\ py' \end{bmatrix}, \ A(x) = \begin{bmatrix} 0 & 1/p \\ q & 0 \end{bmatrix}, \ B(x) = \begin{bmatrix} 0 & 0 \\ -r & 0 \end{bmatrix},$$

any solution $y(x, \lambda)$ to $Ly = \lambda ry$ is, for fixed x in (a, b), analytic in the complex variable λ for $|\lambda| < \infty$ as is $y'(x, \lambda)$ by Theorem 8.4 in Chapter 1 of [9] and the application to linear systems that follows the theorem. The same conclusion follows when applied to the differential equation $\tilde{L}y = \lambda \tilde{r} y$ for $\tilde{a} < x < \tilde{b}$ for a fixed $\tilde{a} < a$, $\tilde{b} > b$, and $\tilde{L}y = -(\tilde{p}y')' + \tilde{q}y$ where $\tilde{p}$, $\tilde{q}$, and $\tilde{r}$ extend p, q, and r to be constant on $[\tilde{a}, a]$ and $[b, \tilde{b}]$. Let $y_1(x, \lambda)$ and $y_2(x, \lambda)$ be a basis of solutions to $Ly = \lambda ry$ for $a < x < b$. Let $c = (a + b)/2$ and $\tilde{y}_1(x, \lambda)$ and $\tilde{y}_2(x, \lambda)$ be the solutions to $\tilde{L}y = \lambda \tilde{r} y$ for $\tilde{a} < x < \tilde{b}$ with initial values

$$\tilde{y}_1(c, \lambda) = y_1(c, \lambda), \ \tilde{y}_1'(c, \lambda) = y_1'(c, \lambda),$$
$$\tilde{y}_2(c, \lambda) = y_2(c, \lambda), \ \tilde{y}_2'(c, \lambda) = y_2'(c, \lambda),$$

respectively. By uniqueness of solutions to initial value problems, $\tilde{y}_1(x, \lambda) = y_1(x, \lambda)$ and $\tilde{y}_2(x, \lambda) = y_2(x, \lambda)$ for x in (a, b); hence, for x in $[a, b]$ because $y_1(x, \lambda)$ and $y_2(x, \lambda)$ have continuous extensions to $[a, b]$. From the discussion above, $\tilde{y}_1(x, \lambda)$ and $\tilde{y}_2(x, \lambda)$ for $x = a$ and $x = b$ are analytic functions of λ for $|\lambda| < \infty$. Hence, the same is true for $y_1(x, \lambda)$ and $y_2(x, \lambda)$ at $x = a$ and $x = b$.

Since $y_1(x, \lambda)$ and $y_2(x, \lambda)$ is a basis of solutions to $Ly = \lambda ry$, $y(x, \lambda) = c_1 y_1(x, \lambda) + c_2 y_2(x, \lambda)$ is the general solution to $Ly = \lambda ry$. Consequently, λ is an eigenvalue and

$y(x,\lambda) = c_1 y_1(x,\lambda) + c_2 y_2(x,\lambda)$ is a corresponding eigenfunction if and only if the 2×2 system

$$\begin{cases} c_1 B_1 y_1 + c_2 B_1 y_2 = 0 \\ c_1 B_2 y_1 + c_2 B_2 y_2 = 0 \end{cases}$$

has a nontrivial solution for c_1 and c_2. This happens if and only if

$$d(\lambda) = \begin{vmatrix} B_1 y_1 & B_1 y_2 \\ B_2 y_1 & B_2 y_2 \end{vmatrix} = 0.$$

The determinant $d(\lambda)$ is an analytic function of λ for $|\lambda| < \infty$ because $y_1(x,\lambda)$ and $y_2(x,\lambda)$ are such functions. The alternative in the theorem follows because such an analytic function is either identically equal to zero or has at most a finite number of zeros in any bounded region of the complex plane. See [6] or [28]. Since eigenvalues of a self-adjoint Sturm-Liouville eigenvalue problem are real, the first alternative in the theorem can not occur for self-adjoint problems and the proof is complete. ∎

Example 7. The non self-adjoint Sturm-Liouville eigenvalue problem

$$-y'' = \lambda y, \; y(0) - y(1) = 0, \; y'(0) + y'(1) = 0,$$

has every complex number as an eigenvalue.

Indeed, the differential equation has general solution

$$y = A \cos \sqrt{\lambda} x + B \sin \sqrt{\lambda} x$$

for arbitrary constants A and B and

$$y' = A\left(-\sqrt{\lambda} \sin \sqrt{\lambda} x\right) + B\left(\sqrt{\lambda} \cos \sqrt{\lambda} x\right).$$

The general solution satisfies the boundary conditions if and only if A and B satisfy

$$\begin{cases} A(1 - \cos \sqrt{\lambda}) + B(-\sin \sqrt{\lambda}) = 0 \\ A\sqrt{\lambda}(-\sin \sqrt{\lambda}) + B\sqrt{\lambda}(1 + \cos \sqrt{\lambda}) = 0 \end{cases}.$$

The system has a nontrivial solution for A and B if and only if

$$\sqrt{\lambda}\left(1 - \cos^2 \sqrt{\lambda} - \sin^2 \sqrt{\lambda}\right) = 0$$

which is satisfied for any complex number λ. For such λ the 2×2 system is satisfied by any pair A and B not both zero that satisfy

$$A\left(1 - \cos \sqrt{\lambda}\right) + B\left(-\sin \sqrt{\lambda}\right) = 0$$

and for such A and B, $y = A \cos \sqrt{\lambda} x + B \sin \sqrt{\lambda} x$ is an eigenfunction corresponding to the eigenvalue λ. Thus, every complex number is an eigenvalue.

4.8.2 Separated Boundary Conditions

Recall that a Sturm-Liouville eigenvalue problem with separated boundary conditions,

$$Ly = \lambda r y, \; B_a y = 0, \; B_b y = 0,$$

that is,

$$\begin{cases} -(p(x)y')' + q(x)y = \lambda r(x)y & a < x < b, \\ \alpha y(a) + \beta y'(a) = 0 & |\alpha| + |\beta| \neq 0, \\ \gamma y(b) + \delta y'(b) = 0 & |\gamma| + |\delta| \neq 0, \end{cases} \tag{4.12}$$

is **regular** if $p(x) > 0$ on $[a, b]$, the functions $p(x)$, $q(x)$, and $r(x)$ are real-valued and continuous on $[a, b]$, $r(x) > 0$ is on $[a, b]$, and the coefficients in the boundary conditions are real numbers. Often $r(x) = 1$; as in the case of the eigenvalue problem corresponding to a Sturm-Liouville boundary value problem.

By Theorem 113 a regular eigenvalue problem with separated boundary conditions is self-adjoint. Moreover, if $\lambda = 0$ is not an eigenvalue of the problem, then the Green's function $g(x, s)$ determined by the differential operator and boundary conditions is real-valued and symmetric.

4.8.2.1 Basic Properties

The properties established in this section apply to regular self-adjoint Sturm-Liouville eigenvalue problems with separated boundary conditions.

As explained earlier, we can assume without loss in generality that $\lambda = 0$ is not an eigenvalue of (4.12) so that the differential operator $L = -(py')' + qy$ and boundary conditions $B_a y = 0$ and $B_b y = 0$ have a Green's function $g(x, s)$ and the eigenvalue problem (4.12) is equivalent to the eigenvalue problem

$$y(x) = \lambda \int_a^b g(x, s)r(s)y(s)\, ds$$

for the kernel $g(x, s)r(s)$. Moreover, the Green's function $g(x, s)$ is real-valued and symmetric.

We consider two cases: the case when the weight function $r(x) = 1$ for all x in $[a, b]$ and the case of a general weight function $r(x) > 0$ on $[a, b]$. The first case is included in the second one but it is beneficial to single out the case $r(x) = 1$ because it occurs frequently and the proofs for this case suggest the line of attack for a general weight function.

4.8.2.2 Case 1: Weight Function $r(x) = 1$ for all x in $[a, b]$

In this case, the equivalent integral equation is

$$y(x) = \lambda \int_a^b g(x, s)y(s)\, ds$$

and the kernel $g(x, s)$ is real-valued and symmetric by Theorem 113. Consequently, the integral operator $G: C[a, b] \to C[a, b]$ defined by

$$Gy(x) = \int_a^b g(x, s)y(s)\, ds$$

is a compact self-adjoint operator when $C[a, b]$ is equipped with the 2-norm by Theorem 53. Recall that μ is an eigenvalue of the integral operator G if $Gy = \mu y$ for some $y \neq 0$ in $C[a, b]$. Therefore, the reciprocal $\mu = 1/\lambda$ of an eigenvalue λ of the kernel $g(x, s)$ is a nonzero eigenvalue of the self-adjoint integral operator G and the kernel and integral operator have the same corresponding eigenfunctions. From Section 3.4 any eigenvalue of G is real and eigenfunctions belonging to distinct eigenvalue are orthogonal. This establishes (once again) all but the last assertion in

Lemma 117 *Any eigenvalue λ of the Sturm-Liouville eigenvalue problem (4.12) with $r(x) = 1$ is real and eigenfunctions corresponding to distinct eigenvalues are orthogonal. The eigenspace of λ has a (finite) basis of real-valued orthonormal eigenfunctions.*

The final assertion follows from Lemma 55 and the fact that all the data in the problem is real-valued.

The Hilbert-Schmidt theorem (Theorem 60) and its corollaries applied to the integral operator G significantly extend the foregoing initial observations.

Theorem 118 *The regular Sturm-Liouville eigenvalue problem (4.12) with $r(x) = 1$ has an infinite sequence of eigenvalues and eigenfunctions with the following properties:*

1. Each eigenvalue is real and simple (has both algebraic and geometric multiplicity 1). The set of magnitudes of the eigenvalues is unbounded and at most a finite number of the eigenvalues are negative. Consequently, the eigenvalues can be listed as

$$\lambda_1 < \lambda_2 < \cdots < \lambda_n < \cdots$$

and $\lambda_n \to \infty$ as $n \to \infty$.
2. The eigenfunctions $\{\phi_n\}_{n=1}^{\infty}$ corresponding to the eigenvalues $\{\lambda_n\}_{n=1}^{\infty}$ can be chosen real-valued and orthonormal,

$$\langle \phi_m, \phi_n \rangle = \delta_{mn},$$

where δ_{mn} is the Kronecker delta.
3. For each continuous function f on $[a, b]$, the unique solution y to the regular Sturm-Liouville boundary value problem $Ly = f$, $B_a y = 0$, and $B_b y = 0$ can be expressed by

$$y(x) = \sum_{n=1}^{\infty} \langle y, \phi_n \rangle \phi_n(x)$$

where the series is absolutely and uniformly convergent on $[a, b]$.

Proof. We will use the notation and observations made in the second paragraph of Section 4.8.2.1. Any eigenvalue λ of (4.12) with $r(x) = 1$ is real by the previous lemma. Each eigenvalue is simple: if z and w are eigenfunctions corresponding to λ, then z and w satisfy the homogeneous Sturm-Liouville differential equation

$$-(py')' + (q - \lambda r)y = 0$$

on $[a, b]$ and

$$\begin{cases} \alpha z(a) + \beta z'(a) = 0 \\ \alpha w(a) + \beta w'(a) = 0 \end{cases} \quad \text{with} \quad |\alpha| + |\beta| > 0.$$

Consequently, the determinant of the 2×2 system must be zero; that is, the Wronskian $W_{z,w}(a) = 0$. It follows that z and w are linearly dependent on $[a, b]$ and that the geometric multiplicity of λ is 1. Furthermore, the algebraic multiplicity also is 1 because the Green's function is self-adjoint; see Lemma 57.

We establish next that the Sturm-Liouville eigenvalue problem has an infinite number of eigenvalues. The proof is by contradiction. Since $G \neq 0$ is a self-adjoint compact integral operator on $C[a, b]$, it has at least one nonzero eigenvalue, say μ, by Theorem 59. Consequently, $\lambda = 1/\mu$ is an eigenvalue of the kernel $g(x, s)$ and the Sturm-Liouville eigenvalue problem has at least one eigenvalue (and corresponding eigenfunction). If the Sturm-Liouville eigenvalue problem has only a finite number of eigenvalues, say $\lambda_1, \ldots, \lambda_N$, then G has only a finite number of nonzero eigenvalues $\mu_n = 1/\lambda_n$ for $n = 1, 2, \ldots, N$ and corresponding orthonormal eigenfunctions ϕ_n for $n = 1, 2, \dots, N$. By the Hilbert-Schmidt theorem

$$Gf(x) = \sum_{n=1}^{N} \langle Gf, \phi_n \rangle \phi_n(x)$$

for all f in $C[a, b]$. Since the unique solution to $Ly = f$, $B_a y = 0$, and $B_b y = 0$ is

$$y(x) = \int_a^b g(x, s) f(s)\, ds = Gf(x),$$

it follows that $LGf(x) = f(x)$ and

$$f(x) = \sum_{n=1}^{N} \langle Gf, \phi_n \rangle L\phi_n(x) = \sum_{n=1}^{N} \langle f, G\phi_n \rangle \lambda_n \phi_n(x) = \sum_{n=1}^{N} \langle f, \phi_n \rangle \phi_n(x)$$

because $G\phi_n = \mu_n \phi_n$ and $\mu_n \lambda_n = 1$. Since $f(x)$ can be any continuous function on $[a, b]$, this equation says that $\{\phi_n\}_{n=1}^{N}$ is a basis for $C[a, b]$, which is impossible because, for example, the functions $1, x, x^2, \dots, x^m$ are linearly independent for every positive integer m. This contradiction establishes that the Sturm-Liouville eigenvalue problem has an infinite number of eigenvalues λ_n and corresponding eigenfunctions ϕ_n. Since λ_n is an eigenvalue of the symmetric kernel $g(x, s)$, the corresponding eigenfunction ϕ_n can be chosen real-valued by Corollary 62 of the Hilbert-Schmidt theorem. Since each eigenvalue is simple, the corresponding real-valued eigenfunctions ϕ_n belong to distinct eigenvalues and are orthogonal; hence, they can be chosen orthonormal.

At this point, we have established Property 2 of the theorem and that there are an infinite number of eigenvalues, each of which is real and simple. We turn now to the assertion that only a finite number of the eigenvalues are negative. We will establish this assertion for separated boundary conditions whose coefficients satisfy $\alpha\beta \leq 0$ and $\gamma\delta \geq 0$, which are the separated boundary conditions that occur most often in applications. The interested reader can find the general result established in [5] or [10]. Let λ be an eigenvalue of (4.12) with $r(x) = 1$ and y be a corresponding real-valued eigenfunction, normalized by

$$\int_a^b y(x)^2\, dx = 1.$$

Multiply the differential equation in (4.12) with $r(x) = 1$ by y and integrate by parts to find

$$\begin{aligned}\lambda \int_a^b y(x)^2\, dx &= \int_a^b y(x)\, d\big(-p(x)y'(x)\big) + \int_a^b q(x)y(x)^2\, dx \\ &= -p(x)y(x)y'(x)\big|_a^b + \int_a^b \left(p(x)y'(x)^2 + q(x)y(x)^2\right) dx.\end{aligned}$$

The restrictions $\alpha\beta \leq 0$ and $\gamma\delta \geq 0$ on the boundary conditions imply that $y(b)y'(b) \leq 0$ and $y(a)y'(a) \geq 0$ so that $-p(x)y(x)y'(x)\big|_a^b \geq 0$; hence,

$$\lambda \geq \int_a^b q(x)y(x)^2\, dx \geq \min_{a \leq x \leq b} q(x) = Q.$$

Thus, the eigenvalues are bounded below by Q. By the Hilbert-Schmidt theorem, the eigenvalues $\mu_n = 1/\lambda_n$ of the integral operator G satisfy $|\mu_n| \to 0$ as $n \to \infty$, and, hence, $|\lambda_n| \to \infty$ as $n \to \infty$. It follows that at most a finite number of the eigenvalues λ_n can be negative because the eigenvalues are bounded below by Q. This completes the proof of Property 1 of the theorem.

We have established all but the last assertion in the theorem. To complete the proof we apply the Hilbert-Schmidt Theorem once more. Since the Green's function $g(x, s)$ is continuous, for each continuous function f on $[a, b]$, the Hilbert-Schmidt expansion

$$Gf(x) = \sum_{n=1}^{\infty} \langle Gf, \phi_n \rangle \phi_n(x)$$

holds with absolute and uniform convergence on $[a, b]$ by the first corollary to the Hilbert-Schmidt theorem. Property 3 follows at once because the unique solution to $Ly = f$, $B_a y = 0$, and $B_b y = 0$ is $y(x) = Gf(x)$. ■

An important interpretation of the third conclusion in the theorem is

Corollary 119 *If y satisfies the boundary conditions $B_a y = 0$ and $B_b y = 0$ and y is in the domain of the Sturm-Liouville differential operator L, then y has the absolutely and uniformly convergent eigenfunction expansion*

$$y(x) = \sum_{n=0}^{\infty} \langle y, \phi_n \rangle \phi_n(x)$$

for x in $[a, b]$.

Proof. Define $f = Ly$ and apply the third part of the theorem. ■

Example 1b. (continued) Let a and l be positive. We found that the eigenvalue problem for

$$Ly = -y'' + ay, \; y(0) = 0, \; y(l) = 0$$

has eigenvalues $\lambda_n = a + (n\pi/l)^2$ and corresponding eigenfunctions the nonzero multiples of $\sin(n\pi x/l)$ for $n = 1, 2, 3, \ldots$. Since $\phi_n(x) = \sqrt{2/l} \sin(n\pi x/l)$ has 2-norm 1 and $\langle \phi_m, \phi_n \rangle = 0$ if $m \neq n$, either by a direct calculation or the fact that the eigenvalue problem is self-adjoint, $\{\phi_n\}_{n=1}^{\infty}$ is an orthonormal set of eigenfunctions for the eigenvalue problem. Since (see Section 4.7) the domain D of L is $C^2[a, b]$, it follows from the corollary that any twice continuously differentiable function y has an absolutely and uniformly convergent eigenfunction expansion on $[0, l]$. Since

$$\langle y, \phi_n \rangle = \sqrt{\frac{2}{l}} \int_0^l y(x) \sin(n\pi x/l)\, dx$$

the eigenfunction expansion is

$$y(x) = \sum_{n=1}^{\infty} \left(\frac{2}{l} \int_0^l y(x) \sin(n\pi x/l)\, dx \right) \sin(n\pi x/l)\, dx,$$

which is just a Fourier sine series for y on $[0, l]$.

We observed in Chapter 1 that many Sturm-Liouville eigenvalue problems that arise in applications have all positive eigenvalues. When separation of variables leads to such an eigenvalue problem, this is a consequence of the fact that the underlying partial differential equations and boundary conditions that describe the physical situation include mechanisms that oppose arbitrarily large responses of the system. The natural eigenfunction expansions of the solutions would not have this property if there were any negative eigenvalues. The next theorem covers most such cases.

Theorem 120 *If $r(x) = 1$, $q \geq 0$, and $\alpha\beta \leq 0$, $\gamma\delta \geq 0$ in the regular eigenvalue problem (4.12), then all the eigenvalues are positive, except when $\alpha = 0$, $\gamma = 0$, and $q \equiv 0$, in which case the eigenvalue problem is*

$$-(py')' = \lambda y, \; y'(a) = 0, \; y'(b) = 0,$$

zero is an eigenvalue, and all other eigenvalues are positive.

Proof. Let λ be an eigenvalue and $y \neq 0$ be a corresponding real-valued eigenfunction. Multiply $Ly = \lambda y$ by y and integrate by parts to obtain

$$\lambda \int_a^b y^2 \, dx = \int_a^b y d(-py') + \int_a^b qy^2 \, dx$$

$$= -p(b)y(b)y'(b) + p(a)y(a)y'(a) + \int_a^b \left(py'^2 + qy^2\right) dx.$$

By the assumptions on the boundary conditions $y(b)y'(b) \leq 0$ and $y(a)y'(a) \geq 0$ so each of the three terms on the right is nonnegative. Hence all the eigenvalues are nonnegative.

Furthermore, zero is an eigenvalue if and only if

$$y(a)y'(a) = 0, \; y(b)y'(b) = 0 \quad \text{and} \quad \int_a^b \left(py'^2 + qy^2\right) dx = 0.$$

Since $p > 0$ on (a, b) and $q \geq 0$ on (a, b), these conditions hold if and only if $y' = 0$ on $[a, b]$, in which case the corresponding eigenfunction $y = k$ a nonzero constant and

$$\alpha k, \; \gamma k = 0, \quad \text{and} \quad k \int_a^b q \, dx = 0,$$

where the first two conditions follows from the boundary condition at $x = a$ and $x = b$. These conditions hold if and only if

$$\alpha = 0, \; \gamma = 0, \quad \text{and} \quad q = 0 \text{ on } [a, b]$$

because $k \neq 0$. Thus, all the eigenvalues are positive except possibly for the case when $\alpha = 0$, $\gamma = 0$ and $q = 0$ on $[a, b]$ when the eigenvalue problem reduces to

$$-(py')' = \lambda y, \; y'(a) = 0, \; y'(b) = 0,$$

a problem for which $\lambda = 0$ is clearly an eigenvalue. For this problem any eigenvalue satisfies

$$\lambda \int_a^b y^2 \, dx = \int_a^b py'^2 \, dx.$$

If $\lambda \neq 0$ the right member must be positive; hence, $\lambda > 0$. ■

The boundary conditions $y'(a) = 0$ and $y'(b) = 0$ are called **Neumann boundary conditions** and arise in transport problems involving a quantity (heat, chemicals, ...) in situations in which none of the quantity passes the boundary of the domain of interest.

A regular Sturm-Liouville eigenvalue problem has at most a finite number of negative eigenvalues, as we have seen. The following corollary establishes this result again but from a somewhat different perspective.

Corollary 121 (of Theorem 120) *If $r(x) = 1$, $\alpha\beta \leq 0$, and $\gamma\delta \geq 0$ in the regular eigenvalue problem (4.12), then at most a finite number of the eigenvalues are negative.*

Proof. There is a positive constant c such that $\hat{q}(x) = q(x) + c > 0$ on $[a, b]$ because $q(x)$ is bounded on $[a, b]$. Consequently, all the eigenvalues of the eigenvalue problem $\hat{L}y = \hat{\lambda}y$, $B_a y = 0$, $B_b y = 0$, where $\hat{L}y = -(py')' + \hat{q}y$, are positive. Since $Ly = \lambda y$ if and only if $\hat{L}y = \hat{\lambda}y$ where $\hat{\lambda} = \lambda + c$, it follows that all eigenvalues of $Ly = \lambda y$, $B_a y = 0$, $B_b y = 0$ satisfy $\lambda + c = \hat{\lambda} > 0$. Thus, $\lambda > -c$. Since the magnitudes of the eigenvalues λ tend to infinity, only a finite number of the eigenvalues can be negative. ■

Example 1b, 3b, 4b (continued). In the first two examples, we calculated the eigenvalues explicitly and found that they were all positive. In Example 4b we assumed the eigenvalues were real and showed that the eigenvalues were positive using the maximum principle. The same conclusion can be reached from Theorem 120 for all three examples without first solving for the eigenvalues. This observation is useful even when the exact values of the eigenvalues can be found because it is helpful to know that the eigenvalues are real and to know their sign before performing any manipulations.

Example 2b (continued). In this example, we calculated the eigenvalues explicitly and found that there may be a finite number of negative eigenvalues, depending on the value of $a < 0$ in the differential operator $Ly = -y'' + ay$. The fact that at most a finite number of eigenvalues can be negative follows from the corollary to Theorem 120.

Recall from Section 3.4 that a symmetric kernel $k(x, s)$ is positive definite if all its eigenvalues are positive.

Theorem 122 *Except for the special Neumann problem in Theorem 120, the Green's functions of all other eigenvalue problems covered by the theorem are positive definite and each such Green's function can be expressed as*

$$g(x, s) = \sum_{n=1}^{\infty} \frac{\phi_n(x)\phi_n(s)}{\lambda_n}$$

where the series converges absolutely and uniformly on $[a, b] \times [a, b]$. *Here* $\{\lambda_n\}$ *is the sequence of eigenvalues of the Green's function and* $\{\phi_n\}$ *is the corresponding sequence of real-valued orthonormal eigenfunctions.*

Proof. The expansion is a direct application of Mercer's theorem in Section 3.4. ■

Example 1a,b. (continued) The boundary value problem $Ly = -y'' + ay = f$, $y(0) = 0$, $y(l) = 0$ with $a > 0$ and $l > 0$ has Green's function

$$g(x, s) = \frac{1}{\sqrt{a}\sinh\sqrt{a}l}\begin{cases} \sinh\sqrt{a}x\sinh\sqrt{a}(l-s), & 0 \le x \le s \le l \\ \sinh\sqrt{a}s\sinh\sqrt{a}(l-x), & 0 \le s \le x \le l \end{cases}.$$

The corresponding eigenvalue problem has eigenvalues $\lambda_n = a + (n\pi/l)^2$ and corresponding orthonormal eigenfunctions $\phi_n(x) = \sqrt{2/l}\sin(n\pi x/l)$ for $n = 1, 2, 3, \ldots$. By Theorem 122

$$g(x, s) = \frac{2}{l}\sum_{n=1}^{\infty} \frac{\sin(n\pi x/l)\sin(n\pi s/l)}{a + (n\pi/l)^2}.$$

In particular, when $s = l/2$ we obtain the Fourier sine series expansion

$$\begin{aligned} \sinh\sqrt{a}x &= \frac{2\sqrt{a}\sinh\sqrt{a}l}{l\sinh(\sqrt{a}l/2)}\sum_{n=1}^{\infty} \frac{\sin(n\pi x/l)\sin(n\pi/2)}{a + (n\pi/l)^2} \\ &= \frac{4}{l}\sqrt{a}\cosh(\sqrt{a}l/2)\sum_{n=1}^{\infty} \frac{\sin(n\pi/2)}{a + (n\pi/l)^2}\sin(n\pi x/l) \end{aligned}$$

for $0 \le x \le l/2$.

4.8.2.3 Case 2: $r(x)$ is a General Weight Function

As we have seen, the self-adjoint Sturm-Liouville eigenvalue problem $Ly = \lambda ry$, $B_a y = 0$, $B_b y = 0$ is equivalent to the integral equation eigenvalue problem

$$y(x) = \lambda \int_a^b g(x, s) r(s) y(s)\ ds. \tag{4.13}$$

Although $g(x, s)$ is real-valued and symmetric, the kernel $g(x, s)r(s)$ is not symmetric, except when $r(x)$ is a constant. Nevertheless, the reasoning used in the case when $r(x) = 1$ can be adjusted to handle a general weight function $r(x)$ by means of a symmetrization process: if λ, y is an eigenvalue, eigenfunction pair for (4.12) so that (4.13) is the equivalent integral equation, then λ and y satisfies

$$\sqrt{r(x)}y(x) = \lambda \int_a^b \sqrt{r(x)}g(x, s)\sqrt{r(s)}\sqrt{r(s)}y(s)\ ds$$

or

$$z(x) = \lambda \int_a^b k(x, s) z(s)\ ds \tag{4.14}$$

where

$$k(x, s) = \sqrt{r(x)}g(x, s)\sqrt{r(s)} \quad \text{and} \quad z(x) = \sqrt{r(x)}y(x).$$

Conversely, if the pair λ, z satisfies the integral equation (4.14), then the pair λ, $y = z/\sqrt{r}$ satisfies (4.13). Thus, the two eigenvalue problems (4.13) and (4.14) are equivalent. The kernel $g(x, s)r(s)$ is called **symmetrizable** because of this equivalence.

Since the kernel $k(x, s)$ is real-valued, continuous, and symmetric, the corresponding integral operator $K : C[a, b] \to C[a, b]$ defined by

$$Kz(x) = \int_a^b k(x, s) z(s)\ ds$$

is a compact self-adjoint linear operator when $C[a, b]$ is equipped with the 2-norm. Thus, we can adjust the reasoning used in the case when $r(x) = 1$ to establish the corresponding results for Sturm-Liouville eigenvalue problems with general weight functions. The details and modified results follow.

By the Hilbert-Schmidt theorem and its corollaries applied to K, the kernel $k(x, s)$ in (4.14) has nonzero eigenvalues $\{\lambda_n\}$, the reciprocals of the nonzero eigenvalues of K, where each eigenvalue is listed to multiplicity and corresponding real-valued orthonormal eigenfunctions $\{\psi_n\}$. The eigenvalues are all real because K is self-adjoint. The sequence $\{\lambda_n\}$ is infinite. Assuming otherwise leads to a contradiction: if $k(x, s)$ has only a finite number of eigenvalues, say $\{\lambda_n\}_{n=1}^N$ and corresponding real-valued orthonormal eigenfunctions $\{\psi_n\}_{n=1}^N$, then by Part 5 of the Hilbert-Schmidt theorem

$$Kf(x) = \sum_{n=1}^{N} \lambda_n^{-1} \langle f, \psi_n \rangle \psi_n(x)$$

for all x in $[a, b]$ and all continuous functions f on $[a, b]$. The displayed equation can be expressed as

$$\int_a^b \left(k(x, s) - \sum_{n=1}^{N} \lambda_n^{-1} \psi_n(x) \psi_n(s) \right) f(s)\ ds = 0$$

for all x in $[a, b]$ and for all continuous functions f on $[a, b]$. It follows that

$$k(x, s) = \sum_{n=1}^{N} \lambda_n^{-1} \psi_n(x)\psi_n(s)$$

by the Corollary 20. Since $k(x, s) = \sqrt{r(x)}g(x, s)\sqrt{r(s)}$, and $\psi_n(x) = \sqrt{r(x)}\phi_n(x)$ where λ_n, ϕ_n is an eigenvalue, eigenfunction pair for $Ly = \lambda ry$, $B_a y = 0$, $B_b y = 0$, the last displayed equation yields

$$g(x, s) = \sum_{n=1}^{N} \lambda_n^{-1} \phi_n(x)\phi_n(s)$$

for all x and all s in $[a, b]$. Consequently, for any continuous function f on $[a, b]$,

$$Gf(x) = \sum_{n=1}^{N} \lambda_n^{-1} \langle f, \phi_n \rangle \phi_n(x).$$

Since $Ly = f$, $B_a y = 0$, $B_b y = 0$ has unique solution $y(x) = Gf(x)$, $LGf(x) = Ly(x) = f(x)$ and

$$f(x) = LGf(x) = \sum_{n=1}^{N} \lambda_n^{-1} \langle f, \phi_n \rangle L\phi_n(x) = \sum_{n=1}^{N} \langle f, \phi_n \rangle \phi_n(x).$$

Just as in the Case $r(x) = 1$, it follows $C[a, b]$ is spanned by the finite set of eigenfunctions $\{\phi_n\}_{n=1}^{N}$, which is a contradiction. So the kernel $k(x, s)$ has an infinite sequence of eigenvalues $\{\lambda_n\}_{n=1}^{\infty}$ and corresponding real-valued orthonormal eigenfunctions $\{\psi_n\}_{n=1}^{\infty}$.

By the equivalence of the eigenvalue problem for the kernel $k(x, s)$ and the Sturm-Liouville eigenvalue problem, it follows that λ_n are the eigenvalues and $\phi_n = \psi_n/\sqrt{r}$ are corresponding real-valued eigenfunctions of the eigenvalue problem (4.12). Moreover, each eigenvalue is real, simple, there are at most a finite number of negative eigenvalues, and $\lambda_n \to \infty$ as $n \to \infty$ by virtually the same arguments used in the case when $r(x) = 1$.

The orthogonality of the eigenfunctions ψ_n of K and the relations $\psi_n = \sqrt{r}\phi_n$ translate into the following condition on the eigenfunctions ϕ_n of the Sturm-Liouville eigenvalue problem

$$\delta_{mn} = \int_a^b \psi_m \psi_n \, ds = \int_a^b \phi_m \phi_n r \, ds.$$

Functions ϕ_m and ϕ_n that satisfy

$$\int_a^b \phi_m(x)\phi_n(x)r(x) \, dx = \delta_{mn}$$

are said to be **orthogonal with respect to the weight function** r. This terminology is motivated by the fact that

$$\langle f, g \rangle_r = \int_a^b f(x)\overline{g(x)}r(x) \, dx$$

is an inner product on the space $C[a, b]$ for any weight function $r(s) > 0$ on $[a, b]$.

The foregoing discussion establishes all but Part 3 of

Theorem 123 *The regular Sturm-Liouville eigenvalue problem (4.12) has an infinite sequence of eigenvalues and eigenfunctions with the following properties:*

1. Each eigenvalue is real and simple (has both algebraic and geometric multiplicity 1). *The set of magnitudes of the eigenvalues is unbounded and at most a finite number of the eigenvalues are negative. Consequently, the eigenvalues can be listed as*

$$\lambda_1 < \lambda_2 < \cdots < \lambda_n < \cdots$$

and $\lambda_n \to \infty$ *as* $n \to \infty$.
2. The corresponding eigenfunctions can be chosen real-valued and orthonormal with respect to the weight function r,

$$\langle \phi_m, \phi_n \rangle_r = \int_a^b \phi_m(s)\phi_n(s)r(s)\, ds = \delta_{mn},$$

where δ_{mn} *is the Kronecker delta.*
3. *For each continuous function* f *on* $[a, b]$, *the unique solution* y *to the regular Sturm-Liouville boundary value problem* $Ly = f$, $B_a y = 0$, *and* $B_b y = 0$ *can be expressed by*

$$y(x) = \sum_{n=1}^{\infty} \langle y, \phi_n \rangle_r \phi_n(x)$$

where the series is absolutely and uniformly convergent on $[a, b]$.

Proof. It remains to prove Property 3. To complete the proof we apply the Hilbert-Schmidt Theorem once more. Since the symmetrized Green's function $k(x, s)$ is continuous, for each continuous function h on $[a, b]$, the Hilbert-Schmidt expansion

$$Kh(x) = \sum_{n=1}^{\infty} \langle Kh, \psi_n \rangle \psi_n(x)$$

holds with absolute and uniform convergence on $[a, b]$ by the first corollary to the Hilbert-Schmidt theorem. Since

$$Kh(x) = \int_a^b \sqrt{r(x)}g(x, s)\sqrt{r(s)}h(s)\, ds = \sqrt{r(x)}G(\sqrt{r}h)(x),$$

that is, $Kh = \sqrt{r}G(\sqrt{r}h)$, and

$$\langle Kh, \psi_n \rangle = \langle \sqrt{r}G(\sqrt{r}h), \sqrt{r}\phi_n \rangle = \langle G(\sqrt{r}h), \phi_n \rangle_r,$$

the Hilbert-Schmidt expansion for $Kh(x)$ can be expressed as

$$\sqrt{r(x)}G(\sqrt{r}h)(x) = \sum_{n=1}^{\infty} \langle G(\sqrt{r}h), \phi_n \rangle_r \sqrt{r(x)}\phi_n(x),$$

$$G(\sqrt{r}h)(x) = \sum_{n=1}^{\infty} \langle G(\sqrt{r}h), \phi_n \rangle_r \phi_n(x),$$

with absolute and uniform convergence on $[a, b]$ because $r(x) > 0$ on $[a, b]$. Since $h = f/\sqrt{r}$ is continuous on $[a, b]$ for any continuous f on $[a, b]$,

$$Gf(x) = \sum_{n=1}^{\infty} \langle Gf, \phi_n \rangle_r \phi_n(x)$$

with absolute and uniform convergence on $[a, b]$. If y is the unique solution to $Ly = f$, $B_a y = 0$, and $B_b y = 0$, then $y = Gf$ and Part 3 is established. ■

The remaining results of the last subsection are extended to the case of a general weight function by virtually the same reasoning used there. We simply state the results here.

Theorem 124 *If $q \geq 0$ and $\alpha\beta \leq 0$, $\gamma\delta \geq 0$ in the regular eigenvalue problem (4.12), then all the eigenvalues are positive, except when $\alpha = 0$, $\gamma = 0$, and $q \equiv 0$, in which case the eigenvalue problem is*

$$-(py')' = \lambda ry,\ y'(a) = 0,\ y'(b) = 0,$$

zero is an eigenvalue, and all other eigenvalues are positive.

Corollary 125 *If $\alpha\beta \leq 0$ and $\gamma\delta \geq 0$ in the separated boundary conditions, then at most a finite number of the eigenvalues of the regular eigenvalue problem (4.12) are negative.*

Theorem 126 *Except for the special Neumann problem in Theorem 124, the Green's functions of all other eigenvalue problems covered by the theorem are positive definite and each such Green's function can be expressed as*

$$g(x, s) = \sum_{n=1}^{\infty} \frac{\phi_n(x)\phi_n(s)}{\lambda_n}$$

where the series converges absolutely and uniformly on $[a, b] \times [a, b]$. Here $\{\lambda_n\}$ is the sequence of eigenvalues of the Green's function and $\{\phi_n\}$ is the corresponding sequence of real-valued orthonormal eigenfunctions.

Another approach to the case when the weight function is not identically 1 is by the change of variable

$$\xi = \int_a^x r(s)\ ds \quad \text{for } x \text{ in } [a, b].$$

Since $r(s) > 0$, ξ is an increasing function of x and the change of variable maps $[a, b]$ onto $[A, B]$ where $A = 0$ and $B = \int_a^b r(s)\ ds$. By the fundamental theorem of calculus $d\xi/dx = r(x) > 0$. Consequently by the inverse function rule of differential calculus, x is a differentiable increasing function of ξ with $dx/d\xi = 1/R(\xi)$ where

$$R(\xi) = r(x),\ P(\xi) = p(x),\ Q(\xi) = q(x), \quad \text{and} \quad Y(\xi) = y(x)$$

and ξ and x are corresponding values under the change of variable. If a prime denotes $d/d\xi$ for functions of ξ and d/dx for functions of x, then

$$(py')' = \left(PRY'\right)'R$$

and hence $Ly = \lambda ry$ expressed in terms of ξ as independent variable is

$$\begin{aligned} -\left(PRY'\right)'R + QY &= \lambda RY, \\ L_1 Y &= \lambda Y, \end{aligned}$$

where

$$L_1 Y = -\left(P_1 Y'\right)' + Q_1 Y,\ P_1 = PR, \quad \text{and} \quad Q_1 = Q/R.$$

Evidently, $P_1 > 0$ and P_1 and Q_1 are continuous on $[A, B]$. Moreover,

$$\begin{aligned} \alpha y(a) + \beta y'(a) &= \alpha_1 Y(A) + \beta_1 Y'(A), \\ \gamma y(b) + \delta y'(b) &= \gamma_1 Y(B) + \delta_1 Y'(B), \end{aligned}$$

where $\alpha_1 = \alpha$, $\beta_1 = \beta R(A)$, $\gamma_1 = \gamma$, and $\delta_1 = \delta R(B)$. Thus, the eigenvalue problem $Ly = \lambda ry$, $B_a y = 0$, $B_b y = 0$ expressed in terms of ξ is $L_1 Y = \lambda Y$, $B_{1A} Y = 0$, $B_{1B} Y = 0$ where

$$B_{1A} Y = \alpha_1 Y(A) + \beta_1 Y'(A) \quad \text{and} \quad B_{1B} Y = \gamma_1 Y(B) + \delta_1 Y'(B).$$

Clearly λ, y is an eigenvalue, eigenfunction pair for the original eigenvalue problem if and only if λ, Y is an eigenvalue, eigenfunction pair for the transformed eigenvalue problem which has weight function 1. Note also that the transformation preserves the signs of the pairs p and P_1, q and Q_1, α and α_1, β and β_1, γ and γ_1, and δ and δ_1. Consequently, all of the results established for the case of a general weight function follow from the case of weight function 1 via this transformation. For example, if y is the unique solution to $Ly = f$, $B_a y = 0$, $B_b y = 0$ where f is a given continuous function on $[a, b]$, then by Part 3 of Theorem 118

$$Y(\xi) = \sum_{n=1}^{\infty} \langle Y, \Phi_n \rangle \Phi_n(\xi)$$

where Y is the unique solution to $L_1 Y = F$, $B_{1A} Y = 0$, $B_{1B} Y = 0$, where the series converges absolutely and uniformly on $[A, B]$, $F(\xi) = f(x)$, and λ_n, $\phi_n(x)$ and λ_n, $\Phi_n(\xi)$ are corresponding eigenvalue, eigenfunction pairs. Since

$$\langle Y, \Phi_n \rangle = \int_A^B Y(\xi)\Phi_n(\xi)\, d\xi = \int_a^b y(x)\phi_n(x) \frac{d\xi}{dx}\, dx = \langle y, \phi_n \rangle_r,$$

it follows at once that

$$y(x) = \sum_{n=1}^{\infty} \langle y, \phi_n \rangle_r \phi_n(x)$$

with absolute and uniform convergence on $[a, b]$. This establishes Part 3 of Theorem 123.

4.8.3 Oscillation and Approximation Properties

The principal results of this section apply to the most important class of Sturm-Liouville eigenvalue problems with separated boundary conditions that occur in applications. They establish that for each N, linear combinations of the eigenfunctions $\{\phi_n\}_{n=0}^{N}$ have approximation and interpolation properties strictly analogous to the linear combinations of $\{x^n\}_{n=0}^{N}$, that is, to ordinary polynomials of degree N. These results follow because the Green's functions for such eigenvalue problems are Kellogg kernels.

Recall from Section 3.6.3 that $k(x, s)$ is a Kellogg kernel if $k(x, s)$ is continuous and symmetric on $[a, b] \times [a, b]$ and satisfies:

K1. $\det\,[k(x_i, x_j)]_{n\times n} > 0, \quad a < x_1 < \cdots < x_n < b,$

K2. $\det\,[k(x_i, s_j)]_{n\times n} \geq 0, \quad a \leq x_1 \leq \cdots \leq x_n \leq b, \ a \leq s_1 \leq \cdots \leq s_n \leq b.$

In this context, $k_{[n]}(x, s) = \det\big[k(x_i, s_j)\big]_{n\times n}$, the nth compound kernel of $k(x, s)$, has domain $\Delta_n \times \Delta_n$ and where Δ_n is the simplex

$$\Delta_n = \big\{x = (x_1, \ldots, x_n) : a \leq x_1 \leq \cdots \leq x_n \leq b\big\}.$$

We maintain the notation of previous sections:

$$\begin{aligned} Ly &= -(p(x)y')' + q(x)y, \ a < x < b, \\ B_a y &= \alpha y(a) + \beta y'(a), \\ B_b y &= \gamma y(b) + \delta y'(b). \end{aligned}$$

Theorem 127 *The regular Sturm-Liouville eigenvalue problem*

$$\begin{cases} Ly = \lambda ry, \ a < x < b, \\ \alpha y(a) + \beta y'(a) = 0, \\ \gamma y(b) + \delta y'(b) = 0, \end{cases} \tag{4.15}$$

where $p > 0$, $q \geq 0$, $r > 0$ *are continuous on* $[a, b]$, $\alpha\beta \leq 0$, $\gamma\delta \geq 0$, $|\alpha| + |\beta| > 0$, *and* $|\gamma| + |\delta| > 0$, *has a Green's function* $g(x, s)$ *which is a Kellogg kernel on* $[a, b] \times [a, b]$, *except when* $\alpha = 0$, $\gamma = 0$, *and* $q = 0$ *and no Green's function exists. Consequently, the kernel* $k(x, s) = \sqrt{r(x)}g(x, s)\sqrt{r(s)}$ *also is a Kellogg kernel.*

Proof. The eigenvalue problem is self-adjoint because it has all real data. See Theorem 105. Consequently, when the Green's function exists it is real-valued and symmetric by Theorem 106. By Theorem 124 all the eigenvalues of (4.15) are positive, except when $\alpha = 0$, $\gamma = 0$, and $q = 0$. Hence, the Green's function exists and is positive definite, except when $\alpha = 0$, $\gamma = 0$, and $q = 0$ in which case $\lambda = 0$ is an eigenvalue and there is no Green's function. This implies that, when it exists, $g(x, x) \geq 0$ for $a \leq x \leq b$ by the first paragraph in the proof of Mercer's theorem in Chapter 3. Moreover, since the boundary conditions are separated, by Theorem 96 the Green's function has the form

$$g(x, s) = \begin{cases} u(x)v(s) & \text{for } a \leq x \leq s \leq b \\ u(s)v(x) & \text{for } a \leq s \leq x \leq b \end{cases},$$

where $u(x)$ and $v(x)$ are real-valued and continuously differentiable on $[a, b]$ and satisfy

$$\begin{aligned} Lu = 0, \ \alpha u(a) + \beta u'(a) = 0 \\ Lv = 0, \ \gamma v(a) + \delta v'(a) = 0 \\ p(x)\, W_{u,v}(x) = -1, \end{aligned}$$

for x in $[a, b]$. Consequently, $u(x)v(x) = g(x, x) \geq 0$ for $a \leq x \leq b$. If $u(c) = 0$ for some c with $a < c < b$, then $u'(c) \neq 0$ because otherwise $u = 0$ on $[a, b]$. Thus $u(x)$ is a nontrivial solution to

$$\begin{aligned} Lu = 0, \ a < x < c, \\ \alpha u(a) + \beta u'(a) = 0, \ u(c) = 0. \end{aligned}$$

That is, 0 is an eigenvalue of the eigenvalue problem

$$\begin{aligned} &Lu = \lambda u, \ a < x < c, \\ &\alpha u(a) + \beta u'(a) = 0, \ u(c) = 0, \end{aligned}$$

which contradicts Theorem 120. Thus, $u(x) \neq 0$ on $a < x < b$. Likewise, $v(x) \neq 0$ on $a < x < b$. Since $u(x)v(x) \geq 0$ for $a < x < b$ it follows that

$$u(x)v(x) > 0 \quad \text{for } a < x < b.$$

Furthermore,

$$\frac{d}{dx}\left(\frac{u(x)}{v(x)}\right) = \frac{v(x)u'(x) - u(x)v'(x)}{v(x)^2} = -\frac{W_{u,v}(x)}{v(x)^2} = \frac{1}{p(x)v(x)^2} > 0$$

for $a < x < b$. So

$$\frac{u(x)}{v(x)} \text{ is increasing on } a < x < b.$$

Since $u(x)v(x) > 0$ and $u(x)/v(x)$ is increasing on $a<x<b$, it follows from Corollary 37 that for $x = (x_1, \ldots, x_n)$ and $s = (s_1, \ldots, s_n)$

$$g_{[n]}(x, s) > 0 \quad \text{when } a < x_1,\ s_1 < x_2,\ s_2 < \cdots < x_n,\ s_n < b$$

and the determinant is 0 for all other choices of $a < x_1 < x_2 < \cdots < x_n < b$ and $a < s_1 < s_2 < \cdots < s_n < b$. Since $g(x, s)$ is continuous on $[a, b] \times [a, b]$, it follows that

$$g_{[n]}(x, s) \geq 0$$

for $a \leq x_1 \leq x_2 \leq \cdots \leq x_n \leq b$ and $a \leq s_1 \leq s_2 \leq \cdots \leq s_n < b$ and

$$g_{[n]}(x, x) > 0 \quad \text{for } a < x_1 < x_2 < \cdots < x_n < b.$$

Thus, the Green's function $g(x, s)$ is a Kellogg kernel.

The final assertion in the theorem follows directly from

$$k_{[n]}(x, s) = \det\left[\sqrt{r(x_i)}g(x_i, s_j)\sqrt{r(s_j)}\right] = \prod_{i=1}^{n}\sqrt{r(x_i)}k_{[n]}(x, s)\prod_{j=1}^{n}\sqrt{r(s_j)}.$$

■

Theorem 128 *The eigenvalues of the regular Sturm-Liouville eigenvalue problem*

$$\begin{cases} Ly = \lambda ry, \ a < x < b, \\ \alpha y(a) + \beta y'(a) = 0, \\ \gamma y(b) + \delta y'(b) = 0, \end{cases}$$

where $p>0$, $r>0$, p, q, *and* r *are real-valued and continuous on* $[a, b]$, $\alpha\beta \leq 0$, $\gamma\delta \geq 0$, $|\alpha| + |\beta| > 0$, *and* $|\gamma| + |\delta| > 0$ *are all real, simple, and can be labeled so that*

$$\lambda_0 < \lambda_1 < \cdots < \lambda_n < \cdots$$

with $\lambda_n \to \infty$ *as* $n \to \infty$. *For* $n = 0, 1, 2, \ldots$ *either the first* $n+1$ *(orthonormal, real-valued) eigenfunctions* $\phi_0(x), \phi_1(x), \ldots, \phi_n(x)$ *corresponding to the first* $n+1$ *eigenvalues is a Tchebycheff system on* (a, b) *or* $\phi_0(x), \phi_1(x), \ldots, \phi_{n-1}(x), -\phi_n(x)$ *a Tchebycheff system on* (a, b). *Consequently, the following oscillation and approximation properties hold:*

1. Given any $n+1$ *points in* (a, b) *and any* $n+1$ *values* $b_0, \ldots, b_n$, *there is a unique* ϕ *-polynomial* $\phi(x) = \sum_{i=0}^{n} a_i\phi_i(x)$ *that take on the prescribed values at the given points.*

2. A nontrivial ϕ *-polynomial has at most* n *zeros in* (a, b) *where nonnodal zeros are counted twice and nodal zeros once.*

3. A nontrivial ϕ*-polynomial* $\phi(x) = \sum_{i=m}^{n} a_i\phi_i(x)$ *has at least* m *nodal zeros in* (a, b) *and has at most* n *zeros there, counting zeros as in Property 2.*

4. ϕ_n *has* n *nodal zeros in* (a, b) *and no other zeros there.*

5. The zeros of ϕ_{n-1} *and* ϕ_n *strictly interlace on* (a, b).

Moreover, $\lambda_0 > 0$ *if* $q \geq 0$ *and either* q *is not identically* 0, *or* $\alpha \neq 0$, *or* $\gamma \neq 0$ *and* $\lambda_0 = 0$ *if* $q = 0$, $\alpha = 0$, *and* $\gamma = 0$.

Proof. The stated properties through item 5 hold, with the addition that $\lambda_0 > 0$, for the eigenvalues and corresponding orthonormal eigenfunctions of any Kellogg kernel by Theorems 73 and 74. There is a constant $q_0 > 0$ such that $\tilde{q}(x) = q(x) + r(x)q_0$ is positive on $[a, b]$ because q is bounded and $r>0$ on $[a, b]$. Let $\tilde{L}y = -(py')' + \tilde{q}y$. Then λ, y is an eigenvalue, eigenfunction pair for $Ly = \lambda ry$, $B_a y = 0$, $B_b y = 0$ if and only if $\tilde{\lambda}$, y is an eigenvalue, eigenfunction pair for $\tilde{L}y = \tilde{\lambda} ry$, $B_a y = 0$, $B_b y = 0$ where $\tilde{\lambda} = \lambda + q_0$. The Green's function $\tilde{g}(x, s)$ of the latter eigenvalue problem is a Kellogg kernel as is the kernel $\tilde{k}(x, s) = \sqrt{r(x)}\tilde{g}(x, s)\sqrt{r(s)}$ by the previous theorem. Hence, the eigenvalues of $\tilde{L}y = \tilde{\lambda} ry$, $B_a y = 0$, $B_b y = 0$, equivalently the eigenvalues of the kernel $\tilde{k}(x, s)$, satisfy

$$0 < \tilde{\lambda}_0 < \tilde{\lambda}_1 < \cdots < \tilde{\lambda}_n < \cdots$$

and its eigenfunctions, which are the eigenfunctions of the original eigenvalue problem, have all the stated properties in the theorem. Since $\tilde{\lambda}_n - q_0 = \lambda_n$,

$$-q_0 < \lambda_0 < \lambda_1 < \cdots < \lambda_n < \cdots.$$

We know from the Hilbert-Schmidt theorem that $|\tilde{\lambda}_n| \to \infty$ as $n \to \infty$ because the integral operator $\tilde{K}$ with symmetric kernel $\tilde{k}(x, s)$ is self-adjoint. Hence, $\lambda_n \to \infty$ as $n \to \infty$. The last two assertions of the theorem follow from Theorem 124. ∎

Example 1b (continued) The eigenvalue problem $-y'' + ay = \lambda y$, $y(0) = 0$, $y(l) = 0$ with $a > 0$ and $l > 0$ has eigenvalues $\lambda_n = a + ((n+1)\pi/l)^2$ and corresponding orthonormal eigenfunctions $\phi_n(x) = \sqrt{2/l}\sin((n+1)\pi x/l)$ for $n = 0, 1, 2, 3, \ldots$. These eigenvalues and eigenfunctions satisfies the hypotheses of Theorem 128 and have all the properties asserted in the theorem. Consequently, the functions

$$\sin\frac{\pi x}{l},\ \sin\frac{2\pi x}{l}, \ldots,\ \sin\frac{n\pi x}{l},\ \pm\sin\frac{(n+1)\pi x}{l}$$

form a Tchebycheff system on $(0, l)$, $\sin((n+1)\pi x/l)$ has exactly n nodal zeros in $(0, l)$, namely,

$$\frac{l}{n+1}, \frac{2l}{n+1}, \frac{3l}{n+1}, \ldots, \frac{nl}{n+1},$$

where the list is empty if $n = 0$, and these nodes strictly interlace on with the nodes of $\sin(n\pi x/l)$. The interlacing of the nodes in $(0, l)$ is easy to check directly because

$$\frac{j-1}{n} < \frac{j}{n+1} < \frac{j}{n}$$

for $j = 1, 2, \ldots, n$.

Example 2b (continued) The eigenvalue problem $-y'' + ay = \lambda y$, $y(0) = 0$, $y(l) = 0$ with $a < 0$ and $l > 0$ has eigenvalues $\lambda_n = a + ((n+1)\pi/l)^2$ and corresponding orthonormal eigenfunctions $\phi_n(x) = \sqrt{2/l}\sin((n+1)\pi x/l)$ for $n = 0, 1, 2, 3, \ldots$. These eigenvalues and eigenfunctions satisfy the hypotheses of Theorem 128 and have all the properties asserted in the theorem. In particular, it has the properties discussed in Example 1b. There may be a finite number of negative eigenvalues in this example.

Example 3b. (continued) This is Example 1b with $a = 0$ and the discussion there applies with $a = 0$.

Example 4b. (continued) The eigenvalue problem $-y'' = \lambda y$, $y(0) - y'(0) = 0$, $y(l) + y'(l) = 0$ with $l > 0$ satisfies the hypotheses of Theorem 128 and therefore its eigenvalues and eigenfunctions have all the properties asserted in the theorem. This is of greater interest than in the previous examples where the eigenvalues and eigenfunctions are known explicitly. Now, the eigenvalues are only known as the roots of the equation

$$\tan\sqrt{\lambda}l = \frac{2\sqrt{\lambda}}{\lambda - 1}$$

augmented by the eigenvalue $\lambda = 1$ in the special situation where $l = (2n+1)\pi/2$ for some nonnegative integer n. The theorem guarantees that this equation has only real positive roots, a fact that is not obvious *a priori*, and that the roots, which are the eigenvalues of the problem, can be listed as

$$0 < \lambda_0 < \lambda_1 < \cdots < \lambda_n < \cdots.$$

The corresponding orthonormal eigenfunctions $\phi_n(x)$, which are nonzero multiples of

$$\sqrt{\lambda_n}\cos\sqrt{\lambda_n}x + \sin\sqrt{\lambda_n}x,$$

have the oscillation and approximation properties stated in the theorem.

4.8.4 Rayleigh Quotient

Consider the regular Sturm-Liouville eigenvalue problem

$$Ly = \lambda r y, \; B_a y = 0, \; B_b y = 0,$$

where $Ly = -(py')' + qy$ and $B_a y = \alpha y(a) + \beta y'(a)$ and $B_b y = \gamma y(b) + \delta y'(b)$ specify separated conditions. The eigenvalue problem has an infinite number of simple eigenvalues

$$\lambda_0 < \lambda_1 < \cdots < \lambda_n < \cdots$$

with $\lambda_n \to \infty$ as $n \to \infty$. This we already know. Recall that the domain of L is

$$D = \{y \in C[a,b] : (py')' \in C[a,b]\}.$$

The quotient that appears in the following theorem is the **Rayleigh quotient.** It will be used in Chapter 7 to find upper estimates of the smallest eigenvalue of a Sturm-Liouville eigenvalue problem as part of a shooting method that accurately determines eigenvalues and corresponding eigenfunctions of the problem.

Theorem 129 *With the notation above, the smallest eigenvalue of a regular Sturm-Liouville eigenvalue problem satisfies*

$$\lambda_0 = \min \frac{\langle Ly, y\rangle}{\langle y, y\rangle_r} = \min \frac{-pyy'\big|_a^b + \int_a^b (py'^2 + qy^2)\,dx}{\int_a^b y^2 r\,dx},$$

where the minimum is over all functions $y \neq 0$ in the domain of L that satisfy the boundary conditions $B_a y = 0$ and $B_b y = 0$. Moreover, the minimum is achieved if and only if y is an eigenfunction corresponding to λ_0.

Proof. If y satisfies the boundary conditions $B_a y = 0$ and $B_b y = 0$ and is in the domain of L, then $Ly = f$ for $f = Ly$; hence, by Theorem 123

$$y(x) = \sum_{n=0}^{\infty} \langle y, \phi_n\rangle_r \phi_n(x),$$

where $\phi_n(x)$ are the corresponding orthonormal eigenfunctions with respect to weight function r, and the series converges absolutely and uniformly on $[a, b]$. Consequently,

$$\begin{aligned}\langle Ly, y\rangle &= \left\langle Ly, \sum_{n=0}^{\infty}\langle y, \phi_n\rangle_r\phi_n\right\rangle = \sum_{n=0}^{\infty}\langle y, \phi_n\rangle_r\langle Ly, \phi_n\rangle \\ &= \sum_{n=0}^{\infty}\langle y, \phi_n\rangle_r\langle y, L\phi_n\rangle = \sum_{n=0}^{\infty}\langle y, \phi_n\rangle_r\langle y, \lambda_n r\phi_n\rangle \\ &= \sum_{n=0}^{\infty}\lambda_n|\langle y, \phi_n\rangle_r|^2 \geq \lambda_0\sum_{n=0}^{\infty}|\langle y, \phi_n\rangle_r|^2 = \lambda_0\langle y, y\rangle_r,\end{aligned}$$

where the last equality follows from a similar calculation using $y = \sum_{n=0}^{\infty}\langle y, \phi_n\rangle_r\phi_n$ to evaluate $\langle y, y\rangle_r$. Equality holds above if and only if $\langle y, \phi_n\rangle_r = 0$ for all $n \geq 1$; hence, if and only if $y = \langle y, \phi_0\rangle_r\phi_0$, equivalently, y is an eigenfunction corresponding to λ_0. Thus, for $y \neq 0$,

$$\lambda_0 \leq \frac{\langle Ly, y\rangle}{\langle y, y\rangle_r}$$

with equality if and only if y is an eigenfunction corresponding to λ_0. The first conclusion in the theorem follows. Finally, a familiar integration by parts argument gives

$$\langle Ly, y\rangle = \int_a^b yd(-py') + \int_a^b qy^2\,dx = -pyy'\big|_a^b + \int_a^b (py'^2 + qy^2)\,dx$$

and the second conclusion follows. ∎

The Courant minimax theorem (see [7] and [10]), which generalizes the preceding theorem in the current context, characterizes (theoretically) all the eigenvalues by minimax conditions that involve orthogonality relations, except in the case of the smallest eigenvalue λ_0. The orthogonality relations make it difficult to obtain useful numerical information from the minimax characterization of λ_n when $n > 0$.

4.8.5 Mixed Boundary Conditions

Our standing notation remains in force:

$$\begin{aligned}Ly &= -(p(x)y')' + q(x)y, \\ B_iy &= a_{i1}y(a) + a_{i2}y'(a) + b_{i1}y(b) + b_{i2}y'(b)\end{aligned}$$

for $i = 1, 2$ and where Ly is a regular Sturm-Liouville differential operator so that $p(x) > 0$ and $p(x)$ and $q(x)$ are real-valued and continuous on $[a, b]$. We call the corresponding eigenvalue problem

$$Ly = \lambda ry,\ B_1y = 0,\ B_2y = 0$$

regular if in addition $r(x) > 0$ on $[a, b]$. Recall that λ, y is an eigenvalue, eigenfunction pair if $y \neq 0$ satisfies the differential equation $Ly = \lambda ry$ on (a, b) and satisfies the boundary conditions. Since the problem is regular, y has additional smoothness: it is continuously differentiable on $[a, b]$ and satisfies the differential equation on $[a, b]$. (See Theorem 112.)

We restrict the discussion to mixed boundary conditions that are self-adjoint so that the eigenvalue problem is self-adjoint. This is the case for periodic boundary conditions and anti-periodic boundary conditions which are the mixed boundary conditions of primary interest in applications.

The basic results in Theorem 123 derived from the Hilbert-Schmidt theorem and its corollaries remain true with one exception.

Theorem 130 *A regular self-adjoint Sturm-Liouville eigenvalue problem* $Ly = \lambda ry$, $B_1 y = 0$, $B_2 y = 0$ *with mixed boundary conditions has an infinite sequence of real eigenvalues* $\{\lambda_n\}_{n=1}^{\infty}$ *and a corresponding sequence of eigenfunctions* $\{\phi_n\}_{n=1}^{\infty}$ *with the following properties:*
1. If the eigenvalues $\{\lambda_n\}_{n=1}^{\infty}$ *are listed in order of increasing modulus and if each eigenvalue is repeated a number of times equal to its multiplicity, then*

$$|\lambda_1| \leq |\lambda_2| \leq |\lambda_3| \leq \cdots \leq |\lambda_n| \leq \cdots,$$

$|\lambda_n| \to \infty$ *as* $n \to \infty$, *and each eigenvalue is either simple or has multiplicity.*
2. The corresponding eigenfunctions are orthonormal with weight function r,

$$\langle \phi_m, \phi_n \rangle_r = \int_a^b \phi_m(s)\phi_n(s)r(s)\, ds = \delta_{mn},$$

where δ_{mn} *is the Kronecker delta, and can be chosen real-valued.*
3. For each continuous function f *on* $[a, b]$, *the unique solution* y *to the regular Sturm-Liouville boundary value problem* $Ly = f$, $B_1 y = 0$, *and* $B_2 y = 0$ *can be expressed by*

$$y(x) = \sum_{n=1}^{\infty} \langle y, \phi_n \rangle_r \phi_n(x)$$

where the series is absolutely and uniformly convergent on $[a, b]$.

Proof. The reasoning used to prove Theorem 123 in which the boundary conditions are separated applies here and establishes all conclusions except the multiplicity assertion in Property 1. The eigenvalues need not all be simple as they are in the case of separated boundary conditions. In the case of mixed boundary conditions, each eigenvalue is either simple or has multiplicity 2. For a given eigenvalue λ, the second order differential equation $Ly = \lambda ry$ has at most two linearly independent solutions. So the multiplicity of any eigenvalue is either 1 or 2 and both possibilities can occur. See the example that follows. ■

Example 5. The regular self-adjoint Sturm-Liouville eigenvalue problem

$$-y'' = \lambda y, \ y(0) = y(l), \ y'(0) = y'(l)$$

has eigenvalues $\lambda_n = (2\pi n/l)^2$ for $n = 0, 1, 2, \ldots$ and corresponding eigenfunctions

$$y_0 = A_0, \ y_n = A_n \cos(2\pi n x/l) + B_n \sin(2\pi n x/l) \quad \text{for } n \geq 1$$

where A_n and B_n are constants with $|A_n| + |B_n| \neq 0$. Thus, λ_0 is a simple eigenvalue and all other eigenvalues have multiplicity 2.

The analogue of Theorem 124 for periodic and antiperiodic boundary conditions is

Theorem 131 *If* $p > 0$, $q \geq 0$, $r > 0$ *are continuous on* $[a, b]$ *and* $p(a) = p(b)$ *in the Sturm-Liouville differential equation and* $B_1 y$ *and* $B_2 y$ *specify periodic or antiperiodic boundary conditions, then all the eigenvalues of the eigenvalue problem* $Ly = \lambda ry$, $B_1 y = 0$, $B_2 y = 0$ *are positive, except when* $q = 0$ *and the boundary conditions are periodic in which case the eigenvalue problem is*

$$-(py')' = \lambda ry, \ y(a) = y(b), \ y'(a) = y'(b),$$

zero is an eigenvalue, and all other eigenvalues are positive.

Proof. If λ is an eigenvalue and y is a corresponding real-valued eigenfunction, multiply $Ly = \lambda ry$ by y and integrate by parts to obtain

$$\lambda \int_a^b y^2 r\,dx = -p(b)y(b)y'(b) + p(a)y(a)y'(a) + \int_a^b \left(py'^2 + qy^2\right) dx.$$

For either periodic or antiperiodic boundary conditions this equation reduces to

$$\begin{aligned}\lambda \int_a^b y^2 r\,dx &= \big(p(a) - p(b)\big)y(b)y'(b) + \int_a^b \left(py'^2 + qy^2\right) dx \\ &= \int_a^b \left(py'^2 + qy^2\right) dx.\end{aligned}$$

Hence, $\lambda \geq 0$ and $\lambda > 0$ unless $y' = 0$, $y = k$ a nonzero constant, and $\int_a^b q\,dx = 0$; that is, $\lambda \geq 0$ and $\lambda > 0$ unless $y = k$ a nonzero constant and $q = 0$. The antiperiodic eigenvalue problem can have no constant eigenfunction; hence its eigenvalues are all positive. The periodic eigenvalue problem with q not identically 0 has all positive eigenvalues. The periodic eigenvalue problem with $q = 0$ does have a nonzero constant eigenfunction corresponding to the eigenvalue $\lambda = 0$ and all its other eigenvalues are positive. ■

As we noted earlier for separated boundary conditions, Part 3 of Theorem 130 can be interpreted as an eigenfunction expansion for functions in the domain of L that satisfy the given boundary conditions. Recall that the domain of L is

$$D = \{y \in C[a, b] : (py')' \in C[a, b]\}.$$

If y is any function in the domain of L that satisfies the given boundary conditions, then y solves the Sturm-Liouville boundary value problem $Ly = f$, $B_1 y = 0$, $B_2 y = 0$ where $f = Ly$. Hence, y has the eigenfunction expansion

$$y(x) = \sum_{n=1}^{\infty} \langle y, \phi_n\rangle_r \phi_n(x)$$

with absolute and uniform convergence on $[a, b]$. If p' is continuous, any y in $C^2[a, b]$ that satisfies the boundary conditions is in the domain of L; consequently, it has an absolutely and uniformly convergent eigenfunction expansion. In particular, applying this observation to the l-periodic eigenvalue problem in Example 5, establishes that any twice continuously differentiable, l-periodic function can be expanded in an absolutely and uniformly convergent Fourier series. Indeed, the orthonormal eigenfunctions corresponding to the eigenvalues $\lambda_n = (2\pi n/l)^2$ are $\phi_0(x) = 1$ for $n = 0$ and $\phi_{2n}(x) = \sqrt{2/l}\cos(2\pi nx/l)$, $\phi_{2n-1}(x) = \sqrt{2/l}\sin(2\pi nx/l)$ for $n = 1, 2, 3, \ldots$. Consequently,

$$\begin{aligned}\langle y, \phi_0\rangle\phi_0 &= \frac{a_0}{2}, \\ \langle y, \phi_{2n}\rangle\phi_{2n} &= a_n \cos(2\pi nx/l), \\ \langle y, \phi_{2n-1}\rangle\phi_{2n-1} &= b_n \sin(2\pi nx/l),\end{aligned}$$

where

$$a_n = \frac{2}{l}\int_0^l y(x)\cos(2\pi nx/l)\,dx,$$

$$b_n = \frac{2}{l}\int_0^l y(x)\sin(2\pi nx/l)\,dx.$$

Thus,

$$y(x) = \sum_{n=0}^{\infty}\langle y, \phi_n\rangle\phi_n(x) = \frac{a_0}{2} + \sum_{n=1}^{\infty}(a_n\cos(2\pi nx/l) + b_n\sin(2\pi nx/l))$$

with absolute and uniform convergence.

Chapter 5

Singular Sturm-Liouville Problems - I

Just as in Chapter 4, the concluding section of Chapter 5 on eigenvalues and eigenfunctions of singular Sturm-Liouville problems is its climax. That section, and the corresponding one in Chapter 6, contain results of great practical importance and focus on the types of singularity that occur naturally when separation of variables is used in polar or spherical coordinates. There are two parts of the discussion. First the basic properties of the eigenvalues and eigenfunctions related to their existence, multiplicity, orthogonality, and eigenfunction expansions are established. These results follow from the Hilbert-Schmidt theorem once suitable properties are established for the Green's functions of singular Sturm-Liouville problems. Second the oscillatory and approximation properties of the eigenfunctions are developed from a unified perspective based on Jentzsch's theorem, Schur's theorem, and the Kellogg conditions; see Section 1.11.2 and Section 3.6.2. The reader primarily interested in the spectral results can skim the necessary background results in Chapter 3 and the properties of Green's functions established in this chapter and concentrate on the material on eigenvalue problems in Section 5.5 and its subsections. Readers seeking a fuller account of properties of solutions to singular Sturm-Liouville differential equations, boundary value problems, and Green's functions will find a readable account in the sections following this introduction.

The overall approach to the study of singular Sturm-Liouville problems parallels that used for regular problems, but with appropriate adjustments to accommodate the singularities. Motivated by the examples in Chapter 1, we consider two types of singular problems for the Sturm-Liouville differential equation

$$-(p(x)y'(x))' + q(x)y(x) = f(x), \quad a < x < b.$$

In this chapter, we consider problems in which $p(x)$ has a simple zero at $x = a$ but is otherwise nonzero and where $p(x)$, $q(x)$, and $f(x)$ are continuous on $[a, b]$. Since p (a) $= 0$ the differential equation is singular at $x = a$. In the next chapter we allow $q(x)$ also to be singular at $x = a$.

The Bessel equation of order 0 and parameter λ

$$R'' + \frac{1}{r}R' + \lambda R = 0$$

equivalently,

$$(rR')' + \lambda rR = 0$$

for $0 < r < b$ serves as a model for the singular Sturm-Liouville problems treated in this chapter. This equation arises from separation of variables in the standard wave equation model for the transverse vibrations of a drumhead, where b is the radius of the drum. (See Section 1.4.)

Bessel's equation of order 0 and parameter λ has two linearly independent solutions, the Bessel functions $J_0(r)$ and $Y_0(r)$. The first is bounded on $(0, b)$ (and determines the eigenfunctions in related eigenvalue problems) the second is unbounded on $(0, b)$. The bounded solution $J_0(r)$ is continuous on $[0, b]$; in fact, it is analytic (has a power series expansion with an infinite radius of convergence). The singular Sturm-Liouville problems considered in the chapter have just such a basis of solutions and the bounded solution determines the eigenfunctions of related eigenvalue problems. They also provide the entry to the shooting methods used in Chapter 7 to determine accurate numerical approximations to eigenvalues and eigenfunctions of the singular problems.

The singular behavior occurs at $x = a$ in this chapter. Corresponding results hold if the singular behavior occurs at $x = b$ instead of $x = a$. Those results can be derived in the same way or, more simply, by a change of variable.

The following standing assumptions are in force throughout the chapter:

Standing Assumptions:

1. $p(x)$ is continuous on $[a, b]$, is differentiable at $x = a$, is nonzero on $a < x \leq b$, and satisfies $p(a) = 0$, $p'(a) \neq 0$.
2. $q(x)$ is continuous on $[a, b]$.
3. $f(x)$ is continuous on $[a, b]$.

   All functions may be complex-valued, unless an explicit statement is made to the contrary.

We will sometimes express (1) in an equivalent way: $p(x) = (x - a)\varphi(x)$ where $\varphi(x)$ is continuous on $[a, b]$ and $\varphi(x) \neq 0$ there, in which case $p'(a) = \varphi(a)$.

Often in applications, $p(x) > 0$ on $a < x \leq b$, equivalently $\varphi(x) > 0$ on $[a, b]$.

5.1 Properties of Solutions

In this section we establish the fundamental nature of solutions to the singular Sturm-Liouville differential equation

$$-(p(x)y'(x))' + q(x)y(x) = f(x), \quad a < x < b, \tag{5.1}$$

under our standing assumptions. By a **solution** to (5.1) we mean a function y such that $(p(x)y'(x))'$ exists for each x in (a, b) and (5.1) holds for each x in (a, b). See Section 4.2 for a discussion of this notion of a solution. The first result, Lemma 132 was inspired by the appendix on special functions in [43] by Tychonoff and Samarski and is the key to all the subsequent developments in the chapter and to the numerical procedure in Chapter 7 that is used to find accurate approximations to eigenvalues and eigenfunctions for such singular problems. Tychonoff and Samarski just considered homogeneous Sturm-Liouville differential equations because their interest was only in the existence and qualitative behavior of eigenfunctions. The lemma that follows embraces inhomogeneous as well as homogeneous equations and includes results not found in the appendix of Tychonoff and Samarski.

The proof of the lemma, although not based on deep analytical results, does involve some subtle elements. It establishes properties of solutions that are needed both for developing theoretical results in this chapter and for establishing convergence of the numerical procedures in Chapter 7.

Lemma 132 *Every solution $y(x)$ to the differential equation (5.1) that is bounded on $a < x < b$ satisfies*

$$\lim_{x \to a} p(x)y'(x) = 0,$$

$$y'(x) = \frac{1}{p(x)} \int_a^x (q(\xi)y(\xi) - f(\xi))\, d\xi \quad \textit{for } a < x < b, \tag{5.2}$$

and $y(x)$ extends to a continuously differentiable function on $[a, b]$ that satisfies the boundary condition

$$q(a)y(a) - p'(a)y'(a) = f(a).$$

Moreover, the extended function also satisfies the differential equation in (5.1) at $x = a$ and at $x = b$.

Proof. Let $y(x)$ be a bounded solution to (5.1). Integration of the differential equation in (5.1) between limits x and c with $a < x$, $c < b$, gives

$$p(x)y'(x) = Q(x) \tag{5.3}$$

where

$$Q(x) = p(c)y'(c) - \int_x^c (q(\xi)y(\xi) - f(\xi))\, d\xi \quad \text{for } a < x < b. \tag{5.4}$$

Since the integrand is bounded and continuous on $a < x < b$, $Q(x)$ is uniformly continuous on (a, b) and extends to a continuous function on $a \leq x \leq b$ by Proposition 7. Hence,

$$\lim_{x \to a} p(x)y'(x) = \lim_{x \to a} Q(x) = Q(a).$$

where the extended function still is denoted by Q. Now from (5.3) for $a < x < b$,

$$y(x) = y(c) - \int_x^c \frac{Q(\xi)}{(\xi - a)\varphi(\xi)}\, d\xi,$$

where $p(x) = (x - a)\varphi(x)$ as in the standing assumptions. Since $y(x)$ is bounded on $a < x < b$ and $Q(x)$ has a limit as x approaches a, it follows that $Q(a) = 0$; hence,

$$\lim_{x \to a} p(x)y'(x) = \lim_{x \to a} Q(x) = Q(a) = 0.$$

Let $x \to a$ in (5.4) and then replace c by x to obtain

$$y'(x) = \frac{1}{p(x)} \int_a^x (q(\xi)y(\xi) - f(\xi))\, d\xi \quad \text{for } a < x < b.$$

It follows that $y'(x)$ is bounded on (a, b), $y(x)$ is uniformly continuous there, and has a unique extension by continuity to a continuous function on $[a, b]$. (See Proposition 7.) Since

$$\frac{d}{dx} \int_a^x (q(\xi)y(\xi) - f(\xi))\, d\xi \bigg|_{x=a} = q(a)y(a) - f(a)$$

by the fundamental theorem of calculus, the expression for $y'(x)$ above and the simplest form of l'Hôpital's rule gives

$$\lim_{x \to a} y'(x) = \frac{q(a)y(a) - f(a)}{p'(a)}.$$

By Lemma 11, $y(x)$ is differentiable at $x = a$,

$$y'(a) = \frac{q(a)y(a) - f(a)}{p'(a)},$$

and the derivative is continuous at $x = a$. Since

$$\frac{y(x) - y(b)}{x - b} = y'(\xi_x)$$

for some ξ_x between x and b, use (5.2) to see that there exists

$$y'(b) = \lim_{x\to b}\frac{y(x)-y(b)}{x-b} = \lim_{\xi_x\to b} y'(\xi_x) = \frac{1}{p(b)}\int_a^b (q(\xi)y(\xi)-f(\xi))\,d\xi.$$

Finally,

$$\lim_{x\to b} y'(x) = \lim_{x\to b}\frac{1}{p(x)}\int_a^x (q(\xi)y(\xi)-f(\xi))\,d\xi = y'(b)$$

which establishes that y' is continuous at $x = b$.

It remains to show that the extended function $y(x)$, which is continuously differentiable on $[a, b]$, also satisfies the differential equation (5.1) at $x = a$ and at $x = b$. Since

$$\frac{p(x)y'(x) - p(a)y'(a)}{x-a} = (py')'(\xi_x) = q(\xi_x)y(\xi_x) - f(\xi_x)$$

for some ξ_x between a and x and ξ_x tends to a as x tends to a, it follows that there exists

$$(py')'(a) = q(a)y(a) - f(a).$$

Thus, y satisfies the differential equation (5.1) at $x = a$. Likewise, y satisfies the differential (5.1) at $x = b$. ■

In view of the lemma, if y is a *bounded* solution of the Sturm-Liouville differential equation (5.1), we may also use y to denote its continuously differentiable extension to the closed interval $[a, b]$.

Lemma 132 suggests how to prove that the differential equation (5.1) has bounded solutions. Since the continuous extension to $[a, b]$ of a bounded solution y of (5.1) also satisfies the differential equation at $x = a$ and $x = b$, integration of (5.2) yields

$$y(x) = y(a) + \int_a^x \left(\frac{1}{p(\xi)}\int_a^\xi (q(\eta)y(\eta)-f(\eta))\,d\eta\right)d\xi$$

for x in $[a, b]$. This suggests introducing the transformation $T_c{:}C[a, b] \to C[a, b]$ defined by

$$T_c y(x) = c + \int_a^x \left(\frac{1}{p(\xi)}\int_a^\xi (q(\eta)y(\eta)-f(\eta))\,d\eta\right)d\xi,$$

where c is a fixed constant and $C[a, b]$ is the Banach space of real or complex valued continuous functions on $[a, b]$. If $y(x)$ is a bounded solution of (5.1), then it (more precisely its continuous extension to $[a, b]$) is a fixed point of the mapping T_c when $c = y(a)$. Conversely, if $y(x)$ is a fixed point of the mapping T_c, then differentiating $y(x) = Ty(x)$ twice, shows that $y(x)$ is a bounded solution of (5.1) with $y(a) = c$, that $y(x)$ is continuously differentiable on $[a, b]$, and that $y(x)$ also satisfies the differential equation at $x = a$ and $x = b$. We shall show that T_c is a contraction mapping on $C[a, b]$ equipped with a suitable norm, apply the contraction mapping fixed point theorem, and thereby establish that T_c has a fixed point and that the differential equation (5.1) has bounded solutions.

Theorem 133 *Fix a real or complex number c and let $C[a, b]$ be the space of real- or complex-valued continuous functions on $[a, b]$. There is a norm on $C[a, b]$ that is equivalent to the maximum norm such that the mapping T_c: $C[a, b] \to C[a, b]$ defined by*

$$T_c y(x) = c + \int_a^x \left(\frac{1}{p(\xi)}\int_a^\xi (q(\eta)y(\eta)-f(\eta))\,d\eta\right)d\xi$$

is a contraction. Consequently, T_c has a unique fixed point y_c in $C[a, b]$.

Proof. For y in $C[a, b]$ define $\|y\|_{\max}= \max_{a\le x\le b}|y(x)|$. We claim the operator T_c is well defined and maps $C[a, b]$ into itself. This essentially amounts to the observation that the improper integral with respect to ξ exists. To confirm this, recall that $p(x) = (x - a)\varphi(x)$ where $\varphi(x)$ is continuous and nonzero on $[a, b]$ and define

$$M = \frac{\|y\|_{\max}\|q\|_{\max}+\|f\|_{\max}}{\min_{a\le x\le b}|\varphi(x)|},$$

fix x in $[a, b]$, and set

$$F(c) = \int_c^x \left(\frac{1}{p(\xi)}\int_a^\xi (q(\eta)y(\eta) - f(\eta))\, d\eta\right) d\xi.$$

(The dependence of F on x is suppressed because x is fixed in this argument.) For $a< c,\ c' \le b$,

$$\begin{aligned}|F(c) - F(c')| &= \left|\int_c^{c'} \left(\frac{1}{p(\xi)}\int_a^\xi (q(\eta)y(\eta) - f(\eta))\, d\eta\right) d\xi\right| \\ &\le M\left|\int_c^{c'} \left(\frac{1}{\xi - a}\int_a^\xi 1\, d\eta\right) d\xi\right| = M|c' - c|.\end{aligned}$$

Thus $F(c)$ is uniformly continuous on $a< c\le b$ and, hence, has a unique extension by continuity to a continuous function on $[a, b]$. The extended function satisfies

$$\begin{aligned}F(a) &= \lim_{c\to a} F(c) = \lim_{c\to a}\int_c^x \left(\frac{1}{p(\xi)}\int_a^\xi (q(\eta)y(\eta) - f(\eta))\, d\eta\right) d\xi \\ &= \int_a^x \left(\frac{1}{p(\xi)}\int_a^\xi (q(\eta)y(\eta) - f(\eta))\, d\eta\right) d\xi\end{aligned}$$

by definition of the improper integral. Thus, the improper integral in question exists and $T_c y(x)$ is well defined. Similarly, for x' and x in $[a, b]$,

$$\begin{aligned}|T_c y(x') - T_c y(x)| &= \left|\int_x^{x'} \left(\frac{1}{p(\xi)}\int_a^\xi (q(\eta)y(\eta) - f(\eta))\, d\eta\right) d\xi\right| \\ &\le M\left|\int_x^{x'} \left(\frac{1}{\xi - a}\int_a^\xi 1\, d\eta\right) d\xi\right| = M|x' - x|.\end{aligned}$$

Thus, $T_c y$ is continuous on $[a, b]$ and $T_c: C[a, b] \to C[a, b]$.

Now T_c is a contraction on $C[a, b]$ with the maximum norm only if $|b - a|$ is suitably small. This technical difficulty can be overcome in various ways. One way is to replace the maximum norm by the norm on $C[a, b]$ defined by

$$\|y\|_B= \max_{a\le x\le b} e^{-B(x-a)}|y(x)| \quad \text{where} \quad B = \frac{\|q\|_{\max}}{\min_{a\le x\le b}|\varphi(x)|},$$

as we did in Chapter 4. Since $e^{-B(b-a)}\|y\|_{\max}\le \|y\|_B\le \|y\|_{\max}$, the new norm is equivalent to the maximum norm and $C[a, b]$ equipped with $\|y\|_B$ is a Banach space. For any

y and z in $C[a, b]$,

$$
\begin{aligned}
|T_c y(x) - T_c z(x)| &= \left| \int_a^x \left(\frac{1}{p(\xi)} \int_a^\xi q(\eta) e^{B(\eta-a)} e^{-B(\eta-a)} (y(\eta) - z(\eta))\, d\eta \right) d\xi \right| \\
&\le B\|y - z\|_B \int_a^x \left(\frac{1}{\xi - a} \int_a^\xi e^{B(\eta-a)}\, d\eta \right) d\xi \\
&\le B\|y - z\|_B \int_a^x \left(\frac{1}{\xi - a} e^{B(\xi-a)} \int_a^\xi 1\, d\eta \right) d\xi \\
&= B\|y - z\|_B \frac{e^{B(x-a)} - 1}{B}.
\end{aligned}
$$

Hence,

$$
\begin{aligned}
e^{-B(x-a)} |T_c y(x) - T_c z(x)| &\le \left(1 - e^{-B(x-a)}\right) \|y - z\|_B, \\
\|T_c y - T_c z\|_B &\le \left(1 - e^{-B(b-a)}\right) \|y - z\|_B,
\end{aligned}
$$

and T_c is a contraction on $C[a, b]$. By the contraction mapping theorem, T_c has a unique fixed point y_c in $C[a, b]$. ∎

Corollary 134 *The singular differential equation (5.1) has nontrivial bounded solutions. If $p(x)$, $q(x)$, and $f(x)$ are real-valued, then (5.1) has a real-valued, nontrivial bounded solution.*

Proof. The fixed point y_c of T_c is a bounded solution of (5.1) and is nontrivial when $c \neq 0$. Assume $p(x)$, $q(x)$, and $f(x)$ are real-valued and that $y = y_1 + iy_2$ is a nontrivial bounded solution to (5.1), where y_1 and y_2 are real-valued. Substitute y into (5.1) to find that y_1 is a bounded solution to (5.1). If y_1 is nontrivial, the desired conclusion follows. If $y_1 = 0$, then $f = 0$ and y_2 is nontrivial and satisfies $(py_2')' + qy_2 = 0$. So y_2 is a real-valued nontrivial bounded solution of (5.1). ∎

Corollary 135 *The only bounded solution to $-(p(x)y'(x))' + q(x)y(x) = 0$, $a < x < b$, $y(a) = 0$ is the identically zero solution.*

Proof. The continuous extension of a bounded solution y to the given problem is a fixed point of the contraction mapping $T_0 : C[a, b] \to C[a, b]$ when $f = 0$. The zero function is clearly the unique fixed point of T_0 when $f = 0$. Thus, $y = 0$. ∎

The case with c any constant and f any continuous function also is important. In this case, the unique fixed point y_c of T_c is the unique bounded solution to

$$
\begin{cases} -(p(x)y'(x))' + q(x)y(x) = f(x), & a < x < b, \\ \qquad y(a) = c, \end{cases} \tag{5.5}
$$

equivalently, is the unique bounded solution to the initial value problem

$$
\begin{cases} -(p(x)y'(x))' + q(x)y(x) = f(x), & a < x < b, \\ y(a) = c,\ y'(a) = (q(a)c - f(a))/p'(a), \end{cases} \tag{5.6}
$$

where the second initial condition follows from Lemma 132. (More precisely, the solution is $y(x) = y_c(x)$ for $a \le x < b$ and $y_c(x)$ is the continuously differentiable extension to $[a, b]$ of the solution and satisfies the differential equation at $x = a$ and $x = b$.)

With the foregoing theorem in hand, we can easily describe the nature of solutions y to the singular differential equation (5.1). Let y_c be the unique fixed point of T_c. Evidently, $y_c(a) = c$. Hence, $y_c(a) \neq 0$ when $c \neq 0$ and two differentiations of $y_c = T_c y_c$ show that y_c is a bounded nontrivial solution to (5.1). On the other hand, as we saw in the remarks motivating consideration of T_c, any bounded solution $y(x)$ to (5.1) with $y(a) = c$ (more properly, the continuous extension of $y(x)$ to $[a, b]$) is a fixed point of T_c; thus, $y = y_c$ because the fixed point is unique. Consequently, if y is any bounded solution of (5.1) with $y(a) \neq 0$, then $y/y(a)$ is the unique fixed point of T_1; that is, $y = y(a)y_1$. It follows that all bounded solutions $y(x)$ to (5.1) with $y(a) \neq 0$ are nonzero multiples of each other.

We summarize and extend this discussion in the following theorem.

Theorem 136 *If $p(x)$ is continuous on $[a, b]$, is differentiable at $x = a$, is nonzero on $a < x \leq b$, and satisfies $p(a) = 0$, $p'(a) \neq 0$; $q(x)$ is continuous on $[a, b]$; and $f(x)$ is continuous on $[a, b]$ then*

(a) (5.1) has a nontrivial bounded solution $u(x)$ that is continuously differentiable on $[a, b]$, and for any such solution $u(a) \neq 0$, $q(a)u(a) - p'(a)u'(a) = f(a)$, and $u(x)$ satisfies the differential equation at $x = a$ and $x = b$;

(b) all nontrivial bounded solutions of (5.1) are nonzero multiples of each other;

(c) solutions $v(x)$ to (5.1) that are linearly independent of a given bounded nontrivial solution $u(x)$ to (5.1) exist and any such solution $v(x)$ is continuously differentiable on $(a, b]$, satisfies the differential equation at $x = b$, and becomes logarithmically infinite as x approaches a; that is,

$$\lim_{x \to a} \frac{v(x)}{\ln(x - a)} = \frac{C}{\varphi(a)u(a)}$$

for some $C \neq 0$.

(d) If $p(x)$, $q(x)$, and $f(x)$ are real-valued, then the solutions $u(x)$ and $v(x)$ can be chosen real-valued.

Proof. The fixed point y_c of T_c is a bounded solution of (5.1) and has all the properties asserted in (a) for any choice of $c \neq 0$ by Lemma 132. So (a) holds for $u = y_c$ for any $c \neq 0$. We have already established (b).

(c) Let c be the midpoint of $[a, b]$. By Theorem 83 there is a unique solution $v(x)$ to the initial value problem for (5.1) with $v(c) = -\overline{u'(c)}$ and $v'(c) = \overline{u(c)}$. The solution $v(x)$ is independent of $u(x)$ because $W_{u,v}(c) = |u(c)|^2 + |u'(c)|^2 \neq 0$. Furthermore, since the differential equation in the initial value problem is regular on the interval (c, b), $v(x)$ extends to a continuously differentiable function on $[c, b]$ and satisfies the differential equation on that interval by Theorem 85. So there exists a solution $v(x)$ to $-(py')' + qy = f$ on $a < x \leq b$ that is linearly independent of $u(x)$.

Now, let $v(x)$ be *any* solution of (5.1) that is independent of $u(x)$. By Lemma 86

$$p(x)(u(x)v'(x) - u'(x)v(x)) = C,$$

with $C \neq 0$ determined by the two independent solutions. Express $p(x)$ as $p(x) = (x - a)\varphi(x)$ where $\varphi(x) \neq 0$ is continuous on $[a, b]$. Since $u(a) \neq 0$ and $\varphi(\xi)$ and $u(\xi)$ are continuous at a, given any $\varepsilon > 0$ there is an $x_0 > a$ such that $u(x) \neq 0$ on $[a, x_0]$ and

$$\left| \frac{Cu(x)}{\varphi(\xi)u(\xi)^2} - \frac{C}{\varphi(a)u(a)} \right| < \frac{\varepsilon}{2} \quad \text{for } a \leq x, \xi \leq x_0.$$

Fix such an x_0. For $a < x \leq x_0$,

$$\frac{d}{dx}\left(\frac{v(x)}{u(x)}\right) = \frac{u(x)v'(x) - u'(x)v(x)}{u(x)^2} = \frac{C}{p(x)u(x)^2}$$

and, hence,

$$v(x) = u(x)\left[\frac{v(x_0)}{u(x_0)} - \int_x^{x_0} \frac{C}{p(\xi)u(\xi)^2}\, d\xi\right].$$

Use the mean value theorem for integrals (Theorem 15) to find

$$v(x) = u(x)\left[\frac{v(x_0)}{u(x_0)} - \frac{C}{\varphi(\xi_x)u(\xi_x)^2}\left(\ln(x_0 - a) - \ln(x - a)\right)\right]$$

for some ξ_x with $x < \xi_x < x_0$. It follows that

$$\frac{v(x)}{\ln(x-a)} - \frac{C}{\varphi(a)u(a)} = I + II$$

where

$$I = \frac{u(x)}{\ln(x-a)}\left[\frac{v(x_0)}{u(x_0)} - \frac{C}{\varphi(\xi_x)u(\xi_x)^2}\ln(x_0 - a)\right]$$

and

$$II = \frac{Cu(x)}{\varphi(\xi_x)u(\xi_x)^2} - \frac{C}{\varphi(a)u(a)}.$$

Since $\lim_{x\to a} I = 0$, there is a $\delta' > 0$ such that $|I| < \varepsilon/2$ if $a < x < a + \delta'$. By earlier choices $|II| < \varepsilon/2$ for $a \leq x$, $\xi_x \leq x_0$. If $\delta = \min(\delta', x_0 - a)$, then

$$a < x < a + \delta \Rightarrow \left|\frac{v(x)}{\ln(x-a)} - \frac{C}{\varphi(a)u(a)}\right| < \varepsilon.$$

Consequently, there exists

$$\lim_{x\to a}\frac{v(x)}{\ln(x-a)} = \frac{C}{\varphi(a)u(a)} \neq 0$$

and the logarithmic growth of $v(x)$ as x approaches a is established.

(d) When $p(x)$, $q(x)$, and $f(x)$ are real-valued $u(x)$ can be chosen real-valued by Corollary 134 and $v(x)$ is real-valued because the initial value problem that determines it has only real data. ■

The continuous dependence result in the next theorem will be needed in Chapter 7 as part of the convergence analysis for a numerical method that accurately evaluates eigenvalues and eigenfunctions. The result we need follows from a reprise of the proof of Theorem 133 in which the operator T_c is replaced by a family of operators $T_{c,\mu}$ that depend on a parameter μ and turn out to be contractions with a uniform contraction constant (independent of μ).

Theorem 137 *Let $c \neq 0$ be fixed, μ be a real parameter that varies in the closed bounded interval I, and $q_\mu(x) = q(x,\mu)$ be a family of continuous functions on $[a, b] \times I$ such that*

$$\lim_{\mu\to\mu_0} \| q_\mu - q_{\mu_0} \|_{\max} = 0$$

for each μ_0 in I; that is, the map that takes μ to q_μ is continuous as a map from I into $C[a, b]$ equipped with the maximum norm. Under the standing assumptions of the chapter, for each μ in I the initial value problem

$$\begin{cases} -(p(x)y'(x))' + q_\mu(x)y(x) = f(x), \ \ a \le x \le b, \\ y(a) = c, \ \ y'(a) = (q_\mu(a)y(a) - f(a))/p'(a), \end{cases} \qquad (5.7)$$

has a unique solution, denoted by $y_\mu(x)$, and given any $\varepsilon > 0$ there is a $\delta > 0$ such that

$$|\mu - \mu_0| < \delta \Rightarrow |y_\mu(x) - y_{\mu_0}(x)| < \varepsilon \quad \text{for } a \le x \le b$$

and

$$|\mu - \mu_0| < \delta \Rightarrow |y'_\mu(x) - y'_{\mu_0}(x)| < \varepsilon \quad \text{for } a \le x \le b.$$

Proof. Let $y(x)$ be a continuous function on $[a, b]$ and define

$$T_{c,\mu}y(x) = c + \int_a^x \left(\frac{1}{p(\xi)} \int_a^\xi (q_\mu(\eta)y(\eta) - f(\eta))\, d\eta \right) d\xi;$$

that is, $T_{c,\mu}$ is the operator T_c used in the proof of Theorem 133 with q replaced by q_μ. Now repeat the proof of Theorem 133 replacing $q(x)$ by $q_\mu(x)$ and in the expressions defining M and B,

$$M = \frac{\|y\|_{\max}\|q\|_{\max} + \|f\|_{\max}}{\min_{a\le x\le b}|\varphi(x)|},$$

$$B = \frac{\|q\|_{\max}}{\min_{a\le x\le b}|\varphi(x)|},$$

interpret $\|q\|_{\max}$ to be the maximum of $q(x, \mu)$ over $[a, b] \times I$,

$$\|q\|_{\max} = \max_{\substack{a\le x\le b \\ \mu \text{ in } I}} |q(x, \mu)|,$$

to find that B is a constant independent of μ in I and the operators $T_{c,\mu} : C[a, b] \to C[a, b]$ satisfy

$$\|T_{c,\mu}y - T_{c,\mu}z\|_B \le \left(1 - e^{-B(b-a)}\right)\|y - z\|_B,$$

where $\|y\|_B = \max e^{-B(x-a)}|y(x)|$ is a norm on $C[a, b]$ that is equivalent to the maximum norm. That is, $T_{c,\mu}$ for μ in I is a family of contractions with a uniform (independent of μ) contraction constant $1 - e^{-B(b-a)}$. Let $y_{c,\mu}$ be the unique fixed point of $T_{c,\mu}$. By Theorem 45 the correspondence μ to $y_{c,\mu}$ from I to $C[a, b]$ equipped with the maximum norm is continuous. That is,

$$\mu \to \mu_0 \Rightarrow \left\| y_{c,\mu} - y_{c,\mu_0} \right\|_{\max} \to 0.$$

Just as in the discussion prior to and following Theorem 133, the existence of a fixed point $y_{c,\mu}$ in $C[a, b]$ to $T_{c,\mu}$ is equivalent to the assertion that $y_\mu(x) = y_{c,\mu}(x)$ is the unique solution to (5.7). See (5.6). Consequently, (5.7) has a unique solution $y_\mu(x)$ and $y_\mu(x)$ converges uniformly to $y_{\mu_0}(x)$ on $[a, b]$ as μ tends to μ_0 by the continuous dependence result just established.

It remains to show that $y'_\mu(x)$ converges uniformly to $y'_{\mu_0}(x)$ on $[a, b]$ as μ tends to μ_0. Differentiate $y_{c,\mu} = T_{c,\mu}y_{c,\mu}$ where $y_{c,\mu} = y_\mu$ to obtain

$$y'_\mu(x) = \frac{1}{p(x)} \int_a^x (q_\mu(\eta)y_\mu(\eta) - f(\eta))\, d\eta$$

and

$$\begin{aligned} y'_\mu(x) - y'_{\mu_0}(x) &= \frac{1}{p(x)}\int_a^x (q_\mu(\eta)y_\mu(\eta) - f(\eta))\,d\eta \\ &\quad - \frac{1}{p(x)}\int_a^x (q_{\mu_0}(\eta)y_{\mu_0}(\eta) - f(\eta))\,d\eta \\ &= \frac{1}{p(x)}\int_a^x (q_\mu(\eta)(y_\mu(\eta) - y_{\mu_0}(\eta)))\,d\eta \\ &\quad - \frac{1}{p(x)}\int_a^x ((q_{\mu_0}(\eta) - q_\mu(\eta))y_{\mu_0}(\eta))\,d\eta. \end{aligned}$$

for x in $[a,\, b]$. Now

$$\begin{aligned} \left|\frac{1}{p(x)}\int_a^x (q_\mu(\eta)(y_\mu(\eta) - y_{\mu_0}(\eta)))\,d\eta\right| &\le \frac{\|y_\mu - y_{\mu_0}\|_{\max}}{\min_{a\le x\le b}|\varphi(x)|}\frac{1}{x-a}\left|\int_a^x q_\mu(\eta)\,d\eta\right| \\ &\le \frac{\|q\|_{\max}}{\min_{a\le x\le b}|\varphi(x)|}\|y_\mu - y_{\mu_0}\|_{\max}, \end{aligned}$$

where $\|q\|_{\max} = \max_{a\le x\le b,\,\mu \text{ in } I}|q(x,\mu)|$. Consequently, the left member of the inequality tends to 0 uniformly on $[a,\, b]$ as μ tends to μ_0. Similarly,

$$\begin{aligned} \left|\frac{1}{p(x)}\int_a^x ((q_{\mu_0}(\eta) - q_\mu(\eta))y_{\mu_0}(\eta))\,d\eta\right| &\le \frac{\|q_\mu - q_{\mu_0}\|_{\max}}{\min_{a\le x\le b}|\varphi(x)|}\frac{1}{x-a}\left|\int_a^x y_{\mu_0}(\eta)\,d\eta\right| \\ &\le \frac{\|y_{\mu_0}\|_{\max}}{\min_{a\le x\le b}|\varphi(x)|}\|q_\mu - q_{\mu_0}\|_{\max}, \end{aligned}$$

where $\|q_\mu - q_{\mu_0}\|_{\max} = \max_{a\le x\le b,\,\mu \text{ in } I}|q(x,\mu) - q(x,\mu_0)|$. Again, the left member of the inequality tends to 0 uniformly on $[a,\, b]$ as μ tends to μ_0. Combining these estimates establishes that $y'_\mu(x)$ converges uniformly to $y'_{\mu_0}(x)$ on $[a,\, b]$ as μ tends to μ_0, which is the final conclusion of the theorem. ■

5.2 Initial Value Problems

The initial value problem corresponding to the singular Sturm-Liouville differential equation (5.1) is

$$\begin{cases} -(p(x)y'(x))' + q(x)y(x) = f(x), & a < x < b, \\ y(c) = c_0, \;\; y'(c) = c_1, \end{cases} \tag{5.8}$$

where c is fixed in $[a,\, b]$ and c_0 and c_1 are given constants. If $a < c < b$, then clearly a solution y to the initial value problem is continuous on its domain $a < x < b$. If $c = a$ (respectively, $c = b$) the initial conditions imply that a solution is continuous at $x = a$ (respectively, $x = b$) and hence is continuous on its domain $a \le x < b$ (respectively, $a < x \le b$).

There are two cases to consider: $c = a$ and $a < c \le b$. Let $c = a$. If the initial value problem has a solution y, then the initial conditions imply that y is continuous at $x = a$ and, hence, bounded on $a \le x \le a'$ for some a' with $a < a' < b$. Since y satisfies the regular Sturm-Liouville differential equation $-(py')' + qy = f$ on $a' < x < b$, it extends to a continuous function on $[a',\, b]$ by Theorem 85. Hence, y is a bounded solution to $-(py')' + qy = f$ on $a < x < b$ and (its continuous extension to $[a,\, b]$) is the unique fixed point of the contraction mapping $T_{c_0} : C[a, b] \to C[a, b]$ in Theorem 133. Thus, if $c = a$ and a solution y to the initial value problem exists, it must be y_{c_0}, the unique fixed point of T_{c_0}. Since $y = y_{c_0}$ and y_{c_0} salsifies $q(a)y_{c_0}(a) - p'(s)y'_{c_0}(a) = f(a)$, c_0 and c_1 must satisfy $q(a)c_0 - p'(s)c_1 = f(a)$ if the initial value problem has a solution. Conversely, if this condition is satisfied the initial value problem has a solution; see (5.6).

Now assume $a < c \leq b$. For positive integers n such that $a + 1/n < c$, the regular initial value problem

$$\begin{cases} -(p(x)y'(x))' + q(x)y(x) = f(x), & a + 1/n \leq x \leq b, \\ y(c) = c_0, \ y'(c) = c_1, \end{cases}$$

has a unique solution $y_n(x)$ for $a + 1/n \leq x \leq b$ by Theorem 81. Suppose x in $a < x \leq b$ lies in both the domain of $y_n(x)$ and $y_m(x)$ and label the solutions so that $n > m$. Since $y_n(x)$ solves the initial value problem on $a + 1/m \leq x \leq b$ and the solution is unique, $y_n(x) = y_m(x)$. Consequently, if x is in $a < x \leq b$ and n satisfies $a + 1/n < x \leq b$, then $y(x) = y_n(x)$ is a well defined function on $a < x \leq b$ and solves the initial value problem

$$\begin{cases} -(p(x)y'(x))' + q(x)y(x) = f(x), & a < x \leq b, \\ y(c) = c_0, \ y'(c) = c_1, \end{cases}.$$

Consequently $y(x)$ also solves (5.8) and has the added property that it is continuously differentiable on $a < x \leq b$ and satisfies the differential equation at $x = b$. If z is also a solution to (5.8), then y and z both solve the regular initial value

$$\begin{cases} -\big(p(x)w'(x)\big)' + q(x)w(x) = f(x), & a + 1/n < x < b, \\ y(c) = c_0, \ y'(c) = c_1, \end{cases}.$$

Since this problem has a unique solution by Theorem 82, $z = y$ on $a + 1/n < x < b$ for every n. Consequently, $z = y$ on $a < x < b$ and equality also holds when $x = b$ when the initial data is given at $c = b$. This shows that (5.8) has a unique solution. In summary:

Theorem 138 *Under the standing assumptions, if $a < c \leq b$, the initial value problem (5.8) has a unique solution that extends to a continuously differentiable function on $a < x \leq b$ and satisfies the differential equation there. If $c = a$, the initial value problem has a solution if and only if c_0 and c_1 satisfy $q(a)c_0 - p'(a)c_1 = f(a)$, in which case the solution is unique, extends to a continuously differentiable function on $[a, b]$, and satisfies the differential equation at $x = a$ and $x = b$.*

As we have observed for regular initial value problems, if $p(x)$, $q(x)$, $f(x)$, c_0, and c_1 are all real-valued, then the unique solution is real-valued.

5.3 Boundary Value Problems

The standing assumptions of the chapter remain in force:

(1) $p(x)$ is continuous on $[a, b]$, is differentiable at $x = a$, is nonzero on $a < x \leq b$, and satisfies $p(a) = 0$, $p'(a) \neq 0$.
(2) $q(x)$ is continuous on $[a, b]$.
(3) $f(x)$ is continuous on $[a, b]$.

We will sometimes express (1) in an equivalent way: $p(x) = (x-a)\varphi(x)$ where $\varphi(x)$ is continuous on $[a, b]$ and $\varphi(x) \neq 0$ there, in which case $p'(a) = \varphi(a)$.

As in Chapter 4, the Sturm-Liouville differential operator is

$$Ly = -(p(x)y'(x))' + q(x)y(x)$$

and $B_b y = \gamma y(b) + \delta y'(b)$ is a linear boundary form, where γ and δ are real or complex numbers with γ and δ not both zero. Boundary conditions at $x = b$ are specified by $B_b y = c_b$, where c_b is a given real or complex number.

The singular Sturm-Liouville boundary value problem associated with the singular differential equation is

$$\begin{cases} -(p(x)y'(x))' + q(x)y(x) = f(x) & \text{for } a < x < b \\ |y(a)| < \infty, \quad \gamma y(b) + \delta y'(b) = c_b. \end{cases} \tag{5.9}$$

The corresponding homogeneous problem is

$$\begin{cases} -(p(x)y'(x))' + q(x)y(x) = 0 & \text{for } a < x < b \\ |y(a)| < \infty, \quad \gamma y(b) + \delta y'(b) = 0 \end{cases} \tag{5.10}$$

Here $|y(a)| < \infty$ is a common shorthand notation which means that $y(x)$ is bounded for $x > a$ and near a, and, hence, bounded on $a < x \leq b$.

As for a regular problem, a **solution** $y(x)$ to (5.9) or (5.10) is a function that satisfies the Sturm-Liouville differential equation on $a < x < b$, satisfies the given boundary conditions, and is continuous on $[a, b]$.

We discussed the reason for the continuity assumption for regular problems in Section 4.5. Some further elaboration is needed here. The formulation of the boundary condition at $x = a$, namely that $|y(a)| < \infty$, is suggested by physical considerations in which such boundary value problems arise. The boundary condition $|y(a)| < \infty$ can in principle allow quite wild behavior of a function that satisfies the singular Sturm-Liouville differential equation as x approaches a. Under our standing assumptions, this does not happen for solutions of the singular differential equation in (5.9). By Lemma 132 any bounded solution $y(x)$ to the differential equation on $a < x < b$ extends to continuous function near $x = a$, equivalently, $\lim_{x\to a} y(x)$ exists in which case defining $y(a)$ to be this limit gives the extension of y to a continuous function near $x = a$. Thus, the requirement that solutions to (5.9) or (5.10) include the continuity requirement is natural in the context of our standing assumptions and makes it explicit that the bounded solutions of interest have limiting values as x approaches a. Furthermore, a solution to (5.9) or (5.10) automatically has additional smoothness.

Lemma 139 *A solution $y(x)$ to (5.9) or (5.10) is continuously differentiable on $[a, b]$ and satisfies the differential equation at $x = a$ and $x = b$.*

Proof. By Lemma 132 any bounded solution y to the differential equation in (5.9) or (5.10) extends to a continuously differentiable function on $[a, b]$ and satisfies the differential equation at $x = a$ and at $x = b$. ■

The next result extends to singular problems a convenient criterion for the existence and uniqueness of solutions to regular Sturm-Liouville boundary value problems.

Theorem 140 *The Sturm-Liouville boundary value problem (5.9) has a unique solution for every choice of $f(x)$ if and only if the corresponding homogeneous problem (5.10) has only the trivial solution.*

Proof. The only if assertion follows immediately.

If (5.10) has only the trivial solution, then any solution $y(x)$ to (5.9) is unique. It remains to prove that (5.9) has a solution for every choice of $f(x)$. By Theorem 136 there is a nontrivial bounded solution $u(x)$ in $C^1[a, b]$ to

$$-(p(x)u'(x))' + q(x)u(x) = 0 \quad \text{for } a \leq x \leq b,$$

and a solution $v(x)$ in $C^1(a, b]$ to

$$-(p(x)v'(x))' + q(x)v(x) = 0 \quad \text{for } a < x \leq b$$

that is logarithmically unbounded as $x \to a$. Note that

$$\gamma u(b) + \delta u'(b) \neq 0;$$

otherwise, $u(x)$ would be a nontrivial solution to (5.10). Let $z(x)$ be the unique bounded solution to the initial value problem (5.6) with $c = 0$,

$$-(p(x)z'(x))' + q(x)z(x) = f(x) \quad \text{for } a < x < b,$$
$$z(a) = 0, \; z'(a) = -f(a)/\varphi(a).$$

Recall that $z(x)$ extends to be continuously differentiable on $[a, b]$ and satisfies the differential equation at $x = a$ and $x = b$. Clearly $y(x)$ will be a solution of (5.9) if and only if $w(x) = y(x) - z(x)$ satisfies

$$\begin{cases} -(p(x)w'(x))' + q(x)w(x) = 0 \quad \text{for } a < x < b, \\ |w(a)| < \infty, \; \gamma w(b) + \delta w'(b) = c_b - \gamma z(b) - \delta z'(b) \end{cases}.$$

The general solution to the homogeneous differential equation for w is

$$w(x) = c_1 u(x) + c_2 v(x)$$

for arbitrary constants c_1 and c_2. The only bounded solutions are $w(x) = c_1 u(x)$. Such a solution will satisfy the boundary condition at $x = b$ if and only if

$$c_1\big(\gamma u(b) + \delta u'(b)\big) = c_b - \gamma z(b) - \delta z'(b),$$

$$c_1 = \frac{c_b - z(b) - \delta z'(b)}{\gamma u(b) + \delta u'(b)}.$$

With this choice of c_1, $y(x) = c_1 u(x) + z(x)$ is a solution of (5.9). ∎

The boundary value problem (5.9) can be reduced to two closely related problems: if y_1 and y_2 solve

$$\begin{cases} -(p(x)y_1'(x))' + q(x)y_1(x) = f(x) \quad \text{for } a < x < b \\ |y_1(a)| < \infty, \; \gamma y_1(b) + \delta y_1'(b) = 0 \end{cases}$$

and

$$\begin{cases} -(p(x)y_2'(x))' + q(x)y_2(x) = 0 \quad \text{for } a < x < b \\ |y_2(a)| < \infty, \; \gamma y_2(b) + \delta y_2'(b) = c_b \end{cases}$$

respectively, then $y = y_1 + y_2$ solves (5.9). If (5.10) has only the trivial solution, both of these auxiliary problems have unique solutions. If fact, if u in $C^1[a, b]$ is a nontrivial bounded solution to the homogeneous differential equation, then exactly as in the proof of the previous theorem, $\gamma u(b) + \delta u'(b) \neq 0$ and the second auxiliary problem has solution

$$y_2 = cu \text{ where } c = \frac{c_b}{\gamma u(b) + \delta u'(b)}.$$

The solution to the first auxiliary problem, which is (5.9) with $c_b = 0$, can be conveniently expressed in term of a Green's function, as we show in the next section.

5.4 Green's Functions

The motivational argument in Section 1.10 used for regular Sturm-Liouville boundary value problems shows that it is reasonable to expect that the solution to (5.9) with $c_b = 0$ can be expressed in terms of a Green's function $g(x, s)$ by

$$y(x) = \int_a^b g(x, s)f(s)\, ds.$$

Specifically, $g(x, s)$ is a **Green's function** for the singular Sturm-Liouville problem (5.9) with $c_b = 0$ if $g(x, s)$ is defined and continuous on $[a, b] \times [a, b]\backslash\{(a, a)\}$, the square $a \le x,\ s \le b$ with the point (a, a) removed, and

$$y(x) = \int_a^b g(x, s)f(s)\, ds, \quad a \le x \le b,$$

uniquely solves (5.9) with $c_b = 0$ for every continuous function $f(x)$ on $[a, b]$.

We will show that a Green's function is unique if it exists and, given existence, that the integral

$$\int_a^b g(x, s)f(s)\, ds$$

is a continuous function of x on $[a, b]$.

We will find that, when the Green's function exists, the integral above is an ordinary Riemann integral for each x with $a < x \le b$ and is a convergent improper Riemann integral when $x = a$.

The Green's function is defined through the boundary value problem (5.9) with $c_b = 0$; however, once the Green's function has been found, it can be used to express the solution to the boundary value problem also when $c_b \neq 0$. That representation is given later in the chapter.

The Green's function representation of solutions has many uses. Once the Green's function is found, the representation makes it possible to investigate how different forcing terms $f(x)$ effect the behavior of the solution. Also, properties of the solution that are not apparent from the boundary value problem itself often can be deduced from the Green's function representation and properties of the Green's function.

Theorem 141 *If the singular boundary value problem (5.9) with $c_b = 0$ has a Green's function, then the Green's function is unique.*

Proof. Let $h(x, s)$ and $g(x, s)$ both satisfy the defining conditions of a Green's function and set $k(x, s) = h(x, s) - g(x, s)$ for (x, s) in $[a, b] \times [a, b]\backslash\{(a, a)\}$. Then

$$\int_a^b h(x, s)f(s)\, ds = y(x) = \int_a^b g(x, s)f(s)\, ds,$$

where $y(x)$ is the unique solution to (5.9) with $c_b = 0$ and right member $f(x)$ in the differential equation. So

$$\int_a^b k(x, s)f(s)\, ds = 0$$

for every continuous $f(x)$ on $[a, b]$.

If $a < x \le b$, $k(x, s)$ is continuous on $a \le s \le b$ and by Corollary 20 $k(x, s) = 0$ for $a < x \le b$, and $a \le s \le b$. If $x = a$ the integral is improper and by the version of Corollary 20 for such integrals, $k(a, s) = 0$ for $a < s \le b$. Thus, $k(x, s) = h(x, s) - g(x, s) = 0$ for all (x, s) in $[a, b] \times [a, b] \backslash \{(a, a)\}$ and a Green's function is unique if it exists. ■

Theorem 142 *The singular boundary value problem (5.9) with $c_b = 0$ has a Green's function $g(x, s)$ if and only if (5.10) has only the trivial solution.*

Proof. If the Green's function exists, then clearly the only solution to (5.10) is the trivial solution.

Assume that the only solution to (5.10) is the trivial solution. By Theorem 140, for each continuous function $f(x)$ on $[a, b]$, (5.9) with $c_b = 0$ has a unique solution $y(x)$ that is defined and continuous on $a \le x \le b$. By Theorem 136 there is a nontrivial solution $u(x)$ in $C^1[a, b]$ to

$$Lu = 0 \quad \text{for } a \le x \le b.$$

Moreover, any such u satisfies

$$\gamma u(b) + \delta u'(b) \ne 0$$

because otherwise $u(x)$ would be a nontrivial solution to (5.10). Also, there is a nontrivial solution $v(x)$ in $C^1(a, b]$ to

$$\begin{aligned} Lv &= 0, \quad a < x \le b, \\ \gamma v(b) + \delta v'(b) &= 0. \end{aligned}$$

One way to establish the existence of $v(x)$ is as follows. Let $v_1(x)$ be a solution to the differential equation (Theorem 136(c)) that becomes logarithmically infinite as x approaches a. Then $v = c_1 u + c_2 v_1$ will satisfy the given conditions if

$$c_1(\gamma u(b) + \delta u'(b)) + c_2(\gamma v_1(b) + \delta v_1'(b)) = 0.$$

Set $c_2 = -1$ and $c_1 = (\gamma v_1(b) + \delta v_1'(b))/(\gamma u(b) + \delta u'(b))$ to obtain a solution $v(x)$ with the required properties.

Next we show that if $u(x)$ is any nontrivial bounded solution to $Lu = 0$ on $[a, b]$ and $v(x)$ is any nontrivial solution to $Lv = 0$ on $(a, b]$, $\gamma v(b) + \delta v'(b) = 0$, then $u(x)$ and $v(x)$ are linearly independent on $a < x \le b$. Indeed, if $\gamma \ne 0$, then $v'(b) \ne 0$ (otherwise, $v(b) = v'(b) = 0$ and v would be the trivial solution) and

$$\begin{aligned} \begin{vmatrix} u(b) & v(b) \\ u'(b) & v'(b) \end{vmatrix} &= \gamma^{-1} \begin{vmatrix} \gamma u(b) + \delta u'(b) & \gamma v(b) + \delta v'(b) \\ u'(b) & v'(b) \end{vmatrix} \\ &= \gamma^{-1}(\gamma u(b) + \delta u'(b))v'(b) \end{aligned}$$

while if $\gamma = 0$, then $v(b) \ne 0$ and

$$\begin{aligned} \begin{vmatrix} u(b) & v(b) \\ u'(b) & v'(b) \end{vmatrix} &= \delta^{-1} \begin{vmatrix} u(b) & v(b) \\ \gamma u(b) + \delta u'(b) & \gamma v(b) + \delta v'(b) \end{vmatrix} \\ &= -\delta^{-1}(\gamma u(b) + \delta u'(b))v(b). \end{aligned}$$

In either case, the Wronskian of u and v is nonzero at $x = b$ and $u(x)$ and $v(x)$ are linearly independent on $a < x \le b$.

Since u and v are linearly independent on $(a, b]$,

$$p(u'v - uv')(x) = C \quad \text{for } a < x \le b$$

for some constant $C \neq 0$ by Lemma 86. Replace v by v/C to obtain a new pair of functions, still denoted by u and v, that satisfy

$$\begin{aligned} &Lu = 0, \quad a \le x \le b, \\ &|u(a)| < \infty, \\ &Lv = 0, \quad a < x \le b, \\ &\gamma v(b) + \delta v'(b) = 0, \end{aligned}$$

and

$$p\big(u'v - uv'\big)(x) = 1 \quad \text{for } a < x \le b.$$

Apply Lagrange's identity (Lemma 80)

$$wLz - zLw = (p(zw' - z'w))'$$

with $z = y$ the solution to (5.9) with $c_b = 0$ and with $w = u$ and with $w = v$ respectively to obtain

$$\begin{aligned} uf &= (p(yu' - y'u))' \\ vf &= (p(yv' - y'v))'. \end{aligned}$$

For $a < a_1 < x \le b$, integrate to find

$$\int_{a_1}^{x} u(s)f(s)\, ds = p(yu' - y'u)(x) - p(yu' - y'u)(a_1),$$

$$\int_{x}^{b} v(s)f(s)\, ds = p(yv' - y'v)(b) - p(yv' - y'v)(x).$$

The boundary condition satisfied by v and y at $x = b$ yield the 2×2 system

$$\begin{cases} y(b)\gamma + y'(b)\delta &= 0 \\ v(b)\gamma + v'(b)\delta &= 0 \end{cases}$$

with γ and δ not both 0. Hence, the determinant of the system $W_{y,v}(b)$ is 0 and

$$p(yv' - y'v)(b) = 0.$$

We have already established in Lemma 132 that

$$\lim_{a_1 \to a} p(a_1)u'(a_1) = 0$$

and

$$\lim_{a_1 \to a} p(a_1)y'(a_1) = 0$$

because y is a bounded solution to the singular Sturm-Liouville differential equation. Consequently, let a_1 tend to a to obtain

$$\int_{a}^{x} u(s)f(s)\, ds = p(yu' - y'u)(x),$$

$$\int_{x}^{b} v(s)f(s)\, ds = -p(yv' - y'v)(x).$$

Multiply the first equation by $v\ (x)$, the second by $u(x)$, and add to get

$$v(x)\int_a^x u(s)f(s)\ ds + u(x)\int_x^b v(s)f(s)\ ds = p(u'v - uv')(x)y(x).$$

Since $p(u'v - uv')(x) = 1$,

$$y(x) = v(x)\int_a^x u(s)f(s)\ ds + u(x)\int_x^b v(s)f(s)\ ds$$

for $a < x \le b$, or, more compactly,

$$y(x) = \int_a^b g(x, s)f(s)\ ds \quad \text{for } a < x \le b, \tag{5.11}$$

where

$$g(x, s) = \begin{cases} v(x)u(s) & \text{for } a \le s \le x \le b \quad \text{and} \quad (x, s) \ne (a, a) \\ u(x)v(s) & \text{for } a \le x \le s \le b \quad \text{and} \quad (x, s) \ne (a, a) \end{cases}.$$

Clearly $g(x, s)$ is continuous on $[a, b] \times [a, b]$ with the point (a, a) removed.

Assertion: If $f(x)$ is continuous on $[a, b]$, then the integral on the right side of (5.11) is a continuous function of x on the closed interval $[a, b]$.

Assume the assertion and let x approach a in (5.11) to obtain

$$y(a) = \lim_{x \to a} y(x) = \lim_{x \to a}\int_a^b g(x, s)f(s)\ ds = \int_a^b g(a, s)f(s)\ ds$$

because the solution y also is continuous on $[a, b]$. This shows that (5.11) also holds when $x = a$ and establishes that $g(x, s)$ is the Green's function for (5.9) with $c_b = 0$.

To complete the proof we must establish the assertion. For $x > a$, the integrand $g(x, s)f(s)$ is continuous for $a \le s \le b$ and $\int_a^b g(x, s)f(s)\ ds$ exists as an ordinary Riemann integral. For $x = a$, the integrand is $g(a, s)f(s) = u(a)v(s)f(s)$ for $a < s \le b$. So $g(a, s)f(s)$ is continuous on $a < s \le b$ and by Theorem 136(c)

$$\lim_{s \to a}\frac{v(s)}{\ln(s - a)} = \frac{C}{\varphi(a)u(a)}$$

for some constant $C \ne 0$. It follows that

$$|g(a, s)f(s)| \le \|u\|_{\max}\|f\|_{\max}M|\ln(s - a)|$$

for some constant M and $a < s \le b$. By the basic comparison test for improper integrals (Proposition 17), the improper integral of $g(a, s)f(s)$ converges and

$$\int_a^b g(a, s)f(s)\ ds = \lim_{x \to a}\int_x^b g(a, s)f(s)\ ds.$$

Thus, $\int_a^b g(x, s)f(s)\ ds$ is defined for all x in $[a, b]$. For $a < x \le b$,

$$\int_a^b g(x, s)f(s)\ ds = v(x)\int_a^x u(s)f(s)\ ds + u(x)\int_x^b v(s)f(s)\ ds$$

which shows that the integral on the left is continuous for $a < x \leq b$. It remains to show that it is continuous at $x = a$. Since

$$\left| v(x) \int_a^x u(s) f(s)\, ds \right| \leq \left| \frac{v(x)}{\ln(x-a)} \right| \|u\|_{\max} \|f\|_{\max} |x-a| |\ln(x-a)|,$$

$\lim_{x \to a} |v(x)/\ln(x-a)|$ exists and is finite, and $|x-a| \ln|x-a| \to 0$ as $x \to a$, there exists

$$\lim_{x \to a} v(x) \int_a^x u(s) f(s)\, ds = 0.$$

It follows that there exists

$$\lim_{x \to a} \int_a^b g(x,s) f(s)\, ds = 0 + \lim_{x \to a} u(x) \int_x^b v(s) f(s)\, ds$$

$$= u(a) \int_a^b v(s) f(s)\, ds = \int_a^b g(a,s) f(s)\, ds$$

because

$$\lim_{x \to a} \int_x^b v(s) f(s)\, ds = \int_a^b v(s) f(s)\, ds$$

by the convergence of the improper integral. Thus,

$$\lim_{x \to a} \int_a^b g(x,s) f(s)\, ds = \int_a^b g(a,s) f(s)\, ds$$

and continuity at $x = a$ is established. ■

The function u, v_1, and v that occur in the foregoing proof can be chosen real-valued if the data $p(x)$, $q(x)$, γ, and δ are real-valued. See Theorem 136 and the comment following Theorem 138. The reasoning used in the proof of the previous theorem leads to the following characterization of and means to construct the Green's function.

Theorem 143 *The singular Sturm-Liouville boundary value problem (5.9) with $c_b = 0$ has a Green's function $g(x,s)$ if and only if there exist functions $u(x)$ in $C^1[a,b]$ and $v(x)$ in $C^1(a,b]$ such that*

$$\begin{cases} Lu = 0 \quad \text{for } a \leq x \leq b \\ |u(a)| < \infty, \end{cases} \tag{5.12}$$

$$\begin{cases} Lv = 0 \quad \text{for } a < x \leq b \\ \gamma v(b) + \delta v'(b) = 0 \end{cases}, \tag{5.13}$$

and

$$p(x)\, W_{u,v}(x) = -1 \quad \text{for } a < x \leq b, \tag{5.14}$$

in which case

$$g(x,s) = \begin{cases} v(x)u(s) & \text{for } a \leq s \leq x \leq b \text{ and } (x,s) \neq (a,a) \\ u(x)v(s) & \text{for } a \leq x \leq s \leq b \text{ and } (x,s) \neq (a,a) \end{cases} \tag{5.15}$$

and (5.9) with $c_b = 0$ has the unique solution

$$y(x) = \int_a^b g(x,s) f(s)\, ds \quad \text{for } a \leq x \leq b.$$

Moreover, $u(x)$ and $v(x)$ can be chosen real-valued when $p(x)$, $q(x)$, γ, and δ are real-valued, in which case the Green's functions is real-valued and symmetric; that is, $g(x,s)$ is a symmetric kernel and the corresponding integral operator is self-adjoint.

Proof. Assume that (5.9) with $c_b = 0$ has a Green's function $g(x, s)$. Then (5.10) has only the trivial solution and the proof of Theorem 142 shows there are functions $u(x)$ and $v\ (x)$ that satisfy (5.12), (5.13), and (5.14) and that the Green's function is given by (5.15).

Conversely, assume there are functions $u(x)$ and $v\ (x)$ that satisfy (5.12), (5.13), and (5.14). Define $g(x, s)$ by (5.15). Clearly $g(x, s)$ is continuous on $[a, b] \times [a, b]\backslash\{(a, a)\}$. We will show that $y(x)$ defined by

$$y(x) = \int_a^b g(x, s)f(s)\, ds \quad \text{for } a \leq x \leq b \tag{5.16}$$

is the unique solution to (5.9) with $c_b = 0$ for every continuous function $f(x)$ on $[a, b]$. This will establish that $g(x, s)$ is the Green's function for (5.9) with $c_b = 0$. To this end, first observe that $y(x)$ is continuous on $[a, b]$ by the assertion established in the proof of the last theorem. Consequently, $|y(a)| < \infty$. Second, express $y(x)$ as

$$y(x) = v(x)\int_a^x u(s)f(s)\, ds + u(x)\int_x^b v(s)f(s)\, ds \quad \text{for } a < x \leq b \tag{5.17}$$

and differentiate to obtain

$$y'(x) = v'(x)\int_a^x u(s)f(s)\, ds + u'(x)\int_x^b v(s)f(s)\, ds \quad \text{for } a < x \leq b.$$

Consequently, for $a < x \leq b$,

$$p(x)y'(x) = p(x)v'(x)\int_a^x u(s)f(s)\, ds + p(x)u'(x)\int_x^b v(s)f(s)\, ds,$$

$$\begin{aligned} -(p(x)y'(x))' &= -p(x)v'(x)u(x)f(x) - (p(x)v'(x))'\int_a^x u(s)f(s)\, ds \\ &\quad + p(x)u'(x)v(x)f(x) - (p(x)u'(x))'\int_x^b v(s)f(s)\, ds, \end{aligned}$$

and

$$q(x)y(x) = q(x)v(x)\int_a^x u(s)f(s)\, ds + q(x)u(x)\int_x^b v(s)f(s)\, ds,$$

Since $p(u'v - v'u) = 1$, it follows that

$$Ly(x) = f(x) + Lv(x)\int_a^x u(s)f(s)\, ds + Lu(x)\int_x^b v(s)f(s)\, ds = f(x).$$

So $y(x)$ satisfies the differential equation in (5.9) with $c_b = 0$. Since

$$y(b) = v(b)\int_a^b u(s)f(s)\, ds,$$

$$y'(b) = v'(b)\int_a^b u(s)f(s)\, ds,$$

and $v\ (x)$ satisfies the boundary condition at $x = b$ so does $y(x)$.

Thus, $y(x)$ is a solution to (5.9) with $c_b = 0$. It remains to show that it is the only solution, equivalently, that (5.10) has only the trivial solution. Suppose that (5.10) were to have a nontrivial solution z. By Theorem 136 u and z are nonzero multiples of each other because they are both nontrivial bounded solutions of the singular differential equation in (5.10). Hence,

u satisfies $\gamma u(b) + \delta u'(b) = 0$ because z satisfies this boundary condition. Then u and v satisfy the 2×2 system of equations

$$\begin{cases} \gamma u(b) + \delta u'(b) = 0 \\ \gamma v(b) + \delta v'(b) = 0 \end{cases} \quad \text{with} \quad |\gamma| + |\delta| \neq 0.$$

It follows that the determinant of the system $W_{u,v}(b) = 0$, which contradicts $p(b)\, W_{u,v}(b) = -1$. Consequently, (5.10) can have only the trivial solution and

$$y(x) = \int_a^b g(x, s) f(s)\, ds \quad \text{for } a \leq x \leq b$$

is the unique solution to (5.9) with $c_b = 0$. ■

The following corollary will be needed later when we take up singular Sturm-Liouville eigenvalue problems.

Corollary 144 *The Green's function $g(x, s)$ determined by the singular Sturm-Liouville differential operator $Ly = -(py')' + qy$ and the boundary conditions $|y(a)| < \infty$, $\gamma y(b) + \delta y'(b) = 0$ has the form*

$$g(x, s) = h(x, s) \ln(\max(x, s) - a)$$

for (x, s) in $[a, b] \times [a, b] \backslash \{(a, a)\}$ where $h(x, s) = h(s, x)$ is continuous on $[a, b] \times [a, b]$ and $h(a, a) \neq 0$. Consequently, there is a constant $M < \infty$ such that

$$\int_a^b |g(x, s)|^2\, ds \leq M$$

for all x in $[a, b]$. Moreover, if $p(x)$, $q(x)$, γ, and δ are real-valued, then $h(x, s)$ can be chosen real-valued and $g(x, s)$ is a symmetric kernel.

Proof. The two-part formula for $g(x, s)$ in Theorem 143 can be expressed as

$$g(x, s) = u(\min(x, s)) v(\max(x, s))$$

for (x, s) in $[a, b] \times [a, b] \backslash \{(a, a)\}$. Evidently $g(x, s)$ is continuous on $[a, b] \times [a, b] \backslash (a, a)$. From Theorem 136

$$\lim_{x \to a} \frac{v(x)}{\ln(x - a)} = m$$

with $0 < |m| < \infty$ and $u(a) \neq 0$ because the bounded solution u is nontrivial. Define

$$w(x) = \begin{cases} v(x)/\ln(x - a) & \text{for } a < x \leq b \\ m & \text{for } x = a \end{cases}.$$

Then $w(x)$ is continuous on $[a, b]$ and, for (x, s) in $[a, b] \times [a, b] \backslash \{(a, a)\}$,

$$\begin{aligned} g(x, s) &= u(\min(x, s)) w(\max(x, s)) \ln(\max(x, s) - a) \\ &= h(x, s) \ln(\max(x, s) - a) \end{aligned}$$

where

$$h(x, s) = u(\min(x, s)) w(\max(x, s))$$

is continuous on $[a, b] \times [a, b]$, $h(x, s) = h(s, x)$ and $h(a, a) = u(a)w(a) \neq 0$. The first assertion in the corollary is established. The second assertion follows from the first and (2.2). The final conclusion follows because u and v can be chosen real-valued when the data is real-valued. ∎

Example 1. Determine when it exists and find the Green's function for the singular Sturm-Liouville boundary value problem

$$-(xy')' - xy = f(x), \quad 0 < x < l,$$
$$|y(0)| \leq \infty, \quad \gamma y(l) + \delta y'(l) = 0,$$

where $(xy')' + xy = 0$ is the Bessel's equation of order 0.

The corresponding homogeneous equation has the Bessel functions $J_0(x)$ and $Y_0(x)$ as linearly independent solutions. Since $J_0(x)$ is bounded on $[0, l]$, we can choose $u = J_0(x)$ in Theorem 143. Since $Y_0(x)$ is unbounded, the corresponding homogeneous problem will have only the trivial solution if and only if

$$\gamma J_0(l) + \delta J_0'(l) \neq 0.$$

The Green's function exists if and only if this inequality is satisfied. We seek a solution v in Theorem 143 of the form $v = cJ_0(x) + Y_0(x)$. Such a v satisfies the boundary condition at $x = l$ if

$$c = -\frac{\gamma Y_0(l) + \delta Y_0'(l)}{\gamma J_0(l) + \delta J_0'(l)}.$$

With this choice for c the Green's function is

$$g(x, s) = \begin{cases} J_0(x)\tilde{Y}_0(s) & \text{for } 0 \leq x \leq s \leq l \text{ and } (x, s) \neq (a, a) \\ \tilde{Y}_0(x)J_0(s) & \text{for } 0 \leq s \leq x \leq l \text{ and } (x, s) \neq (a, a) \end{cases}$$

where $\tilde{Y}_0(x) = cJ_0(x) + Y_0(x)$.

A closely related example involves the modified Bessel functions.

Example 2. Determine when it exists and find the Green's function for the singular Sturm-Liouville boundary value problem

$$-(xy')' + xy = f(x), \quad 0 < x < l,$$
$$|y(0)| \leq \infty, \quad \gamma y(l) + \delta y'(l) = 0,$$

where $(xy')' - xy = 0$ is the modified Bessel's equation of order 0.

The corresponding homogeneous equation has the modified Bessel functions $I_0(x)$ and $K_0(x)$ as linearly independent solutions. Since $I_0(x)$ is bounded on $[0, l]$, we can choose $u = I_0(x)$ in Theorem 143. Since $K_0(x)$ is unbounded, the corresponding homogeneous problem will have only the trivial solution if and only if

$$\gamma I_0(l) + \delta I_0'(l) \neq 0.$$

The Green's function exists if and only if this inequality is satisfied. We seek a solution v in Theorem 143 of the form $v = cI_0(x) + K_0(x)$. Such a v satisfies the boundary condition at $x = l$ if

$$c = -\frac{\gamma K_0(l) + \delta K_0'(l)}{\gamma I_0(l) + \delta I_0'(l)}.$$

With this choice for c the Green's function is

$$g(x,s) = \begin{cases} I_0(x)\tilde{K}_0(s) & \text{for } 0 \le x \le s \le l \quad \text{and} \quad (x,s) \ne (a,a) \\ \tilde{K}_0(x)I_0(s) & \text{for } 0 \le s \le x \le l \quad \text{and} \quad (x,s) \ne (a,a) \end{cases}$$

where $\tilde{K}_0(s)(x) = cI_0(x) + K_0(x)$.

The Green's function $g(x,s)$ for $Ly = f$, $|y(a)| < \infty$, $\gamma y(b) + \delta y'(b) = 0$ has the following properties (when it exists) and these properties characterize the Green's function, in strict analogy to the regular case:

1. $g(x,s)$ is continuous on $[a,b] \times [a,b] \backslash \{(a,a)\}$, the square with the point (a,a) removed, has continuous partial derivatives on the upper triangle $(x \le s)$ and on the lower triangle $(s \le x)$ of $[a,b] \times [a,b] \backslash \{(a,a)\}$, and there exists

$$\lim_{(x,s)\to(a,a)} \frac{g(x,s)}{\ln(\max(x,s) - a)} = l$$

   where $0 < |l| < \infty$.

2. $g(x,s)$, regarded as a function of x for fixed s in $[a,b]$, satisfies the differential equation $Ly = 0$ for $x \ne s$ in (a,b).

3. $g(x,s)$, regarded as a function of x for fixed s in (a,b), satisfies the boundary conditions of the problem.

4. $g(x,s)$, regarded as a function of x for fixed s in (a,b), has a jump in its derivative with respect to x at $x = s$ given by

$$\frac{\partial g}{\partial x}(s+,s) - \frac{\partial g}{\partial x}(s-,s) = -\frac{1}{p(s)}.$$

A direct verification confirms that the Green's function in Theorem 143 has the four properties. Once we establish that the four properties characterize the Green's function, $g(x,s)$ must be the function in Theorem 143. Since that function satisfies $g(x,s) = g(s,x)$, Properties 1–4 hold with the roles of x and s interchanged.

The next lemma verifies that the Green's function in Theorem 143 has Property 1. We leave the verification of Properties 2, 3, and 4 to the reader. The lemma also includes results needed in the next theorem which establishes that Properties 1–4 characterize the Green's function.

Lemma 145 *(a) The Green's function $g(x,s)$ in Theorem 143 has Property 1.*

(b) Any function $g(x,s)$ with Property 1 has the form

$$g(x,s) = h(x,s)\ln(\max(x,s) - a)$$

on $[a,b] \times [a,b] \backslash \{(a,a)\}$ where $h(x,s)$ is continuous on $[a,b] \times [a,b]$ and $h(a,a) \ne 0$.
(b) If $g(x,s)$ has Property 1 and $f(x)$ is any continuous function on $[a,b]$, then

$$\int_a^b g(x,s)f(x)\,dx$$

is a continuous function of s on $[a,b]$.

Proof. (a) By Corollary 144,

$$g(x, s) = h(x, s)\ln(\max(x, s) - a)$$

on $[a, b] \times [a, b] \backslash (a, a)$ where $h(x, s)$ is continuous on $[a, b] \times [a, b]$ and $h(a, a) \neq 0$. Hence,

$$\lim_{(x,s)\to(a,a)} \frac{g(x, s)}{\ln(\max(x, s) - a)} = h(a, a) \neq 0.$$

(b) If $g(x, s)$ has Property 1, then

$$h(x, s) = \begin{cases} g(x, s)/\ln(\max(x, s) - a) & \text{for } (x, s) \text{ in } [a, b] \times [a, b]\backslash\{(a, a)\} \\ l & \text{for } (x, s) = (a, a) \end{cases}$$

is continuous on $[a, b] \times [a, b]$, $h(a, a) = l \neq 0$, and (b) is established.

(c) Consider

$$y(s) = \int_a^b g(x, s)f(x)\, dx$$

for $a \leq s \leq b$. By Property 1, the integrand is continuous for each s in $a < s \leq b$ and the integral exists as a proper Riemann integral for such s. Since

$$\lim_{x\to a} \frac{g(x, a)}{\ln(x - a)} = l$$

from Property 1 and f is continuous on $[a, b]$, it follows that

$$|g(x, a)f(x)| \leq (|l| + 1)\|f\|_{\max}|\ln(x - a)|$$

for $a < x \leq b'$ and some b' with $a < b' < b$. Consequently (see Proposition 17), the integral defining $y(a)$ exists as a convergent improper Riemann integral

$$y(a) = \int_a^b g(x, a)f(x)\, dx = \lim_{a'\to a} \int_{a'}^b g(x, a)f(x)\, dx$$

and $y(x)$ is well defined for x in $[a, b]$.

It remains to show that $y(x)$ is continuous on $[a, b]$. Since the integrand $g(x, s)f(x)$ is continuous on $[a, b] \times [a', b]$ for any a' with $a < a' < b$, it follows from Proposition 18 that $y(s)$ is continuous on $[a', b]$. Since $a' > a$ can be chosen arbitrarily, it follows that $y(s)$ is continuous on $a < s \leq b$.

Finally, we establish that $y(s)$ is continuous at $s = a$. From (b), for $a < s < b$,

$$\begin{aligned} y(s) &= \int_a^b g(x, s)f(x)dx = \int_a^b h(x, s)\ln(\max(x, s) - a)f(x)dx \\ &= \int_a^b (h(x, s) - h(x, a))\ln(\max(x, s) - a)f(x)dx \\ &\quad + \int_a^b h(x, a)\ln(\max(x, s) - a)f(x)dx \\ &= I + II. \end{aligned}$$

We claim that, $I \to 0$ as $s \to a$ and that

$$II \to \int_a^b h(x, a)\ln(x - a)\, f(x)\, dx = \int_a^b g(x, a)\, f(x)\, dx = y(a)$$

as $s \to a$ which establishes the continuity of $y(s)$ at $s = a$.

To establish that $I \to 0$ as $s \to a$, let $\varepsilon > 0$ be given. By the uniform continuity of $h(x, s)$ on $[a, b] \times [a, b]$ there is a $\delta > 0$ such that

$$|h(x, s) - h(x, a)| < \varepsilon$$

for all $a \leq x \leq b$ and $a < s < a + \delta$. Therefore, for $a < s < a + \delta$,

$$|I| \leq \varepsilon \, \|f\|_{\max} \int_a^b |\ln(\max(x, s) - a)| \, dx.$$

This inequality implies that $I \to 0$ as $s \to a$ because the integral on the right is a convergent improper integral that is bounded as s varies in $[a, b]$. See the basic comparison test (Proposition 17) and the examples that follow it.

To show that $II \to \int_a^b h(x, a) \ln(x - a) f(x) \, dx$ as $s \to a$, express II as

$$\begin{aligned} II &= \int_a^s h(x, a) \ln(\max(x, s) - a) f(x) dx \\ &\quad + \int_s^b h(x, a) \ln(\max(x, s) - a) f(x) dx \\ &= \int_a^s h(x, a) \ln(s - a) f(x) dx + \int_s^b h(x, a) \ln(x - a) f(x) dx. \end{aligned}$$

The first integral on the right is bounded in absolute value by

$$\|h\|_{\max} \|f\|_{\max} (s - a) |\ln(s - a)|$$

which tends to zero as $s \to a$. Since the improper integral of $|\ln(x - a)|$ over $[a, b]$ converges and $h(x, a) f(x)$ is continuous on $[a, b]$, another application of Proposition 17 implies that the second integral on the right converges to the improper integral $\int_a^b h(x, a) \ln(x - a) f(x) \, dx = y(a)$. Thus, the asserted limit of II as $s \to a$ is established, and (c) of the lemma is proved. ■

Properties 1-4 above characterize the Green's function:

Theorem 146 *If a function $g(x, s)$ exists with Properties 1-4, then $Ly = 0$, $|y(a)| < \infty$, $\gamma y(b) + \delta y'(b) = 0$ has only the trivial solution and $g(x, s)$ is the Green's function for the differential operator Ly and boundary conditions $|y(a)| \leq \infty$, $\gamma y(b) + \delta y'(b) = 0$. Moreover, $g(x, s) = g(s, x)$.*

Proof. As usual, $B_b y = \gamma y(b) + \delta y'(b)$. Fix s with $a < s < b$ and define functions z_1 and z_2 by

$$z_1(x) = g(x, s) \quad \text{for } a \leq x \leq s \text{ and } z_2(x) = g(x, s) \quad \text{for } s \leq x \leq b.$$

By Properties 2 and 3, $z_1(x)$ satisfies $Lz_1 = 0$ on $a < x < s$, $|z_1(a)| < \infty$ and $z_2(x)$ satisfies $Lz_2 = 0$ on $s < x < b$, $B_b z_2 = 0$. By Lemma 132 z_1 extends to a continuously differentiable function on $[a, s]$ and satisfies the differential equation there. Since z_2 satisfies the regular Sturm-Liouville problem $Lz_2 = 0$ on (s, b), $z_2(s) = g(s, s)$, $B_b z_2 = 0$, it extends to a continuously differentiable function on $[s, b]$ and satisfies the differential equation there. We show first that $Ly = 0$, $|y(a)| < \infty$, $B_b y = 0$ has only the trivial solution. Assume the contrary and let $z(x)$ be a nontrivial solution. By Lemma 132 z extends to a continuously differentiable function on $[a, b]$. Since

$$Lz = 0 \quad \text{for } a < x < s, \; |z(a)| < \infty,$$

and

$$Lz_1 = 0 \quad \text{for } a < x < s, \ |z_1(a)| < \infty,$$

by Theorem 136 applied on the interval $[a, s]$, if $z_1(x)$ is nontrivial it is a multiple of $z(x)$. The same is true if $z_1(x)$ is identically zero on $[a, s]$. Thus, $z_1(x) = c_1(s)z(x)$ on $a \leq x \leq s$ for some scalar $c_1(s)$ that depends on the fixed value of s.

Since

$$\gamma z(b) + \delta z'(b) = 0,$$
$$\gamma z_2(b) + \delta z_2'(b) = 0,$$

with $|\gamma| + |\delta| \neq 0$,

$$W_{z,z_2}(b) = 0,$$

and z and z_2 are linearly dependent solutions on $[s, b]$. Thus,

$$d(s)z(x) + d_2(s)z_2(x) = 0$$

for x in $[s, b]$, where $d(s)$ and $d_2(s)$ are scalars, not both 0, whose value depends on the fixed value of s in (a, b). If $d_2(s) = 0$, then $z(x) = 0$ on $[s, b]$ and $z(x)$ solves the initial value problem $Lz = 0$ on (a, b), $z(s) = 0$, $z'(s) = 0$. Thus, $z(x) = 0$ on (a, b) by the uniqueness of solutions to initial value problems. This contradicts the fact that $z(x)$ is a nontrivial solution. Consequently, $d_2(s) \neq 0$ and $z_2(x) = c_2(s)z(x)$ on $s \leq x \leq b$ where $c_2(s) = -d(s)/d_2(s)$. Since $g(x, s)$ is continuous at $x = s$ by Property 1, it follows that

$$c_2(s)z(s) = g(s+, s) = g(s-, s) = c_1(s)z(s).$$

Since z is nontrivial, there exist s_0 in (a, b) where $z(s_0) \neq 0$; Hence, $c_1(s_0) = c_2(s_0)$ and

$$g_x(s_0+, s_0) - g_x(s_0-, s_0) = c_2(s_0)z'(s_0) - c_1(s_0)z'(s_0) = 0,$$

which contradicts the jump condition in Property 4. Hence, $Ly = 0$, $|y(a)| < \infty$, $B_b y = 0$ has only the trivial solution and $Ly = f$, $B_a y = 0$, $B_b y = 0$ has a unique solution y for each function f in $C[a, b]$.

Finally we establish that a function $g(x, s)$ with Properties 1-4 is the Green's function. To this end, for any continuous function f, let y be the unique solution to $Ly = f$, $|y(a)| < \infty$, $B_b y = 0$. Fix s in (a, b), regard $g(x, s)$ as a function of x in $[a, b]$ and let $a < c < r < s < t < b$. By Property 2

$$0 = \int_c^r yLg\,dx = \int_c^r y(-pg')'\,dx + \int_c^r yqg\,dx.$$

Integration by parts gives

$$\begin{aligned}
0 &= -ypg'\big|_c^r + \int_c^r pg'y'\,dx + \int_c^r qygdx \\
&= -ypg'\big|_c^r + py'g\big|_c^r - \int_c^r g(py')'\,dx + \int_c^r qygdx \\
&= (py'g - ypg')\big|_c^r + \int_c^r gLydx \\
&= (py'g - ypg')\big|_c^r + \int_c^r gfdx.
\end{aligned}$$

Thus,

$$-(py'g - ypg')\big|_c^r = \int_c^r gf\,dx.$$

In the same way,

$$-(py'g - ypg')\big|_t^b = \int_t^b gf\,dx.$$

Since s is fixed with $a < s < b$, g is continuous in x, $\lim_{c \to a} p(c)y'(c) = 0$ and $\lim_{c \to a} p(c)g'(c) = 0$ by Lemma 132, and y is continuous on $[a, b]$, the evaluation at the lower limit as $c \to a$ gives 0. Let $r \to s$ to obtain

$$-(py'g - ypg')\big|_{x=s-} = \int_a^s gf\,dx.$$

Since

$$\begin{cases} \gamma g(b) + \delta g'(b) = 0 \\ \gamma y(b) + \delta y'(b) = 0 \end{cases}$$

with $|\gamma| + |\delta| > 0$, the determinant of the 2×2 system is 0 and the contribution to the evaluated term above at $x = b$ is 0. Let $t \to s$ to obtain

$$(py'g - ypg')\big|_{x=s+} = \int_s^b gf\,dx.$$

Combining these results gives

$$(py'g - ypg')\big|_{x=s-}^{s+} = \int_a^b gf\,dx.$$

Since $a < s < b$, p, y', and g are continuous in x near $x = s$ so is $py'\,g$ and, hence,

$$(-ypg')\big|_{x=s-}^{s+} = \int_a^b gf\,dx.$$

By the jump condition (Property 4)

$$-ypg'\big|_{s-}^{s+} = -y(s)p(s)(g_x(s+, s) - g_x(s-, s)) = y(s)$$

and

$$y(s) = \int_a^b g(x, s)f(x)\,dx.$$

for s in (a, b). Since the solution $y(s)$ is continuous on $[a, b]$ and the integral on the right is continuous on $[a, b]$ by Lemma 145, the displayed equality also holds on $[a, b]$. By definition $h(s, x) = g(x, s)$ is the Green's function for the differential operator L and the boundary conditions $|y(a)| < \infty$ and $B_b y = 0$. By uniqueness it must be given by the formula in Theorem 143 which shows that $h(s, x) = h(x, s)$. Thus, $g(x, s)$ is the Green's function and $g(x, s) = g(s, x)$. ∎

If the fully inhomogeneous problem (5.9) has a unique solution, it can be found by adding the solution $\tilde{y}$ to $Ly = 0$, $|y(a)| < \infty$, $B_b y = c_b$ to the Green's function solution of $Ly = f$, $|y(a)| < \infty$, $B_b y = 0$. Alternatively, it can be expressed directly in terms of the Green's function for $Ly = f$, $|y(a)| < \infty$, $B_b y = 0$. Suppose that $Ly = 0$, $|y(a)| < \infty$, $B_b y = 0$ has only the trivial solution so that $Ly = f$, $|y(a)| < \infty$, $B_b y = c_b$ has a unique solution that we will denote by y and let $g(x, s)$ be the Green's function for $Ly = f$, $|y(a)| < \infty$, $B_b y = 0$. Fix x in (a, b), regard $g(x, s)$ as a function of s, denote derivatives with respect to s by primes, use Properties 1–4 with the roles of x and s interchanged, and reason exactly as we did in the foregoing proof to obtain

$$-(py'g - ypg')\big|_{s=c}^{r} = \int_c^r gf\,ds$$

and

$$-(py'g - ypg')\big|_{s=t}^{b} = \int_t^b gf\,ds$$

for $a < c < r < x < t < b$. Let $c \to a$ and then $r \to x$ to obtain

$$-(py'g - ypg')\big|_{s=x-} = \int_a^x gf\,ds.$$

Likewise, let $t \to x$ to get

$$-(py'g - ypg')\big|_{s=x+}^{b} = \int_x^b gf\,ds$$

and combine results to find that

$$-(py'g - ypg')\big|_{s=b} + (py'g - ypg')\big|_{x-}^{x+} = \int_a^b gf\,ds.$$

As before, p, y', and g are continuous in s for s near x so that

$$(-ypg')\big|_{s=x-}^{x+} = (py'g - ypg')\big|_{s=b} + \int_a^b gf\,ds.$$

By the jump condition (Property 4 with the roles of x and s interchanged)

$$(-ypg')\big|_{s=x-}^{x+} = -y(x)p(x)(g_s(x, x+) - g_s(x, x-)) = y(x).$$

Thus,

$$y(x) = p(y'g - yg')\big|_{s=b} + \int_a^b gf\,ds.$$

Since y satisfies an inhomogeneous boundary condition at $x = b$ instead of the corresponding homogeneous boundary condition, the evaluation at $x = b$ is different from before. At $x = b$ the functions y and g satisfy

$$\begin{cases} \gamma y(b) + \delta y'(b) = c_b \\ \gamma g(b) + \delta g'(b) = 0 \end{cases}$$

and, by the standard elimination process,

$$\gamma\Delta(x, b) = -c_b g'(b) \quad \text{and} \quad \delta\Delta(x, b) = c_b g(b),$$

where

$$\Delta(x, s) = y'(s)g(x, s) - y(s)g'(x, s)$$

and primes indicates derivatives with respect to s. Using these results in the formula for $y(x)$ above yields

Theorem 147 *If $g(x, s)$ is the Green's function determined by the Sturm-Liouville differential operator $Ly = -(py')' + qy$ and the separated boundary conditions $|y(a)| < \infty$, $\gamma y(b) + \delta y'(b) = 0$, then the Sturm-Liouville boundary value problem (5.9) has the unique solution*

$$y(x) = p(b)\Delta(x, b) + \int_a^b g(x, s)f(s)\, ds,$$

where

$$\Delta(x, b) = \begin{cases} -c_b g_s(x, b)/\gamma & \text{if } \gamma \neq 0 \\ c_b g(x, b)/\delta & \text{if } \gamma = 0 \end{cases}$$

for x in $[a, b]$.

5.5 Eigenvalue Problems

We surveyed several singular Sturm-Liouville eigenvalue problems in Chapter 1. Those problems are representative of the vast majority of singular eigenvalue problems that come up in applications. The given data and coefficient functions are all real-valued. Consequently, in our treatment of eigenvalue problems, we adjust the standing assumptions for the chapter as follows:

Standing Assumptions for Eigenvalue Problems

(1) $p(x) \geq 0$ is continuous on $[a, b]$, is differentiable at $x = a$, is nonzero on $a < x \leq b$, and satisfies $p(a) = 0$, $p'(a) \neq 0$.

(2) $q(x)$ is continuous on $[a, b]$.

(3) $q(x)$ is real-valued and γ and δ are real numbers not both zero.

(4) The weight function $r(x)$ in an eigenvalue problem is continuous on $[a, b]$ and either $r(x) > 0$ on $[a, b]$ or $r(x) = (x - a)^m \rho(x)$ where $m > 0$ and $\rho(x) > 0$ is continuous on $a \leq x \leq b$.

We will sometimes express (1) in an equivalent way: $p(x) = (x - a)\varphi(x)$ where $\varphi(x)$ is continuous on $[a, b]$ and $\varphi(x) > 0$ there, in which case $p'(a) = \varphi(a)$.

The eigenvalue problem for a singular Sturm-Liouville differential equation is

$$\begin{cases} -(p(x)y')' + q(x)y = \lambda r(x)y, & a < x < b, \\ |y(a)| < \infty, & \\ \gamma y(b) + \delta y'(b) = 0, & |\gamma| + |\delta| \neq 0, \end{cases} \tag{5.18}$$

or, more compactly,

$$Ly = \lambda ry, \ |y(a)| < \infty, \ B_b y = 0,$$

where $Ly = -(py')' + qy$ and $B_b y = \gamma y(b) + \delta y'(b)$.

The weight function $r(x)$ determines an inner product on $C[a, b]$ by

$$\langle y, z\rangle_r = \int_a^b y(x)\overline{z(x)}r(x)\, dx.$$

The functions y and z are orthogonal with respect to the weight function r if $\langle y, z\rangle_r = 0$.

The example with Bessel's equation of order 0 and parameter λ at the start of the chapter involves a weight function with a simple zero at 0.

Just as in Chapter 4, a real or complex number λ is an **eigenvalue** of a Sturm-Liouville eigenvalue problem and a real or complex-valued function $y \neq 0$ is a corresponding **eigenfunction** if (5.18) is satisfied for the pair λ and y. We also say the eigenfunction y belongs to the eigenvalue λ. When we say y satisfies (5.18), we mean that y satisfies the differential equation on (a, b), satisfies the given boundary conditions, and is continuous on $[a, b]$. The rationale for the continuity requirement is the same as for solutions to singular boundary value problems; see Section 5.3. As for boundary value problems, this definition implies further smoothness for y. Indeed, simply regard y as a solution to $Ly = f$ where $f = \lambda ry$ and apply Lemma 132 to obtain

Lemma 148 *If $y(x)$ is an eigenfunction of (5.18), then $y(x)$ is continuously differentiable on $[a, b]$ and satisfies the Sturm-Liouville differential equation at every point in $[a, b]$. Moreover, $y(x)$ satisfies the initial condition*

$$p'(a)y'(a) = (q(a) - \lambda r(a))y(a).$$

The final conclusion in the lemma will play an essential role in the shooting method used in Chapter 7 to find accurate numerical approximations of eigenvalues and eigenfunctions.

The domain of the differential operator L is

$$D = \{y \in C[a, b] : Ly \in C[a, b]\} = \{y \in C[a, b] : (py')' \in C[a, b]\}$$

exactly as for the regular problems in Chapter 4. This choice for the domain of L guarantees that any eigenfunction y is in the domain of L. The usual integration by parts argument, first integrating from a' to b where $a < a' < b$, using $\lim_{x\to a} p(x)y'(x) = 0$, $\lim_{x\to a} p(x)z'(x) = 0$, and letting $a' \to a$, gives

$$\langle Ly, z\rangle = \langle y, Lz\rangle$$

for all y and z in the domain of L that satisfy the given boundary conditions $|y(a)| < \infty$, $B_b y = 0$, $|z(a)| < \infty$, $B_b z = 0$. (The limit relations at $x = a$ hold for y and z because they satisfy the singular Sturm-Liouville differential equations $-(py')' = f$ and $-(pz')' = g$ on (a, b) where $f = -(py')'$ and $g = -(pz')'$ are continuous functions on $[a, b]$.) As usual,

$$\langle y, z\rangle = \int_a^b y(x)\overline{z(x)}\,dx.$$

Since $\langle Ly, z\rangle = \langle y, Lz\rangle$ for all y and z in the domain of L that satisfy the given boundary conditions, the eigenvalue problem (5.18) is self-adjoint and we have

Lemma 149 *Any eigenvalue of the self-adjoint Sturm-Liouville eigenvalue problem (5.18) is real, and eigenfunctions belonging to distinct eigenvalue are orthogonal with respect to the weight function r. Each eigenvalue has a corresponding real-valued eigenfunction.*

Proof. If $Ly = \lambda ry$ with $y \neq 0$, then

$$\langle y, y\rangle_r = \langle \lambda ry, y\rangle = \langle Ly, y\rangle = \langle y, Ly\rangle = \langle y, \lambda ry\rangle = \bar{\lambda}\langle y, y\rangle_r.$$

Since $\langle y, y\rangle_r > 0$, it follows that $\lambda = \bar{\lambda}$ and λ is real. If $Lz = \mu rz$ with $z \neq 0$, then

$$\lambda\langle y, z\rangle_r = \langle \lambda ry, z\rangle = \langle Ly, z\rangle = \langle y, Lz\rangle = \langle y, \mu rz\rangle = \mu\langle y, z\rangle_r$$

because μ is real. If $\lambda \neq \mu$ then $\langle y, z\rangle_r = 0$. If λ is an eigenvalue and y a corresponding eigenfunction, separating the equation $Ly = \lambda ry$ into its real and imaginary parts shows that Rey and Imy both satisfy $Ly = \lambda ry$. Since one of Rey or Imy is nonzero, it is a real-valued eigenfunction. ■

Theorem 150 *The eigenvalue problem (5.18) has at most a finite number of eigenvalues in any bounded region of the complex plane.*

Proof. Since the second order differential equation $Ly = \lambda ry$ for $a < x < b$ is expressible as the first order linear system $Z' = (A(x) + \lambda B(x))Z$ for $a < x < b$ where

$$Z = \begin{bmatrix} y \\ py' \end{bmatrix}, \; A(x) = \begin{bmatrix} 0 & 1/p \\ q & 0 \end{bmatrix}, \; B(x) = \begin{bmatrix} 0 & 0 \\ -r & 0 \end{bmatrix},$$

any solution $y(x, \lambda)$ to $Ly = \lambda ry$ is, for fixed x in (a, b), analytic in the complex variable λ for $|\lambda| < \infty$ as is $y'(x, \lambda)$ by Theorem 8.4 in Chapter 1 of [9] and the following application to linear systems. The same conclusion follows when applied to the differential equation $\tilde{L}y = \lambda \tilde{r}y$ for $a < x < \tilde{b}$ for a fixed $\tilde{b} > b$ and $\tilde{L}y = -(\tilde{p}y')' + \tilde{q}y$ where $\tilde{p}$, $\tilde{q}$, and $\tilde{r}$ extend p, q, and r to be constant on $[b, b']$. Since $Ly = \lambda ry$ is

$$-(py')' + (q - \lambda r)y = 0 \quad \text{for } a < x < b,$$

there is a nontrivial bounded solution $u(x, \lambda)$ to this equation that extends to a continuously differentiable function on $[a, b]$ and an unbounded solution $v(x, \lambda)$ that extends to a continuously differentiable function on $(a, b]$ by Theorem 136. Let $\tilde{u}(x, \lambda)$ and $\tilde{v}(x, \lambda)$ be the solutions to $\tilde{L}y = \lambda \tilde{r}y$ that have, respectively, the same initial data at $c = (a + b)/2$ that $u(x, \lambda)$ and $v(x, \lambda)$ have. The solutions $\tilde{u}(x, \lambda)$ and $\tilde{v}(x, \lambda)$ exist on $(a, \tilde{b})$ and, by uniqueness of solutions to initial value problems, agree with $u(x, \lambda)$ and $v(x, \lambda)$ on (a, b) and, hence, on $(a, b]$ because all four solutions are continuous at $x = b$. Consequently, $u(b, \lambda) = \tilde{u}(b, \lambda)$ and $v(b, \lambda) = \tilde{v}(b, \lambda)$ are analytic functions of λ for $|\lambda| < \infty$.

Every solution to $Ly = \lambda ry$ can be expressed as a linear combination of $u(x, \lambda)$ and $v(x, \lambda)$; therefore, all nontrivial bounded solutions are nonzero multiples of $u(x, \lambda)$ and λ is an eigenvalue of $Ly = \lambda ry$ with corresponding eigenfunction a nonzero multiple of $u(x, \lambda)$ if and only if

$$\gamma u(b, \lambda) + \delta u'(b, \lambda) = 0.$$

The function on the left is analytic in $|\lambda| < \infty$. Such an analytic function is either identically equal to zero or has at most a finite number of zeros in any bounded region of the complex plane. See [6] or [28]. Since the eigenvalues of a self-adjoint Sturm-Liouville eigenvalue problem are real, it follows that the function $\gamma u(b, \lambda) + \delta u'(b, \lambda)$ has at most a finite number of zeros in any bounded region of the complex plane and the proof is complete. ■

5.5.1 Fundamental Properties

We observed in Chapter 1 that many Sturm-Liouville eigenvalue problems that arise in applications have all positive eigenvalues. When separation of variables leads to such an eigenvalue problem, this is a consequence of the fact that the underlying partial differential equations and boundary conditions that describe the physical situation include mechanisms that oppose arbitrarily large responses of the system. The natural eigenfunction expansions of the solutions would not have this property if there were any negative eigenvalues. The next theorem covers most such cases for the singular problems under consideration. A corollary of the theorem establishes that many singular Sturm-Liouville eigenvalue problems have at most a finite number of negative eigenvalues.

Theorem 151 *If $q \geq 0$, and $\gamma\delta \geq 0$ in addition to the standing assumptions, then all the eigenvalues of the eigenvalue problem $Ly = \lambda ry$, $|y(a)| < \infty$, $\gamma y(b) + \delta y'(b) = 0$ are positive, except when $\gamma = 0$ and $q = 0$, in which case the eigenvalue problem is*

$$-(py')' = \lambda ry, \;\; |y(a)| < \infty, \;\; y'(b) = 0,$$

zero is an eigenvalue, and all other eigenvalues are positive.

Proof. Let λ be an eigenvalue and $y \neq 0$ be a corresponding real-valued eigenfunction. Recall that y is continuously differentiable on $[a, b]$, Multiply $Ly = \lambda ry$ by y and integrate by parts to obtain

$$\lambda \int_a^b y^2 r \, ds = \int_a^b y d(-py') + \int_a^b qy^2 \, ds$$
$$= -p(b)y(b)y'(b) + p(a)y(a)y'(a) + \int_a^b \left(py'^2 + qy^2\right) ds.$$

Since $p(a) = 0$,

$$\lambda \int_a^b y^2 r \, ds = -p(b)y(b)y'(b) + \int_a^b \left(py'^2 + qy^2\right) ds.$$

By the assumptions on the boundary conditions $y(b)y'(b) \leq 0$ so each term on the right is nonnegative. Hence all the eigenvalues are nonnegative. Furthermore, zero is an eigenvalue if and only if

$$y(b)y'(b) = 0 \text{ and } \int_a^b \left(py'^2 + qy^2\right) ds = 0.$$

Since $p > 0$ on (a, b) and $q \geq 0$ on (a, b), these conditions hold if and only if $y' = 0$ on $[a, b]$, in which case the corresponding eigenfunction $y = k$ is a nonzero constant and

$$\gamma k = 0 \text{ and } k^2 \int_a^b q \, ds = 0,$$

where the first condition follows from the boundary condition at $x = b$. These conditions hold if and only if

$$\gamma = 0 \text{ and } q = 0 \text{ on } [a, b]$$

because $k \neq 0$. Thus, all the eigenvalues are positive except possibly for the case when $\gamma = 0$ and $q = 0$ on $[a, b]$ when the eigenvalue problem reduces to

$$-(py')' = \lambda ry, \ |y(a)| < \infty, \ y'(b) = 0,$$

a problem for which $\lambda = 0$ is clearly an eigenvalue. For this problem any eigenvalue satisfies

$$\lambda \int_a^b y^2 r \, ds = \int_a^b py'^2 \, ds.$$

If $\lambda \neq 0$ the right member must be positive; hence, $\lambda > 0$. ■

Corollary 152 *If $\gamma\delta \geq 0$ and $q(a) > 0$ if $r(a) = 0$ in addition to the standing assumptions, then all the eigenvalues of $Ly = \lambda ry$, $|y(a)| < \infty$, $\gamma y(b) + \delta y'(b) = 0$ are real and at most a finite number are negative.*

Proof. The eigenvalues are real because the problem is self-adjoint. For either type of weight function, there is a positive constant c such that $\hat{q}(x) = q(x) + cr(x) > 0$ on $[a, b]$ because $q(x)$ is continuous on $[a, b]$ and $q(a) > 0$ if $r(a) = 0$. Consequently, all the eigenvalues of the eigenvalue problem $\hat{L}y = \hat{\lambda}ry$, $|y(a)| < \infty$, $\gamma y(b) + \delta y'(b) = 0$, where $\hat{L}y = -(py')' + \hat{q}y$, are positive. Since $Ly = \lambda ry$ if and only if $\hat{L}y = \hat{\lambda}ry$ where $\hat{\lambda} = \lambda + c$, it follows that all eigenvalues of $Ly = \lambda ry$, $|y(a)| < \infty, \gamma y(b) + \delta y'(b) = 0$ satisfy $\lambda + c = \hat{\lambda} > 0$. Thus, $\lambda > -c$. By Theorem 150 at most a finite number of eigenvalues lie in the interval $(-c, 0)$. ■

Further fundamental properties of the eigenvalues and eigenfunctions follow from the Hilbert-Schmidt theorem upon recasting (5.18) as an eigenvalue problem for the kernel $g(x, s)r(s)$

$$y(x) = \lambda \int_a^b g(x, s)r(s)y(s)\, ds \tag{5.19}$$

where $g(x, s)$ is the Green's function for the Sturm-Liouville differential operator L with the boundary conditions in (5.18). The kernel $g(x, s)$ is symmetric under the standing assumptions by Theorem 143. The equivalence of the eigenvalue problem (5.18) and the eigenvalue problem (5.19) is established just as for a regular Sturm-Liouville eigenvalue problem. See Section 4.8. Specifically, λ, y is an eigenvalue, eigenfunction pair for the Sturm-Liouville eigenvalue problem (5.18) if and only if λ, y is an eigenvalue, eigenfunction pair for the kernel $g(x, s)r(s)$.

The recasting just described requires that $\lambda = 0$ is not an eigenvalue of (5.18). We can use Theorem 150 to finesse the case when $\lambda = 0$ is an eigenvalue. To this end, let q_0 be a constant, $\tilde{q}(x) = q(x) + q_0 r(x)$, and $\tilde{L}y = -(py')' + \tilde{q}y$. Since

$$Ly = \lambda ry \Leftrightarrow \tilde{L}y = (\lambda + q_0)ry,$$

λ, y is an eigenvalue, eigenfunction pair for (5.18) if and only if $\lambda + q_0$, y is an eigenvalue, eigenfunction pair for the eigenvalue problem $\tilde{L}y = \tilde{\lambda}ry$ with the same boundary conditions as (5.18). By Theorem 150 we can fix q_0 such that 0 is not an eigenvalue for the eigenvalue problem $\tilde{L}y = \tilde{\lambda}y$, $|y(a)| < \infty$, $B_b y = 0$. This problem has a Green's function and any conclusions reached about its eigenvalues and eigenfunctions by means of the equivalent integral equation eigenvalue problem transfer immediately by translation of its eigenvalues to conclusions about the eigenvalues of the original eigenvalue problem. The corresponding eigenfunctions are the same.

In short, we can assume without loss in generality that $\lambda = 0$ is not an eigenvalue of (5.18) and convert it to the equivalent eigenvalue problem (5.19).

If $y(x)$ is continuous on $[a, b]$ and satisfies (5.19), then

$$\sqrt{r(x)}y(x) = \lambda \int_a^b \sqrt{r(x)}g(x, s)\sqrt{r(s)}\sqrt{r(s)}y(s)\, ds$$

and $z(x) = \sqrt{r(x)}y(x)$ is continuous on $[a, b]$ and satisfies

$$z(x) = \lambda \int_a^b k(x, s)z(s)\, ds \tag{5.20}$$

where

$$k(x, s) = \sqrt{r(x)}g(x, s)\sqrt{r(s)}$$

is a mildly singular, symmetric kernel by Corollary 144. (See Section 3.7.3.) Conversely, if $z(x)$ is continuous on $[a, b]$ and satisfies (5.20), then there are two cases to consider according as $r > 0$ on $[a, b]$ or r has a zero at $x = a$ and is positive on $(a, b]$. In the first case, (5.20) implies that

$$\frac{z(x)}{\sqrt{r(x)}} = \lambda \int_a^b g(x, s)r(s)\frac{z(s)}{\sqrt{r(s)}}\, ds$$

for x in $[a, b]$; that is, that $y(x) = z(x)/\sqrt{r(x)}$ satisfies (5.19). Thus, the eigenvalue problems (5.19) and (5.20) are equivalent when $r > 0$ on $[a, b]$.

Now assume $r(x) = (x-a)^m \rho(x)$ where $m > 0$ and $\rho(x) > 0$ and continuous on $[a, b]$. If $z(x)$ is continuous on $[a, b]$ and satisfies (5.20), then

$$\frac{z(x)}{\sqrt{r(x)}} = \lambda \int_a^b g(x, s)\sqrt{r(s)} z(s)\, ds$$

for $a < x \leq b$. Since $g(x, s)$ is mildly singular, the integral on the right is a continuous function on $[a, b]$ by Lemma 145. Therefore, there exists

$$\lim_{x \to a} \frac{z(x)}{\sqrt{r(x)}} = \lambda \int_a^b g(a, s)\sqrt{r(s)} z(s)\, ds.$$

Define $y(x)$ on $[a, b]$ by $y(x) = z(x)/\sqrt{r(x)}$ for $a < x \leq b$ and

$$y(a) = \lambda \int_a^b g(a, s)\sqrt{r(s)} z(s)\, ds.$$

Then $y(x)$ is continuous on $a \leq x \leq b$ and, for $a < x \leq b$,

$$y(x) = \lambda \int_a^b g(x, s)\sqrt{r(s)} z(s)\, ds = \lambda \int_a^b g(x, s) r(s) y(s)\, ds.$$

Equality also holds at $x = a$ by the definition of $y(a)$. In summary, if $z(x)$ is continuous on $[a, b]$ and satisfies (5.20), then $z(x)/\sqrt{r(x)}$ has a unique extension by continuity to a continuous function $y(x)$ on $[a, b]$ that satisfies (5.19). This establishes the equivalence of (5.19) and (5.20) in the case where the weight function r has a zero at $x = a$. Thus, for all weight functions under consideration the two eigenvalue problems are equivalent.

Since the Green's function $g(x, s)$ is a mildly singular symmetric kernel so is $k(x, s)$. Consequently, the integral operator K with kernel $k(x, s)$ is a self-adjoint, compact, bounded, linear operator on $C[a, b]$. See the paragraph preceding Theorem 51. The Hilbert-Schmidt theorem and a line of reasoning similar to that used for regular Sturm-Liouville eigenvalue problems leads to the following properties of the eigenvalues and eigenfunctions of singular Sturm-Liouville eigenvalue problems that occur most frequently in applications.

Theorem 153 *The self-adjoint Sturm-Liouville eigenvalue problem (5.18) with $\gamma\delta \geq 0$ and either $q \geq 0$ or if q assumes negative values $q(a) > 0$ if $r(a) = 0$ has an infinite sequence of real eigenvalues $\{\lambda_n\}_{n=1}^{\infty}$ and a corresponding sequence of real-valued eigenfunctions $\{\phi_n\}_{n=1}^{\infty}$ with the following properties:*

1. Each eigenvalue is simple (has both algebraic and geometric multiplicity 1). Moreover, at most a finite number of the eigenvalues are negative and the sequence of eigenvalues is unbounded; hence, the eigenvalues can be listed as

$$\lambda_1 < \lambda_2 < \cdots < \lambda_n < \cdots$$

with $\lambda_n \to \infty$ as $n \to \infty$.

2. The corresponding eigenfunctions can be chosen real-valued and orthonormal with weight function r,

$$\langle \phi_m, \phi_n \rangle_r = \int_a^b \phi_m(s)\, \phi_n(s) r(s)\, ds = \delta_{mn},$$

where δ_{mn} is the Kronecker delta.

3. If the weight function $r(x)$ is positive and continuous on $[a, b]$, then for each continuous function f on $[a, b]$, the unique solution y to the singular Sturm-Liouville boundary value *problem* $Ly = f$, $|y(a)| < \infty$, $B_b y = 0$ *can be expressed as*

$$y(x) = \sum_{n=1}^{\infty} \langle y, \phi_n \rangle_r \phi_n(x)$$

where the series is absolutely and uniformly convergent on $[a, b]$.
4. If the weight function $r(x) = (x-a)^m \rho(x)$ *with* $m > 0$ *and* $\rho(x)$ *positive and continuous on* $[a, b]$, *then the conclusion in Part 3 holds for continuous functions* f *on* $[a, b]$ *for which* $\lim_{x\to a} f(x)/(x-a)^m$ *exists and is finite.*

Proof. We rely on the discussion and notation that precedes the theorem. In particular, we can assume without loss of generality that zero is not an eigenvalue of the eigenvalue problem. Let K be the self-adjoint, compact, bounded, linear operator on $C[a, b]$ with mildly singular symmetric kernel $k(x, s) = \sqrt{r(x)}g(x, s)\sqrt{r(s)}$, where $g(x, s)$ is the Green's function associated with (5.18). Then λ, $y(x)$ is an eigenvalue, eigenfunction pair for the Sturm-Liouville eigenvalue problem (5.18) if and only if λ, $\sqrt{r(x)}y(x)$ is an eigenfunction, eigenvalue pair for the symmetric kernel $k(x, s)$.

1. Any eigenvalue λ of (5.18) is real because the eigenvalue problem is self-adjoint. If λ is an eigenvalue of (5.18) and $y_1(x)$ and $y_2(x)$ are corresponding eigenfunctions, then $y_1(x)$ and $y_2(x)$ are nontrivial bounded solutions to the singular Sturm-Liouville differential equation

$$-(py')' + (q - \lambda r)y = 0 \quad \text{for } a < x < b.$$

Consequently, by Theorem 136, $y_1(x)$ and $y_2(x)$ are nonzero multiples of each other and the geometric multiplicity of λ is 1. The algebraic multiplicity also is 1 because the kernel $k(x, s)$ is self-adjoint; see Lemma 57.

We establish next that the Sturm-Liouville eigenvalue problem has an infinite number of eigenvalues. The proof is by contradiction. Since $K \neq 0$ is a self-adjoint compact integral operator on $C[a, b]$, it has at least one nonzero eigenvalue, say μ, by Theorem 59. Consequently, $\lambda = 1/\mu$ is an eigenvalue of the kernel $k(x, s)$ and the Sturm-Liouville eigenvalue problem has at least one eigenvalue (and corresponding eigenfunction). Suppose that the Sturm-Liouville eigenvalue problem has only a finite number of eigenvalues, say $\lambda_1, \ldots, \lambda_N$, with corresponding eigenfunctions $\phi_1, \ldots, \phi_N$. By the equivalences above, K has only a finite number of nonzero eigenvalues $\mu_n = 1/\lambda_n$ for $n = 1, 2, \ldots, N$ and corresponding orthonormal eigenfunctions $\psi_n = \sqrt{r}\phi_n$. By the Hilbert-Schmidt theorem

$$Kf(x) = \sum_{n=1}^{N} \langle Kf, \psi_n \rangle \psi_n(x)$$

for all f in $C[a, b]$. Equivalently,

$$\sqrt{r(x)}G(\sqrt{r}f)(x) = \sum_{n=1}^{N} \langle Kf, \psi_n \rangle \sqrt{r(x)}\phi_n(x).$$

Hence,

$$G(\sqrt{r}f)(x) = \sum_{n=1}^{N} \langle Kf, \psi_n \rangle \phi_n(x)$$

for $a < x \leq b$. In fact, equality also holds at $x = a$ because both members of the equality are continuous on $[a, b]$. Since y solves the boundary value problem $Ly = \sqrt{r}f$, $|y(a)| < \infty$,

$\gamma y(b) + \delta y'(b) = 0$ if and only if $y = G(\sqrt{r}f)$, it follows that

$$\sqrt{r}f = Ly = LG(\sqrt{r}f).$$

Consequently,

$$\sqrt{r}f = L\left(\sum_{n=1}^{N} \langle Kf, \psi_n \rangle \phi_n\right) = \sum_{n=1}^{N} \langle Kf, \psi_n \rangle \lambda_n r \phi_n$$

and

$$f(x) = \sum_{n=1}^{N} \lambda_n \langle Kf, \psi_n \rangle \psi_n(x)$$

for $a < x \leq b$ and equality holds on $[a, b]$ as above. Since $f(x)$ can be any continuous function on $[a, b]$, this equation says that $\{\psi_n\}_{n=1}^{N}$ is a basis for $C[a, b]$, which is impossible because, for example, the functions $1, x, x^2, \ldots, x^m$ are linearly independent for every positive integer m. This contradiction establishes that the Sturm-Liouville eigenvalue problem has an infinite number of eigenvalues λ_n and corresponding eigenfunctions ϕ_n.

By the Hilbert-Schmidt theorem, the eigenvalues λ_n of $k(x, s)$ satisfy $|\lambda_n| \to \infty$ as $n \to \infty$. By Theorem 151 and Corollary 152 at most a finite number of the eigenvalues λ_n can be negative. It follows that the eigenvalues can be listed in increasing order as

$$\lambda_1 < \lambda_2 < \cdots < \lambda_n < \cdots$$

and that $\lambda_n \to \infty$ as $n \to \infty$ which completes the proof of Property 1 of the theorem.

2. Since λ_n is an eigenvalue of the symmetric kernel $k(x, s)$, the corresponding eigenfunction ψ_n can be chosen real-valued by Corollary 62 of the Hilbert-Schmidt theorem and the sequence of eigenfunctions $\{\psi_n\}_{n=1}^{\infty}$ can be chosen orthonormal with weight function 1. Then the eigenfunctions ϕ_n of the Sturm-Liouville eigenvalue problem determined by $\psi_n = \sqrt{r}\phi_n$ are orthonormal with weight function r,

$$\langle \phi_m, \phi_n \rangle_r = \langle \sqrt{r}\phi_m, \sqrt{r}\phi_n \rangle = \langle \psi_m, \psi_n \rangle = \delta_{mn}$$

and Property 2 is established.

3. Since $k(x, s)$ is mildly singular, it follows from Corollary 144 that there is a constant $M < \infty$ such that

$$\int_a^b |k(x, s)|^2 \, ds \leq M$$

for all x in $[a, b]$. Consequently, for any continuous function $\tilde{h}$ on $[a, b]$,

$$K\tilde{h}(x) = \sum_{n=1}^{\infty} \langle K\tilde{h}, \psi_n \rangle \, \psi_n(x)$$

with absolute and uniform convergence on $[a, b]$ by Corollary 61 of the Hilbert-Schmidt Theorem. Since

$$K\tilde{h}(x) = \int_a^b \sqrt{r(x)}g(x, s)\sqrt{r(s)}\tilde{h}(s) \, ds = \sqrt{r(x)}\,G(\sqrt{r}\tilde{h})(x),$$

that is, $K\tilde{h} = \sqrt{r}G(\sqrt{r}\tilde{h})$, and

$$\langle K\tilde{h}, \psi_n \rangle = \langle \sqrt{r}G(\sqrt{r}\tilde{h}), \sqrt{r}\phi_n \rangle = \langle G(\sqrt{r}\tilde{h}), \phi_n \rangle_r,$$

it follows that

$$\sqrt{r(x)}\,G(\sqrt{r}\tilde{h})(x) = \sum_{n=1}^{\infty} \langle G(\sqrt{r}\tilde{h}), \phi_n\rangle_r \sqrt{r(x)}\phi_n(x) \tag{5.21}$$

with absolute and uniform convergence on $[a, b]$.

Let $f(x)$ be continuous on $[a, b]$. If the weight function $r(x) > 0$ on $[a, b]$, then $\tilde{h} = f/\sqrt{r}$ is continuous on $[a, b]$. Set $\tilde{h} = f/\sqrt{r}$ in (5.21) to obtain the expansion

$$Gf(x) = \sum_{n=1}^{\infty} \langle Gf, \phi_n\rangle_r \phi_n(x)$$

where the series converges absolutely and uniformly because the cancelled factor $\sqrt{r(x)}$ is positive and continuous on $[a, b]$. If y is the unique solution to $Ly = f$, $|y(a)| < \infty$, and $B_b y = 0$, then $y = Gf$ and Property 3 is established.

4. If the weight function has the form $r(x) = (x-a)^m \rho(x)$ with $m > 0$ and $\rho(x) > 0$ on $[a, b]$, $f(x)$ is continuous on $[a, b]$, and $\lim_{x\to a} f(x)/(x-a)^m$ exists and is finite, then $\tilde{h} = f/\sqrt{r}$ for $a < x \le b$ has a unique extension to a continuous function on $[a, b]$, still denoted by $\tilde{h}$, obtained by defining $\tilde{h}(a) = 0$. Set $\tilde{h} = f/\sqrt{r}$ in (5.21) to obtain the expansion

$$Gf(x) = \sum_{n=1}^{\infty} \langle Gf, \phi_n\rangle_r \phi_n(x)$$

for $a < x \le b$. At this point, we cannot assert that the series is absolutely and uniformly convergent on $[a, b]$ nor that equality holds when $x = a$ because the common factor $\sqrt{r(x)}$ in (5.21) is zero at $x = a$.

We show next that $\sum_{n=1}^{\infty} \langle Gf, \phi_n\rangle_r \phi_n(x)$ is absolutely and uniformly convergent on $[a, b]$. First

$$\langle Gf, \phi_n\rangle_r = \langle Gf, r\phi_n\rangle = \langle Gf, \lambda_n^{-1} L\phi_n\rangle = \lambda_n^{-1}\langle LGf, \phi_n\rangle = \lambda_n^{-1}\langle f, \phi_n\rangle.$$

Second, the function $\tilde{f}(x) = f(x)/r(x)$ for $a < x \le b$ has a unique extension by continuity to a continuous function on $[a, b]$, still denoted by $\tilde{f}$, obtained by defining

$$\tilde{f}(a) = \lim_{x\to a} \frac{f(x)}{r(x)} = \frac{1}{\rho(a)} \lim_{x\to a} \frac{f(x)}{(x-a)^m}$$

and

$$\langle f, \phi_n\rangle = \langle \tilde{f}, \phi_n\rangle_r.$$

Thus, $\langle Gf, \phi_n\rangle_r = \lambda_n^{-1}\langle \tilde{f}, \phi_n\rangle_r$ and

$$\sum_{n=1}^{\infty} \langle Gf, \phi_n\rangle_r \phi_n(x) = \sum_{n=1}^{\infty} \langle \tilde{f}, \phi_n\rangle_r \lambda_n^{-1}\phi_n(x).$$

Since $L\phi_n = \lambda_n r\phi_n$ implies $\lambda_n^{-1}\phi_n = G(r\phi_n)$ it follows that

$$\lambda_n^{-1}\phi_n(x) = G(r\phi_n)(x) = \int_a^b g(x, s)\phi_n(s) r(s)\, ds$$

for x in $[a, b]$. So for each fixed x in $[a, b]$, $\lambda_n^{-1}\phi_n(x)$ is the n-th Fourier coefficient of the function of s, $g(x, s)$, with respect to the family $\{\phi_n\}$ that is orthonormal with weight function r.

By Bessel's inequality and Corollary 144

$$\sum_{n=1}^{\infty} |\lambda_n^{-1}\phi_n(x)|^2 \leq \int_a^b |g(x,s)|^2 r(s)\, ds \leq \|r\|_{\max} \int_a^b |g(x,s)|^2\, ds \leq M'$$

for all x in $[a, b]$ and some constant $M' < \infty$. Also

$$\sum_{n=1}^{\infty} \left| \langle \widetilde{f}, \phi_n \rangle_r \right|^2 \leq \left\| \widetilde{f} \right\|_r^2$$

by Bessel's inequality. Consequently, by the Cauchy-Schwarz inequality

$$\begin{aligned}\sum_{n=N}^{\infty} \left| \langle Gf, \phi_n \rangle_r \phi_n(x) \right| &= \sum_{n=N}^{\infty} \left| \left\langle \widetilde{f}, \phi_n \right\rangle_r \lambda_n^{-1}\phi_n(x) \right| \\ &\leq \left(\sum_{n=N}^{\infty} \left| \left\langle \widetilde{f}, \phi_n \right\rangle_r \right|^2 \right)^{1/2} \left(\sum_{n=N}^{\infty} |\lambda_n^{-1}\phi_n(x)|^2 \right)^{1/2} \\ &\leq M' \left(\sum_{n=N}^{\infty} \left| \left\langle \widetilde{f}, \phi_n \right\rangle_r \right|^2 \right)^{1/2}.\end{aligned}$$

Since the numerical series on the right converges, the absolute and uniform convergence of $\sum_{n=1}^{\infty} |\langle Gf, \phi_n \rangle_r \phi_n(x)|$ on $[a, b]$ is established.

Thus, in addition to the pointwise convergence in

$$Gf(x) = \sum_{n=1}^{\infty} \langle Gf, \phi_n \rangle_r \phi_n(x)$$

for $a < x \leq b$ established earlier, the series on the right converges absolutely and uniformly on $[a, b]$. If y is the unique solution to $Ly = f$, $|y(a)| < \infty$, and $B_b y = 0$, then $y = Gf$ so

$$y(x) = \sum_{n=1}^{\infty} \langle y, \phi_n \rangle_r \phi_n(x)$$

for $a < x \leq b$ and the series on the right converges absolutely and uniformly on $[a, b]$. The left member of the displayed equality is continuous on $[a, b]$ and the same is true of the right member by Theorem 23. Hence,

$$y(a) = \lim_{x \to a} y(x) = \lim_{x \to a} \sum_{n=1}^{\infty} \langle y, \phi_n \rangle_r \phi_n(x) = \sum_{n=1}^{\infty} \langle y, \phi_n \rangle_r \phi_n(a).$$

Thus,

$$y(x) = \sum_{n=1}^{\infty} \langle y, \phi_n \rangle_r \phi_n(x)$$

for $a \leq x \leq b$ and the series converges absolutely and uniformly on $[a, b]$. ■

We mention that most of the conclusions of the theorem hold without the additional assumptions $\gamma\delta \geq 0$ and either $q \geq 0$ or if q assumes negative values $q(a) > 0$ if $r(a) = 0$. Without these assumptions the proof only establishes Property 1 in the weaker form that the eigenvalues are real, simple, and can be listed by increasing absolute value as

$$|\lambda_1| < |\lambda_2| < \cdots < |\lambda_n| < \cdots$$

with $|\lambda_n| \to \infty$ as $n \to \infty$. The proofs of Properties 2, 3, and 4 did not rely on the assumptions $\gamma\delta \geq 0$ and $q(a) > 0$ if $r(a) = 0$.

5.5.2 Oscillation and Approximation Properties

The principal results of this section apply to the most important class of singular Sturm-Liouville eigenvalue problems with separated boundary conditions that occur in applications. They establish that for each N, linear combinations of appropriately chosen eigenfunctions $\{\phi_n\}_{n=0}^{N}$ have approximation and interpolation properties strictly analogous to the linear combinations of $\{x^n\}_{n=0}^{N}$, that is, to ordinary polynomials of degree N. These results follow from the fact that the Green's functions for such eigenvalue problems are mildly singular Kellogg kernels.

Recall from Sections 3.7 and 3.7.3 that a symmetric, mildly singular kernel $k(x, s)$ with domain $[a, b] \times [a, b] \backslash (a, a)$ that satisfies

$$\text{K1.}\quad \det\big[k(x_i, x_j)\big]_{n\times n} > 0, \qquad a < x_1 < \cdots < x_n < b,$$

$$\text{K2.}\quad \det\left[k(x_i, s_j)\right]_{n\times n} \geq 0, \qquad \text{for } x,\, s \text{ in } \left(\Delta_n \times \widetilde{\Delta}_n\right) \cup \left(\widetilde{\Delta}_n \times \Delta_n\right),$$

is called a mildly singular Kellogg kernel. A mildly singular Kellogg kernel $k(x, s)$ and its compound kernels $k_{[n]}(x, s) = \det\left[k(x_i, s_j)\right]_{n\times n}$ with domains $D_n = (\Delta_n \times \widetilde{\Delta}_n) \cup (\widetilde{\Delta}_n \times \Delta_n)$ determine integral operators $K : C[a, b] \to C[a, b]$ and $K_{[n]} : C(D_n) \to C(D_n)$ that are self-adjoint, compact, bounded, linear operators. Here Δ_n is the simplex

$$\Delta_n = \{x = (x_1, \ldots, x_n) : a \leq x_1 \leq \cdots \leq x_n \leq b\}$$

and

$$\widetilde{\Delta}_n = \{x = (x_1, \ldots, x_n) : a < x_1 \leq \cdots \leq x_n \leq b\}.$$

Theorem 154 *In addition to the standing assumptions assume that $q \geq 0$, $\gamma\delta \geq 0$, and either $\gamma \neq 0$ or q is not identically zero. Then the Sturm-Liouville boundary value problem determined by the differential operator Ly and boundary conditions $|y(a)| < \infty$, $\gamma y(b) + \delta y'(b) = 0$ has a Green's function $g(x, s)$ and $g(x, s)$ is a mildly singular Kellogg kernel. Moreover, for any continuous function $r(x)$ on $[a, b]$ that is positive on $a < x < b$, the kernel $k(x, s) = \sqrt{r(x)}g(x, s)\sqrt{r(s)}$ is a mildly singular Kellogg kernel.*

Proof. By Theorem 151 the eigenvalue problem $Ly = \lambda y$, $|y(a)| < \infty$, and $\gamma y(b) + \delta y'(b) = 0$ has only positive eigenvalues. Hence, the Green's function $g(x, s)$ exists and by Theorem 143

$$g(x, s) = \begin{cases} v(x)u(s) & \text{for } a \leq s \leq x \leq b \text{ and } (x, s) \neq (a, a) \\ u(x)v(s) & \text{for } a \leq x \leq s \leq b \text{ and } (x, s) \neq (a, a) \end{cases}$$

where the functions $u(x)$ in $C^1[a, b]$ and $v\,(x)$ in $C^1(a, b]$ are real-valued and satisfy

$$\begin{cases} Lu = 0 \quad \text{for } a < x \leq b \\ |u(a)| < \infty \end{cases},$$

$$\begin{cases} Lv = 0 \quad \text{for } a < x \leq b \\ \gamma v(b) + \delta v'(b) = 0 \end{cases},$$

and

$$p(x)\, W_{u,v}(x) = -1 \quad \text{for } a < x \leq b.$$

We claim that

$$u(x)v(x) > 0 \quad \text{for } a < x < b$$

and

$$\frac{u(x)}{v(x)} \text{ is increasing on } a < x < b.$$

Suppose $v(c) = 0$ for some c with $a < c < b$. Multiply the differential equation $-(pv')' + qv = 0$ by v and integrate by parts to obtain

$$[-v(x)p(x)v'(x)]_c^b + \int_c^b \left(p(x)v'(x)^2 + q(x)v(x)^2\right) dx = 0$$

and

$$-p(b)v(b)v'(b) + \int_c^b \left(p(x)v'(x)^2 + q(x)v(x)^2\right) dx = 0.$$

The boundary condition at $x = b$ implies that $-p(b)v(b)v'(b) \geq 0$. Consequently, $v'(x) = 0$ on $[c, b]$, $v(x) = k$, a constant, on that interval and $k = 0$ because $v(c) = 0$. Consequently, $v(c) = v'(c) = 0$ which implies that $v(x) = 0$ on $a < x < b$. This contradicts the fact that $v(x)$ is not identically zero on $a < x < b$ and establishes that $v(x) \neq 0$ on $a < x < b$.

For $a < x < b$,

$$\frac{d}{dx}\left(\frac{u(x)}{v(x)}\right) = \frac{v(x)u'(x) - u(x)v'(x)}{v(x)^2} = \frac{1}{p(x)v(x)^2} > 0$$

and, hence, $u(x)/v(x)$ is increasing on $a < x \leq b$. Since $u(x)$ is bounded and $v\ (x)$ becomes unbounded as $x \to a$,

$$\lim_{x \to a} \frac{u(x)}{v(x)} = 0.$$

Consequently,

$$\frac{u(x)}{v(x)} > 0 \quad \text{for } a < x \leq b$$

and

$$u(x)v(x) > 0 \quad \text{for } a < x \leq b.$$

Since $u(x)v(x) > 0$ and $u(x)/v(x)$ is increasing on $a < x < b$, it follows from Corollary 37 that

$$g_{[n]}(x, s) > 0 \text{ for } a < x_1,\ s_1 < x_2,\ s_2 < \cdots < x_n,\ s_n < b$$

and is 0 for all other choices of $a < x_1 < x_2 < \cdots < x_n < b$ and $a < s_1 < s_2 < \cdots < s_n < b$. Since $g(x, s)$ is a mildly singular kernel on $[a, b] \times [a, b] \backslash \{(a, a)\}$, each compound kernel $g_{[n]}(x, s)$ is continuous on its domain $(\Delta_n \times \widetilde{\Delta}_n) \cup (\widetilde{\Delta}_n \times \Delta_n)$. It follows that

$$g_{[n]}(x, s) \geq 0$$

for $x = (x_1, \ldots, x_n)$, $s = (s_1, \ldots, s_n)$ in $(\Delta_n \times \widetilde{\Delta}_n) \cup (\widetilde{\Delta}_n \times \Delta_n)$ and that

$$g_{[n]}(x, x) > 0 \text{ for } a < x_1 < x_2 < \cdots < x_n < b.$$

Thus, $g(x, s)$ is a mildly singular Kellogg kernel.

For $x = (x_1, \ldots, x_n)$, $s = (s_1, \ldots, s_n)$ in $(\Delta_n \times \widetilde{\Delta}_n) \cup (\widetilde{\Delta}_n \times \Delta_n)$,

$$k_{[n]}(x, s) = \det\left[\sqrt{r(x_i)}\, g(x_i, s_j) \sqrt{r(s_j)}\right] = \prod_{i=1}^{n} \sqrt{r(x_i)}\, g_{[n]}(x, s) \prod_{j=1}^{n} \sqrt{r(s_j)}$$

and it follows that $k_{[n]}(x, s)$ is a mildly singular Kellogg kernel. ■

Theorem 155 *If, in addition to the standing assumptions, the singular Sturm-Liouville eigenvalue problem*

$$\begin{cases} Ly = \lambda ry, \ a < x < b, \\ |y(a)| < \infty, \quad \gamma y(b) + \delta y'(b) = 0, \end{cases}$$

satisfies $\gamma\delta \geq 0$ *and either* $q \geq 0$ *on* $[a, b]$ *or if* q *changes sign on* $[a, b]$, $q(a) > 0$ *if* $r(a) = 0$, *then the eigenvalues of the singular eigenvalue problem are all real, simple, and can be labeled so that*

$$\lambda_0 < \lambda_1 < \cdots < \lambda_n < \cdots$$

with $\lambda_n \to \infty$ *as* $n \to \infty$. *The corresponding eigenfunctions* $\phi_0(x), \phi_1(x), \ldots, \phi_n(x), \ldots$ *can be chosen orthonormal (with weight function* r*) and such that* $\phi_0(x), \phi_1(x), \ldots, \phi_n(x)$ *is a Tchebycheff system on* (a, b) *for* $n = 0, 1, 2, \ldots$ *Consequently, the following oscillatory and approximation properties hold:*

1. *Given any* $n+1$ *points in* (a, b) *and any* $n+1$ *values* $b_0, \ldots, b_n$, *there is a unique* ϕ*-polynomial* $\phi(x) = \sum_{i=0}^{n} a_i\phi_i(x)$ *that take on the prescribed values at the given points.*
2. *A nontrivial* ϕ*-polynomial has at most* n *zeros in* (a, b) *where nonnodal zeros are counted twice and nodal zeros once.*
3. *A nontrivial* ϕ*-polynomial* $\phi(x) = \sum_{i=m}^{n} a_i\phi_i(x)$ *has at least* m *nodal zeros in* (a, b) *and has at most* n *zeros there, counting zeros as in Property 2.*
4. ϕ_n *has* n *nodal zeros in* (a, b) *and no other zeros there.*
5. *The zeros of* ϕ_{n-1} *and* ϕ_n *strictly interlace on* (a, b).

Proof. Let $\tilde{L}y = -(py')' + \tilde{q}y$, where $\tilde{q} = q + r$. Then λ, y is an eigenvalue, eigenfunction pair for $Ly = \lambda ry$, $|y(a)| < \infty$, $B_b y = 0$ if and only if $\tilde{\lambda}$, y is an eigenvalue, eigenfunction pair for $\tilde{L}y = \tilde{\lambda}ry$, $|y(a)| < \infty$, $B_b y = 0$ where $\tilde{\lambda} = \lambda + 1$. If $q \geq 0$, then $\tilde{q} \geq 0$ and $\tilde{q} \neq 0$ so by Theorem 154 the properties stated in theorem hold for the eigenvalues and eigenfunctions of the kernel $\tilde{k}(x, s) = \sqrt{r(x)}\tilde{g}(x, s)\sqrt{r(s)}$ where $\tilde{g}(x, s)$ is the Green's function for the eigenvalue problem associated with $\tilde{L}$. Since $\tilde{\lambda}$, $\phi(x)$ is an eigenvalue, eigenfunction pair the singular Sturm-Liouville eigenvalue problem for $\tilde{L}y = \tilde{\lambda}ry$ if and only if $\tilde{\lambda}$, $\sqrt{r(x)}\phi(x)$ is an eigenvalue, eigenfunction pair of the kernel $\tilde{k}(x, s)$ by the equivalence established earlier in the chapter, it follows that the properties in the theorem hold for the eigenvalue problem for $\tilde{L}$ and, hence, for the eigenvalue problem for L. Thus, the theorem is established for the case $q \geq 0$.

Suppose q is continuous and changes sign on $[a, b]$. There is a constant $c > 0$ such that $\tilde{q}(x) = q(x) + cr(x)$ is positive on $[a, b]$ because q and r are continuous on $[a, b]$, $r(x) > 0$ on $a < x \leq b$, and $q(a) > 0$ if $r(a) = 0$. Let $\tilde{L}y = -(py')' + \tilde{q}y$. Then λ, y is an eigenvalue, eigenfunction pair for $Ly = \lambda ry$, $|y(a)| < \infty$, $B_b y = 0$ if and only if $\tilde{\lambda}$, y is an eigenvalue, eigenfunction pair for $\tilde{L}y = \tilde{\lambda}ry$, $|y(a)| < \infty$, $B_b y = 0$ where $\tilde{\lambda} = \lambda + c$. Since $\tilde{q} > 0$ on $[a, b]$, we have already established the conclusions of the theorem for the eigenvalue problem for $\tilde{L}$ and the same conclusions follow immediately for the eigenvalue problem for L. ■

Example 3. In Section 1.4 we reviewed the standard wave equation model for the radially symmetric, transverse vibrations $u(x, t)$ of a circular membrane (a drum). Separation of variables led to the eigenvalue problem

$$R'' + \frac{1}{r}R' + \lambda R = 0,\ \ 0 < r < b,\ \ |R(0)| < \infty,\ \ R(b) = 0,$$

for the spatial factor of a separated solution, where b is the radius of the drum head; equivalently,

$$-\left(rR'\right)' = \lambda rR,\ \ 0 < r < b,\ |R(0)| < \infty,\ \ R(b) = 0.$$

This eigenvalue problem satisfies the hypotheses of all the theorems of this section. Hence, the eigenvalue problem has eigenvalues

$$0 < \lambda_0 < \lambda_1 < \cdots < \lambda_n < \cdots$$

and corresponding eigenfunctions $R_n(r)$ that have all the oscillation and interpolation properties in Theorem 155. Moreover, the eigenvalues and eigenfunctions have all the properties in Theorem 153. In particular, the eigenfunctions are orthonormal with weight function r and each twice continuously differentiable function $f(r)$ on $[0, b]$ that satisfies $f(b) = 0$ and for which $f'(0) = 0$ has the eigenfunction expansion

$$f(r) = \sum_{n=0}^{\infty} \langle f, R_n \rangle_r R_n(r)$$

with absolute and uniform convergence on $[0, b]$. The eigenfunction expansion follows directly from Theorem 153 because the function $y = f(r)$ for $0 \le r \le b$ satisfies $Ly = g$ on $(0, b)$, $|y(0)| < \infty$, $y(b) = 0$ where $g(r) = -\left(rf'(r)\right)'$ is continuous on $[0, b]$. The weight function r has a simple zero at zero and

$$\frac{g(r)}{r} = \frac{rf''(r) + f'(r)}{r} = f''(r) + \frac{f'(r) - f'(0)}{r} + \frac{f'(0)}{r}.$$

So $\lim_{r \to 0} g(r)/r$ exists and is finite if and only if $f'(0) = 0$.

Since Bessel's equation of order 0 and parameter λ, $-(rR')' = \lambda rR$, has as bounded solutions only the multiples of $J_0\left(\sqrt{\lambda}r\right)$, it follows that

$$R_n(r) = c_n J_0\left(\sqrt{\lambda_n}r\right)$$

for some constant $c_n \neq 0$. Two points of view are possible here. First, it is well known that the Bessel function $J_0(z)$ has an infinite number of zeros

$$z_0 < z_1 < \cdots < z_n < \cdots$$

that are all positive and tend to infinity as $n \to \infty$. Since $R_n(b) = 0$ for $n = 0, 1, 2, \ldots$, it follows that the eigenvalues of the eigenvalue problem are determined by the zeros of $J_0(z)$ by

$$\lambda_n = \left(\frac{z_n}{b}\right)^2$$

for $n = 0, 1, 2, \ldots$. Second, the results established in this section guarantee that all the eigenvalues λ_n are positive, infinite in number, and satisfy

$$J_0\left(\sqrt{\lambda_n}b\right) = c_n^{-1} R_n(b) = 0.$$

This gives an alternative proof that $J_0(z)$ has an infinite number of positive zeros.

Remark. The condition $f'(0) = 0$ in Example 3 is more natural than may first meet the eye. In a separation of variables solution for a vibrating drum, the drum head might be displaced by $f(r)$ for $0 \leq r \leq b$ at time $t = 0$ and released from rest. Then an eigenfunction expansion as in the example would be needed to fit the initial shape of the drum. Now, the initial shape of the drum is the surface obtained by rotating the graph of $y = f(r)$ for $0 \leq r \leq b$ about the y-axis. The two-dimensional surface obtained will have a singularity (a cusp) over the center of the drum head unless $f'(0) = 0$. Thus, realistic initial shapes for the radially symmetric vibrations of a drum will satisfy this condition. So the limit condition that arose in the proof of Theorem 153 is seen to be physically realistic.

Example 4. As the drum vibrates its rim does not remain at rest as the boundary condition $u(b, t) = 0$ in the model assumes. In reality, the rim vibrates slightly and a more realistic boundary condition is $u(b, t) + \delta u_r(b, t) = 0$ where $\delta > 0$ is a small positive constant. This boundary conditions models a slight elastic restoring force acting along the rim. The corresponding eigenvalue problem for the spatial part of a radially symmetric solution $u(r, t)$ is

$$-(rR')' = \lambda rR, \; 0 < r < b, \; |R(0)| < \infty, \;\; R(b) + \delta R'(b) = 0.$$

Once again, this eigenvalue problem satisfies the hypotheses of all the theorems of this section. Consequently, the discussion in Example 3 carries over to this situation with the single adjustment that the zeros z_n are now the zeros of the function $J_0(z) + \delta J_0'(z)$.

5.5.3 Rayleigh Quotient

Consider the singular Sturm-Liouville eigenvalue problem

$$Ly = \lambda ry, \;\; |y(a)| < \infty, \;\; \gamma y(b) + \delta y'(b) = 0,$$

where $Ly = -(py')' + qy$, $\gamma\delta \geq 0$, and either $q \geq 0$ or if q changes sign on $[a, b]$, $q(a) > 0$ if $r(a) = 0$ so that the conclusions of Theorem 155 hold. The eigenvalue problem has an infinite number of simple eigenvalues

$$\lambda_0 < \lambda_1 < \cdots < \lambda_n < \cdots$$

with $\lambda_n \to \infty$ as $n \to \infty$ and corresponding eigenfunctions $\phi_n(x)$ that are orthonormal with respect to the weight function r. Recall that the domain of L is

$$D = \{y \in C[a, b] : Ly \in C[a, b]\}.$$

The quotient that appears in the following theorem is the **Rayleigh quotient.** It will be used in Chapter 7 to find upper estimates of the smallest eigenvalue of a singular Sturm-Liouville eigenvalue problem as part of a shooting method that accurately determines eigenvalues and corresponding eigenfunctions of the problem.

Theorem 156 *With the notation and assumptions above and with weight function* $r(x) = (x - a)^m \rho(x)$ *where* $m \geq 0$ *and* $\rho(x) > 0$ *and continuous on* $[a, b]$, *the smallest* *eigenvalue of a singular eigenvalue problem* $Ly = \lambda ry$, $|y(a)| < \infty$, $\gamma y(b) + \delta y'(b) = 0$ *satisfies*

$$\lambda_0 = \min \frac{\langle Ly, y\rangle}{\langle y, y\rangle_r} = \min \frac{-p(b)y(b)y'(b) + \int_a^b (py'^2 + qy^2)\, dx}{\int_a^b y^2 r\, dx},$$

where the minimum is over all functions $y \neq 0$ *in the domain of* L *that satisfy the boundary conditions* $|y(a)| < \infty$, $\gamma y(b) + \delta y'(b) = 0$ *and* $\lim_{x\to a} Ly(x)/(x-a)^m$ *exists and is finite. Moreover, the minimum is achieved if and only if* y *is an eigenfunction corresponding to* λ_0.

Remark. Any eigenfunction y satisfies the limit condition of the theorem because $Ly = \lambda ry$. If the weight function is positive on $[a, b]$, that is if $m = 0$, then the limit condition is satisfied for all y in the domain of L because Ly is continuous on $[a, b]$. If $m > 0$ the limit condition further restricts the functions over which the minimum is taken.

Proof. If y satisfies the boundary conditions $|y(a)| < \infty$, $\gamma y(b) + \delta y'(b) = 0$, is in the domain of L, and $\lim_{x\to a} Ly(x)/(x-a)^m$ exists and is finite, then $Ly = f$ for $f = Ly$, f is continuous on $[a, b]$, and $\lim_{x\to a} f(x)/(x-a)^m$ exists and is finite. Consequently, y is continuously differentiable on $[a, b]$ by Lemma 139 and

$$y(x) = \sum_{n=0}^{\infty} \langle y, \phi_n\rangle_r\, \phi_n(x),$$

where the series converges absolutely and uniformly on $[a, b]$ by Theorem 153. Consequently,

$$\begin{aligned}
\langle Ly, y\rangle &= \left\langle Ly, \sum_{n=0}^{\infty} \langle y, \phi_n\rangle_r\, \phi_n \right\rangle = \sum_{n=0}^{\infty} \langle y, \phi_n\rangle_r \langle Ly, \phi_n\rangle \\
&= \sum_{n=0}^{\infty} \langle y, \phi_n\rangle_r \langle y, L\phi_n\rangle = \sum_{n=0}^{\infty} \langle y, \phi_n\rangle_r \langle y, \lambda_n r\phi_n\rangle \\
&= \sum_{n=0}^{\infty} \lambda_n |\langle y, \phi_n\rangle_r|^2 \geq \lambda_0 \sum_{n=0}^{\infty} |\langle y, \phi_n\rangle_r|^2 = \lambda_0 \langle y, y\rangle_r,
\end{aligned}$$

where the last equality follows from a similar calculation using $y = \sum_{n=0}^{\infty} \langle y, \phi_n\rangle_r\, \phi_n$ to evaluate $\langle y, y\rangle_r$. Equality holds above if and only if $\langle y, \phi_n\rangle_r = 0$ for all $n \geq 1$; hence, if and only if $y = \langle y, \phi_0\rangle_r\, \phi_0$, equivalently, y is an eigenfunction corresponding to λ_0. Thus, for $y \neq 0$,

$$\lambda_0 \leq \frac{\langle Ly, y\rangle}{\langle y, y\rangle_r}$$

with equality if and only if y is an eigenfunction corresponding to λ_0. The first conclusion in the theorem follows. Finally, a familiar integration by parts argument gives

$$\begin{aligned}
\langle Ly, y\rangle &= \int_a^b y d(-py') + \int_a^b qy^2\, dx = -pyy'\big|_a^b + \int_a^b (py'^2 + qy^2)\, dx \\
&= -p(b)y(b)y'(b) + \int_a^b (py'^2 + qy^2)\, dx
\end{aligned}$$

because $p(a) = 0$. ■

5.6 Concluding Remarks

We have assumed throughout this chapter that the Sturm-Liouville differential equation is singular only at one endpoint of the interval on which it is defined. This is quite natural from a physical perspective. For example, the standard wave equation model for the vibrations of a circular drum has no physical singularity in the wave equation. However, the method of separation of variables will succeed only in polar coordinates and the use of polar coordinates introduces a single "fake" (nonphysical) singularity at the pole of that system and leads to the singularity in the Sturm-Liouville equation. Since there is no physical singularity, it is reasonable to expect that the solution to the Sturm-Liouville equation is well behaved (at least continuous) at the singularity and makes plausible the mathematical conclusion we reached that the solution $y(x)$ has a limit as x approaches 0 (the singularity) and indeed that $p(x)y'(x)$ has limit 0 as x approaches 0.

Chapter 6

Singular Sturm-Liouville Problems - II

This is the second chapter on singular Sturm-Liouville boundary value problems, eigenvalue problems, and their Green's functions. In Chapter 5 the Sturm-Liouville differential equation

$$-(p(x)y'(x))' + q(x)y(x) = f(x), \quad a < x < b \tag{6.1}$$

was singular because $p(x)$ could vanish at one endpoint of the interval $[a, b]$ while $q(x)$ was continuous there. In this chapter, the Strum-Liouville differential equation is singular in two respects. First, $p(x)$ can vanish at one endpoint of the interval $[a, b]$, say at $x = a$. Second, $q(x)$ also is singular at $x = a$, with a singularity of the form $q(x) = q_1(x)/(x - a)$.

Just as in Chapter 5, the concluding section of Chapter 6 on eigenvalues and eigenfunctions of singular Sturm-Liouville problems is its climax. That section focuses on the type of singularity that occurs naturally when separation of variables is used in polar or spherical coordinates. There are two parts of the discussion. First the basic properties of the eigenvalues and eigenfunctions related to their existence, multiplicity, orthogonality, and eigenfunction expansions are established. These results follow from the Hilbert-Schmidt theorem once suitable properties are established for the Green's functions of singular Sturm-Liouville problems. Second the oscillatory and approximation properties of the eigenfunctions are developed from a unified perspective based on Jentzsch's theorem, Schur's theorem, and the Kellogg conditions; see Section 1.11.2 and Section 3.7. The reader primarily interested in the spectral results can skim the necessary background results in Chapter 3 and the properties of Green's functions established in this chapter and concentrate on the material on eigenvalue problems in Section 6.4 and its subsections. Readers seeking a fuller account of properties of solutions to singular Sturm-Liouville differential equations, boundary value problems, and Green's functions will find a readable account in the sections following this introduction.

The Bessel differential equation of order n and parameter λ, for $n = 1, 2, 3, \ldots$, serves as a motivating example for the singular problems studied in this chapter. That equation is

$$(xy')' - \frac{n^2}{x}y + \lambda xy = 0 \quad 0 < x < b$$

equivalently,

$$-(xy')' + \left(\frac{n^2 - \lambda x^2}{x}\right)y = 0 \quad 0 < x < b.$$

This Bessel equation arises from separation of variables when a reasonable degree of circular or cylindrical symmetry is involved in a model of a wave, diffusion, or steady-state process and polar coordinates are used to separate the spatial variables. Observe that $p(x) = x$ and $q(x) = q_1(x)/(x - a)$ where $q_1(x) = n^2 - \lambda x^2$ in Bessel's equation, $p(x) > 0$ and continuous on $(0, b]$, $q_1(x)$ is continuous on $[0, b]$, and $q_1(0) > 0$.

Although the behavior of the singular Sturm-Liouville differential equations, boundary value problems, and eigenvalue problems treated in this chapter is generally like the behavior encountered in Chapter 5, there are important differences and some basic properties will be developed in a different order here to accommodate the differences and added singular behavior. In particular, it is advantageous to study the homogeneous differential equation first.

When we say a function is bounded for $x > a$ and near a we mean that there is an open interval $a < x < c$ for some $c > a$ on which the function is bounded, and similarly for any other property a function my have near a.

Several arguments needed to accommodate the two singularities in this chapter depend on the order properties of the real numbers. Therefore, we assume the following throughout the chapter.

Standing Assumptions:

(1) $p(x) = (x - a)\,\varphi(x)$ where $\varphi(x)$ is positive and continuous on $[a, b]$.

(2) $q(x) = q_1(x)/(x - a)$ where $q_1(a) > 0$ and $q_1(x)$ is real-valued and continuous on $[a, b]$.

(3) $f(x)$ is real-valued and continuous on $[a, b]$.

(4) γ, δ, and c_b are real numbers with $|\gamma| + |\delta| > 0$.

In Chapter 5, standing assumption (1) was expressed in the equivalent form: $p(x)$ is continuous on $[a, b]$, is differentiable at $x = a$, is nonzero on $a < x \le b$, and satisfies $p(a) = 0$, $p'(a) \ne 0$. It follows at once from $p(x) = (x - a)\,\varphi(x)$ that $p(a) = 0$ and that $p'(a) = \varphi(a)$. In Chapter 6 the factorization $p(x) = (x - a)\,\varphi(x)$ will be used more frequently and hypotheses on $p(x)$ will be stated indirectly through hypotheses on $\varphi(x)$. In particular, we will need to assume $\varphi(x)$ is continuously differentiable to obtain certain key results. The next lemma, which we have not found elsewhere in the literature, helps clarify the relationship between smoothness assumptions on $\varphi(x)$ and smoothness assumptions on $p(x)$.

Lemma 157 *Let $p(x)$ satisfy standing assumption (1) so that $p(x) = (x - a)\,\varphi(x)$ where $\varphi(x)$ is positive and continuous on $[a, b]$ and $p'(a) = \varphi(a)$. Then*
(a) If $\varphi(x)$ is continuously differentiable on $[a, b]$ then $p(x)$ is continuously differentiable on $[a, b]$, $p''(a)$ exists and $p''(a) = 2\varphi'(a)$.
(b) If $p(x)$ is continuously differentiable on $[a, b]$, $p''(x)$ exists for $x \ge a$ and near a and is continuous at $x = a$, then $\varphi(x)$ is continuously differentiable on $[a, b]$ and $\varphi'(a) = p''(a)/2$.
(c) If $p(x)$ is continuously differentiable on $[a, b]$ and $p''(a)$ and $\varphi'(a)$ exist, then $\varphi(x)$ is continuously differentiable on $[a, b]$ and $\varphi'(a) = p''(a)/2$.

Proof. (a) If $\varphi(x)$ is continuously differentiable on $[a, b]$, then $p(x) = (x - a)\,\varphi(x)$ is continuously differentiable on $[a, b]$ and since $\varphi(a) = p'(a)$,

$$p'(x) - p'(a) = (x - a)\,\varphi'(x) + \varphi(x) - \varphi(a),$$

$$\frac{p'(x) - p'(a)}{x - a} = \varphi'(x) + \frac{\varphi(x) - \varphi(a)}{x - a}.$$

Since $\varphi(x)$ is continuously differentiable on $[a, b]$, both terms on the right have limit $\varphi'(a)$ as $x \to a$ and there exists $p''(a) = 2\varphi'(a)$.

(b) Assume $p(x)$ is continuously differentiable on $[a, b]$, $p''(x)$ exists for $x \ge a$ and near a and is continuous at $x = a$. Then $\varphi(x)$ is continuously differentiable on $a < x \le b$ because $\varphi(x) = (x - a)^{-1} p(x)$. Furthermore,

$$\varphi(x) - \varphi(a) = (x - a)^{-1} p(x) - p'(a) = \frac{p(x) - p(a) - p'(a)\,(x - a)}{x - a},$$

$$\frac{\varphi(x) - \varphi(a)}{x - a} = \frac{p(x) - p(a) - p'(a)(x - a)}{(x - a)^2}.$$

By l'Hôpital's rule or Taylor's theorem with remainder, there exists

$$\varphi'(a) = \lim_{x\to a} \frac{\varphi(x) - \varphi(a)}{x - a} = \frac{p''(a)}{2}.$$

Now, it follows from the relation

$$\frac{p'(x) - p'(a)}{x - a} = \varphi'(x) + \frac{\varphi(x) - \varphi(a)}{x - a}$$

established in (a) that there exists

$$\lim_{x\to a} \varphi'(x) = p''(a) - \varphi'(a).$$

Since $\varphi(x)$ is continuous on $[a, b]$, it follows from Lemma 11 that $\lim_{x\to a} \varphi'(x) = \varphi'(a)$. Thus φ' is continuous at $x = a$ (hence is continuous on $[a, b]$) and $\varphi'(a) = p''(a)/2$.

(c) If $p(x)$ is continuously differentiable on $[a, b]$ and $p''(a)$ and $\varphi'(a)$ exist, then from the relation

$$\frac{p'(x) - p'(a)}{x - a} = \varphi'(x) + \frac{\varphi(x) - \varphi(a)}{x - a}$$

established in (a), there exists

$$\lim_{x\to a} \varphi'(x) = p''(a) - \varphi'(a).$$

Just as in the proof of (b) it follows that φ' is continuous at $x = a$ (hence is continuous on $[a, b]$) and $\varphi'(a) = p''(a)/2$. ■

Since the coefficients in (6.1) are real-valued and the boundary conditions introduced later involve only real data, the real part of any complex-valued solution to a problem under study in this chapter is a real-valued solution to the same problem. The imaginary part is a real-valued solution of the corresponding homogeneous problem. Thus, without loss in generality, we make the

> **Convention:** by a **solution** to the singular Sturm-Liouville differential equation (6.1) we mean a *real-valued* function $y(x)$ such that $(p(x)y'(x))'$ exists for all x in (a, b) and $y(x)$ satisfies (6.1) for all x in (a, b).

For a discussion of this notion of a solution see Section 4.2.

6.1 Properties of Solutions

The standing assumptions on page 250 are in force in this section. A preliminary observation about the behavior of solutions to (6.1) at $x = b$ will be needed later.

Lemma 158 *If $y(x)$ is a solution of (6.1) and is continuous on $a < x \leq b$, then $y(x)$ is continuously differentiable on $a < x \leq b$ and satisfies the differential equation at $x = b$.*

Proof. For any c with $a < c < b$, $y(x)$ is a solution to the regular Sturm-Liouville differential equation $-(py')' + qy = f$ on the interval $c < x < b$ and is continuous on $c \leq x \leq b$. By Lemma 79 $y(x)$ is continuously differentiable on $[c, b]$ and satisfies the differential equation at $x = b$. Since $c > a$ can be chosen arbitrarily, the conclusion of the lemma follows. ∎

Next we establish the fundamental nature of solutions to the homogeneous Sturm-Liouville differential equation

$$-(p(x)y'(x))' + q(x)y(x) = 0, \ a < x < b. \tag{6.2}$$

Several lemmas prepare the way and provide the entree to the principal results of the chapter. They play an essential role both for the Sturm-Liouville boundary value problems and Sturm-Liouville eigenvalue problems that are associated with the singular Sturm-Liouville differential operator $Ly = -(py')' + qy$.

Lemma 159 *If $y(x)$ is a nontrivial solution of the equation (6.2), then y is strictly positive or strictly negative for $x > a$ and near a.*

Proof. Clearly there is a c with $a < c < b$ such that $q(x) > 0$ for $a < x < c$. Suppose that $y(x)$ has more than one zero in $a < x < c$. Let α and β be a pair of such zeros, labeled so that $a < \alpha < \beta < c$. Multiply (6.2) by $y(x)$ and integrate by parts to obtain

$$0 = \int_\alpha^\beta (y(-py')' + qy^2)\,dx = y(-py')|_\alpha^\beta + \int_\alpha^\beta (py'^2 + qy^2)\,dx,$$

$$0 = \int_\alpha^\beta (py'^2 + qy^2)\,dx.$$

Since $q > 0$ on $[\alpha, \beta]$, it follows that $y(x) = 0$ on $[\alpha, \beta]$. Thus, y solves the initial value problem $-(py')' + qy = 0$, $y(\alpha) = 0$, $y'(\alpha) = 0$ on (a, b) and must vanish identically on (a, b) by the uniqueness of solutions to initial value problems, a contradiction to the fact that y is nontrivial. Consequently, $y(x)$ has at most one zero in $a < x < c$ and therefore maintains a strict fixed sign for $x > a$ and near a. ∎

Lemma 160 *If $y(x)$ is a solution to (6.2) that is* bounded *on $a < x < b$, then*

$$\lim_{x \to a} p(x)y'(x) = 0,$$

$$p(x)y'(x) = \int_a^x q(s)y(s)\,ds \quad \text{for } a < x < b, \tag{6.3}$$

$y(x)$ extends to a continuous function on $a \leq x \leq b$, and

$$y(a) = 0.$$

Moreover, $y(x)$ is continuously differentiable on $a < x \leq b$ and satisfies the homogeneous differential equation (6.2) on $a < x \leq b$.

Proof. Fix $c > a$ such that $q(x) > 0$ on $a \leq x \leq c$. By the previous lemma, we can further assume c is chosen so that $y(x)$ is nonzero on $a < x \leq c$. Indeed, without loss in generality, assume that $y(x) > 0$ on $a < x \leq c$. Now integrate (6.2) from x to c to get

$$p(x)y'(x) = Q(x), \quad \text{for } a < x < b, \tag{6.4}$$

where

$$Q(x) = p(c)y'(c) - \int_x^c q(s)y(s)\,ds.$$

Since the integrand is positive for $a < x \leq c$, $Q(x)$ decreases as x decreases in $a < x \leq c$; hence,

$$\lim_{x\to a} p(x)y'(x) = \lim_{x\to a} Q(x) \equiv Q(a)$$

exists, finite or infinite. Now from (6.4) for $a < x \leq c$,

$$y(x) = y(c) - \int_x^c \frac{Q(s)}{p(s)}\,ds = y(c) - \int_x^c \frac{Q(s)}{(s-a)\varphi(s)}\,ds. \tag{6.5}$$

Since $y(x)$ is bounded on $a < x \leq c$ and $Q(s)$ is continuous on $a < s \leq c$ and has a limit (finite or infinite) as s decreases to a, it follows that $Q(s)$ has limit zero as s approaches a. Otherwise the integral on the right would become unbounded as x decreases to a. Thus,

$$\lim_{x\to a} p(x)y'(x) = \lim_{x\to a} Q(x) = Q(a) = 0.$$

Since $\lim_{x\to a} Q(x) = 0$, the definition of Q shows that the improper Riemann integral

$$\int_a^c q(s)y(s)\,ds$$

converges and that

$$p(c)y'(c) = \int_a^c q(s)y(s)\,ds.$$

Since

$$p(x)y'(x) - p(c)y'(c) = \int_c^x q(s)y(s)\,ds$$

for $a < x < b$, it follows that

$$p(x)y'(x) = \int_a^x q(s)y(s)\,ds$$

for $a < x < b$. Consequently, $y'(x) > 0$ on $a < x \leq c$, $y(x)$ is increasing there, and, hence, $\lim_{x\to a} y(x)$ exists and is finite because $y(x) > 0$ on $a < x \leq c$. The definition $y(a) = \lim_{x\to a} y(x)$ extends $y(x)$ to a continuous function on $a \leq x \leq c$. Since the improper integral

$$\int_a^c q(s)y(s)\,ds = \int_a^c \frac{q_1(s)}{s-a}y(s)\,ds$$

converges, $q_1(a) > 0$, and $q_1(s)$ and $y(s)$ are continuous on $a \leq s \leq c$, it follows that $y(a) = 0$.

Since $y'(x)$ exists on $a < x < b$, $y(x)$ is continuous on $a < x < b$ and is also continuous at $x = a$, as we just established. By (6.4) and the fact that Q is bounded on $c \leq x \leq b$ and p has a positive minimum on $c \leq x \leq b$, y' is bounded $c \leq x \leq b$ and by Corollary 8 y has a unique extension by continuity to a continuous function on $c \leq x \leq b$. Thus, y extends to a continuous function on $a \leq x \leq b$.

It remains to prove the last two assertions of the lemma. We show first that $y'(x)$ is continuous on $a < x \leq b$. Since both py' and $1/p$ are continuous on $a < x < b$, their product y' is continuous on $a < x < b$. There is a constant M such that $|q(s)y(s)| \leq M$ for s in (c, b) because q and y are bounded there. From (6.3),

$$|p(x)y'(x) - p(\xi)y'(\xi)| \leq \left|\int_\xi^x |q(s)y(s)|\,ds\right| \leq M|x-\xi|$$

for $c < x$, $\xi < b$ and py' is uniformly continuous on (c, b). By Proposition 7 py' has a unique extension by continuity to a continuous function on $[c, b]$. Let B be the value at $x = b$ of the continuous extension of py'. Since

$$y'(x) = \frac{1}{p(x)} p(x)y'(x)$$

for $c < x < b$, there exists

$$\lim_{x \to b} y'(x) = \lim_{x \to b} \frac{1}{p(x)} \lim_{x \to b} p(x)y'(x) = \frac{B}{p(b)}.$$

Since y is continuous on $[c, b]$, it follows from Lemma 11 that y is differentiable at $x = b$ and that its derivative is continuous there. Thus, $y(x)$ is continuously differentiable on $a < x \leq b$.

Since $y(x)$ is continuous on $a \leq x \leq b$ and (6.3) holds on $a < x \leq b$,

$$\frac{p(x)y'(x) - p(b)y'(b)}{x - b} = \frac{1}{x - b} \int_b^x q(s)y(s)\, ds$$

and the fundamental theorem of calculus or l'Hôpital's rule implies that there exists

$$(py')'(b) = q(b)y(b);$$

thus, the homogeneous differential equation also is satisfied at $x = b$. ■

By the two lemmas just established, any nontrivial bounded solution $y(x)$ to (6.2) has an isolated zero at $x = a$. Note the sharp contrast: all bounded nontrivial solutions to the singular differential equation in Chapter 5 satisfy $y(a) \neq 0$. In Chapter 6 all bounded nontrivial solutions satisfy $y(a) = 0$. This means, in particular, that the spirit of the reasoning used in Chapters 5 to establish the existence of bounded nontrivial solutions cannot be used here because the natural integral operator arising by integration of the differential equation always has $y = 0$ as a fixed point.

The next lemma establishes the nature of a basis of solutions to the homogeneous differential equation (6.2) when (6.2) has a bounded nontrivial solution of the form $y(x) = (x - a)^{\nu} z(x)$ with $z(a) \neq 0$ and $z(x)$ continuous on $[a, b]$. Later we will show that solutions of this form exist with $z(x)$ continuously differentiable on $[a, b]$.

Lemma 161 *If the homogeneous differential equation (6.2) has a nontrivial solution of the form $u(x) = (x - a)^{\nu} z(x)$ where $\nu > 0$, $z(a) \neq 0$, and $z(x)$ is continuous on $[a, b]$, then every solution $v(x)$ that is linearly independent of $u(x)$ is singular at $x = a$; more precisely,*

$$\lim_{x \to a} (x - a)^{\nu} v(x) = -\frac{C}{2\nu\varphi(a)z(a)}$$

where $C \neq 0$ is a constant determined by the two solutions $u(x)$ and $v(x)$; consequently, $v(x) = (x - a)^{-\nu} \tilde{z}(x)$ for $a < x \leq b$ and some continuous function $\tilde{z}(x)$ on $[a, b]$ with $\tilde{z}(a) \neq 0$. Moreover, every bounded solution $y(x)$ to (6.2) is a scalar multiple of $u(x)$.

Proof. Assume $u(x) = (x - a)^{\nu} z(x)$ is a solution of (6.2) as described in the lemma. There exist x_0 with $a < x_0 \leq b$ such that $u(x) \neq 0$ on $a < x \leq x_0$. Let $v(x)$ be a solution of (6.2) that is linearly independent of $u(x)$. By Lemma 86, for $a < x \leq x_0$,

$$\left(\frac{v(x)}{u(x)}\right)' = \frac{u(x)v'(x) - v(x)u'(x)}{u(x)^2} = \frac{C}{p(x)u(x)^2}$$

where $C \neq 0$ is determined by the two linearly independent solutions. For any x_1 with $a < x_1 \leq x_0$,

$$v(x) = u(x)\left[\frac{v(x_1)}{u(x_1)} - \int_x^{x_1} \frac{C}{p(s)u(s)^2}\, ds\right]$$

for $a < x \leq x_1$. By the mean value theorem for integrals (Theorem 15),

$$-\int_x^{x_1} \frac{C}{p(s)u(s)^2}\, ds = \frac{C}{2\nu\varphi(s_x)z(s_x)^2}\left((x_1 - a)^{-2\nu} - (x - a)^{-2\nu}\right)$$

for some s_x between x and x_1. Thus,

$$v(x) = (x - a)^{\nu} z(x)\left[\frac{v(x_1)}{u(x_1)} + \frac{C}{2\nu\varphi(s_x)z(s_x)^2}\left((x_1 - a)^{-2\nu} - (x - a)^{-2\nu}\right)\right],$$

and

$$(x - a)^{\nu} v(x) + \frac{C}{2\nu\varphi(a)z(a)} = (x - a)^{2\nu} z(x)\left(\frac{v(x_1)}{u(x_1)} + \frac{C(x_1 - a)^{-2\nu}}{2\nu\varphi(s_x)z(s_x)^2}\right)$$
$$+ \left(\frac{C}{2\nu\varphi(a)z(a)} - \frac{Cz(x)}{2\nu\varphi(s_x)z(s_x)^2}\right).$$

Since φ and z are continuous on $[a, x_0]$, we can fix x_1 with $a < x_1 < x_0$ sufficiently close to a so that the second summand on right is as near zero as desired. With x_1 so fixed, the first summand on the right has limit zero as x tends to a. It follows that there exists

$$\lim_{x \to a} (x - a)^{\nu} v(x) = -\frac{C}{2\nu\varphi(a)z(a)} \neq 0.$$

Define $\tilde{z}(x) = (x - a)^{\nu} v(x)$ for $a < x \leq b$ and $\tilde{z}(a)$ to be the limit above. Then $\tilde{z}(x)$ is continuous on $[a, b]$, $\tilde{z}(a) \neq 0$, and $v(x) = (x - a)^{-\nu}\, \tilde{z}(x)$ for $a < x \leq b$.

To prove the last assertion in the lemma, let $w(x)$ be a solution of (6.2) such that $u(x)$ and $w(x)$ are linearly independent on (a, b). By the basic existence and uniqueness theorem (Theorem 83) such a $w(x)$ exists and may be chosen so that the Wronskian of $u(x)$ and $w(x)$ at $(a + b)/2$ is 1. The solution $w(x)$ is unbounded on (a, b) because it is linearly independent of $u(x)$. Let $y(x)$ be any bounded solution to (6.2). There are constants c_0 and c_1 such that

$$y(x) = c_0 u(x) + c_1 w(x)$$

for $a < x < b$. Since $y(x)$ and $u(x)$ are bounded on (a, b) and $w(x)$ is unbounded, it follows that $c_1 = 0$; hence,

$$y(x) = c_0 u(x)$$

completing the proof. ∎

We turn now to the proof that the homogeneous singular Sturm-Liouville differential equation (6.2) has nontrivial bounded solutions $y(x)$ of the form $y(x) = (x - a)^{\nu} z(x)$ with $\nu > 0$ and $z(a) \neq 0$. Such a factorization is suggested by corresponding results for Sturm-Liouville problems with regular singular points. Also, the steps in the proof are essential for the numerical procedure we use in the next chapter to find accurate numerical approximations to the eigenvalues and eigenfunctions of the singular Sturm-Liouville eigenvalue problems in this chapter.

The idea behind the proof is to substitute $y(x) = (x-a)^{\nu} z(x)$ into (6.2) and determine a (unique) value for ν that leads to a relatively well behaved, singular initial value problem that determines $z(x)$. The crux of the proof is to show that the initial value problem has a (unique) solution $z(x)$ with desirable smoothness properties at the endpoints of the interval $[a, b]$. It is convenient to start with a slightly more general initial value problem (needed in Chapter 7) that emerges from this process, to add some continuous dependence results (also needed in Chapter 7 for the numerical calculation of eigenvalues and eigenfunctions), and then to make the substitution of $y(x) = (x-a)^{\nu} z(x)$ in (6.2).

Theorem 162 *Let $g(x)$ be continuous on $[a, b]$ and c_0 be a fixed constant. If $\alpha(x)$ and $\beta(x)$ are continuous on $[a, b]$, $\alpha(a) > 0$, and $\alpha'(a)$ exists, then the singular initial value problem*

$$\begin{cases} (x-a)z'' + \alpha(x)z' + \beta(x)z = g(x) \quad \text{for } a < x \leq b \\ z(a) = c_0,\; z'(a) = (g(a) - \beta(a)c_0)/\alpha(a) \end{cases} \tag{6.6}$$

has a unique solution $z(x)$. The solution satisfies

$$\lim_{x \to a} (x-a)z''(x) = 0.$$

Proof. Before proceeding to the proof, we must nail down the meaning of a solution to the singular initial value problem (6.6). By a solution $z(x)$ to (6.6) we mean a continuously differentiable function $z(x)$ on $[a, b]$ that satisfies the differential equation on $a < x \leq b$ and satisfies the given initial conditions. Discussion: since $z(x)$ satisfies the differential equation on $a < x \leq b$, $z'(x)$ is automatically continuous there. The assumption that $z'(x)$ is continuous on $[a, b]$ amounts to the assumption that $z'(x)$ is continuous at $x = a$ and this requirement provides a reasonable connection between the behavior of $z(x)$ on $a < x \leq b$ and the initial values assigned to it at $x = a$.

Let $x \to a$ in the differential equation and use the initial conditions to reach the limit conclusion of the theorem.

For the moment, assume that $z(x)$ is a solution of (6.6) and express the differential equation as

$$z''(x) + \frac{\alpha(x)}{x-a} z'(x) = \frac{g(x) - \beta(x)z(x)}{x-a} \tag{6.7}$$

for $a < x \leq b$. Note that

$$\frac{\alpha(x)}{x-a} = \frac{\alpha(x) - \alpha(a)}{x-a} + \frac{\alpha(a)}{x-a} = \alpha_1(x) + \frac{c}{x-a}$$

where

$$c = \alpha(a) > 0$$

and $\alpha_1(x)$ is continuous on $[a, b]$ with the understanding that $\alpha_1(a) = \alpha'(a)$. The differential equation has as an integrating factor

$$\mu(x) = \exp\left(\int \frac{\alpha(x)}{x-a} dx\right) = \exp\left(\int \alpha_1(x) dx\right) \exp\left(c \int \frac{1}{x-a} dx\right)$$
$$= A(x)(x-a)^c$$

where

$$A(x) = \exp\left(\int_a^x \alpha_1(s)\, ds\right)$$

is positive and continuous on $[a, b]$.

Multiply (6.7) by the integrating factor to get

$$(\mu(x)z'(x))' = \mu(x)\frac{g(x) - \beta(x)z(x)}{x - a}$$

for $a < x \leq b$. Integrate from a' to x for $a < a' < b$ and let $a' \to a$ and use the fact that $z'(x)$ is continuous at $x = a$ and that $\mu(a) = 0$ to get

$$\mu(x)z'(x) = \int_a^x \mu(s)\frac{g(s) - \beta(s)z(s)}{s - a}\, ds.$$

The calculation implies that the improper integral converges, a fact that can be confirmed independently using $\mu(x) = A(x)(x - a)^c$ for $c > 0$. Consequently,

$$z'(x) = \frac{1}{A(x)(x - a)^c}\int_a^x A(s)(s - a)^{c-1}(g(s) - \beta(s)z(s))\, ds.$$

for $a < x \leq b$. Integrate again from a' to x for $a < a' < b$ and let $a' \to a$ to get

$$z(x) = c_0 + \int_a^x \frac{\int_a^t A(s)(s - a)^{c-1}(g(s) - \beta(s)z(s))\, ds}{A(t)(t - a)^c}\, dt. \tag{6.8}$$

The calculation implies that the improper integral with respect to t converges. In summary, if $z(x)$ is a solution of the singular initial value problem (6.6), then $z(x)$ is continuous on $[a, b]$ and is a solution of the singular integral equation (6.8).

Notice that the t-integral in (6.8) is a convergent improper integral for any function $z(x)$ that is continuous on $[a, b]$. One way to see this is to observe that the t-integrand is continuous on $[a, x]$ with the understanding that at $t = a$ it is defined by

$$\begin{aligned}
&\lim_{t\to a}\frac{\int_a^t A(s)(s - a)^{c-1}(g(s) - \beta(s)z(s))\,ds}{A(t)(t - a)^c} \\
&= \lim_{t\to a}\frac{A(t)(t - a)^{c-1}(g(t) - \beta(t)z(t))}{A(t)c(t - a)^{c-1} + A'(t)(t - a)^c} \\
&= \frac{g(a) - \beta(a)z(a)}{c} = \frac{g(a) - \beta(a)c_0}{\alpha(a)}.
\end{aligned}$$

Now assume that $z(x)$ is continuous on $[a, b]$ and satisfies the integral equation (6.8) so that $z(a) = c_0$. The fundamental theorem of calculus and the limit calculation above shows that

$$\lim_{x\to a} z'(x) = \frac{g(a) - \beta(a)c_0}{\alpha(a)}.$$

Since $z(x)$ is continuous on $[a, b]$, it follows from Lemma 11 that $z'(a)$ exists, $z'(a) = (g(a) - \beta(a)c_0)/\alpha(a)$, and $z'(x)$ is continuous at $x = a$. Thus, $z(x)$ satisfies the initial conditions in (6.6) and $z'(x)$ is continuous at $x = a$. Reverse the steps leading to (6.8) to confirm that $z(x)$ satisfies the differential equation in (6.8). In particular, $z(x)$ is continuously differentiable on $a < x \leq b$ and, hence, on $[a, b]$. Thus, $z(x)$ is a solution to (6.6).

In summary, $z(x)$ is a solution of the singular initial value problem (6.6) if and only if $z(x)$ is continuous on $[a, b]$ and is a solution of the integral equation (6.8). Thus, the theorem will be established if we prove that the integral equation (6.8) has a unique continuous solution $z(x)$ on $[a, b]$.

Define a (linear) transformation $T : C[a,b] \to C[a,b]$ by

$$Tz(x) = c_0 + \int_a^x \frac{\int_a^t A(s)(s-a)^{c-1}(g(s) - \beta(s)z(s))\,ds}{A(t)(t-a)^c}\,dt.$$

The transformation maps $C[a,b]$ into itself because the t-integrand is continuous, as noted above. Furthermore, T is a contraction when $C[a,b]$ is equipped with the norm

$$\|z\|_L = \max_{a \le x \le b} e^{-L(x-a)}|z(x)|$$

and a suitable choice is made for $L > 0$. Whatever choice is made for $L > 0$, this norm is equivalent to the maximum norm, $\|z\|_{\max}$. Since

$$Tz(x) - Tw(x) = \int_a^x \left(\frac{1}{A(t)(t-a)^c}\int_a^t A(s)(s-a)^{c-1}\beta(s)(w(s) - z(s))\,ds\right)dt$$

and

$$\begin{aligned}
&\left|\int_a^x \left(\frac{1}{A(t)(t-a)^c}\int_a^t A(s)(s-a)^{c-1}\beta(s)(w(s)-z(s))ds\right)dt\right| \\
&\le \frac{\|\beta\|_{\max}\|A\|_{\max}}{A_{\min}}\int_a^x \left(\frac{1}{(t-a)^c}\int_a^t e^{L(s-a)}(s-a)^{c-1}ds\right)dt\|w-z\|_L \\
&\le \frac{\|\beta\|_{\max}\|A\|_{\max}}{A_{\min}}\int_a^x \left(\frac{e^{L(t-a)}}{(t-a)^c}\int_a^t (s-a)^{c-1}ds\right)dt\|w-z\|_L \\
&= \frac{\|\beta\|_{\max}\|A\|_{\max}}{cLA_{\min}}(e^{L(x-a)} - 1)\|w-z\|_L
\end{aligned}$$

where $A_{\min} = \min_{a \le x \le b} A(x)$, it follows that

$$|Tw(x) - Tz(x)| \le \frac{\|\beta\|_{\max}\|A\|_{\max}}{cLA_{\min}}(e^{L(x-a)} - 1)\|w-z\|_L.$$

Hence,

$$\begin{aligned}
e^{-L(x-a)}|Tw(x) - Tz(x)| &\le \frac{\|\beta\|_{\max}\|A\|_{\max}}{cLA_{\min}}(1 - e^{-L(x-a)})\|w-z\|_L, \\
\|Tw - Tz\|_L &\le \frac{\|\beta\|_{\max}\|A\|_{\max}}{cLA_{\min}}\|w-z\|_L.
\end{aligned}$$

Thus, T is a contraction if we fix $L > \|\beta\|_{\max}\|A\|_{\max}/cA_{\min}$ and, by the contraction mapping theorem, there is a unique fixed point for T; that is, (6.8) has a unique solution z in $C[a,b]$. As noted earlier, the existence of such a fixed point is equivalent to the assertions made in statement of the theorem. ■

The following continuous dependence results will be needed later in the context of the numerical evaluation of eigenvalues and eigenfunctions.

Theorem 163 *Let μ be a real parameter that varies in the closed bounded interval I and $\beta_\mu(x) = \beta(x,\mu)$ be a family of continuous functions on $[a,b] \times I$ such that*

$$\lim_{\mu \to \mu_0} \|\beta_\mu - \beta_{\mu_0}\|_{\max} = 0$$

for each μ_0 in I; that is, the map that takes μ to β_μ is continuous as a map from I into $C[a, b]$ equipped with the maximum norm. Let $z_\mu(x)$ be the unique solution to (6.6) where the coefficient $\beta(x)$ in the differential equation is replaced by $\beta_\mu(x)$. Then given any $\varepsilon > 0$ there is a $\delta > 0$ such that

$$|\mu - \mu_0| < \delta \Longrightarrow |z_\mu(x) - z_{\mu_0}(x)| < \varepsilon \quad \text{for } a \leq x \leq b$$

and

$$|\mu - \mu_0| < \delta \Longrightarrow \left|z'_\mu(x) - z'_{\mu_0}(x)\right| < \varepsilon \quad \text{for } a \leq x \leq b.$$

Proof. The notation introduced in the previous proof will be used here. Let T_μ be obtained from the operator T in the previous proof by replacing β by β_μ for μ in I. Since

$$\|\beta_\mu\|_{\max} = \max_{a \leq x \leq b} |\beta_\mu(x)| \leq \max_{\substack{a \leq x \leq b \\ \mu \text{ in } I}} |\beta(x, \mu)| = B < \infty$$

because $\beta(x, \mu)$ is continuous on the compact set $[a, b] \times I$, the operators T_μ will be contractions, with a contraction constant independent of μ in I, if L in the previous proof is fixed with $L > B\|A\|_{\max}/cA_{\min}$. Consequently, $\{T_\mu\}$ for μ in I is a family of contractions with a uniform contraction constant independent of μ in I and the function z_μ is the unique fixed point of T_μ by the equivalence describe earlier.

Furthermore, for each fixed function z in $C[a, b]$,

$$\begin{aligned} &\left|T_\mu z(x) - T_{\mu_0} z(x)\right| \\ &= \left|\int_a^x \left(\frac{1}{A(t)(t-a)^c} \int_a^t \left(\beta_\mu(s) - \beta_{\mu_0}(s)\right) A(s)(s-a)^{c-1} z(s)\, ds\right) dt\right| \\ &\leq \frac{\|A\|_{\max}\|z\|_{\max}(b-a)}{cA_{\min}} \left\|\beta_\mu - \beta_{\mu_0}\right\|_{\max} \end{aligned}$$

for $a \leq x \leq b$. Consequently, for fixed z in $C[a, b]$,

$$\left\| T_\mu z - T_{\mu_0} z\right\|_{\max} \leq \frac{\|A\|_{\max}\|z\|_{\max}(b-a)}{cA_{\min}} \left\|\beta_\mu - \beta_{\mu_0}\right\|_{\max}.$$

and the map μ to $T_\mu z$ is continuous from I into $C[a, b]$ because the map μ to β_μ is continuous from I into $C[a, b]$. By Theorem 45 the map μ to z_μ from I to $C[a, b]$ is continuous and the first conclusion in the theorem is established.

Replace β by β_μ and differentiate (6.8) to obtain

$$z'_\mu(x) = \frac{1}{A(x)(x-a)^c} \int_a^x A(s)(s-a)^{c-1}\left(-\beta_\mu(s) z_\mu(s) + g(s)\right) ds$$

for $a < x \leq b$. Hence,

$$\begin{aligned} &\left|z'_\mu(x) - z'_{\mu_0}(x)\right| \\ &\leq \frac{1}{A(x)(x-a)^c} \int_a^x A(s)(s-a)^{c-1} \left|\beta_\mu(s)\left(z_\mu(s) - z_{\mu_0}(s)\right)\right| ds \\ &\quad + \frac{1}{A(x)(x-a)^c} \int_a^x A(s)(s-a)^{c-1} \left|\left(\beta_\mu(s) - \beta_{\mu_0}(s)\right) z_{\mu_0}(s)\right| ds \\ &\leq \frac{\|A\|_{\max}(x-a)^c}{A_{\min} c (x-a)^c} \left\{\left\|\beta_\mu\right\|_{\max} \left\|z_\mu - z_{\mu_0}\right\|_{\max} + \left\|\beta_\mu - \beta_{\mu_0}\right\|_{\max} \left\|z_{\mu_0}\right\|_{\max}\right\}, \end{aligned}$$

and

$$\left|z'_\mu(x) - z'_{\mu_0}(x)\right| \le \frac{\|A\|_{\max}}{cA_{\min}}\left\{B\|z_\mu - z_{\mu_0}\|_{\max} + \left\|\beta_\mu - \beta_{\mu_0}\right\|_{\max}\|z_{\mu_0}\|_{\max}\right\}$$

for $a < x \le b$. The inequality also holds for $x = a$ because the left member is continuous at $x = a$. Since $\|z_\mu - z_{\mu_0}\|_{\max}$ and $\|\beta_\mu - \beta_{\mu_0}\|_{\max}$ tend to 0 as μ tends to μ_0 the final conclusion of the theorem follows. ■

We can now establish that nontrivial bounded solutions exist to (6.2) under reasonably mild conditions with the aid of Theorem 162.

Theorem 164 *Assume that $\varphi(x)$ is continuously differentiable on $[a, b]$ and $q'_1(a)$ exists in addition to the standing assumptions. Then:*
(a) The homogeneous equation (6.2) has a nontrivial bounded solution of the form

$$u(x) = (x - a)^\nu z(x)$$

where

$$\nu = \sqrt{\frac{q_1(a)}{\varphi(a)}} = \sqrt{\frac{q_1(a)}{p'(a)}} > 0,$$

$z(a) = 1$, and $z(x)$ is continuously differentiable on $[a, b]$.
(b) The solution in (a) is unique and every bounded solution to (6.2) is a constant multiple of a solution in (a).
(c) Every solution $v(x)$ to (6.2) that is linearly independent of a bounded solution has the form $v(x) = (x - a)^{-\nu}\tilde{z}(x)$ for $a < x \le b$ where $\tilde{z}(x)$ is a continuous function on $[a, b]$ with $\tilde{z}(a) \ne 0$, and $\tilde{z}''(x)$ is continuous on $a < x < b$;
(d) The functions $u(x)$ and $v(x)$ are continuously differentiable on $a < x \le b$ and satisfy the differential equation (6.2) at $x = b$.

Proof. (a) Assume for the moment that (6.2) has a solution of the form

$$y(x) = (x - a)^\nu z(x)$$

where $z(x)$ and $\nu > 0$ are to be determined. Then

$$\begin{aligned}
py' &= (x - a)\varphi\Big((x - a)^\nu z' + \nu(x - a)^{\nu-1} z\Big) \\
&= (x - a)^{\nu+1}\varphi z' + \nu(x - a)^\nu \varphi z \\
(py')' &= (x - a)^{\nu+1}\varphi z'' + (x - a)^{\nu+1}\varphi' z' + (\nu + 1)(x - a)^\nu \varphi z' \\
&\quad + \nu(x - a)^\nu \varphi z' + \nu(x - a)^\nu \varphi' z + \nu^2(x - a)^{\nu-1}\varphi z \\
&= (x - a)^{\nu+1}\varphi z'' + (x - a)^\nu\big((2\nu + 1)\varphi + (x - a)\varphi'\big)z' \\
&\quad + (x - a)^{\nu-1}\big(\nu^2\varphi + \nu(x - a)\varphi'\big)z
\end{aligned}$$

and

$$\begin{aligned}
qy &= \frac{q_1}{x - a}(x - a)^\nu z \\
&= (x - a)^{\nu-1} q_1 z.
\end{aligned}$$

Consequently,

$$(py')' - qy = (x-a)^{\nu+1}\varphi z'' + (x-a)^{\nu}\big((2\nu+1)\varphi + (x-a)\varphi'\big)z' \\ + (x-a)^{\nu-1}\big(\nu^2\varphi - q_1 + \nu(x-a)\varphi'\big)z$$

or, equivalently,

$$(py')' - qy = (x-a)^{\nu}[(x-a)\varphi z'' + \big((2\nu+1)\varphi + (x-a)\varphi'\big)z' \\ + \Big((x-a)^{-1}\big(\nu^2\varphi - q_1\big) + \nu\varphi'\Big)z].$$

We seek a choice for ν that will remove the singular behavior in the coefficient of z. The ratio

$$\frac{\nu^2\varphi(x) - q_1(x)}{x-a}$$

can have a finite limit at $x = a$ only if

$$\nu^2\varphi(a) - q_1(a) = 0;$$

that is, only if

$$\nu = \sqrt{\frac{q_1(a)}{\varphi(a)}},$$

in which case,

$$\begin{aligned} \lim_{x\to a}\frac{\nu^2\varphi(x) - q_1(x)}{x-a} &= \lim_{x\to a}\frac{q_1(a)\varphi(x) - \varphi(a)q_1(x)}{(x-a)\varphi(a)} \\ & \lim_{x\to a}\frac{q_1(a)\dfrac{\varphi(x)-\varphi(a)}{x-a} - \varphi(a)\dfrac{q_1(x)-q_1(a)}{x-a}}{\varphi(a)} \\ &= \frac{q_1(a)\varphi'(a) - \varphi(a)q_1'(a)}{\varphi(a)}. \end{aligned}$$

Consequently, with the choice of ν just specified, the function

$$q_2(x) = \begin{cases} \dfrac{q_1(a)\varphi'(a) - \varphi(a)q_1'(a)}{\varphi(a)} & \text{for } x = a \\ \dfrac{\nu^2\varphi(x) - q_1(x)}{x-a} & \text{for } a < x \le b \end{cases},$$

is continuous on $[a, b]$ and

$$(py')' - qy = (x-a)^{\nu}\big[(x-a)\varphi z'' + \big((2\nu+1)\varphi + (x-a)\varphi'\big)z' + \big(q_2 + \nu\varphi'\big)z\big]$$

for $a < x < b$. Thus, (6.2) will have a solution of the form

$$y(x) = (x-a)^{\nu} z(x) \quad \text{with} \quad \nu = \sqrt{\frac{q_1(a)}{\varphi(a)}}$$

if and only if $z(x)$ satisfies

$$(x-a)\varphi(x)z'' + \big((2\nu+1)\varphi(x) + (x-a)\varphi'(x)\big)z' + \big(q_2(x) + \nu\varphi'(x)\big)z = 0$$

for $a < x < b$, equivalently, if and only if $z(x)$ satisfies

$$(x-a)z'' + \alpha(x)z' + \beta(x)z = 0, \quad \text{for } a < x < b,$$

where

$$\alpha(x) = \frac{(2\nu+1)\varphi(x) + (x-a)\varphi'(x)}{\varphi(x)} \quad \text{and} \quad \beta(x) = \frac{q_2(x) + \nu\varphi'(x)}{\varphi(x)}. \tag{6.9}$$

Furthermore, if z' is continuous at $x = a$, the differential equation for z implies that

$$\lim_{x \to a} (x-a)z''(x) = A \Longleftrightarrow z'(a) = -\frac{\beta(a)}{\alpha(a)} z(a) - \frac{A}{\alpha(a)}.$$

This equivalence suggests that the most well behaved solution to the differential equation for z, if it exists, is the solution that satisfies $\lim_{x\to a}(x-a)z''(x) = 0$, in which case

$$z'(a) = -\frac{\beta(a)}{\alpha(a)} z(a).$$

This leads us to seek a nontrivial solution to (6.2) of the form $u(x) = (x-a)^\nu z(x)$ where $\nu = \sqrt{q_1(a)/\varphi(a)}$ and where $z(x)$ solves the initial value problem

$$\begin{cases} (x-a)z'' + \alpha(x)z' + \beta(x)z = 0, & a < x \le b, \\ z(a) = 1, \;\; z'(a) = -\dfrac{\beta(a)}{\alpha(a)}. \end{cases}$$

Under the hypothesis of the theorem, $\alpha(x)$ and $\beta(x)$ given by (6.9) are continuous on $[a, b]$, $\alpha(a) = 2\nu + 1 > 0$, and

$$\alpha(x) - \alpha(a) = (x-a)\varphi'(x)/\varphi(x)$$

so there exists

$$\alpha'(a) = \lim_{x\to a} \frac{\alpha(x) - \alpha(a)}{x-a} = \frac{\varphi'(a)}{\varphi(a)}.$$

Hence, the coefficients $\alpha(x)$ and $\beta(x)$ satisfy the hypotheses in Theorem 162 applied with $g(x) = 0$ and $c_0 = 1$. Therefore, the initial value problem above has a unique solution $z(x)$ which is continuously differentiable on $[a, b]$. This completes the proof of (a).

(b) By Lemma 161 every bounded solution to the homogeneous equation (6.2) is a multiple of the solution in (a). Suppose $u_1(x) = (x-a)^\nu z_1(x)$ has the properties in (a). Then $u_1(x)$ is a bounded solution of $-(py')' + qy = 0$ on (a, b) and

$$u_1(x) = c_1 u(x)$$

on (a, b) for some constant c_1. Hence,

$$z_1(x) = c_1 z(x)$$

on (a, b). Let x tend to a to conclude that $c_1 = 1$ and $z_1(x) = z(x)$ on $[a, b]$. Thus, there is only one solution $u(x)$ that satisfies the conditions in (a).

(c) The assertion follows at once from Lemma 161.

(d) Since u and v are continuous on $a < x \le b$, they are both continuously differentiable there and satisfy the differential equation at $x = b$ by Lemma 158. ■

The focus in this section has been on nontrivial bounded solutions of (6.2). However, it is an easy consequence of the results obtained for bounded solutions and the asymptotic behavior of companion unbounded solutions that each such unbounded solution can be expressed as

$$y(x) = (x-a)^{-\nu}\tilde{z}(x)$$

where $\nu = \sqrt{q_1(a)/\varphi(a)}$, $\tilde{z}(a) \neq 0$, and $\tilde{z}$ is continuous on $[a, b]$. In fact, replacing ν by $-\nu$ in the long calculation at the beginning of the proof of Theorem 164 shows that $\tilde{z}(x)$ must be a nontrivial solution of

$$(x-a)\tilde{z}'' + \tilde{\alpha}(x)\tilde{z}' + \tilde{\beta}(x)\tilde{z} = 0 \quad \text{for } a < x \leq b$$

where

$$\tilde{\alpha}(x) = \frac{(-2\nu+1)\varphi(x) + (x-a)\varphi'(x)}{\varphi(x)},$$

$$\tilde{\beta}(x) = \frac{q_2(x) - \nu\varphi'(x) + \lambda r(x)}{\varphi(x)}.$$

Furthermore, the method of proof of Theorem 164 can be applied to the initial value problem

$$\begin{cases} (x-a)\tilde{z}'' + \tilde{\alpha}(x)\tilde{z}' + \tilde{\beta}(x)\tilde{z} = 0 & \text{for } a < x \leq b \\ \tilde{z}(a) = 1, \quad \tilde{z}'(a) = -\dfrac{\tilde{\beta}(a)}{\tilde{\alpha}(a)} \end{cases}$$

provided $\tilde{\alpha}(a) = -2\nu + 1 > 0$, that is provided $\nu < 1/2$, to prove that a solution $\tilde{z}(x)$ exists in $C^1[a, b]$ and satisfies the given initial conditions.

6.2 Boundary Value Problems

The standing assumptions on page 250 remain in force in this section. Consequently, all the results established in the previous section are available for use here. As noted earlier, since all the data is real-valued, we can restrict our attention to *real-valued* solutions of the boundary value problems that follow.

The Sturm-Liouville boundary value problem associated with the singular differential equation (6.1) is

$$\begin{cases} -\left(p(x)y'(x)\right)' + q(x)y(x) = f(x) & \text{for } a < x < b \\ |y(a)| < \infty, \ \gamma y(b) + \delta y'(b) = c_b. \end{cases} \tag{6.10}$$

The corresponding homogeneous problem is

$$\begin{cases} -\left(p(x)y'(x)\right)' + q(x)y(x) = 0 & \text{for } a < x < b \\ |y(a)| < \infty, \ \gamma y(b) + \delta y'(b) = 0 \end{cases} \tag{6.11}$$

The notation $|y(a)| < \infty$ means that y is bounded for $x > a$ and near a, just as in Chapter 5. As usual, $Ly = -\left(py'\right)' + qy$.

A function $y(x)$ is a **solution** to (6.10) if it satisfies the singular differential equation on $a < x < b$, satisfies the boundary condition at $x = b$, and is continuous on $a \leq x \leq b$. As always, $y(x)$ is a solution to the differential equation if $\left(p(x)y'(x)\right)'$ exists for each x in $a < x < b$ and $y(x)$ satisfies the differential equation there. See Section 4.2 for a discussion of this notion of a solution. We discussed the reason for the continuity assumption for singular problems in Chapter 5. Essentially the same remarks apply here. The formulation of the boundary condition at $x = a$, namely that $|y(a)| < \infty$, is suggested by physical considerations in which such boundary value problems arise. The boundary condition $|y(a)| < \infty$ can in principle allow quite wild behavior of a function that satisfies the singular Sturm-Liouville differential equation as x approaches a. Under our standing assumptions, this does not happen for solutions of the singular homogeneous differential equation in (6.11). By Lemma 160 any bounded solution $y(x)$ to the homogeneous differential equation on $a < x < b$ extends to continuous function near $x = a$; in fact $\lim_{x\to a} y(x) = 0$ and defining $y(a) = 0$ gives the extension of y to a continuous function near $x = a$. We will show later that whenever (6.10) has a unique solution $y(x)$ the same is true; that is, $\lim_{x\to a} y(x) = 0$ and setting $y(a) = 0$ gives the extension of y to a continuous function near $x = a$. Thus, it is natural include the continuity requirement in the context of our standing assumptions and makes it explicit that the bounded solutions of interest have limiting values as x approaches a.

We start with two lemmas that are useful in the study of Sturm-Liouville boundary value problems and eigenvalue problems. The first is a direct consequence of Lemma 158.

Lemma 165 *If $y(x)$ is a solution of the singular Sturm-Liouville boundary value problem (6.10), then $y(x)$ is continuously differentiable on $a < x \leq b$ and satisfies the differential equation at $x = b$.*

Lemma 166 *The following results hold.*

(a) There is a nontrivial bounded solution u to $Ly = 0$ on $a < x \leq b$ such that

$$u(x) = (x-a)^{\nu} z(x) \quad \text{with } z \text{ in } C^1[a,b],\ z(a) \neq 0,\ \nu = \sqrt{\frac{q_1(a)}{\varphi(a)}} > 0.$$

(b) If u in (a) satisfies $\gamma u(b) + \delta u'(b) \neq 0$, there is a nontrivial continuously differentiable function $v(x)$ on $a < x \leq b$ that satisfies

$$\begin{aligned} Lv &= 0, \quad a < x \leq b, \\ \gamma v(b) + \delta v'(b) &= c_b. \end{aligned}$$

(c) Any bounded nontrivial solution u to $Ly = 0$ on $a < x \leq b$ that satisfies $\gamma u(b) + \delta u'(b) \neq 0$ and any nontrivial solution v to $Ly = 0$ on $a < x \leq b$ that satisfies $\gamma v(b) + \delta v'(b) = 0$ are linearly independent on $a < x \leq b$ and v has the form

$$v(x) = (x-a)^{-\nu} \tilde{z}(x) \quad \text{with } \tilde{z} \in C[a,b],\ \tilde{z}(a) \neq 0, \text{ and } \nu = \sqrt{\frac{q_1(a)}{\varphi(a)}}.$$

Proof. (a) By Theorem 164 the Sturm-Liouville differential equation $Ly = 0$, $a < x < b$ has a nontrivial bounded solution $u(x)$ with the properties in (a).

(b) By the basic existence and uniqueness theorem for initial value problems,

$$\begin{aligned} Lw &= 0, \quad a < x < b, \\ w(c) &= -u'(c),\ w'(c) = u(c), \end{aligned}$$

where $c = (a+b)/2$, has a unique solution on $a < x < b$. Moreover, w extends to a continuously differentiable function on $[c, b]$ that satisfies the differential equation there by Theorem 85. Since $W_{u,w}(c) = u(c)^2 + u'(c)^2 > 0$, the solutions u and w are linearly independent on $a < x < b$. Hence, the differential equation $Ly = 0$ for $a < x \leq b$ has general solution $v = c_1 u + c_2 w$ and v will be nontrivial and satisfy the boundary condition at $x = b$ if

$$c_1\big(\gamma u(b) + \delta u'(b)\big) + c_2\big(\gamma w(b) + \delta w'(b)\big) = c_b$$

can be satisfied with c_1 and c_2 not both zero. The choices $c_2 = -1$ and $c_1 = \big(c_b + \gamma w(b) + \delta w'(b)\big)/\big(\gamma u(b) + \delta u'(b)\big)$ do the job.

(c) Next we show that if $\gamma u(b) + \delta u'(b) \neq 0$ where $u(x)$ is any nontrivial bounded solution to $Lu = 0$ on $a < x \leq b$ and $v(x)$ is any nontrivial solution to $Lv = 0$ on $a < x \leq b$ with $\gamma v(b) + \delta v'(b) = 0$, then $u(x)$ and $v(x)$ are linearly independent on $a < x \leq b$. Indeed, if $\gamma \neq 0$, then $v'(b) \neq 0$ (otherwise, $v(b) = v'(b) = 0$ and $v(x)$ would be the trivial solution) and

$$\begin{aligned}\begin{vmatrix} u(b) & v(b) \\ u'(b) & v'(b) \end{vmatrix} &= \gamma^{-1} \begin{vmatrix} \gamma u(b) + \delta u'(b) & \gamma v(b) + \delta v'(b) \\ u'(b) & v'(b) \end{vmatrix} \\ &= \gamma^{-1}\big(\gamma u(b) + \delta u'(b)\big) v'(b)\end{aligned}$$

while if $\gamma = 0$, then $v(b) \neq 0$ and

$$\begin{aligned}\begin{vmatrix} u(b) & v(b) \\ u'(b) & v'(b) \end{vmatrix} &= \delta^{-1} \begin{vmatrix} u(b) & v(b) \\ \gamma u(b) + \delta u'(b) & \gamma v(b) + \delta v'(b) \end{vmatrix} \\ &= -\delta^{-1}\big(\gamma u(b) + \delta u'(b)\big) v(b).\end{aligned}$$

In either case, the Wronskian of $u(x)$ and $v(x)$ is nonzero at $x = b$ and $u(x)$ and $v(x)$ are linearly independent on $a < x \leq b$. The final conclusion of the lemma on the form of v follows from Theorem 164. ∎

The foregoing lemma prepares the way to establish the basic connection between the inhomogeneous and homogeneous Sturm-Liouville boundary value problems. The proof of the theorem that follows essentially constructs the Green's function for the inhomogeneous problem when $c_b = 0$. However, a discussion of Green's functions is deferred until the next section.

Theorem 167 *The singular Sturm-Liouville boundary value problem (6.10) has a unique solution for every function $f(x)$ that is continuous on $[a, b]$ if and only if the corresponding homogeneous problem (6.11) has only the trivial solution.*

Proof. If (6.10) has a unique solution for every choice of $f(x)$, then (6.11) has a unique solution. Clearly $y = 0$ is a solution and, hence, is the only solution to the homogeneous boundary value problem.

Conversely, assume (6.11) has only the trivial solution. By Lemma 166(a) there is a nontrivial bounded solution u to $Lu = 0$ on $a < x \leq b$. Since u is nontrivial and bounded, $\gamma u(b) + \delta u'(b) \neq 0$; otherwise u would be a nontrivial solution to (6.11). By Lemma 166(b) there is a nontrivial solution v to $Lv = 0$ on $a < x \leq b$ with $\gamma v(b) + \delta v'(b) = 0$. The solutions u and v to $Ly = 0$ on $a < x \leq b$ are linearly independent,

$$\begin{aligned} u(x) &= (x-a)^{\nu} z(x) \quad \text{with z in } C^1[a, b],\ z(a) \neq 0, \\ v(x) &= (x-a)^{-\nu} \tilde{z}(x) \quad \text{with } \tilde{z} \text{ in } C[a, b],\ \tilde{z}(a) \neq 0, \end{aligned}$$

and $\nu = \sqrt{q_1(a)/\varphi(a)} > 0$ by Lemma 166. Also

$$p(x)\big(u'(x)v(x) - u(x)v'(x)\big) = C \neq 0$$

by Lemma 86. Replace v by v/C to obtain a pair of solutions, still denoted by u and v, such that

$$\begin{aligned} &Lu = 0, \quad a < x \leq b, \\ &|u(a)| < \infty, \end{aligned}$$

$$Lv = 0, \quad a < x \le b,$$
$$\gamma v(b) + \delta v'(b) = 0,$$
$$p(u'v - uv') = 1 \quad \text{for } a < x \le b,$$
$$u(x) = (x-a)^{\nu} z(x) \quad \text{with } z \text{ in } C^1[a,b],\ z(a) \neq 0,$$

and

$$v(x) = (x-a)^{-\nu} \tilde{z}(x) \quad \text{with } \tilde{z} \text{ in } C[a,b],\ \tilde{z}(a) \neq 0.$$

The solutions u and v to $Ly=0$ together with some plausible reasoning will lead to a solution formula for (6.10) when $c_b=0$. Once that formula is obtained we will check directly that the formula does in fact solve (6.10) when $c_b=0$. So for the moment, assume that (6.10) when $c_b=0$ has a solution y with the property that $\lim_{x\to a} py' = 0$. Apply Lemma 80 (Lagrange's identity) with $z=u$ and y the solution to $Ly=f$, $|y(a)| < \infty$, $\gamma y(b) + \delta y'(b) = 0$ to obtain

$$\int_a^x -uf\, ds = p(uy' - yu')\big|_a^x.$$

Since $\lim_{x\to a} pu' = 0$ by Lemma 160 and $\lim_{x\to a} py' = 0$ by assumption,

$$\int_a^x -uf\, ds = p(x)\big(u(x)y'(x) - y(x)u'(x)\big).$$

In the same way, replace z by v in Lagrange's identity to get

$$\int_x^b -vf\, ds = p(vy' - yv')\big|_x^b.$$

Since γ and δ are not both zero and

$$\gamma v(b) + \delta v'(b) = 0$$
$$\gamma y(b) + \delta y'(b) = 0$$

the determinant of the 2×2 system is zero, the evaluation at the upper limit b gives 0, and

$$-\int_x^b vf\, ds = -p(x)\big(v(x)y'(x) - y(x)v'(x)\big).$$

Thus,

$$\int_a^x uf\, ds = p(x)\big(-u(x)y'(x) + y(x)u'(x)\big)$$

and

$$\int_x^b vf\, ds = p(x)\big(v(x)y'(x) - y(x)v'(x)\big).$$

Multiply the last equation by $u(x)$, the equation above it by $v(x)$, and add to eliminate $y'(x)$ and obtain

$$v(x)\int_a^x uf\, ds + u(x)\int_x^b vf\, ds = y(x)p(x)\big(v(x)u'(x) - u(x)v'(x)\big).$$

Since u and v were normalized to satisfy $p(u'v - uv') = 1$,

$$y(x) = v(x)\int_a^x u(s)f(s)\, ds + u(x)\int_x^b v(s)f(s)\, ds$$

for $a < x \leq b$. We extend the definition of y to a function defined on $[a, b]$ by setting $y(a) = 0$. This is the unique extension of y to a continuous function on $[a, b]$, as will be confirmed shortly. The formula above was obtained under the assumption that a solution to (6.10) with $c_b = 0$ did exist and had the additional property that $\lim_{x \to a} py' = 0$. We now check that

$$y(x) = \begin{cases} 0 & \text{for } x = a \\ v(x) \int_a^x u(s)f(s)\,ds + u(x) \int_x^b v(s)f(s)\,ds & \text{for } a < x \leq b \end{cases} \tag{6.12}$$

is the solution of (6.10) with $c_b = 0$ (without the additional assumption that $\lim_{x \to a} py' = 0$.) Differentiate

$$y(x) = v(x) \int_a^x u(s)f(s)\,ds + u(x) \int_x^b v(s)f(s)\,ds \quad \text{for } a < x \leq b \tag{6.13}$$

to obtain

$$y'(x) = v'(x) \int_a^x u(s)f(s)\,ds + u'(x) \int_x^b v(s)f(s)\,ds \tag{6.14}$$

for $a < x \leq b$. Consequently,

$$p(x)y'(x) = p(x)v'(x) \int_a^x u(s)f(s)\,ds + p(x)u'(x) \int_x^b v(s)f(s)\,ds,$$

$$-\big(p(x)y'(x)\big)' = -p(x)v'(x)u(x)f(x) - \big(p(x)v'(x)\big)' \int_a^x u(s)f(s)\,ds$$
$$+ p(x)u'(x)v(x)f(x) - \big(p(x)u'(x)\big)' \int_x^b v(s)f(s)\,ds,$$

and

$$q(x)y(x) = q(x)v(x) \int_a^x u(s)f(s)\,ds + q(x)u(x) \int_x^b v(s)f(s)\,ds,$$

Since $p(u'v - v'u) = 1$, addition of these two equations gives

$$Ly(x) = f(x) + Lv(x) \int_a^x u(s)f(s)\,ds + Lu(x) \int_x^b v(s)f(s)\,ds = f(x)$$

for $a < x \leq b$. So $y(x)$ satisfies the differential equation in (6.10) and also satisfies the differential equation at $x = b$.

From (6.13) and (6.14)

$$y(b) = v(b) \int_a^b u(s)f(s)\,ds,$$

$$y'(b) = v'(b) \int_a^b u(s)f(s)\,ds,$$

and $y(x)$ satisfies the boundary condition at $x = b$ because $v(x)$ does.

Equation (6.13) shows that $y(x)$ is continuous for $a < x \leq b$. It remains to show that $y(x)$ is continuous at $x = a$ to establish that (6.12) is the solution to (6.10) with $c_b = 0$. We must show that

$$\lim_{x \to a} y(x) = 0.$$

Since

$$y(x) = v(x) \int_a^x u(s)f(s)\,ds + u(x) \int_x^b v(s)f(s)\,ds$$

for $a < x \leq b$, the existence of the asserted limit follows from

$$L_1 = \lim_{x\to a} v(x) \int_a^x u(s)f(s)\,ds = 0 \quad \text{and} \quad L_2 = \lim_{x\to a} u(x) \int_x^b v(s)f(s)\,ds = 0.$$

To establish the limits we use the mean value theorem for integrals and the continuity of z, $\tilde{z}$, and f on $[a, b]$. For some s_x between a and x,

$$\begin{aligned} L_1 &= \lim_{x\to a} (x-a)^{-\nu}\,\tilde{z}(x)\,z(s_x)\,f(s_x) \int_a^x (s-a)^{\nu}\,ds \\ &= \lim_{x\to a} \frac{\tilde{z}(x)\,z(s_x)\,f(s_x)\,(x-a)}{\nu+1} = 0. \end{aligned}$$

Likewise, for some s_x between x and b,

$$u(x) \int_x^b v(s)f(s)\,ds = z(x)\tilde{z}(s_x)f(s_x)(x-a)^{\nu} \int_x^b (s-a)^{-\nu}\,ds$$

and

$$(x-a)^{\nu} \int_x^b (s-a)^{-\nu}\,ds = \left\{ \begin{array}{ll} (x-a)^{\nu} \left[\dfrac{(b-a)^{-\nu+1} - (x-a)^{-\nu+1}}{-\nu+1} \right] & \text{if } \nu \neq 1 \\ (x-a)[\ln(b-a) - \ln(x-a)] & \text{if } \nu = 1 \end{array} \right\},$$

which has limit 0 as x approaches a; consequently, there exists

$$L_2 = \lim_{x\to a} u(x) \int_x^b v(s)f(s)\,ds = 0$$

and

$$\lim_{x\to a} y(x) = 0.$$

Thus, $y(x)$ is a continuous function on $[a, b]$ and solves the boundary value problem (6.10).

We have proven that (6.10) with $c_b = 0$ has a unique solution, say $y_1(x)$ when the corresponding homogeneous boundary value problem has only the trivial solution. Under the same assumption, the boundary value problem

$$\begin{aligned} -(py')' + qy = 0, \quad & a < x < b, \\ |y(a)| < \infty, \ & \gamma y(b) + \delta y'(b) = c_b \end{aligned}$$

has a unique solution. Indeed, the general solution to the homogeneous differential equation is $y_2 = c_1 u + c_2 v$, with u and v as above. The solution y_2 will satisfy the boundary condition at $x = a$ if $c_2 = 0$ and will satisfy the boundary condition at $x = b$ if $c_1 = c_b/(\gamma u(b) + \delta u'(b))$. With these choices for c_1 and c_2, y_2 solves the boundary value problem above and $y = y_1(x) + y_2(x)$ solves the inhomogeneous boundary value problem (6.10). ■

If $y(x)$ is a solution of the homogeneous boundary value problem (6.11), then $\lim_{x\to a} p(x)y'(x) = 0$ by Lemma 160. Here is a companion result that will be needed later when we study Green's functions.

Theorem 168 *If the homogeneous boundary value problem (6.11) has only the trivial solution and y is the unique solution to the inhomogeneous boundary value problem (6.10), then* $\lim_{x\to a} p(x)y'(x) = 0$.

Proof. With $u(x) = (x-a)^\nu z(x)$ and $v(x) = (x-a)^{-\nu}\tilde{z}(x)$ as in the proof of Theorem 167, from (6.14)

$$\begin{aligned} p(x)y'(x) &= p(x)v'(x)\int_a^x u(s)f(s)\,ds + p(x)u'(x)\int_x^b v(s)f(s)\,ds \\ &= I + II \end{aligned}$$

for $a < x \le b$. We claim that I and II have limit 0 as x approaches a. Consider I. For $x > a$ and near a, $u(x) \neq 0$, $pv' = \big(pu'v - 1\big)/u$ because $p(u'v - uv') = 1$, and

$$\begin{aligned} I &= \frac{p(x)u'(x)v(x) - 1}{u(x)}\int_a^x u(s)f(s)\,ds \\ &= \frac{p(x)u'(x)v(x)}{u(x)}\int_a^x u(s)f(s)\,ds - \frac{1}{u(x)}\int_a^x u(s)f(s)\,ds \\ &= A - B. \end{aligned}$$

A and B both have limit 0 as x tends to a: by the mean value theorem for integrals

$$\int_a^x u(s)f(s)\,ds = z(\xi_x)f(\xi_x)\int_a^x (s-a)^\nu\,ds = z(\xi_x)f(\xi_x)\frac{(x-a)^{\nu+1}}{\nu+1}$$

for some ξ_x between a and x and

$$B = \frac{1}{u(x)}\int_a^x u(s)f(s)\,ds = \frac{1}{(x-a)^\nu z(x)}z(\xi_x)f(\xi_x)\frac{(x-a)^{\nu+1}}{\nu+1} \to 0$$

as $x \to a$. Next

$$\begin{aligned} \frac{p(x)u'(x)v(x)}{u(x)} &= \frac{(x-a)\varphi(x)\Big[(x-a)^\nu z'(x) + \nu(x-a)^{\nu-1}z(x)\Big]v(x)}{(x-a)^\nu z(x)} \\ &= \frac{\varphi(x)\big[(x-a)z'(x) + \nu z(x)\big]v(x)}{z(x)} \\ &= \frac{\varphi(x)\big[(x-a)z'(x) + \nu z(x)\big]\tilde{z}(x)}{z(x)}(x-a)^{-\nu} \end{aligned}$$

and

$$\begin{aligned} A &= \frac{p(x)u'(x)v(x)}{u(x)}\int_a^x u(s)f(s)\,ds \\ &= \frac{\varphi(x)\big[(x-a)z'(x) + \nu z(x)\big]\tilde{z}(x)}{z(x)}(x-a)^{-\nu}z(\xi_x)f(\xi_x)\frac{(x-a)^{\nu+1}}{\nu+1} \end{aligned}$$

which again has limit 0 as x tends to a. Thus $I \to 0$ as $x \to a$.

Consider II. By the mean value theorem for integrals

$$\int_x^b v(s)f(s)\,ds = \tilde{z}(\xi_x)f(\xi_x)\int_x^b (s-a)^{-\nu}\,ds$$

for some ξ_x between x and b and

$$\begin{aligned} p(x)u'(x) &= (x-a)\varphi(x)\Big[(x-a)^\nu z'(x) + \nu(x-a)^{\nu-1}z(x)\Big] \\ &= \varphi(x)\big[(x-a)z'(x) + \nu z(x)\big](x-a)^\nu. \end{aligned}$$

So

$$II = \varphi(x)\big[(x-a)z'(x) + \nu z(x)\big]\tilde{z}(\xi_x)f(\xi_x)(x-a)^{\nu}\int_x^b (s-a)^{-\nu}\,ds.$$

The limit as x approaches a of the last two factors on the right is 0 as we saw near the end of the preceding proof. Thus, $II \to 0$ as $x \to a$. Combine results to find that there exists

$$\lim_{x\to a} p(x)y'(x) = 0.$$

■

Not surprisingly, the smoothness of a solution to (6.10) at the singularity depends on whether $q(x)$ is continuous at $x = a$ or has a pole at $x = a$. In the first case, Chapter 5, the solution $y(x)$ is continuously differentiable on $[a, b]$. In the second case, Chapter 6, we have shown that the solution $y(x)$ is continuous on $[a, b]$ with $y(a) = 0$ but no further smoothness can be guaranteed at $x = a$.

Example 1. Consider the Sturm-Liouville boundary value problem

$$\begin{cases} -(xy')' + \dfrac{1-4x^2}{4x}y = \dfrac{2}{\pi} - \dfrac{1}{2\pi} + \dfrac{2}{\pi}x^2 & \text{for } 0 < x < \dfrac{\pi}{2}\cdot \\ |y(0)| < \infty,\ y(\pi/2) = 0 \end{cases}$$

The singular differential equation in the corresponding homogeneous problem

$$\begin{cases} (xy')' - \dfrac{1/4}{x}y + xy = 0 & \text{for } 0 < x < \pi/2 \\ |y(0)| < \infty,\ y(\pi/2) = 0 \end{cases},$$

is Bessel's equation of order 1/2. Its general solution is

$$y = c_1 J_{1/2}(x) + c_2 J_{-1/2}(x) = c_1\sqrt{\frac{2}{\pi}}x^{-1/2}\sin x + c_2\sqrt{\frac{2}{\pi}}x^{-1/2}\cos x.$$

Apparently, the only solution to the homogeneous boundary value problem is the trivial solution so the given boundary value problem has a unique solution. In this case, the inhomogeneous problem was chosen to have solution

$$y = \frac{\pi}{2}J_{1/2}(x) - \frac{2}{\pi}x = \sqrt{\frac{\pi}{2}}x^{-1/2}\sin x - \frac{2}{\pi}x$$

for $0 < x \le \pi/2$. The solution clearly extends to a continuous function on $[0, \pi/2]$ with $y(0) = 0$. Evidently the solution is not differentiable at $x = 0$.

6.3 Green's Functions

The standing assumptions on page 250 remain in force throughout this section, and, as noted earlier, since all the data is real-valued, we can restrict our attention to *real-valued* solutions of the boundary value problems that follow.

When a unique solution exists, the motivational argument in Section 1.10 used for regular Sturm-Liouville boundary value problems shows that it is reasonable to expect that the solution to (6.10) with $c_b = 0$ can be expressed in terms of a Green's function $g(x, s)$ by

$$y(x) = \int_a^b g(x, s)f(s)\,ds.$$

Specifically, $g(x, s)$ is a **Green's function** for the singular Sturm-Liouville problem (6.10) with $c_b = 0$ if $g(x, s)$ is defined and continuous on the square $a \leq x, s \leq b$ with the point (a, a) removed and

$$y(x) = \int_a^b g(x, s)f(s)\,ds, \quad a \leq x \leq b,$$

uniquely solves (6.10) with $c_b = 0$ for every function $f(x)$ that is continuous on $[a, b]$. In contrast to Chapter 5 where the Green's function has a logarithmic singularity as (x, s) approaches (a, a), the Green's function in Chapter 6 remains bounded on its domain. However, there is no continuous extension of the Green's function to the full square $[a, b] \times [a, b]$. These assertions will be established as we go along. In this situation, we could append (a, a) to the domain of the Green's function and define $g(a, a)$ arbitrarily to obtain a Green's function defined on the full square, but it seems more natural not to do this.

The integral

$$\int_a^b g(x, s)f(s)\,ds$$

exists as an ordinary Riemann integral for $a < x \leq b$ because the integrand $g(x, s)f(s)$ is a continuous function of s in $[a, b]$. When $x = a$ the integrand $g(a, s)f(s)$ is only defined on $a < s \leq b$ and the integral is interpreted as an improper Riemann integral

$$\int_a^b g(a, s)f(s)\,ds = \lim_{a' \to a} \int_{a'}^b g(a, s)f(s)\,ds.$$

We will establish shortly that the improper integral converges. In fact, the limit is 0.

The Green's function is defined through the boundary value problem (6.10) with $c_b = 0$; however, once the Green's function has been found, it can be used to express the solution to the boundary value problem also when $c_b \neq 0$. That representation is given later in the chapter.

Once the Green's function is found, the representation makes it possible to investigate how different forcing terms $f(x)$ effect behavior of the solution. Also, properties of the solution that are not apparent from the boundary value problem itself often can be deduced from the Green's function representation and properties of the Green's function.

Theorem 169 *If the singular boundary value problem (6.10) with $c_b = 0$ has a Green's function, then the Green's function is unique and must be real-valued.*

Proof. The uniqueness proof is the same as for Theorem 141. If $g(x, s) = g_1(x, s) + ig_2(x, s)$ where g_1 and g_2 are real-valued, then separating

$$y(x) = \int_a^b g(x, s)f(s)\,ds,$$

into real and imaginary parts and using the fact that a solution $y(x)$ is real-valued gives

$$0 = \int_a^b g_2(x, s)f(s)\,ds$$

for every continuous $f(x)$ on $[a, b]$. By the version of Corollary 20 for improper integrals it follows that $g_2(x, s) = 0$ on $[a, b] \times [a, b] \backslash \{(a, a)\}$ and $g(x, s) = g_1(x, s)$ is real-valued. ■

Theorem 170 *The Sturm-Liouville boundary value problem (6.10) with* $c_b = 0$ *has a Green's function* $g(x, s)$ *if and only if the corresponding homogeneous problem (6.11) has only the trivial solution.*

Proof. If there is a Green's function $g(x, s)$, then

$$y(x) = \int_a^b g(x, s)f(s)\, ds$$

is the unique solution to (6.10) with $c_b = 0$. In particular, if $f = 0$, the unique solution is $y = 0$; that is, (6.11) has only the trivial solution.

Conversely, if (6.11) has only the trivial solution, then the reasoning used in the proof of Theorem 167 shows that the boundary value problem (6.10) with $c_b = 0$ and $f(x)$ a given continuous function on $[a, b]$ has the unique solution

$$y(x) = \begin{cases} 0 & \text{for } x = a \\ v(x) \int_a^x u(s)f(s)\, ds + u(x) \int_x^b v(s)f(s)\, ds & \text{for } a < x \leq b \end{cases}$$

where u and v are the functions used in the proof of Theorem 167. Those functions satisfy

$$Lu = 0, \quad a < x \leq b,$$
$$|u(a)| < \infty,$$

$$Lv = 0, \quad a < x \leq b,$$
$$\gamma v(b) + \delta v'(b) = 0,$$

$$p(u'v - uv') = 1 \quad \text{for } a < x \leq b,$$

$$u(x) = (x - a)^{\nu} z(x) \quad \text{with } z \text{ in } C^1[a, b], \quad \text{and} \quad z(a) \neq 0,$$

and

$$v(x) = (x - a)^{-\nu} \tilde{z}(x) \quad \text{with } \tilde{z} \in C[a, b], \quad \text{and} \quad \tilde{z}(a) \neq 0,$$

where $\nu = \sqrt{q_1(a)/\varphi(a)} > 0$.

Define

$$g(x, s) = \begin{cases} u(x)v(s) & \text{for } a \leq x \leq s \leq b \ \text{ and } \ (x, s) \neq (a, a) \\ v(x)u(s) & \text{for } a \leq s \leq x \leq b \ \text{ and } \ (x, s) \neq (a, a) \end{cases} \tag{6.15}$$

We will show that this is the Green's function for (6.10) with $c_b = 0$. The function $g(x, s)$ is clearly continuous on $[a, b] \times [a, b] \setminus \{(a, a)\}$. It follows directly from the definition of $g(x, s)$ that

$$\int_a^b g(x, s)f(s)\, ds = v(x) \int_a^x u(s)f(s)\, ds + u(x) \int_x^b v(s)f(s)\, ds$$

for $a < x \leq b$ and that

$$\int_a^b g(a, s)f(s)\, ds = \lim_{a' \to a} \int_{a'}^b u(a)v(s)f(s)\, ds = 0$$

because $u(a) = 0$. Consequently, the two-part formula above for the solution $y(x)$ of (6.10) with $c_b = 0$ and a given right member $f(x)$ can be expressed as

$$y(x) = \int_a^b g(x, s)f(s)\, ds$$

for $a \leq x \leq b$ and $g(x, s)$ is the Green's function for (6.10) with $c_b = 0$. ■

Theorem 171 *The singular Sturm-Liouville boundary value problem (6.10) with $c_b = 0$ has a Green's function $g(x, s)$ if and only if there exist functions u and v with u continuous on $a \leq x \leq b$, v continuously differentiable on $a < x \leq b$,*

$$\begin{cases} Lu = 0 \quad \text{for } a < x < b \\ |u(a)| < \infty \end{cases}, \tag{6.16}$$

$$\begin{cases} Lv = 0 \quad \text{for } a < x < b \\ \gamma v(b) + \delta v'(b) = 0 \end{cases}, \tag{6.17}$$

and

$$p(x)\, W_{u,v}(x) = -1 \quad \text{for } a < x < b, \tag{6.18}$$

in which case

$$g(x, s) = \begin{cases} u(x)v(s) & \text{for } a \leq x \leq s \leq b \quad \text{and} \quad (x, s) \neq (a, a) \\ v(x)u(s) & \text{for } a \leq s \leq x \leq b \quad \text{and} \quad (x, s) \neq (a, a) \end{cases} \tag{6.19}$$

and (6.10) with $c_b = 0$ has the unique solution

$$y(x) = \int_a^b g(x, s) f(s)\, ds \quad \text{for } a \leq x \leq b. \tag{6.20}$$

Proof. If the Green's function exists, the corresponding homogeneous boundary value problem has only the trivial solution and the proofs of Theorem 167 and 170 establish that functions $u(x)$ and $v(x)$ exist with the stated continuity and differentiability properties, that satisfy (6.16), (6.17), (6.18), and that the Green's function is given by (6.19).

Assume now that functions $u(x)$ and $v(x)$ exist that satisfy (6.16), (6.17), and (6.18). By Lemma 160, $u(x)$ extends to a continuous function on $[a, b]$, is continuously differentiable on $a < x \leq b$, and satisfies the singular Sturm-Liouville differential equation at $x = b$. By Theorem 164 parts (a) and (b), $u(x) = (x - a)^{\nu} z(x)$ where $z(a) \neq 0$, $z(x)$ is continuously differentiable on $[a, b]$, and $\nu = \sqrt{q_1(a)/\varphi(a)}$. By parts (c) and (d) of that theorem, $v(x) = (x - a)^{-\nu} \tilde{z}(x)$ for $a < x \leq b$ where $\tilde{z}(x)$ is a continuous function on $[a, b]$ with $\tilde{z}(a) \neq 0$, and $v(x)$ satisfies the singular differential equation at $x = b$.

With these properties of $u(x)$ and $v(x)$ established and with $g(x, s)$ defined by (6.19), the reasoning used in the proof of Theorem 170 shows that $g(x, s)$ is the Green's function for (6.10) with $c_b = 0$. ∎

The following corollary will be needed later when we study Sturm-Liouville eigenvalue problems. See Section 3.7 for the definition of a mildly singular kernel.

Corollary 172 *The Green's function $g(x, s)$ determined by the singular Sturm-Liouville differential operator $Ly = -(py')' + qy$ and the boundary conditions $|y(a)| < \infty$, $\gamma y(b) + \delta y'(b) = 0$ is a mildly singular, symmetric kernel. Indeed,*

$$g(x, s) = h(x, s)(\min(x, s) - a)^{\nu}(\max(x, s) - a)^{-\nu}$$

on $[a, b] \times [a, b] \backslash \{(a, a)\}$ where $h(x, s) = h(s, x)$ is real-valued and continuous on $[a, b] \times [a, b]$, $h(a, a) \neq 0$, and $\nu = \sqrt{q_1(a)/\varphi(a)}$. Consequently, there is a constant $M < \infty$ such that

$$\int_a^b |g(x, s)|^2\, ds \leq M$$

for all x in $[a, b]$.

Proof. The two-part formula for $g(x, s)$ in Theorem 171 can be expressed as

$$g(x, s) = u(\min(x, s))v(\max(x, s))$$

for (x, s) in $[a, b] \times [a, b] \backslash \{(a, a)\}$. Clearly $g(x, s) = g(s, x)$ and $g(x, s)$ is real-valued because u and v are. So $g(x, s)$ is a symmetric kernel. Since $u(x) = (x - a)^{\nu} z(x)$ where $z(a) \neq 0$, $z(x)$ is continuously differentiable on $[a, b]$, and $v(x) = (x - a)^{-\nu} \tilde{z}(x)$ for $a < x \leq b$ where $\tilde{z}(x)$ is a continuous function on $[a, b]$ with $\tilde{z}(a) \neq 0$, $g(x, s)$ is continuous on $[a, b] \times [a, b] \backslash \{(a, a)\}$ and

$$\begin{aligned} g(x, s) &= z(\min(x, s))\tilde{z}(\max(x, s))(\min(x, s) - a)^{\nu}(\max(x, s) - a)^{-\nu} \\ &= h(x, s)(\min(x, s) - a)^{\nu}(\max(x, s) - a)^{-\nu} \end{aligned}$$

where $h(x, s) = z(\min(x, s))\tilde{z}(\max(x, s)) = h(s, x)$ is continuous on $[a, b] \times [a, b]$ and $h(a, a) = z(a)\tilde{z}(a) \neq 0$. Since

$$0 \leq \frac{\min(x, s) - a}{\max(x, s) - a} \leq 1$$

on $[a, b] \times [a, b] \backslash \{(a, a)\}$, $g(x, s)$ is bounded and continuous there. Let (x, s) tend to (a, a) along the line $s - a = m(x - a)$ with slope m, $0 < m < 1$, that lies in the lower triangle of $[a, b] \times [a, b]$ to find that $g(x, s)$ tends to $m^{\nu} h(a, a)$ along that line. Since $h(a, a) \neq 0$, $g(x, s)$ can have no continuous extension to the full square $[a, b] \times [a, b]$. Thus, $g(x, s)$ is a mildly singular kernel. The first assertion in the corollary is established.

The second assertion follows from the first because $|g(x, s)| \leq |h(x, s)|$ on $[a, b] \times [a, b] \backslash \{(a, a)\}$ and for x in $[a, b]$,

$$\int_a^b |g(x, s)|^2 \, ds \leq \int_a^b |h(x, s)|^2 \, ds \leq \|h\|_{\max}^2 (b - a)$$

where $\|h\|_{\max} = \max_{a \leq x, s \leq b} |h(x, s)|$. ■

Example 2. Determine when it exists and find the Green's function for the singular Sturm-Liouville boundary value problem

$$-(xy')' + \frac{n^2}{x} y - xy = f(x), \quad 0 < x < l,$$

$$|y(0)| \leq \infty, \quad \gamma y(l) + \delta y'(l) = 0,$$

where the differential equation is Bessel's equation of integral order $n \geq 1$. The corresponding homogeneous equation has the Bessel functions $J_n(x)$ and $Y_n(x)$ as linearly independent solutions. Since $J_n(x)$ is bounded on $[0, l]$, we can choose $u = J_n(x)$ in Theorem 171. Since $Y_n(x)$ is unbounded, the corresponding homogeneous problem will have only the trivial solution if and only if

$$\gamma J_n(l) + \delta J_n'(l) \neq 0.$$

The Green's function exists if and only if this inequality is satisfied. We seek a solution v in Theorem 171 of the form $v = cJ_n(x) + Y_n(x)$. Such a v satisfies the boundary condition at $x = l$ if

$$c = -\frac{\gamma Y_n(l) + \delta Y_n'(l)}{\gamma J_n(l) + \delta J_n'(l)}.$$

With this choice for c the Green's function is

$$g(x, s) = \begin{cases} J_n(x)\tilde{Y}_n(s) & \text{for } 0 \leq x \leq s \leq l \text{ and } (x, s) \neq (a, a) \\ \tilde{Y}_n(x)J_n(s) & \text{for } 0 \leq s \leq x \leq l \text{ and } (x, s) \neq (a, a) \end{cases}$$

where $\tilde{Y}_n(x) = cJ_n(x) + Y_n(x)$.

A closely related example involves the modified Bessel functions.

Example 3. Determine when it exists and find the Green's function for the singular Sturm-Liouville boundary value problem

$$-(xy')' + \frac{n^2}{x}y + xy = f(x), \quad 0 < x < l,$$
$$|y(0)| \leq \infty, \quad \gamma y(l) + \delta y'(l) = 0,$$

where the differential equation is the modified Bessel's equation of order n.

The corresponding homogeneous equation has the modified Bessel functions $I_n(x)$ and $K_n(x)$ as linearly independent solutions. Since $I_n(x)$ is bounded on $[0, l]$, we can choose $u = I_n(x)$ in Theorem 171. Since $K_n(x)$ is unbounded, the corresponding homogeneous problem will have only the trivial solution if and only if

$$\gamma I_n(l) + \delta I_n'(l) \neq 0.$$

The Green's function exists if and only if this inequality is satisfied. We seek a solution v in Theorem 171 of the form $v = cI_n(x) + K_n(x)$. Such a v satisfies the boundary condition at $x = l$ if

$$c = -\frac{\gamma K_n(l) + \delta K_n(l)}{\gamma I_n(l) + \delta I_n'(l)}.$$

With this choice for c the Green's function is

$$g(x, s) = \begin{cases} I_n(x)\tilde{K}_n(s) & \text{for } 0 \leq x \leq s \leq l \quad \text{and} \quad (x, s) \neq (a, a) \\ \tilde{K}_n(x)I_n(s) & \text{for } 0 \leq s \leq x \leq l \quad \text{and} \quad (x, s) \neq (a, a) \end{cases}$$

where $\tilde{K}_n(x) = cI_n(x) + K_n(x)$.

The Green's function $g(x, s)$ for $Ly = f$, $|y(a)| < \infty$, $\gamma y(b) + \delta y'(b) = 0$ has the following properties (when it exists) and these properties characterize the Green's function, in strict analogy to the regular case:

1. $g(x, s)$ is a bounded, continuous function on $[a, b] \times [a, b] \backslash \{(a, a)\}$ of the form
$$g(x, s) = h(x, s)(\min(x, s) - a)^{\nu}(\max(x, s) - a)^{-\nu}$$
where $h(x, s)$ is continuous on $[a, b] \times [a, b]$, $h(a, a) \neq 0$, and $\nu > 0$. Moreover, $g(x, s)$ has continuous partial derivatives on the upper triangle $(a < x \leq s \leq b)$ and on the lower triangle $(a < s \leq x \leq b)$ of $[a, b] \times [a, b] \backslash \{(a, a)\}$.
2. $g(x, s)$, regarded as a function of x for fixed s in $[a, b]$, satisfies the differential equation $Ly = 0$ for $x \neq s$ in (a, b).
3. $g(x, s)$, regarded as a function of x for fixed s in (a, b), satisfies the boundary conditions of the problem.
4. $g(x, s)$, regarded as a function of x for fixed s in (a, b), has a jump in its derivative with respect to x at $x = s$ given by
$$\frac{\partial g}{\partial x}(s+, s) - \frac{\partial g}{\partial x}(s-, s) = -\frac{1}{p(s)}.$$

A direct verification confirms that the Green's function in Theorem 171 has the four properties. The next lemma will be used in the proof that Properties 1-4 characterize the Green's function and also confirms that the Green's function has Property 1. We leave the verification of Properties 2, 3, and 4 to the reader. Once we establish that the four roperties characterize the Green's function, $g(x, s)$ must be the function in Theorem 171. Since that function satisfies $g(x, s) = g(s, x)$, Properties 1-4 hold with the roles of x and s interchanged.

Lemma 173 *(a) The Green's function $g(x, s)$ in Theorem 171 has Property 1.*

(b) If $g(x, s)$ is any function that has the form

$$g(x, s) = h(x, s)(\min(x, s) - a)^{\nu}(\max(x, s) - a)^{-\nu}$$

on $[a, b] \times [a, b] \backslash \{(a, a)\}$ where $h(x, s)$ is continuous on $[a, b] \times [a, b]$ and $h(a, a) \neq 0$, and $f(x)$ is any continuous function on $[a, b]$, then

$$\int_a^b g(x, s) f(x)\, dx$$

is a continuous function of s on $[a, b]$.

Proof. (a) From the proof of Corollary 172,

$$g(x, s) = h(x, s)(\min(x, s) - a)^{\nu}(\max(x, s) - a)^{-\nu}$$

on $[a, b] \times [a, b] \backslash \{(a, a)\}$ where $h(x, s) = z(\min(x, s))\tilde{z}(\max(x, s))$, $z(x)$ is continuously differentiable on $[a, b]$, $z(a) \neq 0$, $\tilde{z}(x)$ is continuous on $[a, b]$ and continuously differentiable on $a < x \leq b$, $\tilde{z}(a) \neq 0$, and $\nu > 0$. Thus, $g(x, s)$ has the required form and has continuous partial derivatives on the indicated triangles in $[a, b] \times [a, b] \backslash \{(a, a)\}$.

(b) Let

$$y(s) = \int_a^b g(x, s) f(x)\, dx$$

for $a \leq s \leq b$. Observe first that the integral defining $y(x)$ exists for each x in $[a, b]$. The integrand is a continuous function of x for each s in $a < s \leq b$ and the integral exists as a proper Riemann integral for such s. If $s = a$ the integrand is only defined on $a < x \leq b$ and is continuous there, the integral is improper, and

$$\begin{aligned} y(a) &= \int_a^b g(x, a) f(x)\, dx = \lim_{a' \to a} \int_{a'}^b g(x, a) f(x)\, dx \\ &= \lim_{a' \to a} \int_{a'}^b h(x, a)(\min(x, a) - a)^{\nu}(\max(x, a) - a)^{-\nu} f(x)\, dx \\ &= \lim_{a' \to a} \int_{a'}^b h(x, a)(0)(x - a)^{-\nu} f(x)\, dx = 0. \end{aligned}$$

It remains to show that $y(x)$ is continuous on $[a, b]$. Since the integrand $g(x, s)f(x)$ is continuous on $[a, b] \times [a', b]$ for any a' with $a < a' < b$, it follows from Proposition 18 that $y(s)$ is continuous on $[a', b]$. Since $a' > a$ can be chosen arbitrarily, it follows that $y(s)$ is continuous on $a < s \leq b$.

Finally, we establish that $y(s)$ is continuous at $s = a$. For $a < s < b$,

$$\begin{aligned}
y(s) &= \int_a^b g(x,s)f(x)\,dx \\
&= \int_a^b h(x,s)(\min(x,s)-a)^{\nu}(\max(x,s)-a)^{-\nu} f(x)\,dx \\
&= \int_a^b (h(x,s)-h(x,a))(\min(x,s)-a)^{\nu}(\max(x,s)-a)^{-\nu} f(x)\,dx \\
&\quad + \int_a^b h(x,a)(\min(x,s)-a)^{\nu}(\max(x,s)-a)^{-\nu} f(x)\,dx \\
&= I + II.
\end{aligned}$$

We claim that, $I \to 0$ and $II \to 0$ as $s \to a$; hence,

$$\lim_{s \to a} y(s) = 0 = y(a)$$

and $y(s)$ is continuous at $s = a$.

To establish that $I \to 0$ as $s \to a$, let $\varepsilon > 0$ be given. By the uniform continuity of $h(x,s)$ on $[a,b] \times [a,b]$ there is a $\delta > 0$ such that

$$|h(x,s) - h(x,a)| < \varepsilon$$

for all $a \leq x \leq b$ and $a < s < a + \delta$. Therefore, for $a < s < a + \delta$,

$$|I| \leq \varepsilon \|f\|_{\max} \int_a^b \left|(\min(x,s)-a)^{\nu}(\max(x,s)-a)^{-\nu}\right| dx \leq \varepsilon \|f\|_{\max}(b-a).$$

Hence, $I \to 0$ as $s \to a$.

To show that $II \to 0$ as $s \to a$, express II as

$$\begin{aligned}
II &= \int_a^s h(x,a)(\min(x,s)-a)^{\nu}(\max(x,s)-a)^{-\nu} f(x)\,dx \\
&\quad + \int_s^b h(x,a)(\min(x,s)-a)^{\nu}(\max(x,s)-a)^{-\nu} f(x)\,dx \\
&= \int_a^s h(x,a)(x-a)^{\nu}(s-a)^{-\nu} f(x)\,dx \\
&\quad + \int_s^b h(x,a)(s-a)^{\nu}(x-a)^{-\nu} f(x)\,dx.
\end{aligned}$$

The first summand of II is bounded in absolute value by

$$\|h\|_{\max}\|f\|_{\max}(s-a)$$

because $\left|(x-a)^{\nu}(s-a)^{-\nu}\right| \leq 1$ and has limit 0 as s tends to a. By the mean value theorem of integral calculus (Theorem 15), the second summand can be expressed as

$$\int_s^b h(x,a)(s-a)^{\nu}(x-a)^{-\nu} f(x)\,dx = h(\xi_x,a)f(\xi_x)(s-a)^{\nu} \int_s^b (x-a)^{-\nu}\,dx$$

for some ξ_x between s and b. Since

$$(s-a)^{\nu} \int_s^b (x-a)^{-\nu}\,dx = \begin{cases} (s-a)^{\nu} \dfrac{(b-a)^{-\nu+1} - (s-a)^{-\nu+1}}{-\nu+1} & \text{if } \nu \neq 1 \\ (s-a)(\ln(b-a) - \ln(s-a)) & \text{if } \nu = 1 \end{cases}$$

has limit 0 as $s \to a$, the second summand in II has limit 0 as $s \to a$. Thus, $II \to 0$ as $s \to a$ and (b) of the lemma is proved. ■

Properties 1-4 above characterize the Green's function.

Theorem 174 *If a function $g(x, s)$ exists with Properties 1-4, then $Ly = 0$, $|y(a)| < \infty$, $\gamma y(b) + \delta y'(b) = 0$ has only the trivial solution and $g(x, s)$ is the Green's function for the differential operator Ly and boundary conditions $|y(a)| \leq \infty$, $\gamma y(b) + \delta y'(b) = 0$. Moreover, $g(x, s) = g(s, x)$.*

Proof. Let $B_b y = \gamma y(b) + \delta y'(b)$. Fix s with $a < s < b$ and define functions z_1 and z_2 by

$$z_1(x) = g(x, s) \quad \text{for } a \leq x \leq s \quad \text{and} \quad z_2(x) = g(x, s) \quad \text{for } s \leq x \leq b.$$

Both $z_1(x)$ and $z_2(x)$ are continuous on their domains by Property 1. By Properties 2 and 3, $z_1(x)$ satisfies $Lz_1 = 0$ on $a < x < s$, $|z_1(a)| < \infty$ and $z_2(x)$ satisfies $Lz_2 = 0$ on $s < x < b$, $B_b z_2 = 0$. By Lemma 165 z_1 is a continuously differentiable on $(a, s]$ and satisfies the differential equation there. Since z_2 satisfies the regular Sturm-Liouville problem $Lz_2 = 0$ on (s, b), $z_2(s) = g(s, s)$, $B_b z_2 = 0$, it is continuously differentiable on $[s, b]$ and satisfies the differential equation there.

We show first that $Ly = 0$, $|y(a)| < \infty$, $B_b y = 0$ has only the trivial solution. Assume the contrary and let $z(x)$ be a nontrivial solution. Since

$$Lz = 0 \quad \text{for } a < x < s, \ |z(a)| < \infty,$$

and

$$Lz_1 = 0 \quad \text{for } a < x < s, \ |z_1(a)| < \infty,$$

by Theorem 164(a, b) applied on the interval $[a, s]$, $z_1(x)$ is a multiple of $z(x)$. Thus, $z_1(x) = c_1(s)z(x)$ on $a \leq x \leq s$ for some scalar $c_1(s)$ that depends on the fixed value of s.

Since

$$\gamma z(b) + \delta z'(b) = 0,$$
$$\gamma z_2(b) + \delta z_2'(b) = 0,$$

and $|\gamma| + |\delta| \neq 0$, the determinant of the 2×2 system $W_{z,z_2}(b) = 0$ and z and z_2 are linearly dependent solutions on $[s, b]$. Thus,

$$d(s)z(x) + d_2(s)z_2(x) = 0$$

for x in $[s, b]$, where $d(s)$ and $d_2(s)$ are scalars, not both 0, whose value depends on the fixed value of s in (a, b). If $d_2(s) = 0$, then $z(x) = 0$ on $[s, b]$ and $z(x)$ solves the initial value problem $Lz = 0$ on (a, b), $z(s) = 0$, $z'(s) = 0$. Thus, $z(x) = 0$ on (a, b) by the uniqueness of solutions to initial value problems. This contradicts the fact that $z(x)$ is a nontrivial solution. Consequently, $d_2(s) \neq 0$ and $z_2(x) = c_2(s)z(x)$ on $s \leq x \leq b$ where $c_2(s) = -d(s)/d_2(s)$. Thus,

$$c_2(s)z(s) = z_2(s) = g(s, s) = z_1(s) = c_1(s)z(s).$$

Since z is nontrivial, there exist s_0 in (a, b) where $z(s_0) \neq 0$; hence, $c_1(s_0) = c_2(s_0)$ and

$$g_x(s_0+, s_0) - g_x(s_0-, s_0) = c_2(s_0)z'(s_0) - c_1(s_0)z'(s_0) = 0,$$

which contradicts the jump condition in Property 4. Hence, $Ly = 0$, $|y(a)| < \infty$, $B_b y = 0$ has only the trivial solution and $Ly = f$, $B_a y = 0$, $B_b y = 0$ has a unique solution y for each function f in $C[a, b]$.

Finally we establish that a function $g(x, s)$ with Properties 1-4 is the Green's function. To this end, for any continuous function f, let y be the unique solution to $Ly = f$, $|y(a)| < \infty$, $B_b y = 0$, which exists by Theorem 167. Fix s in (a, b), regard $g(x, s)$ as a function of x in $[a, b]$ and let $a < c < r < s < t < b$. By Property 2

$$0 = \int_c^r yLg\,dx = \int_c^r y(-pg')'\,dx + \int_c^r yqg\,dx.$$

Integration by parts gives

$$\begin{aligned} 0 &= -ypg'\big|_c^r + \int_c^r pg'y'\,dx + \int_c^r qyg\,dx \\ &= -ypg'\big|_c^r + py'g\big|_c^r - \int_c^r g(py')'\,dx + \int_c^r qyg\,dx \\ &= (py'g - ypg')\big|_c^r + \int_c^r gLy\,dx \\ &= (py'g - ypg')\big|_c^r + \int_c^r gf\,dx. \end{aligned}$$

Thus,

$$-(py'g - ypg')\big|_c^r = \int_c^r gf\,dx.$$

In the same way,

$$-(py'g - ypg')\big|_t^b = \int_t^b gf\,dx.$$

Since s is fixed with $a < s < b$, g is continuous in x, $\lim_{c\to a} p(c)y'(c) = 0$ by Theorem 168, $\lim_{c\to a} p(c)g'(c) = 0$ by Lemma 160, and y is continuous on $a \le x \le b$, the evaluation at the lower limit as $c \to a$ gives 0. Let $r \to s$ to obtain

$$-(py'g - ypg')\big|_{x=s-} = \int_a^s gf\,dx.$$

Since

$$\begin{cases} \gamma y(b) + \delta y'(b) = 0 \\ \gamma g(b) + \delta g'(b) = 0 \end{cases}$$

with $|\gamma| + |\delta| > 0$, the determinant of the 2×2 system is 0 and the contribution to the evaluated term above at $x = b$ is 0. Let $t \to s$ to obtain

$$(py'g - ypg')\big|_{x=s+} = \int_s^b gf\,dx.$$

Combining these results gives

$$(py'g - ypg')\big|_{x=s-}^{s+} = \int_a^b gf\,dx.$$

Since $a < s < b$, p, y', and g are continuous in x near $x = s$. Hence, $py'g$ is continuous near $x = s$ and

$$(-ypg')\big|_{x=s-}^{s+} = \int_a^b gf\,dx.$$

By the jump condition (Property 4)

$$-ypg'\Big|_{s-}^{s+} = -y(s)p(s)\big(g_x(s+, s) - g_x(s-, s)\big) = y(s)$$

and

$$y(s) = \int_a^b g(x, s)f(x)\, dx.$$

for s in (a, b). Since $y(s)$ is continuous on $[a, b]$ and the integral on the right is continuous on $[a, b]$ by Lemma 173, the equality also holds at $s = a$ and $s = b$. By definition $g(x, s)$ is the Green's function for the differential operator $Ly = -(py')' + qy$ and the boundary conditions $|y(a)| < \infty$ and $B_b y = 0$. By uniqueness it must be given by the formula in Theorem 171 which shows that $g(s, x) = g(x, s)$. ■

If the fully inhomogeneous problem (6.10) has a unique solution, it can be expressed directly in terms of the Green's function for $Ly = f$, $|y(a)| < \infty$, $B_b y = 0$. Suppose that $Ly = 0$, $|y(a)| < \infty$, $B_b y = 0$ has only the trivial solution so that $Ly = f$, $|y(a)| < \infty$, $B_b y = c_b$ has a unique solution that we will denote by y and let $g(x, s)$ be the Green's function for $Ly = f$, $|y(a)| < \infty$, $B_b y = 0$. Fix x in (a, b), regard $g(x, s)$ as a function of s, denote derivatives with respect to s by primes, and use Properties 1-4 with the roles of x and s interchanged exactly as we did in the foregoing proof to obtain

$$-(py'g - ypg')\Big|_{s=c}^{r} = \int_c^r gf\, ds$$

and

$$-(py'g - ypg')\Big|_{s=t}^{b} = \int_t^b gf\, ds$$

for $a < c < r < x < t < b$. Let $c \to a$ and then $r \to x$ to obtain

$$-(py'g - ypg')\Big|_{s=x-} = \int_a^x gf\, ds.$$

Likewise, let $t \to x$ to get

$$-(py'g - ypg')\Big|_{s=x+}^{b} = \int_x^b gf\, ds$$

and combine results to find that

$$-(py'g - ypg')\Big|_{s=b} + (py'g - ypg')\Big|_{x-}^{x+} = \int_a^b gf\, ds.$$

As before, p, y', and g are continuous in s for s near x so that

$$(-ypg')\Big|_{s=x-}^{x+} = (py'g - ypg')\Big|_{s=b} + \int_a^b gf\, ds.$$

By the jump condition (Property 4 with the roles of x and s interchanged)

$$(-ypg')\Big|_{s=x-}^{x+} = -y(x)p(x)\big(g_s(x, x+) - g_s(x, x-)\big) = y(x).$$

Thus,

$$y(x) = p(y'g - yg')\Big|_{s=b} + \int_a^b gf\, ds.$$

Since y satisfies an inhomogeneous boundary condition at $x = b$ instead of the corresponding homogeneous boundary condition, the evaluation at $x = b$ is different from before. At $x = b$ the functions y and g satisfy

$$\begin{cases} \gamma y(b) + \delta y'(b) = c_b \\ \gamma g(b) + \delta g'(b) = 0 \end{cases}$$

and, by the standard elimination process,

$$\gamma \Delta(x, b) = -c_b g'(b) \quad \text{and} \quad \delta \Delta(x, b) = c_b g(b),$$

where

$$\Delta(x, s) = y'(s) g(x, s) - y(s) g'(x, s)$$

and primes indicates derivatives with respect to s. Using these results in the formula for $y(x)$ above yields

Theorem 175 *If $g(x, s)$ is the Green's function determined by the Sturm-Liouville differential operator $Ly = -(py')' + qy$ and the separated boundary conditions $|y(a)| < \infty$, $\gamma y(b) + \delta y'(b) = 0$, then the Sturm-Liouville boundary value problem (6.10) has the unique solution*

$$y(x) = p(b)\Delta(x, b) + \int_a^b g(x, s)\, f(s)\, ds,$$

where

$$\Delta(x, b) = \begin{cases} -c_b g_s(x, b)/\gamma & \text{if } \gamma \neq 0 \\ c_b g(x, b)/\delta & \text{if } \gamma = 0 \end{cases}$$

for x in $[a, b]$.

Proof. The formula for $y(x)$ was established for $a < x < b$. Both members of the formula are continuous on the closed interval $[a, b]$; therefore, the formula also holds at $x = a$ and $x = b$. ■

6.4 Eigenvalue Problems

The standing assumptions on page 250 remain in force throughout this section and are augmented by an assumption about the weight functions that may occur in the eigenvalue problems:

> A weight function $r(x)$ is a continuous function on $[a, b]$ and either $r(x) > 0$ on $[a, b]$ or $r(x) = (x - a)^m \rho(x)$ where $m > 0$ is a real number and $\rho(x) > 0$ on $a \leq x \leq b$.

As usual, $C[a, b]$ is the space of continuous functions on $[a, b]$ equipped with the maximum norm and

$$\langle y, z \rangle = \int_a^b y(x)\overline{z(x)}\, dx$$

is an inner product on $C[a, b]$. The weight function $r(x)$ also determines an inner product on $C[a, b]$ by

$$\langle y, z \rangle_r = \int_a^b y(x)\overline{z(x)} r(x)\, dx.$$

The functions y and z are orthogonal with respect to the weight function r if $\langle y, z\rangle_r = 0$.

All of the foregoing assumptions are satisfied by the eigenvalue problem for Bessel's equation of order $n > 0$,

$$\begin{cases} -(xy')' + (n^2/x)y = \lambda xy \quad 0 < x < b & 0 < x < b, \\ |y(0)| < \infty, & \\ \gamma y(b) + \delta y'(b) = 0, & |\gamma| + |\delta| \neq 0. \end{cases}$$

which serves as a model for the type of eigenvalue problems that follow.

The eigenvalue problem for a singular Sturm-Liouville differential equation is

$$\begin{cases} -(p(x)y')' + q(x)y = \lambda r(x)y, & a < x < b, \\ |y(a)| < \infty,\ \gamma y(b) + \delta y'(b) = 0, & \end{cases} \tag{6.21}$$

or, more compactly,

$$Ly = \lambda ry,\ |y(a)| < \infty,\ B_b y = 0,$$

where $Ly = -(py')' + qy$, and $B_b y = \gamma y(b) + \delta y'(b)$.

The eigenvalue problem for Bessel's equation of order n and parameter λ involves a weight function with a simple zero at 0.

Just as in Chapter 5, a real or complex number λ is an **eigenvalue** of a Sturm-Liouville eigenvalue problem and a real or complex-valued function $y \neq 0$ is a corresponding **eigenfunction** if (6.21) is satisfied for the pair λ and y. We also say the eigenfunction y belongs to the eigenvalue λ. When we say y satisfies (6.21), we mean that y satisfies the differential equation on (a, b), satisfies the given boundary conditions, and is continuous on $[a, b]$. The rationale for the continuity requirement is the same as for solutions to singular boundary value problems; see Section 6.2. As for boundary value problems, this definition implies further smoothness for y. Since an eigenfunction y is a bounded solution of

$$-(py')' + (q - \lambda r)y = 0$$

on (a, b) that is continuous on $[a, b]$, Lemma 160 implies

Lemma 176 *If $y(x)$ is an eigenfunction of (6.21), then $y(x)$ is continuous on $[a, b]$, $y(a) = 0$, $\lim_{x\to a} p(x)y'(x) = 0$, and $y(x)$ is continuously differentiable on $a < x \leq b$ and satisfies the Sturm-Liouville differential equation there.*

If y is an eigenfunction of (6.21), then $\lim_{x\to a} p(x)y'(x) = 0$ and y is continuous on $[a, b]$. Since $Ly = \lambda ry$ for $a < x < b$ and the right member is continuous on $[a, b]$, it follows that Ly, which is defined initially for $a < x < b$, is continuous on that interval and has a unique extension by continuity to a continuous function on $[a, b]$. We denote the extended function by Ly for simplicity. Thus, for the study of eigenvalue problems, it is natural to take the domain of L to be the set

$$D = \left\{ y \in C[a, b] : Ly \in C[a, b] \quad \text{and} \quad \lim_{x\to a} p(x)y'(x) = 0 \right\},$$

with a slight abuse of notation: $Ly \in C[a, b]$ means Ly is continuous on (a, b) and has a unique extension by continuity to the closed interval $[a, b]$, with the extended function still denoted by Ly. The domain of L is an inner product space with the usual inner product $\langle y, z\rangle$.

Lemma 177 *(a) Every eigenfunction y of (6.21) is in the domain of L.*
(b) If y and z are in the domain of L and satisfy the boundary conditions $B_b y = 0$ and $B_b z = 0$, respectively, then $\langle Ly, z\rangle = \langle y, Lz\rangle$.

Proof. (a) Clearly any eigenfunction y of (6.21) is in the domain of L by the previous lemma and the observations following it.

(b) Let y and z be in the domain D of L and satisfy the boundary conditions $B_b y=0$ and $B_b z=0$. For $a<c<b$ the usual integration by parts argument gives

$$\int_c^b (Ly)\bar{z}\,dx = p(\bar{z}'y-\bar{z}y')\Big|_c^b + \int_c^b y\overline{(Lz)}\,dx$$

Since y and z satisfy the same separated boundary conditions at $x=b$, the contribution at the upper limit is 0 by a now familiar argument. Since y and z are in the domain of L, the contribution at the lower limit tends to 0 as $c\to a$. Let $c\to a$ to obtain

$$\int_a^b (Ly)\bar{z}\,dx = \int_a^b y\overline{(Lz)}\,dx.$$

Thus, $\langle Ly,\, z\rangle = \langle y,\, Lz\rangle$. ■

Since $\langle Ly,\, z\rangle = \langle y,\, Lz\rangle$ for all y and z in the domain of L that satisfy the given boundary conditions, the eigenvalue problem (6.21) is self-adjoint and we have

Lemma 178 *Any eigenvalue of the self-adjoint Sturm-Liouville eigenvalue problem (6.21) is real and eigenfunctions belonging to distinct eigenvalue are orthogonal with respect to the weight function r. Each eigenvalue has a corresponding real-valued eigenfunction.*

Proof. If $Ly=\lambda ry$ with $y\neq 0$, then

$$\lambda\langle y,\, y\rangle_r = \langle \lambda ry,\, y\rangle = \langle Ly,\, y\rangle = \langle y,\, Ly\rangle = \langle y,\, \lambda ry\rangle = \bar{\lambda}\langle y,\, y\rangle_r.$$

Since $\langle y,\, y\rangle_r > 0$, it follows that $\lambda=\bar{\lambda}$ and λ is real. If $Lz=\mu rz$ with $z\neq 0$, then

$$\lambda\langle y,\, z\rangle_r = \langle \lambda ry,\, z\rangle = \langle Ly,\, z\rangle = \langle y,\, Lz\rangle = \langle y,\, \mu rz\rangle = \mu\langle y,\, z\rangle_r$$

because μ is real. If $\lambda\neq\mu$ then $\langle y,\, z\rangle_r = 0$. Since any eigenvalue of (6.21) is real, both the real and imaginary parts of any eigenfunction satisfy all the conditions in the eigenvalue problem and at least one of them is not identically zero. Consequently, each eigenvalue has a corresponding real-valued eigenfunction. ■

Theorem 179 *The eigenvalue problem (6.21) has at most a finite number of eigenvalues in any bounded region of the complex plane.*

Proof. Since the second order differential equation $Ly=\lambda ry$ for $a<x<b$ is expressible as the first order linear system $Z'=(A(x)+\lambda B(x))Z$ for $a<x<b$ where

$$Z=\begin{bmatrix} y \\ py' \end{bmatrix},\quad A(x)=\begin{bmatrix} 0 & 1/p \\ q & 0 \end{bmatrix},\quad B(x)=\begin{bmatrix} 0 & 0 \\ -r & 0 \end{bmatrix},$$

any solution $y(x,\lambda)$ to $Ly=\lambda ry$ is, for fixed x in (a,b), analytic in the complex variable λ for $|\lambda|<\infty$ as is $y'(x,\lambda)$ by Theorem 8.4 in Chapter 1 of [9] and the following application to linear systems. The same conclusion follows when applied to the differential equation $\hat{L}y=\lambda\hat{r}y$ for $a<x<\hat{b}$ for a fixed $\hat{b}>b$ and $\hat{L}y=-(\hat{p}y')'+\hat{q}y$ where $\hat{p}$, $\hat{q}$, and $\hat{r}$ extend p, q, and r to be constant on $[b,\,\hat{b}]$. Since $Ly=\lambda ry$ is

$$-(py')'+(q-\lambda r)y=0 \quad \text{for } a<x<b,$$

there is a nontrivial bounded solution $u(x, \lambda)$ to this equation that extends to a continuous function on $[a, b]$ that is continuously differentiable on $(a, b]$ and an unbounded solution $v(x, \lambda)$ that extends to a continuously differentiable function on $(a, b]$ by Theorem 164. Let $\hat{u}(x, \lambda)$ and $\hat{v}(x, \lambda)$ be the solutions to $\hat{L}y = \lambda\hat{r}y$ that have, respectively, the same initial data at $c = (a + b)/2$ that $u(x, \lambda)$ and $v(x, \lambda)$ have. The solutions $\hat{u}(x, \lambda)$ and $\hat{v}(x, \lambda)$ exist on $(a, \hat{b})$ and, by uniqueness of solutions to initial value problems, agree with $u(x, \lambda)$ and $v(x, \lambda)$ on (a, b) and, hence, on $(a, b]$ because all four solutions are continuous at $x = b$. Consequently, $u(b, \lambda) = \hat{u}(b, \lambda)$ and $v(b, \lambda) = \hat{v}(b, \lambda)$ are analytic functions of λ for $|\lambda| < \infty$.

Every solution to $Ly = \lambda ry$ can be expressed as a linear combination of $u(x, \lambda)$ and $v(x, \lambda)$; therefore, all nontrivial bounded solutions are nonzero multiples of $u(x, \lambda)$ and λ is an eigenvalue of $Ly = \lambda ry$ with corresponding eigenfunction a nonzero multiple of $u(x, \lambda)$ if and only if

$$\gamma u(b, \lambda) + \delta u'(b, \lambda) = 0.$$

The function on the left is analytic in $|\lambda| < \infty$. Such an analytic function is either identically equal to zero or has at most a finite number of zeros in any bounded region of the complex plane. See [6] or [28]. Since the eigenvalues of a self-adjoint Sturm-Liouville eigenvalue problem are real, it follows that the function $\gamma u(b, \lambda) + \delta u'(b, \lambda)$ has at most a finite number of zeros in any bounded region of the complex plane and the proof is complete. ■

6.4.1 Fundamental Properties

We observed in Chapter 1 that many Sturm-Liouville eigenvalue problems that arise in applications have all positive eigenvalues. When separation of variables leads to such an eigenvalue problem, this is a consequence of the fact that the underlying partial differential equations and boundary conditions that describe the physical situation include mechanisms that oppose arbitrarily large responses of the system. The natural eigenfunction expansions of the solutions would not have this property if there were any negative eigenvalues. The next theorem covers most such cases for the singular problems under consideration. A corollary of the theorem establishes that many singular Sturm-Liouville eigenvalue problems have at most a finite number of negative eigenvalues.

Theorem 180 *If $q \geq 0$ on $a < x \leq b$ and $\gamma\delta \geq 0$ in addition to the standing assumptions, then all the eigenvalues of the eigenvalue problem $Ly = \lambda ry$, $|y(a)| < \infty$, $\gamma y(b) + \delta y'(b) = 0$ are positive.*

Proof. Let λ be an eigenvalue and $y \neq 0$ be a corresponding real-valued eigenfunction. Multiply $Ly = \lambda ry$ by y and integrate by parts to obtain

$$\lambda \int_c^b y^2 r\, dx = \int_c^b y d(-py') + \int_c^b qy^2\, dx$$

$$= -p(b)y(b)y'(b) + p(c)y(c)y'(c) + \int_c^b (py'^2 + qy^2)\, dx$$

for any c with $a < c < b$. Since $\lim_{c \to a} p(c)y'(c) = 0$ by Lemma 160 and the integral on the left converges as $c \to a$ because the integrand is continuous on $[a, b]$, the integral on the right converges and

$$\lambda \int_a^b y^2 r\, dx = -p(b)y(b)y'(b) + \int_a^b (py'^2 + qy^2)\, dx.$$

By the assumption on the boundary condition at $x = b$, $y(b)y'(b) \leq 0$. So each term on the right is nonnegative and all the eigenvalues are nonnegative. If zero were an eigenvalue, then $y' = 0$

on $a < x < b$ because $p > 0$ there and $y = k$ on $[a, b]$ for some nonzero constant k. Since $y(a) = 0$ for any eigenfunction, $k = 0$ and we have reached a contradiction. Thus, all the eigenvalues are positive. ■

Corollary 181 *If $\gamma\delta \geq 0$ in addition to the standing assumptions, then at most a finite number of the eigenvalues of $Ly = \lambda ry$, $|y(a)| < \infty$, $\gamma y(b) + \delta y'(b) = 0$ are negative.*

Proof. For either type of weight function, there is a positive constant c such that

$$\hat{q}(x) = q(x) + cr(x) = \frac{q_1(x) + cr(x)(x-a)}{x-a} = \frac{\hat{q}_1(x)}{x-a} > 0$$

on $a < x \leq b$ because $q_1(x)$ is continuous on $[a, b]$, $q_1(a) > 0$, $r(x)$ is continuous on $[a, b]$ and positive on $a < x \leq b$. Since $\hat{q}(x)$ also satisfies the standing assumptions, all the eigenvalues of the eigenvalue problem $\hat{L}y = \hat{\lambda} ry$, $|y(a)| < \infty$, $\gamma y(b) + \delta y'(b) = 0$, where $\hat{L}y = -(py')' + \hat{q}y$, are positive. Since $Ly = \lambda ry$ if and only if $\hat{L}y = \hat{\lambda} ry$ where $\hat{\lambda} = \lambda + c$, it follows that all eigenvalues of $Ly = \lambda ry$, $|y(a)| < \infty$, $\gamma y(b) + \delta y'(b) = 0$ satisfy $\lambda + c = \hat{\lambda} > 0$. Thus, $\lambda > -c$. By Theorem 179 there are at most a finite number of eigenvalues in the interval $[-c, 0]$. ■

Further properties of the eigenvalues and eigenfunctions follow from the Hilbert-Schmidt theorem (Chapter 3) upon recasting the eigenvalue problem (6.21) as an eigenvalue problem for the kernel $g(x, s)r(s)$,

$$y(x) = \lambda \int_a^b g(x, s)r(s)y(s)\, ds, \tag{6.22}$$

where $g(x, s)$ is the Green's function for the Sturm-Liouville differential operator L with the boundary conditions in (6.21). The equivalence of the eigenvalue problem (6.21) and the eigenvalue problem (6.22) is established just as for a regular Sturm-Liouville eigenvalue problem. See Section 4.8.

The recasting just described requires that $\lambda = 0$ is not an eigenvalue of (6.21). We can use Theorem 179 to finesse the case when $\lambda = 0$ is an eigenvalue. To this end, let q_0 be a constant, $\hat{q}(x) = q(x) + q_0 r(x)$, and $\hat{L}y = -(py')' + \hat{q}y$. Since

$$Ly = \lambda ry \Longleftrightarrow \hat{L}y = (\lambda + q_0)ry,$$

λ, y is an eigenvalue, eigenfunction pair for (6.21) if and only if $\lambda + q_0$, y is an eigenvalue, eigenfunction pair for the eigenvalue problem $\hat{L}y = \hat{\lambda} ry$ with the same boundary conditions as (6.21). By Theorem 179 we can fix q_0 such that 0 is not an eigenvalue for the eigenvalue problem $\hat{L}y = \hat{\lambda} ry$, $|y(a)| < \infty$, $B_b y = 0$. This problem has a Green's function and any conclusions reached about its eigenvalues and eigenfunctions by means of the equivalent integral equation eigenvalue problem transfer immediately by translation of its eigenvalues to conclusions about the eigenvalues of the original eigenvalue problem. The corresponding eigenfunctions are the same.

In short, we can assume without loss of generality that $\lambda = 0$ is not an eigenvalue of (6.21) and convert it to the equivalent eigenvalue problem (6.22).

If $y(x)$ is continuous on $[a, b]$ and satisfies (6.22), then

$$\sqrt{r(x)}y(x) = \lambda \int_a^b \sqrt{r(x)}g(x, s)\sqrt{r(s)}\sqrt{r(s)}y(s)\, ds$$

and $z(x) = \sqrt{r(x)}y(x)$ is continuous on $[a, b]$ and satisfies

$$z(x) = \lambda \int_a^b k(x, s)z(s)\, ds \tag{6.23}$$

where

$$k(x, s) = \sqrt{r(x)}g(x, s)\sqrt{r(s)}$$

is a mildly singular, symmetric kernel by Corollary 172. Conversely, if $z(x)$ is continuous on $[a, b]$ and satisfies (6.23), then there are two cases to consider according as $r > 0$ on $[a, b]$ or r has a zero at $x = a$ and is positive on $(a, b]$. In the first case, (6.23) implies that

$$\frac{z(x)}{\sqrt{r(x)}} = \lambda \int_a^b g(x, s)r(s)\frac{z(s)}{\sqrt{r(s)}},\, ds$$

for x in $[a, b]$; that is, that $y(x) = z(x)/\sqrt{r(x)}$ satisfies (6.22). Thus, the eigenvalue problems (6.22) and (6.23) are equivalent when $r > 0$ on $[a, b]$.

Now assume $r(x) = (x - a)^m \rho(x)$ where $m > 0$ and $\rho(x) > 0$ on $[a, b]$. If $z(x)$ is continuous on $[a, b]$ and satisfies (6.23), then

$$\frac{z(x)}{\sqrt{r(x)}} = \int_a^b g(x, s)\sqrt{r(s)}z(s)\, ds$$

for $a < x \leq b$. Since $g(x, s)$ is mildly singular, the integral on the right is a continuous function on $[a, b]$ by Lemma 173. Therefore, there exists

$$\lim_{x \to a} \frac{z(x)}{\sqrt{r(x)}} = \lambda \int_a^b g(a, s)\sqrt{r(s)}z(s)\, ds.$$

Define $y(x)$ on $[a, b]$ by $y(x) = z(x)/\sqrt{r(x)}$ for $a < x \leq b$ and

$$y(a) = \lambda \int_a^b g(a, s)\sqrt{r(s)}z(s)\, ds.$$

Then $y(x)$ is continuous on $a \leq x \leq b$ and $r(s)y(s) = \sqrt{r(s)}z(s)$ on $[a, b]$ because $r(a) = 0$. For $a < x \leq b$,

$$y(x) = \lambda \int_a^b g(x, s)\sqrt{r(s)}z(s)\, ds = \lambda \int_a^b g(x, s)r(s)y(s)\, ds$$

and equality also holds at $x = a$ by the definition of $y(a)$. In summary, if $z(x)$ is continuous on $[a, b]$ and satisfies (6.23), then $z(x)/\sqrt{r(x)}$ has a unique extension by continuity to a continuous function $y(x)$ on $[a, b]$ that satisfies (6.22). This establishes the equivalence of (6.22) and (6.23) in the case where the weight function r has a zero at $x = a$. Thus, for all weight functions under consideration the two eigenvalue problems are equivalent.

Since the Green's function $g(x, s)$ is a mildly singular symmetric kernel so is $k(x, s)$. Consequently, the integral operator K with kernel $k(x, s)$ is a self-adjoint, compact, bounded, linear operator on $C[a, b]$. (See the paragraph that precedes Theorem 51.) The Hilbert-Schmidt theorem, its corollaries, and a line of reasoning similar to that used for regular Sturm-Liouville eigenvalue problems lead to the following properties of the eigenvalues and eigenfunctions of the singular Sturm-Liouville eigenvalue problem (6.21) when the boundary condition at $x = b$ satisfies $\gamma\delta \geq 0$, the most frequently occurring case in applications.

Theorem 182 *The Sturm-Liouville eigenvalue problem (6.21) with $\gamma\delta \geq 0$ has an infinite sequence of real eigenvalues $\{\lambda_n\}_{n=1}^{\infty}$ and a corresponding sequence of real-valued eigenfunctions $\{\phi_n\}_{n=1}^{\infty}$ with the following properties:*

1. Each eigenvalue is simple (has both algebraic and geometric multiplicity 1). Moreover, at most a finite number of the eigenvalues are negative and the sequence of eigenvalues is unbounded; hence, the eigenvalues can be listed as

$$\lambda_1 < \lambda_2 < \cdots < \lambda_n < \cdots$$

with $\lambda_n \to \infty$ as $n \to \infty$.
2. The corresponding eigenfunctions can be chosen real-valued and orthonormal with weight function r,

$$\langle \phi_m, \phi_n \rangle_r = \int_a^b \phi_m(s)\phi_n(s)r(s)\, ds = \delta_{mn},$$

where δ_{mn} is the Kronecker delta.
3. If the weight function $r(x)$ is positive and continuous on $[a, b]$, then for each continuous function f on $[a, b]$, the unique solution y to the singular Sturm-Liouville boundary value problem $Ly = f$, $|y(a)| < \infty$, $\gamma y(b) + \delta y'(b) = 0$ can be expressed as

$$y(x) = \sum_{n=1}^{\infty} \langle y, \phi_n \rangle_r \phi_n(x)$$

where the series is absolutely and uniformly convergent on $[a, b]$.
4. If the weight function $r(x) = (x - a)^m \rho(x)$ with $m > 0$ and $\rho(x)$ positive and continuous on $[a, b]$, then the conclusion in Part 3 holds for continuous functions f on $[a, b]$ for which $\lim_{x\to a} f(x)/(x - a)^m$ exists and is finite.

Proof. We rely on the discussion and notation that precedes the theorem. In particular, we can assume without loss in generality that zero is not an eigenvalue of the eigenvalue problem. Let K be the self-adjoint, compact, bounded, linear operator on $C[a, b]$ with mildly singular symmetric kernel $k(x, s) = \sqrt{r(x)}g(x, s)\sqrt{r(s)}$, where $g(x, s)$ is the Green's function associated with (6.21). Then λ, $y(x)$ is an eigenvalue, eigenfunction pair for the Sturm-Liouville eigenvalue problem (6.21) if and only if λ, $\sqrt{r(x)}y(x)$ is an eigenfunction, eigenvalue pair for the symmetric kernel $k(x, s)$.

1. Any eigenvalue λ of (6.21) is real because the eigenvalue problem is self-adjoint. If λ is an eigenvalue of (6.21) and $y_1(x)$ and $y_2(x)$ are corresponding eigenfunctions, then $y_1(x)$ and $y_2(x)$ are nontrivial bounded solutions to the singular Sturm-Liouville differential equation

$$-(py')' + (q - \lambda r)y = 0 \text{ for } a < x < b.$$

Consequently, by Theorem 164, $y_1(x)$ and $y_2(x)$ are nonzero multiples of each other and the geometric multiplicity of λ is 1. The algebraic multiplicity also is 1 because the kernel $k(x, s)$ is self-adjoint; see Lemma 57.

We establish next that the Sturm-Liouville eigenvalue problem has an infinite number of eigenvalues. The proof is by contradiction. Since $K \neq 0$ is a self-adjoint compact integral operator on $C[a, b]$, it has at least one nonzero eigenvalue, say μ, by Theorem 59. Consequently, $\lambda = 1/\mu$ is an eigenvalue of the kernel $k(x, s)$ and the Sturm-Liouville eigenvalue problem has at least one eigenvalue (and corresponding eigenfunction). Suppose that the Sturm-Liouville eigenvalue problem has only a finite number of eigenvalues, say $\lambda_1, \ldots, \lambda_N$, with corresponding eigenfunctions $\phi_1, \ldots, \phi_N$. By the equivalences above, K has only a finite number of nonzero eigenvalues $\mu_n = 1/\lambda_n$ for $n = 1, 2, \ldots, N$ and corresponding orthonormal eigenfunctions $\psi_n = \sqrt{r}\phi_n$. By the Hilbert-Schmidt theorem

$$Kf(x) = \sum_{n=1}^{N} \langle Kf, \psi_n \rangle \psi_n(x)$$

for all f in $C[a, b]$. Equivalently,

$$\sqrt{r(x)}\,G(\sqrt{r}f)(x) = \sum_{n=1}^{N} \langle Kf, \psi_n \rangle \sqrt{r(x)}\phi_n(x).$$

Hence,

$$G(\sqrt{r}f)(x) = \sum_{n=1}^{N} \langle Kf, \psi_n \rangle \phi_n(x)$$

for $a < x \leq b$. In fact, equality also holds at $x = a$ because both members of the equality are continuous on $[a, b]$. Since y solves the boundary value problem $Ly = \sqrt{r}f$, $|y(a)| < \infty$, $\gamma y(b) + \delta y'(b) = 0$ if and only if $y = G(\sqrt{r}f)$, it follows that

$$\sqrt{r}f = Ly = LG(\sqrt{r}f).$$

Consequently,

$$\sqrt{r}f = L\left(\sum_{n=1}^{N} \langle Kf, \psi_n \rangle \phi_n \right) = \sum_{n=1}^{N} \langle Kf, \psi_n \rangle \lambda_n r \phi_n$$

and

$$f(x) = \sum_{n=1}^{N} \lambda_n \langle Kf, \psi_n \rangle \psi_n(x)$$

for $a < x \leq b$ and equality holds on $[a, b]$ as above. Since $f(x)$ can be any continuous function on $[a, b]$, this equation says that $\{\psi_n\}_{n=1}^{N}$ is a basis for $C[a, b]$, which is impossible because, for example, the functions $1, x, x^2, \ldots, x^m$ are linearly independent for every positive integer m. This contradiction establishes that the Sturm-Liouville eigenvalue problem has an infinite number of eigenvalues λ_n and corresponding eigenfunctions ϕ_n.

By the Hilbert-Schmidt theorem, the eigenvalues λ_n of $k(x, s)$ satisfy $|\lambda_n| \to \infty$ as $n \to \infty$. By Corollary 181 at most a finite number of the eigenvalues λ_n can be negative. It follows that the eigenvalues can be listed in increasing order as

$$\lambda_1 < \lambda_2 < \cdots < \lambda_n < \cdots$$

and that $\lambda_n \to \infty$ as $n \to \infty$ which complete the proof of Property 1 of the theorem.

2. Since λ_n is an eigenvalue of the symmetric kernel $k(x, s)$, the corresponding eigenfunction ψ_n can be chosen real-valued by Corollary 62 of the Hilbert-Schmidt theorem and the sequence of eigenfunctions $\{\psi_n\}_{n=1}^{\infty}$ can be chosen orthonormal with weight function 1. Then the eigenfunctions ϕ_n of the Sturm-Liouville eigenvalue problem determined by $\psi_n = \sqrt{r}\phi_n$ are real-valued and orthonormal with weight function r,

$$\langle \phi_m, \phi_n \rangle_r = \langle \sqrt{r}\phi_m, \sqrt{r}\phi_n \rangle = \langle \psi_m, \psi_n \rangle = \delta_{mn}$$

and Property 2 is established.

3. Since $k(x, s)$ is mildly singular, it follows from Corollary 172 that there is a constant $M < \infty$ such that

$$\int_a^b |k(x, s)|^2 ds \leq M$$

for all x in $[a, b]$. Consequently, for any continuous function $\tilde{h}$ on $[a, b]$,

$$K\tilde{h}(x) = \sum_{n=1}^{\infty} \langle K\tilde{h}, \psi_n \rangle \psi_n(x)$$

with absolute and uniform convergence on $[a, b]$ by Corollary 61 of the Hilbert-Schmidt Theorem. Since

$$K\tilde{h}(x) = \int_a^b \sqrt{r(x)}\, g(x, s) \sqrt{r(s)}\, \tilde{h}(s)\, ds = \sqrt{r(x)}\, G(\sqrt{r}\tilde{h})(x),$$

that is, $K\tilde{h} = \sqrt{r}G(\sqrt{r}\tilde{h})$, and

$$\langle K\tilde{h}, \psi_n \rangle = \langle \sqrt{r}G(\sqrt{r}\tilde{h}), \sqrt{r}\phi_n \rangle = \langle G(\sqrt{r}\tilde{h}), \phi_n \rangle_r,$$

it follows that

$$\sqrt{r(x)}\, G(\sqrt{r}\tilde{h})(x) = \sum_{n=1}^{\infty} \langle G(\sqrt{r}\tilde{h}), \phi_n \rangle_r \sqrt{r(x)}\phi_n(x) \tag{6.24}$$

with absolute and uniform convergence on $[a, b]$.

Let $f(x)$ be continuous on $[a, b]$. If the weight function $r(x) > 0$ on $[a, b]$, then $\tilde{h} = f/\sqrt{r}$ is continuous on $[a, b]$. Set $\tilde{h} = f/\sqrt{r}$ in (6.24) to obtain the expansion

$$Gf(x) = \sum_{n=1}^{\infty} \langle Gf, \phi_n \rangle_r \phi_n(x)$$

where the series converges absolutely and uniformly because the cancelled factor $\sqrt{r(x)}$ is positive and continuous on $[a, b]$. If y is the unique solution to $Ly = f$, $|y(a)| < \infty$, and $B_b y = 0$, then $y = Gf$ and Property 3 is established.

4. If the weight function has the form $r(x) = (x - a)^m \rho(x)$ with $m > 0$ and $\rho(x) > 0$ on $[a, b]$, $f(x)$ is continuous on $[a, b]$, and $\lim_{x \to a} f(x)/(x - a)^m$ exists and is finite, then $\tilde{h} = f/\sqrt{r}$ for $a < x \le b$ has a unique extension to a continuous function on $[a, b]$, still denoted by $\tilde{h}$, obtained by defining $\tilde{h}(a) = 0$. Set $\tilde{h} = f/\sqrt{r}$ in (6.24) to obtain the expansion

$$Gf(x) = \sum_{n=1}^{\infty} \langle Gf, \phi_n \rangle_r \phi_n(x)$$

for $a < x \le b$. At this point, we cannot assert that the series is absolutely and uniformly convergent on $[a, b]$ nor that equality holds when $x = a$ because the common factor $\sqrt{r(x)}$ in (6.24) is zero at $x = a$.

We show next that $\sum_{n=1}^{\infty} \langle Gf, \phi_n \rangle_r \phi_n(x)$ is absolutely and uniformly convergent on $[a, b]$. First

$$\langle Gf, \phi_n \rangle_r = \langle Gf, r\phi_n \rangle = \langle Gf, \lambda_n^{-1} L\phi_n \rangle = \lambda_n^{-1} \langle LGf, \phi_n \rangle = \lambda_n^{-1} \langle f, \phi_n \rangle.$$

Second, the function $\widetilde{f}(x) = f(x)/r(x)$ for $a < x \le b$ has a unique extension by continuity to a continuous function on $[a, b]$, still denoted by $\widetilde{f}$, obtained by defining

$$\widetilde{f}(a) = \lim_{x \to a} \frac{f(x)}{r(x)} = \frac{1}{\rho(a)} \lim_{x \to a} \frac{f(x)}{(x - a)^m}$$

and

$$\langle f, \phi_n \rangle = \left\langle \tilde{f}, \phi_n \right\rangle_r.$$

Thus, $\langle Gf, \phi_n \rangle_r = \lambda_n^{-1} \left\langle \tilde{f}, \phi_n \right\rangle_r$ and

$$\sum_{n=1}^{\infty} \langle Gf, \phi_n \rangle_r \phi_n(x) = \sum_{n=1}^{\infty} \left\langle \tilde{f}, \phi_n \right\rangle_r \lambda_n^{-1} \phi_n(x).$$

Since $L\phi_n = \lambda_n r \phi_n$ implies $\lambda_n^{-1} \phi_n = G(r\phi_n)$ it follows that

$$\lambda_n^{-1} \phi_n(x) = G(r\phi_n)(x) = \int_a^b g(x, s)\, \phi_n(s) r(s)\, ds$$

for x in $[a, b]$. So, for each fixed x in $[a, b]$, $\lambda_n^{-1} \phi_n(x)$ is the n-th Fourier coefficient of the function of s, $g(x, s)$, with respect to the family $\{\phi_n\}$ that is orthonormal with weight function r. By Bessel's inequality and Corollary 172

$$\sum_{n=1}^{\infty} |\lambda_n^{-1} \phi_n(x)|^2 \le \int_a^b |g(x, s)|^2 r(s)\, ds \le \|r\|_{\max} \int_a^b |g(x, s)|^2\, ds \le M'$$

for all x in $[a, b]$ and some constant $M' < \infty$. Also

$$\sum_{n=1}^{\infty} \left| \left\langle \tilde{f}, \phi_n \right\rangle_r \right|^2 \le \left\| \tilde{f} \right\|_r^2$$

by Bessel's inequality. Consequently, by the Cauchy-Schwarz inequality

$$\begin{aligned}
\sum_{n=N}^{\infty} |\langle Gf, \phi_n \rangle_r \phi_n(x)| &= \sum_{n=N}^{\infty} \left| \left\langle \tilde{f}, \phi_n \right\rangle_r \lambda_n^{-1} \phi_n(x) \right| \\
&\le \left(\sum_{n=N}^{\infty} \left| \left\langle \tilde{f}, \phi_n \right\rangle_r \right|^2 \right)^{1/2} \left(\sum_{n=N}^{\infty} \left| \lambda_n^{-1} \phi_n(x) \right|^2 \right)^{1/2} \\
&\le M' \left(\sum_{n=N}^{\infty} \left| \left\langle \tilde{f}, \phi_n \right\rangle_r \right|^2 \right)^{1/2}.
\end{aligned}$$

Since the numerical series on the right converges, the absolute and uniform convergence of $\sum_{n=1}^{\infty} |\langle Gf, \phi_n \rangle_r \phi_n(x)|$ on $[a, b]$ is established.

Thus, in addition to the pointwise convergence in

$$Gf(x) = \sum_{n=1}^{\infty} \langle Gf, \phi_n \rangle_r \phi_n(x)$$

for $a < x \le b$ established earlier, the series on the right converges absolutely and uniformly on $[a, b]$. If y is the unique solution to $Ly = f$, $|y(a)| < \infty$, and $B_b y = 0$, then $y = Gf$ so

$$y(x) = \sum_{n=1}^{\infty} \langle y, \phi_n \rangle_r \phi_n(x)$$

for $a < x \le b$ and the series on the right converges absolutely and uniformly on $[a, b]$. The left member of the displayed equality is continuous on $[a, b]$ and the same is true of the right member by Theorem 23. Hence,

$$y(a) = \lim_{x \to a} y(x) = \lim_{x \to a} \sum_{n=1}^{\infty} \langle y, \phi_n \rangle_r \, \phi_n(x) = \sum_{n=1}^{\infty} \langle y, \phi_n \rangle_r \, \phi_n(a).$$

Thus,

$$y(x) = \sum_{n=1}^{\infty} \langle y, \phi_n \rangle_r \, \phi_n(x)$$

for $a \le x \le b$ and the series converges absolutely and uniformly on $[a, b]$. ∎

We mention that most of the conclusions of the theorem hold without the assumption $\gamma\delta \ge 0$. Without this assumption the proof only establishes Property 1 in the weaker form that the eigenvalues are real, simple, and can be listed by increasing absolute value as

$$|\lambda_1| < |\lambda_2| < \cdots < |\lambda_n| < \cdots$$

with $|\lambda_n| \to \infty$ as $n \to \infty$. The proofs of Properties 2, 3, and 4 did not rely on the assumptions $\gamma\delta \ge 0$.

6.4.2 Oscillation and Approximation Properties

The principal results of this section apply to the most important class of singular Sturm-Liouville eigenvalue problems with separated boundary conditions that occur in applications. They establish that for each N, linear combinations of appropriately chosen eigenfunctions $\{\phi_n\}_{n=0}^{N}$ have approximation and interpolation properties strictly analogous to the linear combinations of $\{x^n\}_{n=0}^{N}$, that is, to ordinary polynomials of degree N. These results follow from the fact that the Green's functions for such eigenvalue problems are mildly singular Kellogg kernels.

Recall from Section 3.7.3 that a symmetric, mildly singular kernel $k(x, s)$ with domain $[a, b] \times [a, b] \backslash \{(a, a)\}$ that satisfies

K1. $\det\left[k(x_i, x_j)\right]_{n\times n} > 0, \quad a < x_1 < \cdots < x_n < b,$

K2. $\det\left[k(x_i, s_j)\right]_{n\times n} \ge 0, \quad \text{for } x, s \text{ in} (\Delta_n \times \tilde{\Delta}_n) \cup (\tilde{\Delta}_n \times \Delta_n),$

is called a mildly singular Kellogg kernel. A mildly singular Kellogg kernel $k(x, s)$ and its compound kernels $k_{[n]}(x, s) = \det\left[k(x_i, s_j)\right]_{n\times n}$ with domains $D_n = (\Delta_n \times \widetilde{\Delta}_n) \cup (\widetilde{\Delta}_n \times \Delta_n)$ determine integral operators $K\colon C[a, b] \to C[a, b]$ and $K_{[n]} : C(D_n) \to C(D_n)$ that are self-adjoint, compact, bounded, linear operators. Here Δ_n is the simplex

$$\Delta_n = \{x = (x_1, \ldots, x_n) : a \le x_1 \le \cdots \le x_n \le b\}$$

and

$$\widetilde{\Delta}_n = \{x = (x_1, \ldots, x_n) : a < x_1 \le \cdots \le x_n \le b\}.$$

Theorem 183 *If the singular Sturm-Liouville eigenvalue problem*

$$\begin{cases} Ly = \lambda r y, \quad a < x < b, \\ |y(a)| < \infty, \quad \gamma y(b) + \delta y'(b) = 0, \end{cases}$$

satisfies $q \ge 0$ for $a < x \le b$ and $\gamma\delta \ge 0$ in addition to the standing assumptions, then Ly together with the given boundary conditions has a Green's function $g(x, s)$. The Green's function is a mildly singular Kellogg kernel as is the kernel $k(x, s) = \sqrt{r(x)} g(x, s) \sqrt{r(s)}$.

Proof. By Theorem 180 all the eigenvalues of the eigenvalue problem are positive. Hence, the Green's function $g(x, s)$ exists and by Theorem 171

$$g(x,s) = \begin{cases} v(x)u(s) & \text{for } a \le s \le x \le b \quad \text{and} \quad (x,s) \ne (a,a) \\ u(x)v(s) & \text{for } a \le x \le s \le b \quad \text{and} \quad (x,s) \ne (a,a) \end{cases}$$

where $u(x)$ is real-valued and continuous on $[a, b]$, $v(x)$ is real-valued and continuously differentiable on $(a, b]$, $v(x)$ becomes unbounded as $x \to a$, and u and v satisfy

$$\begin{cases} Lu = 0 \quad \text{for } a < x \le b \\ |u(a)| < \infty \end{cases},$$

$$\begin{cases} Lv = 0 \quad \text{for } a < x \le b \\ \gamma v(b) + \delta v'(b) = 0 \end{cases},$$

and

$$p(x)\, W_{u,v}(x) = -1 \quad \text{for } a < x \le b.$$

We claim that

$$u(x)v(x) > 0 \quad \text{for } a < x < b$$

and

$$\frac{u(x)}{v(x)} \text{ is increasing on } a < x < b.$$

Suppose $v(c) = 0$ for some c with $a < c < b$. Multiply the differential equation $-(pv')' + qv = 0$ by v and integrate by parts to obtain

$$[-v(x)p(x)v'(x)]_c^b + \int_c^b \left(p(x)v'(x)^2 + q(x)v(x)^2\right) dx = 0$$

and

$$-p(b)v(b)v'(b) + \int_c^b \left(p(x)v'(x)^2 + q(x)v(x)^2\right) dx = 0.$$

The boundary condition at $x = b$ implies that $-p(b)v(b)v'(b) \ge 0$. Consequently, $v'(x) = 0$ on $[c, b]$, $v(x) = k$, a constant, on that interval and $k = 0$ because $v(c) = 0$. Consequently, $v(c) = v'(c) = 0$ which implies that $v(x) = 0$ on $a < x < b$. This contradicts the fact that $v(x)$ is not identically zero on $a < x < b$ and establishes that $v(x) \ne 0$ on $a < x < b$.

For $a < x < b$,

$$\frac{d}{dx}\left(\frac{u(x)}{v(x)}\right) = \frac{v(x)u'(x) - u(x)v'(x)}{v(x)^2} = \frac{1}{p(x)v(x)^2} > 0$$

and, hence, $u(x)/v(x)$ is increasing on $a < x \le b$. Since $u(x)$ is bounded and $v(x)$ becomes unbounded as $x \to a$,

$$\lim_{x \to a} \frac{u(x)}{v(x)} = 0.$$

Consequently,

$$\frac{u(x)}{v(x)} > 0 \quad \text{for } a < x \le b$$

and

$$u(x)v(x) > 0 \quad \text{for } a < x \le b.$$

Since $u(x)v(x) > 0$ and $u(x)/v(x)$ is increasing on $a < x < b$, it follows from Corollary 37 that

$$g_{[n]}(x, s) > 0 \quad \text{for } a < x_1,\, s_1 < x_2,\, s_2 < \cdots < x_n,\, s_n < b$$

and is 0 for all other choices of $a < x_1 < x_2 < \cdots < x_n < b$ and $a < s_1 < s_2 < \cdots < s_n < b$. Since $g(x, s)$ is a mildly singular kernel on $[a, b] \times [a, b] \backslash \{(a, a)\}$, each compound kernel $g_{[n]}(x, s)$ is continuous on its domain $(\Delta_n \times \widetilde{\Delta}_n) \cup (\widetilde{\Delta}_n \times \Delta_n)$. It follows that

$$g_{[n]}(x, s) \geq 0$$

for $x = (x_1, \ldots, x_n)$, $s = (s_1, \ldots, s_n)$ in $\left(\Delta_n \times \widetilde{\Delta}_n\right) \cup \left(\widetilde{\Delta}_n \times \Delta_n\right)$ and that

$$g_{[n]}(x, x) > 0 \quad \text{for } a < x_1 < x_2 < \cdots < x_n < b.$$

Thus, $g(x, s)$ is a mildly singular Kellogg kernel.

For $x = (x_1, \ldots, x_n)$, $s = (s_1, \ldots, s_n)$ in $\left(\Delta_n \times \widetilde{\Delta}_n\right) \cup \left(\widetilde{\Delta}_n \times \Delta_n\right)$,

$$k_{[n]}(x, s) = \det\left[\sqrt{r(x_i)}\,g(x_i, s_j)\sqrt{r(s_j)}\right] = \prod_{i=1}^{n} \sqrt{r(x_i)}\,g_{[n]}(x, s) \prod_{j=1}^{n} \sqrt{r(s_j)}$$

and it follows that $k_{[n]}(x, s)$ is a mildly singular Kellogg kernel. ■

Theorem 184 *If, in addition to the standing assumptions, the singular Sturm-Liouville eigenvalue problem*

$$\begin{cases} Ly = \lambda r y, & a < x < b, \\ |y(a)| < \infty, & \gamma y(b) + \delta y'(b) = 0, \end{cases}$$

satisfies $\gamma\delta \geq 0$, *then the eigenvalues of the singular eigenvalue problem are all real, simple, and can be labeled so that*

$$\lambda_0 < \lambda_1 < \cdots < \lambda_n < \cdots$$

with $\lambda_n \to \infty$ *as* $n \to \infty$. *The corresponding eigenfunctions* $\phi_0(x), \phi_1(x), \ldots, \phi_n(x), \ldots$ *can be chosen orthonormal (with weight function* r*) and such that* $\phi_0(x)$, $\phi_1(x)$, $\ldots$, $\phi_n(x)$ *form a Tchebycheff system on* (a, b) *for each* $n = 0, 1, 2, \ldots$. *Consequently, the following oscillatory and approximation properties hold:*

1. *Given any* $n+1$ *points in* (a, b) *and any* $n+1$ *values* $b_0, \ldots, b_n$, *there is a unique* ϕ *-polynomial* $\phi(x) = \sum_{i=0}^{n} a_i\phi_i(x)$ *that takes on the prescribed values at the given points.*
2. *A nontrivial* ϕ*-polynomial has at most* n *zeros in* (a, b) *where nonnodal zeros are counted twice and nodal zeros once.*
3. *A nontrivial* ϕ*-polynomial* $\phi(x) = \sum_{i=m}^{n} a_i\phi_i(x)$ *has at least* m *nodal zeros in* (a, b) *and has at most* n *zeros there, counting zeros as in Property 2.*
4. ϕ_n *has* n *nodal zeros in* (a, b) *and no other zeros there.*
5. *The zeros of* ϕ_{n-1} *and* ϕ_n *strictly interlace on* (a, b).

Proof. The desired conclusions follow from the equivalence established earlier: λ, $\phi(x)$ is an eigenvalue, eigenfunction pair for the singular Sturm-Liouville eigenvalue problem if and only if λ, $\sqrt{r(x)}\phi(x)$ is an eigenvalue, eigenfunction pair of the kernel $k(x, s) = \sqrt{r(x)}\,g(x, s)\sqrt{r(s)}$. The stated properties hold with $\lambda_0 > 0$ for the eigenvalues and corresponding eigenfunctions of any mildly singular Kellogg kernel. See Section 3.7.3. Exactly as in the proof of Corollary 181 there is a constant $c > 0$ such that

$$\hat{q}(x) = q(x) + cr(x) = \frac{q_1(x) + cr(x)(x - a)}{x - a} = \frac{\hat{q}_1(x)}{x - a} > 0$$

on $a < x \leq b$ and $\hat{q}(x)$ also satisfies the standing assumptions. The eigenvalue problem $\hat{L}y = \hat{\lambda}ry$, $|y(a)| < \infty$, $\gamma y(b) + \delta y'(b) = 0$, where $\hat{L}y = -(py')' + \hat{q}y$, has all positive eigenvalues and a Green's function that is a mildly singular Kellogg kernel. Hence, all the conclusions in the theorem hold for the eigenvalues $\hat{\lambda}_n$ and eigenfunctions $\hat{\phi}_n$ of the eigenvalue problem for $\hat{L}$ with $\hat{\lambda}_0 > 0$.

Since λ, y is an eigenvalue, eigenfunction pair for $Ly = \lambda ry$ if and only if $\hat{\lambda} = \lambda + c$, y is an eigenvalue, eigenfunction pair for $\hat{L}y = \hat{\lambda}ry$, all the eigenvalues of $Ly = \lambda ry$, $|y(a)| < \infty$, $\gamma y(b) + \delta y'(b) = 0$ satisfy $\lambda + c = \hat{\lambda} > 0$ and λ and $\hat{\lambda}$ have the same corresponding eigenfunctions. The eigenvalues $\hat{\lambda}_n$ can be listed as

$$0 < \hat{\lambda}_0 < \hat{\lambda}_1 < \cdots < \hat{\lambda}_n < \cdots$$

and the corresponding eigenfunctions $\hat{\phi}_n(x) = \phi_n(x)$ can be chosen to have the properties listed in the theorem. Since $\lambda_n + c = \tilde{\lambda}_n$,

$$\lambda_0 < \lambda_1 < \cdots < \lambda_n < \cdots$$

and the theorem is established. ■

Example 4. In Section 1.5.3 we gave a model for the temperature $u(r, \theta, t)$ in a circular plate of radius b with insulated top and bottom and whose outer edge is held at temperature zero. Two separation constants were used, $-\lambda$ and μ. Physically realistic separated solutions, $u = T(t)\Theta(\theta)R(r)$, to the heat equation and homogeneous boundary condition are determined by $T(t) = T_\lambda(t) = e^{-\lambda t}$ and the eigenvalue problems

$$\Theta'' + \mu\Theta = 0,\ \Theta(0) = \Theta(2\pi),\ \Theta'(0) = \Theta'(2\pi)$$

and

$$-(rR')' + \left(\frac{\mu}{r} - \lambda r\right)R = 0,\ |R(0)| < \infty,\ R(b) = 0.$$

The periodic boundary conditions reflect the fact that polar angles that differ by a multiple of 2π mark the same point in the plate. It follows that $\mu = n^2$ for $n = 0, 1, 2, \ldots$ and

$$\Theta = \Theta_n(\theta) = a_n \cos n\theta + b_n \sin n\theta$$

where a_n and b_n are arbitrary constants. For each n, the corresponding eigenvalue problem for $R = R_n$ is

$$-(rR_n')' + \left(\frac{n^2}{r} - \lambda r\right)R_n = 0,\ |R_n(0)| < \infty,\ R_n(b) = 0.$$

We concentrate on this singular eigenvalue problem. The differential equation is Bessel's equation of order n with parameter λ. The eigenvalue problem for the parameter λ can be expressed as

$$-(rR_n')' + \frac{n^2}{r}R_n = \lambda rR_n,\ |R_n(0)| < \infty,\ R_n(b) = 0$$

and the differential operator $L_n y = -(ry')' + (n^2/r)y$ satisfies all the assumptions made in this chapter. Hence, the eigenvalue problem has eigenvalues

$$0 < \lambda_{n0} < \lambda_{n1} < \cdots < \lambda_{nm} < \cdots$$

and corresponding eigenfunctions $R_{nm}(r)$ that have all the oscillation and interpolation properties in Theorem 184. Moreover, the eigenvalues and eigenfunctions have all the properties in Theorem 182. In particular, the eigenfunctions are orthonormal with weight function r and each twice continuously differentiable function $f(r)$ on $[0, b]$ that satisfies $f(b) = 0$ and for which $\lim_{r\to 0}(L_n f(r))/r$ exists and is finite has the eigenfunction expansion

$$f(r) = \sum_{m=0}^{\infty} \langle f, R_{nm}\rangle_r \, R_{nm}(r)$$

with absolute and uniform convergence on $[0, b]$. The meaning of the limit condition will be discussed in a remark following the example. The eigenfunction expansion follows directly from Theorem 182(4) because f is the solution to $L_n y = g$ for $g = L_n f$.

Since Bessel's equation of order n and parameter λ has as bounded solutions only the multiples of $J_n(\sqrt{\lambda} r)$, it follows that

$$R_{nm}(r) = c_{nm} J_n\left(\sqrt{\lambda_{nm}}\, r\right)$$

for some constant $c_m \neq 0$. Two points of view are possible here. First, it is well known that the Bessel function $J_n(z)$ has an infinite number of zeros

$$z_{n0} < z_{n1} < \cdots < z_{nm} < \cdots$$

that are all positive and tend to infinity as $m \to \infty$. Since $R_{nm}(b) = 0$ for $m = 0, 1, 2, \ldots,$ it follows that the eigenvalues of the eigenvalue problem are determined by the zeros of $J_n(z)$ by

$$\lambda_{nm} = \left(\frac{z_{nm}}{b}\right)^2$$

for $m = 0, 1, 2, \ldots$. Second, the results established in this chapter guarantee that all the eigenvalues λ_{nm} are positive, infinite in number, and satisfy

$$J_n\left(\sqrt{\lambda_{nm}}\, b\right) = c_{nm}^{-1} R_{nm}(b) = 0.$$

This give an alternative proof that $J_n(z)$ has an infinite number of positive zeros.

Remark. The absolute and uniform convergence of the eigenfunction expansions for the functions $f(r)$ in Example 4 requires that $\lim_{r\to 0}(L_n f(r))/r$. To better understand the limit condition, observe that

$$\frac{L_n f(r)}{r} = \frac{-(rf'(r))' + (n^2/r)f(r)}{r} = -f''(r) - \frac{f'(r) - f'(0)}{r} - \frac{f'(0)}{r} + \frac{n^2}{r^2} f(r)$$

and

$$\frac{n^2}{r^2} f(r) = \frac{n^2}{r^2}\left(f(0) + f'(0)r + \frac{1}{2} f''(\rho_r) r^2\right)$$

for some ρ_r between 0 and r by Taylor's theorem with remainder. Consequently,

$$\frac{L_n f(r)}{r} = -f''(r) + \frac{n^2}{2} f''(\rho_r) - \frac{f'(r) - f'(0)}{r} - \frac{(n^2 - 1)f'(0)}{r} + \frac{n^2 f(0)}{r^2}$$

and $(L_n f(r))/r$ has a finite limit, namely $(n^2 - 4)f''(0)/2$, if and only if ($n = 1$ or $f'(0) = 0$) and ($n = 0$ or $f(0) = 0$). That is,

if $n = 0$ the series converges absolutely and uniformly to f if $f'(0) = 0$;

if $n = 1$ the series converges absolutely and uniformly to f if $f(0) = 0$;

$n \geq 2$ the series converges absolutely and uniformly to f if $f'(0) = f'(0) = 0$.

These conditions should seem reasonable because for $n=0$ the derivative of each eigenfunction $c_{0m}J_0(\sqrt{\lambda_{0m}}r)$ is 0 at $r=0$; for $n=1$ each eigenfunction $c_{1m}J_1(\sqrt{\lambda_{1m}}r)$ is 0 at $r=0$; and for $n\geq 2$ each eigenfunction $c_{nm}J_n(\sqrt{\lambda_{nm}}r)$ and its derivative is 0 at $r=0$.

Example 5. If instead of maintaining temperature zero on the edge of the circular plate in Example 4, we had modeled heat transfer across the boundary according to Newton's law of cooling, then the boundary condition in the model would be $\gamma u(b,\theta,t)+\delta u_r(b,\theta,t)=0$ where γ and δ are positive constants and the resulting singular eigenvalue problem becomes

$$-(rR_n')'+\frac{n^2}{r}R_n=\lambda rR_n,\ 0<r<b,$$

$$|R_n(0)|<\infty,\ \gamma R_n(b)+\delta R_n'(b)=0.$$

Once again, this eigenvalue problem satisfies the hypotheses of all the theorems of this chapter. Consequently, the discussion in Example 4 carries over to this situation with the single adjustment that the zeros z_{nm}, for $m=0,\ 1,\ 2,\ldots$, are now the zeros of the function $\gamma J_n(z)+\delta J_n'(z)$.

6.4.3 Rayleigh Quotient

Consider the singular Sturm-Liouville eigenvalue problem

$$Ly=\lambda ry,\ |y(a)|<\infty,\ \gamma y(b)+\delta y'(b)=0,$$

where $Ly=-(py')'+qy$ and $\gamma\delta\geq 0$ so that the conclusions of Theorem 184 hold. The eigenvalue problem has an infinite number of simple eigenvalues

$$\lambda_0<\lambda_1<\cdots<\lambda_n<\cdots$$

with $\lambda_n\to\infty$ as $n\to\infty$ and corresponding real-valued eigenfunctions $\phi_n(x)$ that are orthonormal with respect to the weight function r. Recall that the domain of L is

$$D=\left\{y\in C[a,b]: Ly\in C[a,b]\quad\text{and}\quad \lim_{x\to a}p(x)y'(x)=0\right\},$$

with the slight abuse of notation: $Ly\in C[a,b]$ means Ly is continuous on (a,b) and has a unique extension by continuity to the closed interval $[a,b]$, with the extended function still denoted by Ly. (See the paragraph following Lemma 176.)

The quotient that appears in the following theorem is the **Rayleigh quotient.** It will be used in Chapter 7 to find upper estimates of the smallest eigenvalue of a Sturm-Liouville eigenvalue problem as part of a shooting method that accurately determines eigenvalues and corresponding eigenfunctions of the problem.

Theorem 185 *With the notation and assumptions above and with weight function $r(x)=(x-a)^m\rho(x)$ where $m\geq 0$ and $\rho(x)>0$ and continuous on $[a,b]$, the smallest eigenvalue of the singular eigenvalue problem $Ly=\lambda ry$, $|y(a)|<\infty$, $\gamma y(b)+\delta y'(b)=0$ satisfies*

$$\lambda_0=\min\frac{\langle Ly,y\rangle}{\langle y,y\rangle_r}=\min\frac{-p(b)y(b)y'(b)+\int_a^b(py'^2+qy^2)\,dx}{\int_a^b y^2r\,dx},$$

where the minimum is over all functions $y\neq 0$ in the domain of L that satisfy the boundary conditions $|y(a)|<\infty$, $\gamma y(b)+\delta y'(b)=0$ and $\lim_{x\to a}Ly(x)/(x-a)^m$ exists and is finite. Moreover, the minimum is achieved if and only if y is an eigenfunction corresponding to λ_0.

Remark. Any eigenfunction y satisfies the limit condition of the theorem because $Ly = \lambda ry$. If the weight function is positive on $[a, b]$, that is if $m = 0$, then the limit condition is satisfied for all y in the domain of L because Ly is continuous on $[a, b]$. If $m > 0$ the limit condition further restricts the functions over which the minimum is taken.

Proof. If y satisfies the boundary conditions $|y(a)| < \infty, \gamma y(b) + \delta y'(b) = 0$, is in the domain of L, and $\lim_{x\to a} Ly(x)/(x-a)^m$ exists and is finite, then $Ly = f$ for $f = Ly$, f is continuous on $[a, b]$, and $\lim_{x\to a} f(x)/(x-a)^m$ exists and is finite. Consequently, y is continuous on $[a, b]$ and

$$y(x) = \sum_{n=0}^{\infty} \langle y, \phi_n \rangle_r \phi_n(x),$$

where the series converges absolutely and uniformly on $[a, b]$ by Theorem 182. Consequently,

$$\begin{aligned}\langle Ly, y \rangle &= \left\langle Ly, \sum_{n=0}^{\infty} \langle y, \phi_n \rangle_r \phi_n \right\rangle = \sum_{n=0}^{\infty} \langle y, \phi_n \rangle_r \langle Ly, \phi_n \rangle \\ &= \sum_{n=0}^{\infty} \langle y, \phi_n \rangle_r \langle y, L\phi_n \rangle = \sum_{n=0}^{\infty} \langle y, \phi_n \rangle_r \langle y, \lambda_n r \phi_n \rangle \\ &= \sum_{n=0}^{\infty} \lambda_n \left| \langle y, \phi_n \rangle_r \right|^2 \geq \lambda_0 \sum_{n=0}^{\infty} \left| \langle y, \phi_n \rangle_r \right|^2 = \lambda_0 \langle y, y \rangle_r,\end{aligned}$$

where the last equality follows from a similar calculation using $y = \sum_{n=0}^{\infty} \langle y, \phi_n \rangle_r \phi_n$ to evaluate $\langle y, y \rangle_r$. Equality holds above if and only if $\langle y, \phi_n \rangle_r = 0$ for all $n \geq 1$; hence, if and only if $y = \langle y, \phi_0 \rangle_r \phi_0$, equivalently, y is an eigenfunction corresponding to λ_0. Thus, for $y \neq 0$,

$$\lambda_0 \leq \frac{\langle Ly, y \rangle}{\langle y, y \rangle_r}$$

with equality if and only if y is an eigenfunction corresponding to λ_0. The first conclusion in the theorem follows. Finally, a familiar integration by parts argument gives

$$\begin{aligned}\int_c^b yLy\,dx &= \int_c^b yd(-py') + \int_c^b qy^2\,dx \\ &-pyy'\Big|_c^b + \int_c^b (py'^2 + qy^2)\,dx.\end{aligned}$$

Now

$$\lim_{c\to a} -pyy'\Big|_c^b = -p(b)y(b)y'(b)$$

and

$$\lim_{c\to a} \int_c^b yLy\,dx = \int_a^b yLy\,dx = \langle Ly, y \rangle$$

because y is in the domain of L. Consequently, letting $c \to a$ in the integrations by parts formula above shows that the improper integral $\int_a^b (py'^2 + qy^2)\,dx$ exists, that

$$\langle Ly, y \rangle = -p(b)y(b)y'(b) + \int_a^b (py'^2 + qy^2)\,dx,$$

and the second conclusion follows. ∎

6.5 Concluding Remarks

We have assumed throughout this chapter that the Sturm-Liouville differential equation is singular only at one endpoint of the interval on which it is defined. This is quite natural from a physical perspective. For example, the standard wave equation model for the vibrations of a circular drum has no physical singularity in the wave equation. However, the method of separation of variables will succeed only in polar coordinates and the use of polar coordinates introduces a single "fake" (nonphysical) singularity at the pole of that system and leads to the singularity in the Sturm-Liouville equation. Since there is no physical singularity, it is reasonable to expect that the solution to the Sturm-Liouville equation is well behaved (at least continuous) at the singularity and makes plausible the mathematical conclusion we reached that the solution $y(x)$ has a limit as x approaches 0 (the singularity) and indeed that $p(x)y'(x)$ has limit 0 as x approaches 0.

Chapter 7

Approximation of Eigenvalues and Eigenfunctions

In this chapter we develop shooting methods for the numerical determination of eigenvalues and eigenvectors of the regular and singular Sturm-Liouville eigenvalue problems with separated boundary conditions treated in Chapters 4, 5, and 6. Other methods based on finite differences or finite elements reduce the eigenvalue problem to a matrix eigenvalue problem. We do not cover such methods because they have been extensively studied elsewhere. Useful references include Isaacson and Keller [21], Stoer and Bulirsch [41], Strang and Fix [42], and Collatz [7], a compendium of numerical techniques together with many examples.

That a shooting method can be used to approximate eigenvalues and eigenvectors of regular problems in not new. See for example Chapter 8, Section 7.3 in [21]. What is new is the accompanying convergence analysis, both for the regular problems in Chapter 4 and for the singular problems in Chapters 5 and 6. The eigenvalues provide the decay or growth rates associated with the physical process that is modeled. These rates are of great importance and often hard to discern from numerical solutions based on finite differences or finite elements.

The shooting methods can be used in principle to find all the eigenvalues and eigenfunctions of the Sturm-Liouville problems. Two important features of the methods are: (1) No roundoff errors accumulate when several eigenvalues and eigenfunctions are determined numerically because each eigenvalue and eigenfunction is found independently of the others. (2) The methods handle both regular and singular problems with equal ease.

We have used the methods to find the first four or five eigenvalues and corresponding eigenfunctions of several test problems and problems for which exact solutions are not known. The accuracy achieved is gratifying, as will be confirmed by numerical results presented later in this chapter. How to code the shooting algorithm and the programming language to use is a matter of personal preference. The examples presented here were obtained using MATLAB and some of its standard packages. An advantage of this approach is that MATLAB has many well-tested, robust, adaptive codes. Alternatively, the reader can easily code the shooting method in any convenient programming language, with the advantage that the coder has full control over all details of program execution.

We will use the notation introduced in Chapters 4, 5, and 6 throughout this chapter. All the eigenvalues under consideration are real and simple, as we established in Chapters 4, 5, and 6, because the Sturm-Liouville eigenvalue problems have separated boundary conditions and are self-adjoint. Furthermore, there are at most a finite number of negative eigenvalues. We established this result for the eigenvalue problems that occur most frequently in applications in Corollary 125, Corollary 152, and Corollary 181. A more general result for regular eigenvalue problems can be found in [5] or [10] where it is shown that for large n, $\lambda_n = n^2\pi^2/(b-a)^2 + O(1)$, independent of the boundary conditions imposed and where $O(1)$ is a bounded function of n. For singular eigenvalue problems we restrict consideration to the classes of problems covered by Corollary 152 and Corollary 181. Consequently, we can always list the eigenvalues as

$$\lambda_0 < \lambda_1 < \cdots < \lambda_n < \cdots$$

throughout the chapter.

Before proceeding further, it is useful to make some observations about the traditional shooting method for solving boundary value problems for ordinary differential equations and the shooting method used here to solve eigenvalue problems.

Assume that the regular Sturm-Liouville boundary value problem

$$\begin{cases} -(p(x)y')' + q(x)y = f(x), & 0 < x < 1, \\ \qquad y(0) = 0, \quad y(1) = 0, \end{cases} \tag{7.1}$$

where $p(x)$, $q(x)$, and $f(x)$ are continuous on $[0, 1]$ has a unique solution, equivalently 0 is not an eigenvalue of the associated eigenvalue problem. To solve the boundary value problem (theoretically or numerically) solve instead the initial value problem

$$\begin{cases} -(p(x)u')' + q(x)u = f(x), & 0 < x < 1, \\ \qquad u(0) = 0, \quad u'(0) = s, \end{cases} \tag{7.2}$$

where s is a parameter that is to be determined so that the solution to the initial value problem $u = u(x) = u(x, s)$ also solves the boundary value problem. The initial value problem is linear and, hence, has a solution u that extends across the entire interval $[0, 1]$. The solution u will solve the boundary value problem if and only if u satisfies the equation

$$u(1, s) = 0.$$

The initial value problem for u has solution $u(x, s) = u_p(x) + sv(x)$ where u_p is the particular solution satisfying $Lu_p = f$, $u_p(0) = 0$, $u_p'(0) = 0$, v satisfies $Lv = 0$, $v(0) = 0$, $v'(0) = 1$, and $Ly = -(py')' + qy$. The function $u(1, s)$ is linear in s. Since the boundary value problem has a unique solution, with say $y'(0) = \sigma$, and solutions to the initial value problem are unique, $u(x, \sigma) = y(x)$ and $u(1, \sigma) = 0$. In fact σ is the unique zero of the function $u(1, s)$. Indeed, if τ is any zero of $u(1, s)$, then $u(1, \tau) = 0$ and $u(x, \tau)$ solves the boundary value problem (7.1). Consequently, $u(x, \sigma)$ and $u(x, \tau)$ both solve (7.1). Since the solution to (7.1) is unique, $u(x, \sigma) = u(x, \tau)$ and $\tau = u'(0, \tau) = u'(0, \sigma) = \sigma$. Since the linear function $u(1, s)$ has σ as its unique root, Newton's method converges in one step to that root; consequently, given any initial guess s_0,

$$\sigma = s_0 - \frac{u(1, s_0)}{\partial u(1, s_0)/\partial s}.$$

Thus, the unique solution $u(x, \sigma)$ of the boundary value problem can be approximated by choosing an arbitrary initial condition s_0, using an initial value problem routine to solve (7.2), calculating σ from the Newton step above, and then solving (7.2) with $s = \sigma$ to obtain a numerical approximation to the solution of the boundary value problem (7.1).

A shooting method for solving eigenvalue problems is similar in spirit but leads to a nonlinear equation that must be solved by a root finding method, typically Newton's method or the bisection method. A simple regular eigenvalue problem illustrates the key ideas:

$$\begin{cases} -y'' = \lambda y, & 0 < x < 1, \\ y(0) = 0, & y(1) = 0. \end{cases} \tag{7.3}$$

If λ is an eigenvalue and $y = y(x) = y(x, \lambda)$ is a corresponding eigenfunction, then $y'(0) \neq 0$. We normalize the eigenfunction by requiring that $y'(0) = 1$. Furthermore, all the eigenvalues of (7.3) are positive and simple from general results in Chapter 4. (These assertions are easily established directly for the problem at hand.)

This time the shooting parameter λ is in the differential equation. The simple eigenvalue problem (7.3) can be solved directly; however, the direct approach in not available for more general eigenvalue problems whereas the following approach is: consider the initial value problem

$$\begin{cases} -u'' = \lambda u, & 0 < x < 1, \\ u(0) = 0, & u'(0) = 1, \end{cases} \tag{7.4}$$

whose unique solution $u = u(x) = u(x, \lambda)$ extends across the entire interval $[0, 1]$. Consequently, λ will be an eigenvalue of (7.3) if and only if the solution to the initial value problem satisfies

$$u(1, \lambda) = 0,$$

in which case $y(x) = u(x, \lambda)$ is the corresponding normalized eigenfunction. Solution of the initial value problem (7.4) yields

$$u(x, \lambda) = \frac{\sin \sqrt{\lambda} x}{\sqrt{\lambda}}.$$

Consequently, λ is an eigenvalue of the eigenvalue problem (7.3) if and only if

$$u(1, \lambda) = \frac{\sin \sqrt{\lambda}}{\sqrt{\lambda}} = 0.$$

Since $u(1, \lambda) = 0$ if and only if $\lambda = (n\pi)^2$ for $n = 1, 2, 3, \ldots$, these values of λ are the eigenvalues of (7.3) and

$$y(x) = \frac{\sin \sqrt{\lambda} x}{\sqrt{\lambda}}$$

are the corresponding normalized eigenfunctions.

The line of reasoning just applied to the simple eigenvalue problem (7.3) and the companion initial value problem (7.4) can be used to find accurate numerical approximations to the eigenvalues and normalized eigenfunctions of the regular Sturm-Liouville problems in Chapter 4. Natural variants of the shooting method yield corresponding results for the singular Sturm-Liouville problems in Chapters 5 and 6.

The foregoing discussion provides perspective on the general developments that follow.

7.1 Regular Problems

In this section we develop a shooting method for the numerical determination of eigenvalues and eigenfunctions of the regular Sturm-Liouville eigenvalue problems with separated boundary conditions of Chapter 4:

$$Ly = \lambda r y, \; B_a y = 0, \; B_b y = 0, \tag{7.5}$$

where

$$\begin{aligned} Ly &= -(p(x)y')' + q(x)y, \text{ for } a < x < b, \\ B_a y &= \alpha y(a) + \beta y'(a), \; |\alpha| + |\beta| > 0, \\ B_b y &= \gamma y(b) + \delta y'(b), \; |\gamma| + |\delta| > 0. \end{aligned}$$

Since the eigenvalue problem is regular,

(1) $p(x) > 0$ and $r(x) > 0$ on $[a, b]$.

(2) $p(x)$, $q(x)$, and $r(x)$ are continuous on $[a, b]$.

(3) α, β, γ and δ are real numbers.

Although the differential equation is only assumed to hold on $a < x < b$, an eigenfunction of a regular Sturm-Liouville eigenvalue problem is actually continuously differentiable on the closed interval $[a, b]$ and the differential equation is satisfied on that interval by Theorem 112.

The shooting method requires an additional smoothness assumption because it uses a variational equation associated with $Ly = \lambda ry$. The following assumption also holds throughout the section on regular eigenvalue problems:

(4) $p(x)$, $q(x)$, and $r(x)$ are continuously differentiable on $[a, b]$.

7.1.1 The Shooting Method

There is a natural conceptual and effective computational approach for finding the eigenvalues and eigenfunctions of a Sturm-Liouville problem via initial value problems and a shooting method. To set the stage and introduce some notation, we briefly outline the approach for an eigenvalue problem with Dirichlet boundary conditions,

$$\begin{aligned} &-(py')' + (q - \lambda r)y = 0, \quad a \le x \le b, \\ &y(a) = 0, \quad y(b) = 0. \end{aligned}$$

If λ is an eigenvalue and $y = y(x) = y(x, \lambda)$ is a corresponding eigenfunction, then $y'(a) \neq 0$ and there is a uniquely determined normalized eigenfunction with $y'(a) = 1$. The normalized eigenfunction solves the initial value problem

$$\begin{aligned} &-(pu)' + (q - \lambda r)u = 0, \quad a \le x \le b, \\ &u(a) = 0, \quad u'(a) = 1, \end{aligned}$$

and satisfies the equation $u(b) = u(b, \lambda) = 0$. Conversely, consider the foregoing initial value problem, depending on the parameter λ. If the u-initial value problem has a solution $u = u(x) = u(x, \lambda)$ that satisfies $u(b, \lambda) = 0$, then λ is an eigenvalue of the Sturm-Liouville problem and $y(x) = u(x, \lambda)$ is the corresponding normalized eigenfunction.

Thus, determining the eigenvalues and eigenfunctions of the given eigenvalue problem is equivalent conceptually and numerically to solving the initial value problem and then determining the eigenvalues and eigenfunctions through the zeros of the equation $u(b, \lambda) = 0$. The essence of an algorithm for this process follows:

Step 1. Determine an initial guess (approximate value) of an eigenvalue of interest. Even better, determine an interval that contains the eigenvalue.

Step 2. Solve the initial value problem

$$\begin{aligned} &-(pu)' + (q - \lambda r)u = 0, \quad a \le x \le b, \\ &u(a) = 0, \quad u'(a) = 1. \end{aligned}$$

for $u = u(x) = u(x, \lambda)$.

Step 3. If $u(b, \lambda) = 0$ (or is zero to within an acceptable error) STOP; λ is an eigenvalue (approximate eigenvalue) and $u(x, \lambda)$ is a corresponding eigenfunction (approximate eigenfunction). ELSE

Step 4. Use a root finder to update the current estimate of λ as a root of $u(b, \lambda) = 0$ and GO TO Step 1 with the updated λ.

So, λ is the shooting parameter. We shall show in the next two sections that a numerical implementation of this approach using the bisection method or Newton's method enables one to determine as accurately as desired a finite number of eigenvalues and corresponding eigenfunctions of the Sturm-Liouville problems (7.5). Then we discuss how to choose an initial guess in Step 1 and how to recognize which eigenvalue and eigenfunction has been found, based on theoretical results and on the numerical output obtained. Thus, if the eigenvalues are listed as

$$\lambda_0 < \lambda_1 < \cdots < \lambda_n < \cdots,$$

the shooting method can be used to approximate accurately any finite number of eigenvalues and corresponding eigenfunctions and to determine which eigenvalues in the list have been found.

7.1.2 Bisection Method and Convergence

Consider the Sturm-Liouville eigenvalue problem (7.5) expressed as

$$\begin{cases} -(p(x)y'(x))' + (q(x) - \lambda r(x))y(x) = 0, & a \leq x \leq b, \\ \alpha y(a) + \beta y'(a) = 0, \quad \gamma y(b) + \delta y'(b) = 0, \end{cases} \tag{7.6}$$

where the coefficients and constants satisfy our standing assumptions (1)–(4), and the related initial value problem

$$\begin{cases} -(p(x)u'(x))' + (q(x) - \lambda r(x))u(x) = 0, & a \leq x \leq b, \\ u(a) = -\beta/\sqrt{\alpha^2 + \beta^2}, \ u'(a) = \alpha/\sqrt{\alpha^2 + \beta^2}. \end{cases} \tag{7.7}$$

Since the eigenvalues λ of (7.6) are all simple and the corresponding eigenfunctions satisfy $\alpha y(a) + \beta y'(a) = 0$, the vector $\langle y(a), y'(a)\rangle$ is parallel to $(-\beta, \alpha)$ and there is a unique eigenfunction corresponding to each eigenvalue normalized by

$$y(a) = -\beta/\sqrt{\alpha^2 + \beta^2}, \ y'(a) = \alpha/\sqrt{\alpha^2 + \beta^2}.$$

Notice that the solution u to the initial value problem (7.7) is normalized in the same way and satisfies the boundary condition $\alpha u(a) + \beta u'(a) = 0$. Denote the global solution to the initial value problem by $u = u(x) = u(x, \lambda)$. Since the differential equation is linear, the solution is defined on $a \leq x \leq b$ by Theorem 82, whatever choice is made for λ. Just as in the case of Dirichlet boundary conditions, λ is an eigenvalue of the Sturm-Liouville eigenvalue problem (7.6) if and only if the initial value problem has a solution $u(x, \lambda)$ defined on $a \leq x \leq b$ that satisfies $F(\lambda) = 0$, where

$$F(\lambda) = \gamma u(b, \lambda) + \delta u'(b, \lambda),$$

in which case $y(x) = u(x, \lambda)$ is the corresponding normalized eigenfunction.

We know that $F(\lambda)$ has as its zeros the eigenvalues of the Sturm-Liouville eigenvalue problem. To be able to use the bisection method to find those zeros, we need to know that F changes sign at each eigenvalue. This follows from the next theorem that also plays a key role in the use of Newton's method in the next section. Under the standing assumptions, $u(x, \lambda)$ is continuously differentiable as a function of its variables. This assertion is a consequence of general continuous dependence results for ordinary differential equations. In particular, it is a direct consequence of Theorem 7.5 in Chapter 1 of [9]. Differentiation of the initial value problem (7.7) shows that $w = w(x) = w(x, \lambda) = \partial u(x, \lambda)/\partial\lambda$ satisfies the variational initial value problem

$$\begin{aligned}
&-(p(x)w'(x))' + (q(x) - \lambda r(x))w(x) = r(x)u(x), \quad a \le x \le b,\\
&w(a) = 0, \quad w'(a) = 0.
\end{aligned}$$

It also follows that $F(\lambda) = \gamma u(b, \lambda) + \delta u'(b, \lambda)$ is a continuously differentiable function of λ and that

$$F'(\lambda) = \gamma w(b, \lambda) + \delta w'(b, \lambda).$$

Theorem 186 *Under the standing assumptions (1)–(4), if λ is an eigenvalue of the Sturm-Liouville problem (7.6), then $F(\lambda) = 0$ and*

$$F'(\lambda) = \gamma w(b, \lambda) + \delta w'(b, \lambda) \neq 0.$$

Proof. The desired conclusion follows via variation of parameters. Let u be the solution to the initial value problem (7.7) and v be any solution to

$$-(p(x)v'(x))' + (q(x)v(x) - \lambda r(x))v(x) = 0$$

on $a \le x \le b$. If $\alpha \neq 0$

$$\begin{aligned}
p(a)\, W_{u,v}(a) &= p(a)\begin{vmatrix} u(a) & v(a) \\ u'(a) & v'(a) \end{vmatrix} \\
&= \alpha^{-1} p(a)\begin{vmatrix} \alpha u(a) + \beta u'(a) & \alpha v(a) + \beta v'(a) \\ u'(a) & v'(a) \end{vmatrix} \\
&= -\alpha^{-1} p(a) u'(a)(\alpha v(a) + \beta v'(a)).
\end{aligned}$$

A similar calculation if $\alpha = 0$ leads to

$$p(a)\, W_{u,v}(a) = \begin{cases} \beta^{-1} p(a) u(a)(\alpha v(a) + \beta v'(a)) & \text{if } \alpha = 0, \\ -\alpha^{-1} p(a) u'(a)(\alpha v(a) + \beta v'(a)) & \text{if } \alpha \neq 0. \end{cases}$$

Since $u(a) \neq 0$ when $\alpha = 0$ and $u'(a) \neq 0$ when $\alpha \neq 0$, it follows that $p(a)\, W_{u,v}(a) \neq 0$ and u and v are linearly independent if and only if v does not satisfy the initial condition that u satisfies. Fix such a $v = v(x) = v(x, \lambda)$. Then u and v are linearly independent and a repetition of the foregoing calculation at $x = b$ gives

$$p(b)\, W_{u,v}(b) = \begin{cases} \delta^{-1} p(b) u(b)(\gamma v(b) + \delta v'(b)) & \text{if } \gamma = 0, \\ -\gamma^{-1} p(b) u'(b)(\gamma v(b) + \delta v'(b)) & \text{if } \gamma \neq 0. \end{cases}$$

Since $p(b)\, W_{u,v}(b) = p(a)\, W_{u,v}(a) \neq 0$ by Lemma 86,

$$\gamma v(b, \lambda) + \delta v'(b, \lambda) \neq 0;$$

that is, v does not satisfy the boundary condition in the eigenvalue problem at $x = b$.

By variation of parameters (Theorem 87) with u and v as above and $f(x) = r(x)u(x)$, the variational initial value problem for w has solution

$$w(x) = A(x)u(x) + B(x)v(x),$$

where

$$A(x) = \frac{1}{p(a)\,W_{u,v}(a)} \int_a^x r(s)u(s)v(s)\,ds,$$

$$B(x) = -\frac{1}{p(a)\,W_{u,v}(a)} \int_a^x r(s)u(s)^2\,ds,$$

and the dependence on λ has been suppressed. Note that $B(b,\lambda) \neq 0$. Recall also that the coefficients in the variation of parameters solution are chosen so that

$$w'(x) = A(x)u'(x) + B(x)v'(x).$$

For an eigenvalue λ of the Sturm-Liouville problem,

$$F(\lambda) = \gamma u(b,\lambda) + \delta u'(b,\lambda) = 0$$

because $u = y$, the normalized eigenfunction, by the uniqueness of solutions to initial value problems, and

$$\begin{aligned}
F'(\lambda) &= \frac{\partial}{\partial \lambda}(\gamma u(b,\lambda) + \delta u'(b,\lambda)) = \gamma w(b,\lambda) + \delta w'(b,\lambda) \\
&= \gamma(A(b,\lambda)u(b,\lambda) + B(b,\lambda)v(b,\lambda)) \\
&\quad + \delta(A(b,\lambda)u'(b,\lambda) + B(b,\lambda)v'(b,\lambda)) \\
&= A(b,\lambda)(\gamma u(b,\lambda) + \delta u'(b,\lambda)) + B(b,\lambda)(\gamma v(b,\lambda) + \delta v'(b,\lambda)) \\
&= B(b,\lambda)(\gamma v(b,\lambda) + \delta v'(b,\lambda)) \neq 0
\end{aligned}$$

and the proof is complete. ■

It follows immediately from Theorem 186 that the bisection method can be used in principle to find each eigenvalue and corresponding normalized eigenfunction to any desired accuracy. Moreover, the numerically determined solutions to the u-initial value problem are approximate eigenfunctions that converge uniformly to the normalized eigenfunction corresponding to the given eigenvalue.

Theorem 187 *Under the standing assumptions (1)–(4), if λ is an eigenvalue of the Sturm-Liouville eigenvalue problem (7.6) and $y = y(x,\lambda)$ is the corresponding normalized eigenfunction, then there is an open interval $(\underline{\lambda}, \overline{\lambda})$ containing λ such that $F(\underline{\lambda})F(\overline{\lambda}) < 0$ and the bisection method can be used to generate a sequence of approximate eigenvalues $\lambda^{(n)} \to \lambda$ and corresponding approximate eigenfunctions $u(x,\lambda^{(n)})$ obtained by solving the u-initial value problem (7.7) such that $u(x,\lambda^{(n)}) \to y(x,\lambda)$ and $u'(x,\lambda^{(n)}) \to y'(x,\lambda)$ uniformly on $a \le x \le b$.*

Proof. By Theorem 186, there is a $\delta_1 > 0$ so that λ is the only zero of $F(\mu)$ in the interval $|\mu - \lambda| \le \delta_1$ and $F(\lambda - \delta_1)F(\lambda + \delta_1) < 0$. Thus $\underline{\lambda} = \lambda - \delta_1$ and $\overline{\lambda} = \lambda + \delta_1$ determine an interval of the required type. Consequently, the bisection method can be used to generate a sequence $\lambda^{(n)}$ of approximate zeros of F with $\lambda^{(n)} \to \lambda$ and corresponding solutions $u(x,\lambda^{(n)})$ to the u-initial value problem with parameter $\lambda^{(n)}$. By Theorem 88, given $\varepsilon > 0$ there is a $\delta_2 > 0$ so that $|\mu - \lambda| < \delta_2$ implies that the solution $u(x,\mu)$ to the u-initial value problem with parameter μ satisfies

$$|u(x,\mu) - y(x,\lambda)| < \varepsilon \quad \text{and} \quad |u'(x,\mu) - y'(x,\lambda)| < \varepsilon$$

for $a \le x \le b$. Since $\lambda^{(n)} \to \lambda$ as $n \to \infty$,

$$|u(x,\lambda^{(n)}) - y(x,\lambda)| < \varepsilon \qquad \text{and} \quad |u'(x,\lambda^{(n)}) - y'(x,\lambda)| < \varepsilon$$

for $a \le x \le b$ provided n is sufficiently large. ■

In Section 7.1.4 we discuss how to determine an interval $(\underline{\lambda}, \overline{\lambda})$ as in the theorem when numerical approximations of a certain eigenvalue and/or a corresponding eigenfunction are needed.

7.1.3 Newton's Method and Convergence

Theorem 186 is also a key step in establishing that the eigenvalues and eigenfunctions of the Sturm-Liouville eigenvalue problem (7.6) can be found using Newton's method as the root finder.

As before, $u = u(x) = u(x, \lambda)$ is the unique solution to the initial value problem (7.7) and $F(\lambda) = \gamma u(b, \lambda) + \delta u'(b, \lambda)$ is continuously differentiable.

Fix an eigenvalue λ of the eigenvalue problem (7.6). Since $F(\lambda) = 0$ and $F'(\lambda) \neq 0$, if $\lambda^{(0)}$ is a sufficiently good initial guess of λ, then all the Newton iterates

$$\lambda^{(n+1)} = \lambda^{(n)} - \frac{F(\lambda^{(n)})}{F'(\lambda^{(n)})}$$

are defined and $\lambda^{(n)} \to \lambda$ as $n \to \infty$. See Theorem 46. Now reasoning in the same way as for the bisection method, one obtains the following result.

Theorem 188 *Under the standing assumptions (1)–(4), if λ is an eigenvalue of the Sturm-Liouville eigenvalue problem (7.6) and $y = y(x, \lambda)$ is the corresponding normalized eigenfunction, then given a sufficiently good initial guess $\lambda^{(0)}$ of λ, Newton's method can be used to generate a sequence of approximate eigenvalues $\lambda^{(n)} \to \lambda$ and approximate eigenfunctions $u(x, \lambda^{(n)})$ obtained by solving the u-initial value problem (7.7) such that $u(x, \lambda^{(n)}) \to y(x, \lambda)$ and $u'(x, \lambda^{(n)}) \to y'(x, \lambda)$ uniformly on $a \leq x \leq b$.*

In Section 7.1.4 we discuss how to find a suitable initial guess $\lambda^{(0)}$ of an eigenvalue when numerical approximations of the eigenvalue and/or its corresponding eigenfunction are needed.

7.1.4 Numerical Results

We give several examples to illustrate the convergence results of the previous section. Additional results will be given in Chapter 8. First we discuss some practical aspects of implementing the shooting method and finding particular eigenvalues and eigenfunctions of interest. Often the hardest part of applying an iterative method is getting it started; that is, finding a good enough initial guess so that the process converges (converges numerically) to a desired answer in a reasonable time and with a reasonable expenditure of computing resources. The results of the last section show that the shooting method can be used in principle to determine all the eigenvalues and eigenfunctions of a regular Sturm-Liouville eigenvalue problem with separated boundary conditions using either the bisection method or Newton's method to update the shooting parameter. However, practical implementation of the method requires some educated guesswork to find a suitable pair of starting values for the bisection method or a suitable initial guess for Newton's method. Our experience, using Newton's method as root-finder, is that finding a suitable initial guess leading to the determination of some eigenvalue (perhaps not an eigenvalue of interest) is not difficult. The program we used was interactive and, once an initial value for the shooting parameter λ had been found for which numerical convergence to an eigenvalue was observed, a graph of the corresponding normalized eigenfunction was displayed. With this information in hand, the following properties of regular Sturm-Liouville eigenvalue problems helped to determine further initial guesses (if needed) that determined the eigenvalues and eigenfunctions of interest in a particular application.

SL-1. The eigenvalues are real, simple, and can be ordered as

$$\lambda_0 < \lambda_1 < \lambda_2 < \cdots$$

SL-2. The eigenfunction y_n corresponding to λ_n has exactly n nodal zeros (interior zeros where a sign change occurs) and no other interior zeros. Moreover, the nodes of y_{n-1} and y_n strictly interlace.

SL-3. If $\alpha\beta \leq 0, \gamma\delta \geq 0, p > 0, q \geq 0$, and $r(x) > 0$ on $[a, b]$, then all the eigenvalues are positive, except for the case when the eigenvalue problem is $-(py')' = \lambda ry,\ y'(a) = 0$, $y'(b) = 0$, in which case 0 is an eigenvalue and all the other eigenvalues are positive. (See Theorem 124.)

SL-4. If $\alpha\beta \leq 0,\ \gamma\delta \geq 0,\ p > 0,\ q \geq 0$, and $r(x) > 0$ on $[a, b]$, then $\lambda_0 \geq \min_{a\leq x\leq b}\dfrac{q(x)}{r(x)}$ and $\lambda_0 > 0$ unless the eigenvalue problem is $-(py')' = \lambda ry$, $y'(a) = 0$, $y'(b) = 0$, in which case 0 is an eigenvalue and all the other eigenvalues are positive. (See Theorem 124.)

The first inequality is SL-4 satisfied by λ_0 is a consequence of the standard argument leading to SL-3: With y_0 normalized by $\int_a^b y_0(x)^2 r(x)\,dx = 1$,

$$\lambda_0 = \int_a^b \lambda_0 y_0^2 r\,dx = \int_a^b (-(py_0')'y_0 + qy_0^2)\,dx$$

$$= \int_a^b (p(y_0')^2 + qy_0^2)\,dx - py_0y_0'\Big|_a^b$$

and $-py_0y_0'\big|_a^b$ is nonnegative for the given boundary data; hence,

$$\lambda_0 \geq \int_a^b q(x)y_0(x)^2\,dx \geq \int_a^b \frac{q(x)}{r(x)} y_0(x)^2 r(x)\,dx \geq \min_{a\leq x\leq b}\frac{q(x)}{r(x)}.$$

The lower bound in SL-4 holds with q only assumed to be continuous.

SL-5. If $\alpha\beta \leq 0,\ \gamma\delta \geq 0,\ p > 0$, and $r(x) > 0$ on $[a, b]$, then $\lambda_0 \geq \min_{a\leq x\leq b}\dfrac{q(x)}{r(x)}$.

Let $M = -\min_{a\leq x\leq b} q(x)/r(x)$. Then $Ly = \lambda ry$ with $y \neq 0$ if and only if

$$-(py')' + (q + Mr)y = (\lambda + M)ry$$

with $y \neq 0$ and

$$q + Mr = \left(\frac{q}{r} + M\right)r \geq \left(\min_{a\leq x\leq b}\frac{q(x)}{r(x)} + M\right)r = 0.$$

Consequently, by SL-3 $\lambda + M \geq 0$, equivalently,

$$\lambda \geq -M = \min_{a\leq x\leq b}\frac{q(x)}{r(x)}.$$

SL-6. *(Rayleigh Quotient)*

$$\lambda_0 = \min\frac{\langle Ly, y\rangle}{\langle y, y\rangle_r} = \min R(y),$$

where

$$R(y) = \frac{-pyy'\big|_a^b + \int_a^b (py'^2 + qy^2)\,dx}{\int_a^b y^2 r\,dx}$$

and the minimum is over all functions $y \neq 0$ in the domain of L that satisfy the boundary conditions $B_a y = 0$ and $B_b y = 0$. The minimum is achieved if and only if y is an eigenfunction corresponding to λ_0. (See Theorem 129.)

In Chapter 4 we established SL-1 and SL-2 for boundary conditions with $\alpha\beta \le 0$ and $\gamma\delta \ge 0$, the most important case for scientific and engineering applications. A proof for general separated boundary conditions can be found in [5], [9], or [10].

Let $F(\lambda) = \gamma u(b, \lambda) + \delta u'(b, \lambda)$ be the function introduced in the previous section, where $u(x, \lambda)$ is the solution to (7.7). We established that $F(\lambda)$ is a smooth function of λ for λ real, that it has only simple zeros, and those zeros are the eigenvalues of the Sturm-Liouville eigenvalue problem (7.6). Figure 7.1 illustrates such a function, see Example 1 below, and suggests a practical strategy for implementing the shooting method to determine eigenvalues and eigenfunctions of (7.6).

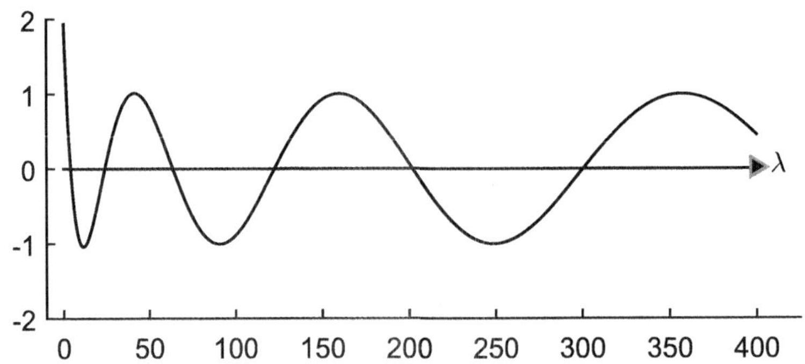

FIGURE 7.1: Graph of $F(\lambda) = \dfrac{\sin\sqrt{\lambda}}{\sqrt{\lambda}} + \cos\sqrt{\lambda}$

We apply the strategy first to determine numerically the smallest eigenvalue λ_0 and its corresponding eigenfunction y_0 using the shooting method and then using this information to systematically search for any additional eigenvalues and eigenfunctions that may be needed. We assume that Newton's method is used in the update step. As we said earlier and as Figure 7.1 suggests, it is not difficult to find a starting value for Newton's method that gives numerical convergence to some eigenvalue λ (perhaps not λ_0) and corresponding normalized eigenfunction $u(x, \lambda)$. A graph of $u(x, \lambda)$ reveals the number of its nodes. If there are, say, seven nodes, then $\lambda = \lambda_7$. Informed by this information, new starting values for Newton's method can be chosen and the shooting algorithm run again. This trial and error method leads to a determination of λ_0 in reasonably short time, usually a minute or two at most. The same approach can be used to locate other desired eigenvalues and eigenfunctions.

The trial and error approach can be refined if finding helpful starting values proves difficult. From SL-5 and SL-6, the smallest eigenvalue λ_0 of the eigenvalue problem satisfies

$$\min_{a \le x \le b} \frac{q(x)}{r(x)} \le \lambda_0 \le R(y) \tag{7.8}$$

for any function $y \neq 0$ in the domain of L that satisfies the boundary conditions $B_a y = 0$ and $B_b y = 0$. There is either a quadratic or cubic polynomial y with this property. Specifically, if y is expressed in powers of $(x - a)$,

$$y(x) = -\beta + \alpha(x - a) + c_2(x - a)^2 + c_3(x - a)^3$$

where

$$c_2 = -\frac{\alpha\delta - \gamma\beta + \alpha(b-a)}{\gamma(b-a)^2 + 2\delta(b-a)} \quad \text{and} \quad c_3 = 0$$

if

$$\gamma(b-a)^2 + 2\delta(b-a) \neq 0,$$

and where

$$c_2 = 0 \quad \text{and} \quad c_3 = -\frac{\alpha\delta - \gamma\beta + \alpha(b-a)}{\gamma(b-a)^3 + 3\delta(b-a)^2} = \frac{\alpha\delta - \gamma\beta + \alpha(b-a)}{\delta(b-a)^2}$$

if

$$\gamma(b-a)^2 + 2\delta(b-a) = 0.$$

Note that $\delta \neq 0$ in this case; else γ and δ would both be zero, a contradiction. The double inequality (7.8) helps to inform a trial and error approach for finding a starting value for the shooting parameter that gives convergence to λ_0. Further help in finding suitable initial guesses for Newton's method in sensitive cases can be found by graphing the function $F(\lambda)$ over some interval with left endpoint at most $\min_{a \le x \le b} q(x)/r(x)$. The standard fourth order ordinary differential equation solvers yield numerical versions of $u(x,\lambda)$ and $u'(x,\lambda)$ for λ at a suitable set of equally spaced points, say, and, hence, a numerical version of $F(\lambda)$. The graph of $F(\lambda)$ can be used to select useful starting values for Newton's method. The same strategies apply when the bisection method is used as the root-finder.

The numerical results in the following examples were obtained by the shooting method, using Newton's method to update the shooting parameter, and following the practical suggestions given earlier for determining starting values. Newton's method was stopped when $|F(\lambda)| < 10^{-6}$ and the change in magnitude of the shooting parameter was less than 10^{-6}. The algorithm was run on a standard desktop computer and convergence was obtained in a matter of seconds, once good starting values were determined.

Example 1. A regular Sturm-Liouville eigenvalue problem of the form

$$\begin{cases} -y'' = \lambda y, & 0 \le x \le 1 \\ y(0) = 0, & y(1) + y'(1) = 0, \end{cases}$$

arises in connection with heat conduction in a laterally insulated rod whose left end is held at temperature 0 and whose right end obeys Newton's law of cooling. All thermal coefficients have been set equal to 1 for simplicity. By SL-3 all the eigenvalues are positive. It is easy to show that the normalized eigenfunctions corresponding to the eigenvalues $\lambda_0, \lambda_1, \ldots$ are

$$y_n(x) = \frac{1}{\sqrt{\lambda_n}} \sin \sqrt{\lambda_n} x,$$

for $n = 0, 1, 2, \ldots$. In this example, the function $F(\lambda)$ of the previous section is

$$F(\lambda) = \frac{\sin\sqrt{\lambda}}{\sqrt{\lambda}} + \cos\sqrt{\lambda}$$

for $\lambda > 0$ and the eigenvalues $\lambda_0, \lambda_1, \ldots$ are its zeros.

The double inequality (7.8) for λ_0 applied with the quadratic polynomial $y = x - 2x^2/3$ yields the bounds

$$0 \le \lambda_0 \le R(y) = \frac{-yy'\big|_0^1 + \int_0^1 y'^2\,dx}{\int_0^1 y^2\,dx} = \frac{1/9 + 7/27}{4/45} = \frac{25}{6} < 4.2.$$

The shooting method produced the following approximations of the first five eigenvalues of this eigenvalue problem. We have found that a flexible doubling and/or halving procedure either of previous initial guesses or previously determined approximate eigenvalues together with a little thought is an effective means for finding an initial guess that results in convergence to a desired eigenvalue. The table shows all the initial guesses used, in the order they were used, to find the first five eigenvalues and corresponding eigenfunctions. The first column shows which eigenvalue was found with the corresponding initial guess.

n	**Initial Guess**	$\lambda_n \approx$	**Check**	**Relative Error**
0	2, 8	4.1159	4.1159	0
1	16, 32	24.1391	24.1393	-8.2852×10^{-6}
2	64	63.6479	63.6591	-1.7594×10^{-4}
3	128	122.8674	122.8892	-1.7740×10^{-4}
6	256	418.8989	418.9868	-2.1457×10^{-4}
4	192	201.8160	201.8513	-1.7488×10^{-4}

Since $25/6 \approx 4.1667$, the Rayleigh quotient of $y = x - 2x^2/3$ is itself a rather good approximation of the smallest eigenvalue. The Check column lists the first five positive zeros of $F(\lambda)$ found using Newton's method. The relative error calculation is made regarding the approximate zeros of $F(\lambda)$ as if they were the exact eigenvalues. Graphs of the corresponding normalized eigenfunctions y_0, y_1, y_2, y_3, and y_4 are shown in Figure 7.2. The graphs show the nodal interlacing property in SL-2.

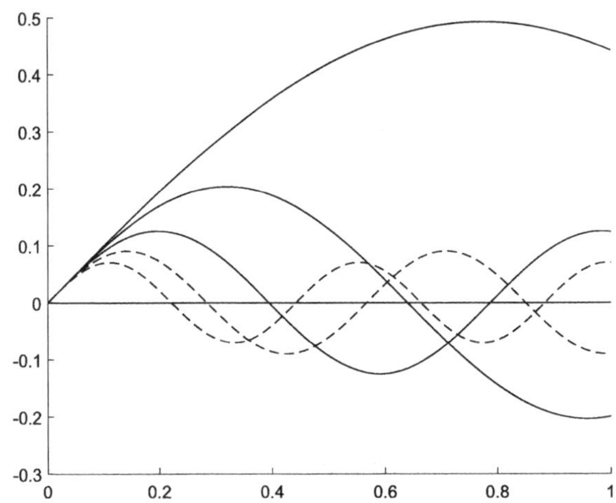

FIGURE 7.2: Eigenfunctions for Example 1

The graph in Figure 7.2 with no nodes (interior zeros at which a sign change occurs) was obtained with initial guess 2. Therefore it must be the eigenfunction y_0 belong to the smallest eigenvalue λ_0. The other eigenvalue-eigenfunction pairs are identified in the same way based on the number of nodes of the eigenfunction. The first initial guess for λ_4, 256, converged to an eigenvalue whose corresponding eigenfunction had 6 nodes; that initial guess gave convergence to λ_6. The next initial guess for λ_4 halfway between 128 and 256 gave convergence to λ_4.

Example 2. Use the shooting method to find the first five eigenvalues and eigenfunctions of the regular Sturm-Liouville eigenvalue problem

$$\begin{cases} -y'' + xy = \lambda(\cos x)y, & 0 \le x \le 1, \\ y(0) = 0, \ y(1) = 0. \end{cases}$$

This eigenvalue problem has weight function $r(x) = \cos x$. The double inequality (7.8) for λ_0 applied with the quadric polynomial $y = x - x^2$ yields the bounds

$$0 \le \lambda_0 \le R(y) = \frac{-yy'\big|_0^1 + \int_0^1 (y'^2 + xy^2)\,dx}{\int_0^1 y^2 \cos x\,dx} = \frac{0 + 7/20}{22\sin 1 - 12(1 + \cos 1)}$$

So $R(y) \approx 12.1807$. The shooting method with the indicated initial guesses lead to the following approximations of the first five eigenvalues. The initial guesses shown were the first ones tried in the search process. The following table shows all the initial guesses used, in the order they were used, to find the first five eigenvalues and corresponding eigenfunctions. The first column shows which eigenvalue was found with the corresponding initial guess.

n	**Initial Guess**	$\lambda_n \approx$
0	11, 20	11.9548
1	40	47.3785
2	80	106.4274
3	160	189.1125
4	320	295.4207

Just as in Example 1, the Rayleigh quotient of $y = x - x^2$ is itself a rather good approximation of the smallest eigenvalue. Graphs of the corresponding normalized eigenfunctions y_0, y_1, y_2, y_3, and y_4 are shown in Figure 7.3.

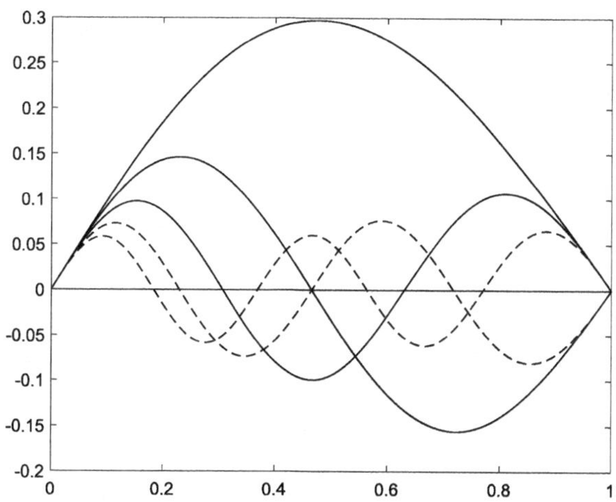

FIGURE 7.3: Eigenfunctions for Example 2

In Example 2 the function $F(\lambda)$ is not known explicitly, although a numerical approximation could be generated using an initial value problem solver, as we mentioned earlier. So what evidence is there, beyond the convergence theory of the last section and the suggestive numerical output, to support the belief that the approximate eigenvalues in the table are accurate?

If the shooting method converges numerically to $\tilde{\lambda}$, then an initial value problem solver can be used to evaluate $F(\tilde{\lambda} - \varepsilon)$ and $F(\tilde{\lambda} + \varepsilon)$ for some $\varepsilon > 0$. If ε can be chosen so that $F(\tilde{\lambda} - \varepsilon)$ and $F(\tilde{\lambda} + \varepsilon)$ are of opposite sign, then $\tilde{\lambda}$ approximates an eigenvalue λ of the eigenvalue problem to within an error of at most ε. A plot of $u(x, \tilde{\lambda})$ will reveal the number of nodes of the approximate eigenfunction and, therefore, which eigenvalue has been approximated to accuracy ε. Since $\tilde{\lambda}$ is almost certainly not exactly an eigenvalue, $F(\tilde{\lambda}) \neq 0$ and hence

$F(\tilde{\lambda}-\varepsilon)$ and $F(\tilde{\lambda}+\varepsilon)$ have the same sign for $\varepsilon > 0$ suitably small. Our experience is that by experimenting with different choices of ε reasonably small a sign change can be detected.

Example 2 (continued) We found $\lambda_0 \approx \tilde{\lambda} = 11.954818$ correctly rounded, where two more digits of the numerical output are shown here. In Example 2, $F(\lambda) = u(1, \lambda)$, where u is the solution of the initial value problem

$$\begin{cases} -u'' + xu = \lambda(\cos x)u, & 0 \le x \le 1, \\ u(0) = 0, \ u'(0) = 1. \end{cases}$$

Numerical experiments with different choices for potential error bounds leads to $F(\tilde{\lambda} - 3 \times 10^{-7}) \approx 2.0 \times 10^{-9}$ and $F(\tilde{\lambda} + 3 \times 10^{-7}) \approx -2.5 \times 10^{-8}$. It follows from the intermediate value theorem that there is an eigenvalue that differs from $\tilde{\lambda}$ by at most 3×10^{-7}. Since $u(x, \tilde{\lambda})$ has no nodes in $0 < x < 1$, we conclude that $|\tilde{\lambda} - \lambda_0| < 3 \times 10^{-7}$. Since $F(\tilde{\lambda} - 2 \times 10^{-7}) \approx -2.5 \times 10^{-9}$ and $F(\tilde{\lambda} + 2 \times 10^{-7}) \approx -2.1 \times 10^{-8}$ we further conclude that $2 \times 10^{-7} < |\tilde{\lambda} - \lambda_0| < 3 \times 10^{-7}$, which shows that the error bound 3×10^{-5} is reasonably sharp.

Another way to test the approximate eigenvalues for accuracy follows. Consider the regular eigenvalue problem

$$\begin{cases} -(py')' + qy = \lambda ry, & a \le x \le b, \\ y(a) = 0, \ y(b) = 0. \end{cases} \tag{7.9}$$

Let λ be an eigenvalue and $y(x)$ be its corresponding normalized eigenfunction, so that $y'(a) = 1$. Define $u(t) = y(x)$, $p_1(t) = p(x)$, $q_1(t) = q(x)$, $r_1(t) = r(x)$ where $t = kx$ for $a \le x \le b$ and k is a given positive constant. Since $\dot{u} = y'/k$ where a dot denotes differentiation with respect to t, u satisfies the differential equation

$$-k^2(p_1\dot{u})^{\cdot} + q_1 u = \lambda r_1 u, \quad ka \le t \le kb$$

as well as the conditions

$$u(ka) = y(a) = 0, \ u(kb) = y(b) = 0, \ \dot{u}(ka) = y'(a)/k = 1/k.$$

Thus, if λ, $y(x)$ is an eigenvalue, normalized eigenfunction pair of (7.9) and the positive constant k is chosen to be $k = \sqrt{\lambda}$, then $u(t)$ satisfies the initial value problem

$$\begin{cases} -(p_1\dot{u})^{\cdot} + (q_1/k^2)u = r_1 u, & ka \le t \le kb, \\ u(ka) = 0, \ \dot{u}(ka) = 1/k \end{cases} \tag{7.10}$$

as well as the condition $u(kb) = 0$. Conversely, if for some $k > 0$, the solution $u(t)$ to this initial value problem also satisfies $u(kb) = 0$, then $\lambda = k^2$, $y(x) = u(kx)$ for $a \le x \le b$ is an eigenvalue, normalized eigenfunction pair for (7.9). These observations lead to the following check on the eigenvalues found by shooting: for each eigenvalue λ of (7.9) found by shooting, solve the initial value problem (7.10), evaluate $u(kb)$, and compare this value to 0. If λ is an exact (approximate) eigenvalue, then $u(kb)$ is exactly (approximately) 0. We chose Dirichlet boundary data in (7.9) for simplicity. The same approach can be used with any separated boundary conditions.

Example 2 (continued) Apply the test above with $k = \sqrt{\lambda}$ to the five eigenvalues λ in the table in Example 2. That is, solve the initial value problem

$$\begin{cases} -(\dot{u})^{\cdot} + (t/k^3)u = (\cos(t/k))u, & 0 \le t \le k, \\ u(0) = 0,\ \dot{u}(0) = 1/k \end{cases}$$

numerically and compare the values for $u(k)$ to 0. The following table shows the comparison and strongly suggests that the numerical approximation of λ_n is quite accurate.

n	$\lambda_n \approx$	$u(k) \approx$
0	11.9548	7.9920×10^{-7}
1	47.3785	3.5285×10^{-8}
2	106.4274	7.1805×10^{-7}
3	189.1125	1.4575×10^{-8}
4	295.4207	9.3689×10^{-8}

Moreover, the graphs of $y(x) = u(kx)$ for $0 \le x \le 1$ have the expected number of nodes in $(0, 1)$, consistent with the fact that $y(x)$ is a corresponding eigenfunction if λ is an eigenvalue. Altogether, these results add considerable confidence to belief that the shooting method has produced accurate approximations to the first five eigenvalues and corresponding normalized eigenfunctions of the eigenvalue problem in Example 2.

Our experience is that finding good starting values for problems with Neumann boundary conditions is more challenging than for other boundary conditions. The next example illustrates what often happens.

Example 3. Use the shooting method to find the first five eigenvalues and eigenfunctions of the regular Sturm-Liouville eigenvalue problem

$$\begin{cases} -y'' + xy = \lambda(\cos x)y, & 0 \le x \le 1, \\ y'(0) = 0,\ y'(1) = 0. \end{cases}$$

As in Example 2, the eigenvalue problem has weight function $r(x) = \cos x$ and the double inequality (7.8) for λ_0 applied with the quadric polynomial $y = 1$ yields the bounds

$$0 \le \lambda_0 \le R(y) = \frac{-yy'\big|_0^1 + \int_0^1 (y'^2 + xy^2)\,dx}{\int_0^1 y^2 \cos x\,dx} = \frac{0 + 1/2}{\sin 1} \approx 0.5942$$

The following table shows all the initial guesses used, in the order they were used, to find the first five eigenvalues and corresponding eigenfunctions. The first column shows which eigenvalue was found with the corresponding initial guess.

n	**Initial Guess**	$\lambda_n \approx$
0	0.3	0.5782
1	10, 20	13.0166
2	40	48.4866
4	80	190.2506
2	60	48.4866
0	70	0.5782
2	50	48.3785
3	90	107.5585

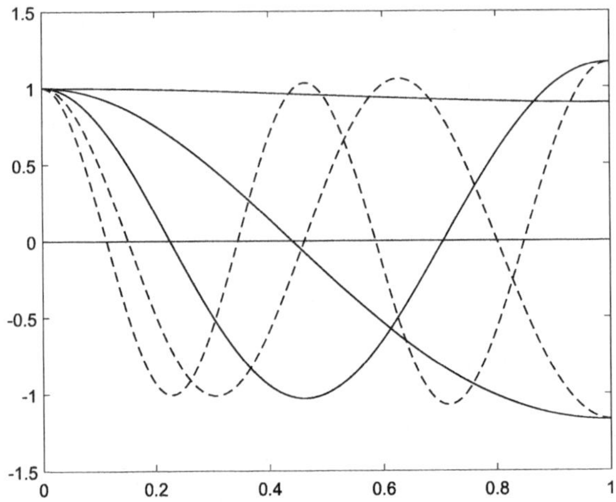

FIGURE 7.4: Eigenfunctions for Example 3

Once again, the Rayleigh quotient of $y=1$ is itself a rather good approximation of the smallest eigenvalue. Figure 7.4 shows the first five corresponding normalized eigenfunctions. The graphs show the nodal interlacing properties in SL-2. The table also illustrates that finding good initial guesses with the strategy suggested in Example 1 proves most challenging when the boundary conditions are of Neumann type.

Example 3 (continued) We found $\lambda_2 \approx \tilde{\lambda} = 48.486581$ correctly rounded, where two more digits of the numerical output are shown here. In Example 3, $F(\lambda) = u'(1, \lambda)$, where u is the solution of the initial value problem

$$\begin{cases} -u'' + xu = \lambda(\cos x)u, & 0 \le x \le 1, \\ u(0) = 1,\ u'(0) = 0. \end{cases}$$

Numerical experiments with different choices for potential error bounds lead to $F(\tilde{\lambda} - 5 \times 10^{-7}) \approx 3.8 \times 10^{-7}$ and $F(\tilde{\lambda} + 5 \times 10^{-7}) \approx -2.0 \times 10^{-8}$. It follows from the intermediate value theorem that there is an eigenvalue that differs from $\tilde{\lambda}$ by at most 5×10^{-7}. Since $u(x, \tilde{\lambda})$ has two nodes in $0 < x < 1$, we conclude that $|\tilde{\lambda} - \lambda_2| < 5 \times 10^{-7}$. Since $F(\tilde{\lambda} - 4 \times 10^{-7}) \approx 3.4 \times 10^{-7}$ and $F(\tilde{\lambda} + 4 \times 10^{-7}) \approx 2.0 \times 10^{-8}$ we further conclude that $4 \times 10^{-7} < |\tilde{\lambda} - \lambda_2| < 5 \times 10^{-7}$, which shows that the error bound 5×10^{-7} is reasonably sharp.

7.2 Singular Problems - I

In this section the shooting method for regular problems is adjusted to cover the singular Sturm-Liouville eigenvalue problems treated in Chapter 5:

$$\begin{cases} -(p(x)y')' + q(x)y = \lambda r(x)y, & a < x < b, \\ |y(a)| < \infty, & \\ \gamma y(b) + \delta y'(b) = 0, & |\gamma| + |\delta| \neq 0, \end{cases} \tag{7.11}$$

The standing assumptions for eigenvalue problems of Chapter 5 remain in force here:

(1) $p(x) \geq 0$ is continuous on $[a, b]$, is differentiable at $x = a$, is nonzero on $a < x \leq b$, and satisfies $p(a) = 0$, $p'(a) \neq 0$.

(2) $q(x)$ is continuous on $[a, b]$.

(3) $p(x)$ and $q(x)$ are real-valued and γ and δ are real numbers.

(4) The weight function $r(x)$ is continuous on $[a, b]$ and either $r(x) > 0$ on $[a, b]$ or $r(x) = (x - a)^m \rho(x)$ where $m > 0$ and $\rho(x) > 0$ is continuous on $a \leq x \leq b$.

Just as for regular problems, the numerical solution procedure will use the variational equation associated with the differential equation in (7.11). Therefore, we further assume that

(5) $p(x)$, $q(x)$, and $r(x)$ are continuously differentiable on $[a, b]$.

By Lemma 148 any eigenfunction $y(x)$ of (7.11) extends to a continuously differentiable function on $[a, b]$, still denoted by $y(x)$, that satisfies the differential equation at $x = a$ and $x = b$ and satisfies the initial condition

$$(q(a) - \lambda r(a))y(a) - p'(a)y'(a) = 0.$$

Since nontrivial bounded solutions of the singular Sturm-Liouville differential equation in (7.11) satisfy $y(a) \neq 0$ by Theorem 136, we normalize eigenfunctions by requiring that

$$y(a) = 1.$$

Consequently, a normalized eigenfunction satisfies the initial conditions

$$y(a) = 1, \;\; y'(a) = \frac{q(a) - \lambda r(a)}{p'(a)}.$$

7.2.1 The Shooting Method

The properties of a normalized eigenfunction to (7.11) lead to a shooting method with shooting parameter λ based on the initial value problem

$$\begin{cases} -(p(x)u'(x))' + (q(x) - \lambda r(x))u(x) = 0, & a \leq x \leq b, \\ u(a) = 1, \; u'(a) = (q(a) - \lambda r(a))/p'(a). \end{cases} \tag{7.12}$$

This initial value problem has a unique solution $u = u(x) = u(x, \lambda)$ that is continuously differentiable on $[a, b]$ by Theorem 138 applied with $c_0 = 1$, $c_1 = (q(a) - \lambda r(a))/p'(a)$, and $f(x) = 0$. If λ is an eigenvalue of (7.11), then the normalized eigenfunction $y(x)$ is the unique solution to (7.12) and has the additional property that $\gamma y(b) + \delta y'(b) = 0$. Conversely if $u(x)$ satisfies (7.12) for some λ for which $\gamma u(b) + \delta u'(b) = 0$, then λ is an eigenvalue of (7.11) and $u(x)$ is the normalized eigenfunction corresponding to λ. In summary, just as for the regular eigenvalue problem, finding an eigenvalue λ and corresponding normalized eigenfunction $y(x)$ is equivalent to finding a value of λ such that (7.12) has a solution $u(x)$ that satisfies $\gamma u(b) + \delta u'(b) = 0$.

Reasoning very much as in the regular case establishes that the shooting method can be used with either the bisection method or Newton's method to determine, as accurately as desired, a finite number of eigenvalues and corresponding eigenfunctions of the singular Sturm-Liouville problems of Chapter 5. The approach just described is based on a different variation of parameters formula from the familiar one for regular problems. We have not seen this formula elsewhere. Consequently, we conclude this section with a statement and proof of the formula for the singular differential equations in Chapter 5.

Theorem 189 (Variation of Parameters) *Fix λ. Let $u(x) = u(x, \lambda)$ be the solution to (7.12), let $v(x) = v(x, \lambda)$ be a solution to the differential equation in (7.12) that is linearly independent of $u(x)$ on $(a, b]$, and let $g(x)$ be continuous on $[a, b]$. Under the standing assumptions (1)–(4), the initial value problem*

$$\begin{cases} -(p(x)z')' + (q(x) - \lambda r(x))z = g(x), & a < x \leq b, \\ z(a) = 0,\ z'(a) = -g(a)/p'(a). \end{cases} \tag{7.13}$$

has the unique solution z that extends to a continuously differentiable on $[a, b]$ and is given by

$$z(x) = \begin{cases} A(x)u(x) + B(x)v(x) & \text{for } a < x \leq b \\ 0 & \text{for } x = a \end{cases} \tag{7.14}$$

where

$$A(x) = \int_a^x \frac{v(s)g(s)}{p(s)\,W_{u,v}(s)}\,ds, \quad B(x) = -\int_a^x \frac{u(s)g(s)}{p(s)\,W_{u,v}(s)}\,ds,$$

$W_{u,v}(x)$ is the Wronskian of u and v, and $p(x)\,W_{u,v}(x) = C$ a nonzero constant on $a < x \leq b$. Moreover, $z(x)$ in (7. 14) also satisfies the differential equation at $x = a$.

Proof. We have already established in Theorem 138 that (7.13) has a unique continuously differentiable solution z on $[a, b]$, but we did not obtain the explicit representation (7.14) for the solution.

It remains to prove that the solution z is given explicitly by (7.14). By Lemma 86 $p(x)\,W_{u,v}(x) = C$ a nonzero constant on $a < x \leq b$. Hence, both improper integrals for $A(x)$ and $B(x)$ converge because $v(x)$ grows logarithmically as $x \to a$ (Theorem 136) and $u(x)g(x)$ is continuous on $[a, b]$. Consequently,

$$\lim_{x \to a} A(x) = 0 \quad \text{and} \quad \lim_{x \to a} B(x) = 0.$$

It is convenient to define $A(a) = 0$ and $B(a) = 0$ so that A and B are continuous on $[a, b]$. The theorem can be established by reasoning much as in the regular case; however, it is easier to simply check directly that the proposed solution formula for $z(x)$ has all the required properties. This follows once we establish:

A. There exists $\lim_{x \to a} z(x) = 0$.

B. There exists $\lim_{x \to a} z'(x) = -g(a)/p'(a)$.

C. $z(x)$ satisfies the differential equation in (7.13).

To prove A note that

$$\lim_{x\to a} z(x) = \lim_{x\to a} A(x)u(x) + \lim_{x\to a} B(x)v(x)$$

because both limits on the right exist: first,

$$\lim_{x\to a} A(x)u(x) = 0 \cdot 1 = 0.$$

Second, by the mean value theorem for integrals,

$$B(x) = -\int_a^x \frac{u(s)g(s)}{C}\, ds = -\frac{u(s_x)g(s_x)}{C}(x-a)$$

for some s_x is between a and x. Since $v(x)$ becomes infinite like $\ln(x-a)$ as $x \to a$ by Theorem 136 and $u(x)g(x)$ is continuous on $[a, b]$,

$$\lim_{x\to a} B(x)v(x) = 0.$$

Combining these results gives

$$\lim_{x\to a} z(x) = 0,$$

which establishes A and the continuity of $z(x)$ at $x = a$. Since $z(x) = A(x)u(x) + B(x)v(x)$ is continuous on $a < x \leq b$, it follows that $z(x)$ is continuous on $[a, b]$.

We establish B in a similar way starting with the observation that

$$A'(x)u(x) + B'(x)v(x) = \frac{v(x)g(x)}{C}u(x) - \frac{u(x)g(x)}{C}v(x) = 0$$

for $a < x \leq b$, and, hence,

$$z'(x) = A(x)u'(x) + B(x)v'(x)$$

there. Now,

$$\lim_{x\to a} A(x)u'(x) = 0 \cdot u'(a) = 0.$$

Since $p(x)\, W_{u,v}(x) = C$,

$$v'(x) = \frac{u'(x)}{u(x)}v(x) + \frac{C}{p(x)u(x)}$$

holds for $x > a$ and near a; hence, using the mean value theorem for integrals,

$$B(x)v'(x) = -\frac{u(s_x)g(s_x)}{C}(x-a)\left(\frac{u'(x)}{u(x)}v(x) + \frac{C}{p(x)u(x)}\right)$$

for some s_x is between a and x. Since

$$\lim_{x\to a} (x-a)v(x) = 0$$

because $v(x)$ becomes logarithmically infinite as x approaches a and

$$\lim_{x\to a} \frac{x-a}{p(x)} = \lim_{x\to a} \frac{x-a}{p(x)-p(a)} = \frac{1}{p'(a)},$$

it follows that

$$\lim_{x\to a} B(x)v'(x) = -\frac{u(a)g(a)}{C}\frac{C}{p'(a)u(a)} = -\frac{g(a)}{p'(a)},$$

which establishes B. The continuity of z on $[a, b]$ and B imply that there exists $z'(a) = -g(a)/p'(a)$ and that z' is continuous at $x = a$ by Lemma 11. The expression $z' = Au' + Bv'$ shows that z' is continuous on $a < x \leq b$. Thus, z is continuously differentiable on $[a, b]$.

To establish C, observe that

$$\begin{aligned} A'(x)(p(x)u'(x)) + B'(x)(p(x)v'(x)) &= \frac{v(x)g(x)}{C}(p(x)u'(x)) \\ &\quad - \frac{u(x)g(x)}{C}(p(x)v'(x)) \\ &= -\frac{g(x)}{C}p(x)\,W_{u,v}(x) = -g(x) \end{aligned}$$

and recall from the proof of B that

$$A'(x)u(x) + B'(x)v(x) = 0.$$

Consequently for $a < x \leq b$,

$$z(x) = A(x)u(x) + B(x)v(x)$$

satisfies

$$\begin{aligned} z'(x) &= A(x)u'(x) + B(x)v'(x), \\ p(x)z'(x) &= A(x)(p(x)u'(x)) + B(x)(p(x)v'(x)), \\ (p(x)z'(x))' &= A(x)(p(x)u'(x))' + B(x)(p(x)v'(x))' - g(x), \\ (q(x) - \lambda r(x))z(x) &= A(x)(q(x) - \lambda r(x))u(x) + B(x)(q(x) - \lambda r(x))v(x) \end{aligned}$$

and, hence,

$$-(p(x)z'(x))' + (q(x) - \lambda r(x))z(x) = A(x) \cdot 0 + B(x) \cdot 0 + g(x),$$

which establishes C for $a < x \leq b$. Finally since

$$-(p(x)z'(x))' = -(q(x) - \lambda r(x))z(x) + g(x)$$

for $a < x \leq b$,

$$\lim_{x \to a}(-(pz')'(x)) = -(q(a) - \lambda r(a))z(a) + g(a)$$

and pz' is continuous on $[a, b]$, there exists

$$-(pz')'(a) = -(q(a) - \lambda r(a))z(a) + g(a)$$

by Lemma 11. So $z(x)$ satisfies the differential equation at $x = a$ and the proof is complete. ■

The proof shows that $\lim_{x \to a}(A(x)u(x) + B(x)v(x)) = 0$. So the variation of parameters solution can be expressed simply as $z(x) = A(x)u(x) + B(x)v(x)$ if $A(x)u(x) + B(x)v(x)$ is interpreted at $x = a$ as its limiting value as x tends to a. Furthermore, the proof also shows that $A(x)$ and $B(x)$ are chosen so that

$$z'(x) = A(x)u'(x) + B(x)v'(x),$$

a result that plays an essential role in the convergence analysis that follows.

7.2.2 Bisection Method and Convergence

The convergence analysis that follows parallels closely that in Section 7.1.2 for regular problems. Recall that λ is an eigenvalue of the eigenvalue problem (7.11) if and only if $u(x) = u(x,\lambda)$ solves the initial value problem (7.12) and also satisfies $\gamma u(b,\lambda) + \delta u'(b,\lambda) = 0$, in which case $u(x)$ is the normalized eigenfunction belonging to λ. Let

$$F(\lambda) = \gamma u(b,\lambda) + \delta u'(b,\lambda).$$

Under our standing assumptions (1)–(5), the solution $u(x,\lambda)$ of (7.12) satisfies a regular Sturm-Liouville differential equation when x is restricted to $[c,b]$ for any fixed $c > a$ in (a,b). It follows from continuous dependence results in Section 7 in Chapter 1 of [9] (see especially Theorem 7.4 and the subsequent material) that the solution $u(x,\lambda)$ is continuously differentiable in x and λ for x in $[c,b]$ and λ in any bounded interval. Since $c > a$ in (a,b) is arbitrary, $u(x,\lambda)$ is continuously differentiable in x and λ for x in $a < x \le b$ and λ in any bounded interval.

Theorem 190 *Under the standing assumptions (1)–(5), if λ is an eigenvalue of the Sturm-Liouville problem (7.11), then $F(\lambda) = 0$ and*

$$F'(\lambda) = \gamma w(b,\lambda) + \delta w'(b,\lambda) \neq 0$$

where $w = w(x,\lambda) = \partial u(x,\lambda)/\partial\lambda$ and $y = u(x,\lambda)$ is the corresponding normalized eigenfunction.

Proof. Let $y(x) = y(x,\lambda)$ be the normalized eigenfunction corresponding to the eigenvalue λ. Then $u(x) = y(x) = y(x,\lambda)$ is the unique solution to (7.12) and $F(\lambda) = 0$ because the eigenfunction satisfies the boundary condition $\gamma u(b,\lambda) + \delta u'(b,\lambda) = 0$.

It remains to show that $F'(\lambda) \neq 0$. Let $v(x) = v(x,\lambda)$ be a solution of the differential equation in (7.12) that is linearly independent of $u(x)$ so that $p(x)\,W_{u,v}(x) = C \neq 0$ on $a < x \le b$, where C is a nonzero constant. Consequently, $W_{u,v}(b) \neq 0$. If $\gamma \neq 0$ in the boundary condition at $x = b$, then $u'(b) \neq 0$ because otherwise the boundary condition implies $u(b) = 0$ in which case $u = 0$, contradicting the fact that u is an eigenfunction. Furthermore,

$$0 \neq W_{u,v}(b) = \begin{vmatrix} u(b) & v(b) \\ u'(b) & v'(b) \end{vmatrix} = \gamma^{-1} \begin{vmatrix} \gamma u(b) + \delta u'(b) & \gamma v(b) + \delta v'(b) \\ u'(b) & v'(b) \end{vmatrix}$$
$$= -\gamma^{-1} u'(b)(\gamma v(b) + \delta v'(b));$$

hence, $\gamma v(b) + \delta v'(b) \neq 0$. Likewise, if $\delta \neq 0$, then $u(b) \neq 0$,

$$W_{u,v}(b) = \delta^{-1} u(b)(\gamma v(b) + \delta v'(b))$$

and $\gamma v(b) + \delta v'(b) \neq 0$. Since one of γ and δ is nonzero, it follows that

$$\gamma v(b) + \delta v'(b) \neq 0.$$

As noted earlier, under our standing assumptions, $u = u(x,\lambda)$ is a continuously differentiable function of both its variables for $a < x \le b$ and λ in any bounded interval. Consequently, we can differentiate the initial value problem for $u = u(x,\lambda)$ with respect to the parameter λ to find that $w = w(x) = w(x,\lambda) = \partial u(x,\lambda)/\partial\lambda$ satisfies the variational initial value problem

$$\begin{cases} -(p(x)w'(x))' + (q(x) - \lambda r(x))w(x) = r(x)u(x), & a < x \le b, \\ w(a) = 0,\ w'(a) = -r(a)/p'(a). \end{cases}$$

Apply Theorem 189 with $g(x) = r(x)u(x)$ to express the solution to the variational initial value problem as

$$w(x) = \begin{cases} A(x)u(x) + B(x)v(x) & \text{for } a < x \leq b \\ 0 & \text{for } x = a \end{cases}$$

where

$$A(x) = \int_a^x \frac{v(s)r(s)u(s)}{C}\,ds \quad \text{and} \quad B(x) = -\int_a^x \frac{r(s)u(s)^2}{C}\,ds \neq 0$$

and the dependence on λ has been suppressed. Recall also that the coefficients in the variation of parameters solution are chosen so that

$$w'(x) = A(x)u'(x) + B(x)v'(x).$$

For an eigenvalue λ of the Sturm-Liouville problem (7.11)

$$F(\lambda) = \gamma u(b,\lambda) + \delta u'(b,\lambda) = 0$$

because $u = y$, the normalized eigenfunction, by the uniqueness of solutions to initial value problems. Furthermore,

$$\begin{aligned} F'(\lambda) &= \frac{\partial}{\partial\lambda}(\gamma u(b,\lambda) + \delta u'(b,\lambda)) = \gamma w(b,\lambda) + \delta w'(b,\lambda) \\ &= \gamma(A(b,\lambda)u(b,\lambda) + B(b,\lambda)v(b,\lambda)) \\ &\quad + \delta(A(b,\lambda)u'(b,\lambda) + B(b,\lambda)v'(b,\lambda)) \\ &= A(b,\lambda)(\gamma u(b,\lambda) + \delta u'(b,\lambda)) + B(b,\lambda)(\gamma v(b,\lambda) + \delta v'(b,\lambda)) \\ &= B(b,\lambda)(\gamma v(b,\lambda) + \delta v'(b,\lambda)) \neq 0 \end{aligned}$$

and the proof is complete. ■

It follows immediately from Theorem 190 that the bisection method can be used to find each eigenvalue to any desired accuracy. Moreover, the numerically determined solutions to the u-initial value problem are approximate eigenfunctions that converge uniformly to the normalized eigenfunction corresponding to the given eigenvalue.

Theorem 191 *Under the standing assumptions (1)–(5), if λ is an eigenvalue of the Sturm-Liouville eigenvalue problem (7.11) and $y = y(x,\lambda)$ is the corresponding normalized eigenfunction, then the bisection method can be used to generate a sequence of approximate eigenvalues $\lambda^{(n)} \to \lambda$ and corresponding approximate eigenfunctions $u(x,\lambda^{(n)})$ obtained by solving the u-initial value problem (7.12) such that $u(x,\lambda^{(n)}) \to y(x,\lambda)$ and $u'(x,\lambda^{(n)}) \to y'(x,\lambda)$ uniformly on $a \leq x \leq b$.*

Proof. By Theorem 190, there is a $\delta_1 > 0$ so that λ is the only zero of $F(\mu)$ in the interval $|\mu - \lambda| < \delta_1$ and F changes sign at λ. Consequently, the bisection method can be used to generate a sequence $\lambda^{(n)}$ of approximate zeros of F with $\lambda^{(n)} \to \lambda$ and corresponding solutions $u(x,\lambda^{(n)})$ to the u-initial value problem with parameter $\lambda^{(n)}$. By Theorem 137, given $\varepsilon > 0$ there is a $\delta_2 > 0$ so that $|\mu - \lambda| < \delta_2$ implies that the solution $u(x,\mu)$ to the u-initial value problem with parameter μ satisfies

$$|u(x,\mu) - y(x,\lambda)| < \varepsilon \quad \text{and} \quad |u'(x,\mu) - y'(x,\lambda)| < \varepsilon$$

for $a \leq x \leq b$. Since $\lambda^{(n)} \to \lambda$ as $n \to \infty$,

$$|u(x,\lambda^{(n)}) - y(x,\lambda)| < \varepsilon \quad \text{and} \quad |u'(x,\lambda^{(n)}) - y'(x,\lambda)| < \varepsilon$$

for $a \leq x \leq b$ provided n is sufficiently large. ■

7.2.3 Newton's Method and Convergence

Theorem 190 is also a key step in establishing that the eigenvalues and eigenfunctions of a Sturm-Liouville eigenvalue problem (7.11) can be found using Newton's method as the root finder.

As in the previous section $u = u(x) = u(x, \lambda)$ is the unique solution to the initial value problem (7.12) and $F(\lambda) = \gamma u(b, \lambda) + \delta u'(b, \lambda)$ is continuously differentiable.

Fix an eigenvalue λ of the eigenvalue problem (7.11). Since $F(\lambda) = 0$ and $F'(\lambda) \neq 0$, if $\lambda^{(0)}$ is a sufficiently good initial guess of λ, then all the Newton iterates

$$\lambda^{(n+1)} = \lambda^{(n)} - \frac{F(\lambda^{(n)})}{F'(\lambda^{(n)})}$$

are defined and $\lambda^{(n)} \to \lambda$ as $n \to \infty$. See Theorem 46. Now reasoning in the same way as for the bisection method, one obtains the following result.

Theorem 192 *Under the standing assumptions (1)–(5), if λ is an eigenvalue of the Sturm-Liouville eigenvalue problem (7.11) and $y = y(x, \lambda)$ is the corresponding normalized eigenfunction, then Newton's method can be used to generate a sequence of approximate eigenvalues $\lambda^{(n)} \to \lambda$ and approximate eigenfunctions $u(x, \lambda^{(n)})$ obtained by solving the u-initial value problem (7.12) such that $u(x, \lambda^{(n)}) \to y(x, \lambda)$ and $u'(x, \lambda^{(n)}) \to y'(x, \lambda)$ uniformly on $a \leq x \leq b$.*

7.2.4 Numerical Results

We give several examples of singular eigenvalue problems of the type considered in Chapter 5 to illustrate the convergence results of the previous section. Additional results will be given in Chapter 8.

Since the shooting method described above for singular eigenvalue problems is an interactive method, just like the method used for regular Sturm-Liouville eigenvalue problems, the discussion in Section 7.1.4 about determining good starting values that lead to numerical convergence of the method applies verbatim to the singular problems. We urge the reader to review that discussion.

We restrict the results in this section to singular eigenvalue problems (7.11) whose boundary conditions satisfy $\gamma\delta \geq 0$, which are the boundary conditions that occur most frequently in applications. The following five properties of singular eigenvalue problems compliment SL-1 to SL-6 for the regular problems in Section 7.1.4. If, in addition to the standing assumptions of this section, the boundary conditions satisfy $\gamma\delta \geq 0$, then the singular eigenvalue problem (7.11) has the following properties:

SL-1. The eigenvalues are real, simple, and can be ordered as

$$\lambda_0 < \lambda_1 < \lambda_2 < \cdots$$

(See Theorem 155.)

SL-2. The eigenfunction y_n corresponding to λ_n has exactly n nodal zeros (interior zeros where a sign change occurs) and no other interior zeros, provided either $q \geq 0$ on $[a, b]$ or, if q changes sign on $[a, b]$, $q(a) > 0$ when $r(a) = 0$. Moreover, the nodes of y_{n-1} and y_n strictly interlace. (See Theorem 155.)

SL-3. If $q \geq 0$ on $[a, b]$, then all the eigenvalues are positive, except for the case when the eigenvalue problem is $-(py')' = \lambda ry$, $|y(a)| < \infty$, $y'(b) = 0$, in which case 0 is an eigenvalue and all the other eigenvalues are positive. (See Theorem 151.)

SL-4. If $q \geq 0$ on $[a, b]$, then

$$\lambda_0 \geq \lim_{a' \to a} \min_{a' \leq x \leq b} \frac{q(x)}{r(x)}$$

and $\lambda_0 > 0$ unless the eigenvalue problem is $-(py')' = \lambda ry$, $|y(a)| < \infty$, $y'(b) = 0$, in which case 0 is an eigenvalue and all the other eigenvalues are positive. (See Theorem 151.)

The inequality in SL-4 satisfied by λ_0 is a consequence of reasoning that parallels the standard argument leading to SL-3 for regular problems. Normalize the eigenfunction y_0 by $\int_a^b y_0(x)^2 r(x)\, dx = 1$ and recall that y_0 is continuously differentiable on $[a, b]$. Consequently,

$$\begin{aligned}\lambda_0 &= \int_a^b \lambda_0 y_0^2 r\, dx = \int_a^b (-(py_0')' y_0 + q y_0^2)\, dx \\ &= \int_a^b (p(y_0')^2 + q y_0^2)\, dx - p y_0 y_0' |_a^b \geq \int_{a'}^b q y_0^2\, dx - p(b) y_0(b) y_0'(b)\end{aligned}$$

because $p(a) = 0$. For $a < a' < b$,

$$\int_a^b q y_0^2\, dx \geq \int_{a'}^b \frac{q}{r} y_0^2 r\, dx \geq \left(\min_{a' \leq x \leq b} \frac{q(x)}{r(x)} \right) \int_{a'}^b y_0^2 r\, dx.$$

Thus,

$$\lambda_0 \geq \left(\min_{a' \leq x \leq b} \frac{q(x)}{r(x)} \right) \int_{a'}^b y_0^2 r\, dx - p(b) y_0(b) y_0'(b).$$

Since $-p(b)y_0(b)y_0'(b) \geq 0$ because $\gamma\delta \geq 0$ and $\min_{a' \leq x \leq b} q(x)/r(x)$ decreases as a' decreases to a, it follows that

$$\lambda_0 \geq \lim_{a' \to a} \left(\min_{a' \leq x \leq b} \frac{q(x)}{r(x)} \right) \int_a^b y_0^2 r\, dx = \lim_{a' \to a} \left(\min_{a' \leq x \leq b} \frac{q(x)}{r(x)} \right).$$

If $q(a) > 0$ and $r(a) = 0$ as is often the case in applications, then the limit in SL-4 is $\min_{a < x \leq b} q(x)/r(x)$ and

$$\lambda_0 \geq \min_{a < x \leq b} \frac{q(x)}{r(x)}.$$

SL-5. *(Rayleigh Quotient)* If the weight function is $r(x) = (x - a)^m \rho(x)$ where $m \geq 0$ and $\rho(x)$ is positive and continuous on $[a, b]$, then

$$\lambda_0 = \min \frac{\langle Ly, y \rangle}{\langle y, y \rangle_r} = \min R(y),$$

where

$$R(y) = \frac{-p(b)y(b)y'(b) + \int_a^b (py'^2 + qy^2)\, dx}{\int_a^b y^2 r\, dx}$$

and the minimum is over all functions $y \neq 0$ in the domain of L that satisfy the boundary conditions $|y(a)| < \infty$, $\gamma y(b) + \delta y'(b) = 0$, and for which $\lim_{x \to a} Ly(x)/(x - a)^m$ exists and is finite. The minimum is achieved if and only if y is an eigenfunction corresponding to λ_0. (See Theorem 156.)

Let $F(\lambda) = \gamma u(b, \lambda) + \delta u'(b, \lambda)$ be the function introduced in the previous section, where $u(x, \lambda)$ is the solution to (7.12). We established that $F(\lambda)$ is a smooth function of λ for λ real, that it has only simple zeros, and those zeros are the eigenvalues of the Sturm-Liouville eigenvalue problem (7.11). Figure 7.5 illustrates such a function, see Example 1 below, and suggests a practical strategy for implementing the shooting method to determine eigenvalues and eigenfunctions of (7.11).

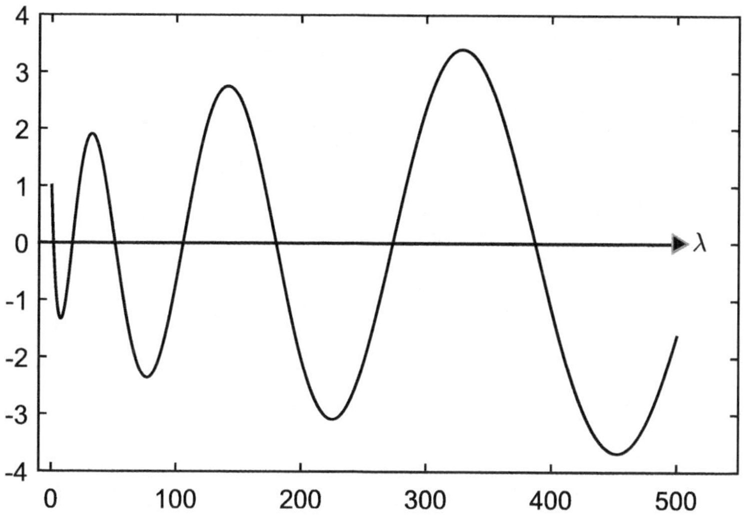

FIGURE 7.5: Graph of $F(\lambda) = J_0(\sqrt{\lambda}) - \sqrt{\lambda}J_1(\sqrt{\lambda})$

We apply the strategy first to determine numerically the smallest eigenvalue λ_0 and its corresponding eigenfunction y_0 using the shooting method and then using this information to systematically search for any additional eigenvalues and eigenfunctions that may be needed. We assume that Newton's method is used in the update step. As we said earlier and as Figure 7.2 suggests, it is not difficult to find a starting value for Newton's method that gives numerical convergence to some eigenvalue λ (perhaps not λ_0) and corresponding normalized eigenfunction $u(x, \lambda)$. A graph of $u(x, \lambda)$ reveals the number of its nodes. If there are, say, seven nodes, then $\lambda = \lambda_7$. Informed by this information, new starting values for Newton's method can be chosen and the shooting algorithm run again. This trial and error method leads to a determination of λ_0 in reasonably short time, usually a minute or two at most. The same approach can be used to locate other desired eigenvalues and eigenfunctions. See also the advice on page 309.

If $q(x) \geq 0$, the trial and error approach can be refined if finding helpful starting values proves difficult. In this case, from SL-4 and SL-5, the smallest eigenvalue λ_0 of the eigenvalue problem satisfies

$$\lim_{a' \to a} \min_{a' \leq x \leq b} \frac{q(x)}{r(x)} \leq \lambda_0 \leq R(y) \tag{7.15}$$

for any function $y \neq 0$ in the domain of L that satisfies the boundary conditions $|y(a)| < \infty$, $\gamma y(b) + \delta y'(b) = 0$ and is such that $\lim_{x \to a} Ly(x)/(x-a)^m$ exists and is finite. A function y with this property is

$$y(x) = (x-a)^{m+1} + c(x-a)^{m+2}$$

where

$$c = -\frac{\gamma(b-a)+\delta(m+1)}{(b-a)(\gamma(b-a)+\delta(m+2))}.$$

It satisfies the boundary condition at $x = b$ and

$$\lim_{x\to a} Ly(x)/(x-a)^m = -(m+1)^2 p'(a).$$

The double inequality (7.15) helps to inform a trial and error approach for finding a starting value for the shooting parameter that gives convergence to λ_0. Further help in finding suitable initial guesses for Newton's method in sensitive cases can be found by graphing the function $F(\lambda)$ over some interval with left endpoint at most $\lim_{a'\to a}\min_{a'\le x\le b} q(x)/r(x)$. The standard fourth order ordinary differential equation solvers yield numerical versions of $u(x,\lambda)$ and $u'(x,\lambda)$ for λ at a suitable set of equally spaced points, say, and, hence, a numerical version of $F(\lambda)$. The graph of $F(\lambda)$ can be used to select useful starting values for Newton's method. The same strategies apply when the bisection method is used as the root-finder.

The numerical results in the following examples were obtained with the shooting method, using Newton's method to update the shooting parameter, and following the practical suggestions given earlier for determining starting values. Newton's method was stopped when $|F(\lambda)| < 10^{-6}$ and the change in magnitude of the shooting parameter was less than 10^{-6}. The algorithm was run on a standard desktop computer and convergence was obtained in a matter of seconds, once good starting values were determined.

Example 1. A singular Sturm-Liouville eigenvalue problem of the form

$$\begin{cases} -(xy')' = \lambda xy, & 0 < x < 1, \\ |y(0)| < \infty, & y(1) + y'(1) = 0. \end{cases}$$

arises in connection with heat conduction in a circular plate with insulated top and bottom and whose circumference obeys Newton's law of cooling. All thermal coefficients have been set equal to 1 by introducing dimensionless variables. By SL-3 all the eigenvalues are positive. The differential equation is Bessel's equation of order 0 and parameter λ. Consequently, the bounded solutions to the differential equation are multiples of $J_0(\sqrt{\lambda}x)$. Since $J_0(0) = 1$ the normalized eigenfunctions are

$$y_n(x) = J_0(\sqrt{\lambda_n}x)$$

where $\lambda_0, \lambda_1, \ldots$ are the zeros of

$$F(\lambda) = J_0(\sqrt{\lambda}) + \sqrt{\lambda}J_0'(\sqrt{\lambda}) = J_0(\sqrt{\lambda}) - \sqrt{\lambda}J_1(\sqrt{\lambda})$$

for $\lambda > 0$.

The double inequality (7.15) for λ_0 applied with the polynomial $y = x^2 - 3x^3/4$ yields the bounds

$$0 \le \lambda_0 \le R(y) = \frac{-xyy'|_{x=1} + \int_0^1 xy'^2\,dx}{\int_0^1 y^2 x\,dx} = \frac{1/16 + 7/160}{61/2688} = \frac{1428}{305} < 4.7.$$

In the table that follows, the initial guess at λ_0 of 2.5 was suggested by the bounds above. The first guess at λ_1 was chosen as about twice $R(y)$ and a flexible interactive doubling and halving procedure either of previous initial guess or previously determined approximate eigenvalues was followed after that to get the eigenvalues λ_1, λ_2, λ_3, and λ_4. The table shows all the initial guesses used, in the order they were used, to find the first five eigenvalues and corresponding eigenfunctions. The first column shows which eigenvalue was found with the corresponding initial guess. The check column records the first five zeros of $F(\lambda) = J_0(\sqrt{\lambda}) - \sqrt{\lambda}J_1(\sqrt{\lambda})$ calculate using Newton's method. The relative error column is calculated using the entries in the check column as proxies for the exact eigenvalue.

n	**Initial Guess**	$\lambda_n \approx$	**Check**	**Relative Error**
0	2.5	1.5770	1.5770	0
1	10	16.6422	16.6421	6.0089×10^{-6}
1	20	16.6422		
2	40	51.2045	51.2055	-1.9529×10^{-5}
4	80	179.4947	179.5171	-1.2478×10^{-4}
3	102	105.4815	105.4931	-1.0996×10^{-4}

Graphs of the corresponding normalized eigenfunctions y_0, y_1, y_2, y_3, and y_4 are shown in Figure 7.6. The graph shows the interlacing of the nodes in SL-2. The initial guess 2.5 produces the approximate eigenvalue 1.5770 in the table and the graph of the corresponding eigenfunction with no nodes in $(0, 1)$. We conclude that 1.5770 is an approximate value of the eigenvalue λ_0. The other eigenvalues are identified in the same manner.

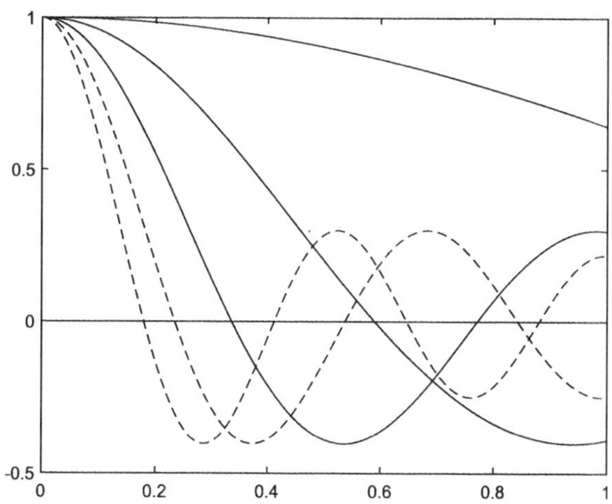

FIGURE 7.6: Eigenfunctions for Example 1

Example 2. Use the shooting method to find the first five eigenvalues and eigenfunctions of the singular Sturm-Liouville eigenvalue problem

$$\begin{cases} -((\sin x)y')' + xy = \lambda(\cos x)y, & 0 < x < 1, \\ |y(0)| < \infty, \; y(1) = 0. \end{cases} \tag{7.16}$$

This eigenvalue problem has weight function $r(x) = \cos x$. The double inequality (7.15) for λ_0 applied with the polynomial $y(x) = x^2 - x^3$ yields the bounds

$$0 \le \lambda_0 \le R(y) = \frac{\int_0^1 ((\sin x)y'^2 + xy^2)\, dx}{\int_0^1 y^2 \cos x\, dx} < 9.83.$$

The shooting method with the indicated initial guesses lead to the following approximations of the first five eigenvalues. The initial guesses shown were the first ones tried in the search process. The following table shows all the initial guesses used, in the order they were used, to find the first five eigenvalues and corresponding eigenfunctions. The first column shows which eigenvalue was found with the corresponding initial guess. The strategy for choosing initial guesses was the same as for Example 1.

n	**Initial Guess**	$\lambda_n \approx$
1	5	8.3131
0	2.5	1.6356
2	16	20.2746
2	40	37.5510
7	74	159.8239
4	57	60.1439

Graphs of the corresponding normalized eigenfunctions y_0, y_1, y_2, y_3, and y_4 are shown in Figure 7.7. The graph shows the interlacing of nodes in SL-2.

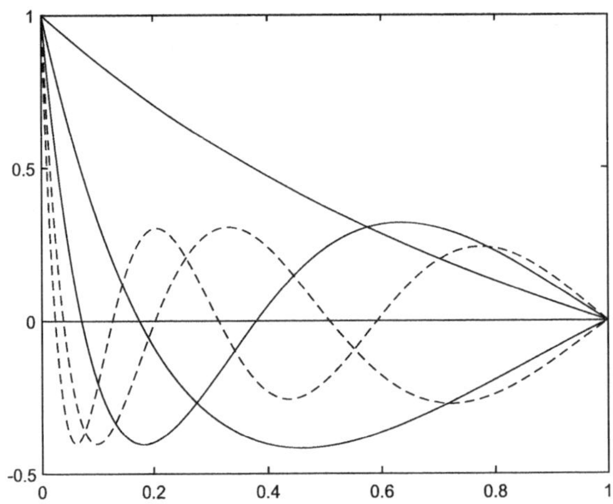

FIGURE 7.7: Eigenfunctions for Example 2

Just as for regular problems, if the shooting method converges numerically to $\tilde{\lambda}$, then an initial value problem solver can be used to evaluate $F(\tilde{\lambda} - \varepsilon)$ and $F(\tilde{\lambda} + \varepsilon)$ for some $\varepsilon > 0$. If ε can be chosen so that $F(\tilde{\lambda} - \varepsilon)$ and $F(\tilde{\lambda} + \varepsilon)$ are of opposite sign, then $\tilde{\lambda}$ approximates an eigenvalue λ of the eigenvalue problem to within an error of at most ε. A plot of $u(x, \tilde{\lambda})$ will reveal the number of nodes of the approximate eigenfunction and, therefore, which eigenvalue has been approximated to accuracy ε. Since $\tilde{\lambda}$ is almost certainly not exactly an eigenvalue, $F(\tilde{\lambda}) \neq 0$ and hence $F(\tilde{\lambda} - \varepsilon)$ and $F(\tilde{\lambda} + \varepsilon)$ have the same sign for $\varepsilon > 0$ suitably small. Our experience is that by experimenting with different choices of ε reasonable small a sign change can be detected.

Example 2 (continued) We found $\lambda_3 \approx \tilde{\lambda} = 37.551007$ correctly rounded, where two more digits of the numerical output are shown here. In Example 2, $F(\lambda) = u(1, \lambda)$, where u is the solution of the initial value problem

$$\begin{cases} -((\sin x)u')' + xu = \lambda(\cos x)u, & 0 \leq x \leq 1, \\ u(0) = 1, \ u'(0) = (0 - \lambda)/1 = -\lambda. \end{cases}$$

Numerical experiments with various choices for a possible error bound $\varepsilon > 0$ yield $F(\tilde{\lambda} - 3 \times 10^{-5}) \approx -9.3 \times 10^{-9}$ and $F(\tilde{\lambda} + 3 \times 10^{-5}) \approx 2.6 \times 10^{-6}$. It follows from the intermediate value theorem that there is an eigenvalue that differs from $\tilde{\lambda}$ by at most 3×10^{-5}. Since the approximate eigenfunction $u(x, \tilde{\lambda})$ has three nodes in $0 < x < 1$, we conclude that $|\tilde{\lambda} - \lambda_3| < 3 \times 10^{-5}$. Since $F(\tilde{\lambda} - 2 \times 10^{-5}) \approx 4.3 \times 10^{-7}$ and $F(\tilde{\lambda} + 2 \times 10^{-5}) \approx 2.2 \times 10^{-6}$ we further conclude that $2 \times 10^{-5} < |\tilde{\lambda} - \lambda_3| < 3 \times 10^{-5}$, which shows that the error bound 3×10^{-5} is reasonably sharp.

Another way to test the approximate eigenvalues for accuracy follows. Consider the singular eigenvalue problem

$$\begin{cases} -(py')' + qy = \lambda ry, & a < x < b, \\ |y(a)| < \infty, \ y(b) = 0. \end{cases} \tag{7.17}$$

Any bounded solution $u(x)$ to the differential equation in (7.17) is (extends to) a continuously differentiable function on $[a, b]$, satisfies the differential equation there, and satisfies the initial condition

$$u'(a) = \frac{q(a) - \lambda r(a)}{p'(a)}$$

by the general results established in Chapter 5.

Let λ be an eigenvalue and $y(x)$ be its corresponding normalized eigenfunction, so that $y(a) = 1$. Define $u(t) = y(x)$, $p_1(t) = p(x)$, $q_1(t) = q(x)$, $r_1(t) = r(x)$ where $t = kx$ for $a \leq x \leq b$ and k is a given positive constant. Since $\dot{u} = y'/k$ where a dot denotes differentiation with respect to t, u satisfies the differential equation

$$-k^2(p_1\dot{u})^{\cdot} + q_1 u = \lambda r_1 u, \qquad ka \leq t \leq kb$$

as well as the conditions

$$u(ka) = y(a) = 1, \ u(kb) = y(b) = 0, \ \dot{u}(ka) = y'(a)/k = \frac{q(a) - \lambda r(a)}{kp'(a)}.$$

Thus, if λ, $y(x)$ is an eigenvalue, normalized eigenfunction pair of (7.17) and the positive constant k is chosen to be $k = \sqrt{\lambda}$, then $u(t)$ satisfies the initial value problem

$$\begin{cases} -(p_1\dot{u})^{\cdot} + (q_1/k^2)u = r_1 u, & ka \leq t \leq kb, \\ u(ka) = 1, \ \dot{u}(ka) = \dfrac{q(a) - k^2 r(a)}{kp'(a)} \end{cases} \tag{7.18}$$

as well as the condition $u(kb) = 0$. Conversely, if for some $k > 0$, the solution $u(t)$ to this initial value problem also satisfies $u(kb) = 0$, then $\lambda = k^2$, $y(x) = u(kx)$ for $a \leq x \leq b$ is an eigenvalue, normalized eigenfunction pair for (7.17). These observations lead to the following check on the eigenvalues found by shooting: for each eigenvalue λ of (7.17) found by shooting, solve the initial value problem (7.18), evaluate $u(kb)$, and compare this value to 0. If λ is an exact (approximate) eigenvalue, then $u(kb)$ is exactly (approximately) 0. We chose Dirichlet boundary data in (7.17) for simplicity. The same approach can be used with any separated boundary conditions.

Example 2 (continued) Apply the test above with $k = \sqrt{\lambda}$ to the five eigenvalues λ in the table in Example 2. That is, solve the initial value problem

$$\begin{cases} -(p_1\dot{u})^{\cdot} + (q_1/k^2)u = r_1 u, & 0 \leq t \leq k, \\ u(0) = 1, \ \dot{u}(0) = \dfrac{0 - k^2(1)}{k(1)} = -k \end{cases}$$

numerically and compare the values for $u(k)$ to 0. Here $p_1(t) = \sin(t/k)$, $q_1(t) = t/k$, $r_1(t) = \cos(t/k)$. The following table show the comparison and strongly suggests that the numerical approximation of λ_n is quite accurate.

n	$\lambda_n \approx$	$u(k) \approx$
0	1.6356	-1.9209×10^{-5}
1	8.3131	-3.8541×10^{-6}
2	20.2746	2.7525×10^{-6}
3	37.5510	-1.2798×10^{-6}
4	60.1439	7.7097×10^{-7}

Moreover, the graphs of $y(x) = u(kx)$ for $0 \le x \le 1$ have the expected number of nodes in $(0, 1)$, consistent with the fact that $y(x)$ is a corresponding eigenfunction if λ is an eigenvalue. Altogether, these results add considerable confidence to belief that the shooting method has produced accurate approximations to the first five eigenvalues and corresponding normalized eigenfunctions of the eigenvalue problem in Example 2.

Example 3. A singular Sturm-Liouville eigenvalue problem of the form

$$\begin{cases} -(xy')' + (\sin \pi x)y = \lambda xy, & 0 < x < 1, \\ |y(0)| < \infty, \ y'(1) = 0. \end{cases}$$

arises in connection with heat conduction in a circular plate with partially insulated top and bottom and an insulated circumference. All thermal coefficients have been set equal to 1 for simplicity. By SL-3 all the eigenvalues are positive. The double inequality (7.15) for λ_0 applied with the polynomial $y = x^2 - 2x^3/3$ yields the bounds

$$0 < \lambda_0 \le R(y) = \frac{-xyy'|_{x=1} + \int_0^1 (xy'^2 + (\sin \pi x)y^2)\, dx}{\int_0^1 xy^2\, dx} \approx \frac{0 + 0.08999}{2/63} < 2.9$$

The following table shows all the initial guesses used, in the order they were used, to find the first five eigenvalues and corresponding eigenfunctions. The first column shows which eigenvalue was found with the corresponding initial guess.

n	**Initial Guess**	$\lambda_n \approx$
0	1.5	1.2212
0	6	1.2212
1	12	16.6360
1	24	16.6360
2	48	51.0938
3	96	105.3506
4	192	179.3574

Figure 7.8 shows the first five corresponding normalized eigenfunctions.

The error in each eigenvalue can be tested using the intermediate value theorem. For example, the shooting method yields the approximation $\lambda_4 \approx \tilde{\lambda} = 179.357373$. In this example $F(\lambda) = u'(1, \lambda)$ where $u(x, \lambda)$ is the solution of the initial value problem

$$\begin{cases} -(xu')' + (\sin \pi x)u = \lambda xu & 0 \le x \le 1, \\ u(0) = 1, \ u'(0) = (0 - \lambda(0))/1 = 0. \end{cases}$$

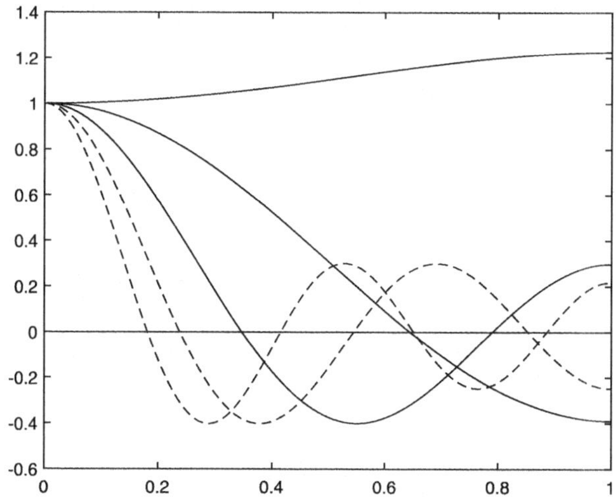

FIGURE 7.8: Eigenfunctions for Example 3

Numerical experiments with different choices for potential error bounds leads to $F(\tilde{\lambda} - 2 \times 10^{-4}) \approx 4.3 \times 10^{-6}$ and $F(\tilde{\lambda} + 2 \times 10^{-4}) \approx -3.9 \times 10^{-5}$. It follows from the intermediate value theorem that there is an eigenvalue that differs from $\tilde{\lambda}$ by at most 2×10^{-4}. Since the approximate eigenfunction $u(x, \tilde{\lambda})$ has four nodes in $0 < x < 1$, we conclude that $|\tilde{\lambda} - \lambda_4| < 2 \times 10^{-4}$. Since $F(\tilde{\lambda} - 10^{-4}) \approx -6.6 \times 10^{-6}$ and $F(\tilde{\lambda} + 10^{-4}) \approx -2.8 \times 10^{-5}$ we further conclude that $10^{-4} < |\tilde{\lambda} - \lambda_4| < 2 \times 10^{-4}$, which shows that the error bound 2×10^{-4} is reasonably sharp.

7.2.5 Concluding Remarks

A few concluding remarks are in order about the numerical implementation of the shooting method for finding eigenvalues and eigenfunction of (7.11) via the initial value problems (7.12) in which λ is the shooting parameter. Since the initial value problem is singular at $x = a$ standard initial value problem solvers do not apply on the full interval $a \leq x \leq b$. We use the initial data in (7.12) and an Euler-like method to extend the initial data to $x = a + \varepsilon$, where $\varepsilon > 0$ is fixed suitably small. The initial value problem is regular on $a + \varepsilon \leq x \leq b$ and the solution u to (7.12) can be extended from $a + \varepsilon$ to b by a standard initial value solver. Recall that $p(x) = (x - a)\varphi(x)$ where $\varphi(x)$ is positive and continuous on $[a, b]$. The Euler-like step during the shooting procedure is done as follows. From the initial data in (7.12)

$$u(a + \varepsilon) = 1 + \varepsilon u'(a) = 1 + \varepsilon(q(a) - \lambda r(a))/p'(a)$$

to first order accuracy in ε. Integrate the differential equation in (7.12) from a to $a + \varepsilon$ and use the fact that the bounded solution u satisfies $p(a)u'(a) = 0$ by Lemma 132 to obtain

$$\begin{aligned} p(a + \varepsilon)u'(a + \varepsilon) &= \int_a^{a+\varepsilon} (q(x) - \lambda r(x))u(x)\, dx \\ &= \varepsilon(q(a + \varepsilon) - \lambda r(a + \varepsilon))u(a + \varepsilon) \end{aligned}$$

to first order accuracy in ε and where the right-hand rule was used to approximate the integral. Since $p(a+\varepsilon) = \varepsilon\varphi(a+\varepsilon)$,

$$u'(a+\varepsilon) = (q(a+\varepsilon) - \lambda r(a+\varepsilon))u(a+\varepsilon)/\varphi(a+\varepsilon)$$

to first order accuracy in ε. A convenient choice for ε in code written for MATLAB and ode45 is $\varepsilon = \varepsilon ps$, the distance from 1.0 to the next larger positive double-precision number (approximately 10^{-16}).

If Newton's method is used in the shooting procedure, then the variational initial value problem of (7.12), its derivative with respect to the shooting parameter λ, is needed. The variational initial value problem also is singular at $x = a$ and first step way from the singularity at $x = a$ can be handled using the same ideas used for u above.

7.3 Singular Problems - II

In this section, the shooting method for the regular problems in Chapter 4 and for the singular problems in Chapter 5 is adjusted to cover the singular Sturm-Liouville problems treated in Chapter 6. The eigenvalue problem for such a singular Sturm-Liouville differential equation is

$$\begin{cases} -(p(x)y')' + q(x)y = \lambda r(x)y, & a < x < b, \\ |y(a)| < \infty, & \\ \gamma y(b) + \delta y'(b) = 0, & |\gamma| + |\delta| \neq 0. \end{cases} \tag{7.19}$$

The shooting method for the singular problems in Chapter 5 used essentially the same strategy as that for regular problems in Chapter 4. Here a different approach is needed. (See the comparison of the two types of singular problems on page 254.) We make the follow assumptions (1)–(5) throughout this section:

Standing Assumptions

(1) $p(x) = (x-a)\varphi(x)$ where $\varphi(x)$ is positive and continuously differentiable on $[a, b]$.

(2) $q(x) = q_1(x)/(x-a)$ where $q_1(x)$ is real-valued and continuous on $[a, b]$.

(3) $q_1(a) > 0$ and $q_1'(a)$ exists.

(4) γ and δ are real numbers with $\gamma\delta \geq 0$ and $|\gamma| + |\delta| > 0$.

Assumptions (1)–(4) guarantee that the principal results established for the singular eigenvalue problems in Chapter 6 hold.

Just as for the regular problems in Chapter 4 and for the singular problems in Chapter 5, the numerical solution procedure will use a variational equation associated with the Sturm-Liouville differential equation. Therefore, we further assume that

(5) $q_1(x)$ and $r(x)$ are continuously differentiable on $[a, b]$.

The differential equation in (7.19) can be expressed as

$$-(p(x)y')' + \tilde{q}(x)y = 0, \quad a < x < b, \tag{7.20}$$

where

$$\tilde{q}(x) = q(x) - \lambda r(x)$$

for $a < x \leq b$. Since

$$\tilde{q}(x) = \frac{q_1(x)}{x-a} - \lambda r(x) = \frac{q_1(x) - (x-a)\lambda r(x)}{x-a} = \frac{\tilde{q}_1(x)}{x-a}, \tag{7.21}$$

the coefficients in the differential equation (7.20) satisfies (1)–(5) with respect to $p(x)$ and $\tilde{q}(x)$. Consequently, it follows from Theorem 164 and its proof that the differential equation (7.20) has a nontrivial bounded solution in $C[a, b]$ for any choice of the parameter λ. One such solution is

$$(x-a)^{\nu} z(x),$$

where

$$\nu = \sqrt{q_1(a)/\varphi(a)}$$

and where z in $C^1[a, b]$ is the unique solution to the initial value problem

$$\begin{cases} (x-a)z'' + \alpha(x)z' + \beta(x)z = 0 \quad \text{for } a < x \leq b \\ z(a) = 1, \; z'(a) = -\beta(a)/\alpha(a) \end{cases} \tag{7.22}$$

where

$$\alpha(x) = \frac{(2\nu + 1)\varphi(x) + (x-a)\varphi'(x)}{\varphi(x)}, \tag{7.23}$$

$$\beta(x) = \frac{\tilde{q}_2(x) + \nu\varphi'(x)}{\varphi(x)}, \tag{7.24}$$

and

$$\tilde{q}_2(x) = \begin{cases} \dfrac{q_1(a)\varphi'(a) - \varphi(a)(q_1'(a) - \lambda r(a))}{\varphi(a)} & \text{for } x = a \\ \dfrac{\nu^2\varphi(x) - (q_1(x) - (x-a)\lambda r(x))}{x-a} & \text{for } a < x \leq b \end{cases}. \tag{7.25}$$

The value of $\tilde{q}_2(x)$ at $x = a$ makes this function continuous on $[a, b]$. See Theorem 164 and its proof where the foregoing results are established.

The parameter λ occurs in the coefficient β so that $\beta = \beta(x) = \beta(x, \lambda)$. We will usually suppress the dependence on λ and just write β or $\beta(x)$. Nevertheless, the solution $z(x)$ depends on λ and we write $z(x) = z(x, \lambda)$ when it is advantageous to explicitly express the dependence on λ.

Since $z(x, \lambda)$ satisfies a nonsingular differential equation when x is restricted to $[c, b]$ for any fixed c with $a < c < b$, it follows from continuous dependence results in [9] that the solution $z(x, \lambda)$ depends smoothly on both x and λ for x in $[c, b]$ and λ in any bounded interval. Since $c > a$ can be chosen arbitrarily, the solution $z(x, \lambda)$ depends smoothly on both x and λ for x in $a < x \leq b$ and λ in any bounded interval. This smoothness will be needed when we discuss the numerical approximation of eigenvalues and eigenfunctions using the bisection method and Newton's method to update an appropriate shooting parameter.

Suppose that λ is an eigenvalue of (7.19) and y is a corresponding eigenfunction. In particular, y is a nontrivial bounded solution of the differential equation (7.20). Since, by Theorem 164, any nontrivial solution of the differential equation in (7.20) is a nonzero multiple of $(x-a)^{\nu} z(x)$, the eigenfunction

$$y(x) = c_0(x-a)^{\nu} z(x)$$

where ν and $z(x)$ are as above and c_0 is a nonzero constant. We normalize the eigenfunction by choosing $c_0 = 1$; thus, the **normalized eigenfunction** of (7.19) corresponding to the eigenvalue λ is

$$y = y(x) = y(x, \lambda) = (x-a)^{\nu} z(x, \lambda)$$

where $\nu = \sqrt{q_1(a)/\varphi(a)} > 0$ and $z(x, \lambda)$ is the unique solution to the initial value problem (7.22).

The normalized eigenfunction $y(x, \lambda)$ also satisfies the boundary condition

$$\gamma y(b, \lambda) + \delta y'(b, \lambda) = 0;$$

equivalently,

$$\tilde{\gamma} z(b, \lambda) + \tilde{\delta} z'(b, \lambda) = 0,$$

where $\tilde{\gamma} = \gamma(b - a) + \delta\nu$ and $\tilde{\delta} = \delta(b - a)$.

7.3.1 The Shooting Method

The properties of a normalized eigenfunction to (7.19) lead to a shooting method with shooting parameter λ based on the initial value problem (7.22). If λ is an eigenvalue of the eigenvalue problem (7.19) and $y(x, \lambda)$ is its normalized eigenfunction, then $y(x) = (x - a)^{\nu} z(x, \lambda)$ where $\nu = \sqrt{q_1(a)/\varphi(a)} > 0$ and $z(x, \lambda)$ is the unique solution to the initial value problem (7.22). Moreover, $\gamma y(b, \lambda) + \delta y'(b, \lambda) = 0$ which is equivalent to $\tilde{\gamma} z(b, \lambda) + \tilde{\delta} z'(b, \lambda) = 0$, where $\tilde{\gamma} = \gamma(b - a) + \delta\nu$ and $\tilde{\delta} = \delta(b - a)$. Conversely, if $z = z(x, \lambda)$ is the unique solution to (7.22) for a value of λ such that $\tilde{\gamma} z(b, \lambda) + \tilde{\delta} z'(b, \lambda) = 0$, then λ is an eigenvalue of the eigenvalue problem (7.19) and $y = (x - a)^{\nu} z(x, \lambda)$ is its corresponding normalized eigenfunction. In summary, λ is an eigenvalue and $y(x, \lambda) = (x - a)^{\nu} z(x, \lambda)$ is its corresponding normalized eigenfunction if and only if λ satisfies $\tilde{\gamma} z(b, \lambda) + \tilde{\delta} z'(b, \lambda) = 0$ and $z(x, \lambda)$ is the unique solution to (7.22).

The shooting method can be used with either the bisection method or Newton's method to find accurate numerical approximations to the eigenvalues and eigenfunctions of (7.19). The convergence analysis for the approach just described is based on a different variation of parameters formula from the familiar one for regular problems or from the variation of parameters formula for the singular problems in Chapter 5. We have not seen this formula elsewhere. Consequently, we conclude this section with a statement and proof of the formula for the singular differential equations in Chapter 6.

Some preliminary observations are needed about the behavior as x approaches a of solutions to the differential equation in (7.22) that are linearly independent of the solution $z(x)$ to the initial value problem (7.22). Express $\alpha(x)/(x - a)$ as

$$\frac{\alpha(x)}{x - a} = \frac{\alpha(x) - \alpha(a)}{x - a} + \frac{\alpha(a)}{x - a} = \alpha_1(x) + \frac{c}{x - a} \tag{7.26}$$

where $c = \alpha(a) = 2\nu + 1 > 0$ and $\alpha_1(x)$ is continuous on $[a, b]$ with the understanding that $\alpha_1(a) = \alpha'(a) = \varphi'(a)/\varphi(a)$. Fix $x_0 > a$ such that $z(x) > 0$ on $[a, x_0]$, and define

$$\tilde{A}(x) = \exp\left(\int_x^{x_0} \alpha_1(s)\, ds\right) > 0$$

for $a \le x \le b$. If $v(x) = v(x, \lambda)$ is a solution of the differential equation in (7.22) that is linearly independent of $z(x, \lambda)$ and $W_{z,v}(x)$ is their Wronskian, then

$$\begin{aligned} W'_{z,v}(x) &= (zv' - z'v)' = zv'' - z''v \\ &= z\left(-\frac{\alpha(x)}{x - a} v' - \frac{\beta(x)}{x - a} v\right) - \left(-\frac{\alpha(x)}{x - a} z' - \frac{\beta(x)}{x - a} z\right) v \\ &= -\frac{\alpha(x)}{x - a} W_{z,v}(x). \end{aligned}$$

Integration and use of (7.26) yields

$$W_{z,v}(x) = W_{z,v}(x_0)\exp\left(\int_x^{x_0} \frac{\alpha(s)}{s-a}\,ds\right) = W_{z,v}(x_0)\tilde{A}(x)\frac{(x_0-a)^c}{(x-a)^c}.$$

Consequently,

$$W_{z,v}(x) = A_0(x)(x-a)^{-c} \quad \text{for } a < x \le b, \tag{7.27}$$

where $c = 2\nu + 1 > 0$ and

$$A_0(x) = W_{z,v}(x_0)(x_0-a)^c\tilde{A}(x)$$

is continuous and nonzero on $[a, b]$.

The expression $W_{z,v}(x) = A_0(x)(x-a)^{-c}$ leads easily to the asymptotic behavior of the solution $v(x)$ as $x \to a$. Since $zv' - z'v = W_{z,v}(x)$, for $a < x < x_0$,

$$\frac{d}{dx}\left(\frac{v(x)}{z(x)}\right) = \frac{W_{z,v}(x)}{z(x)^2} = \frac{A_0(x)(x-a)^{-c}}{z(x)^2},$$

$$\frac{v(x_0)}{z(x_0)} - \frac{v(x)}{z(x)} = \int_x^{x_0} \frac{A_0(s)(s-a)^{-c}}{z(s)^2}\,ds,$$

and

$$v(x) = \frac{v(x_0)}{z(x_0)}z(x) - z(x)\int_x^{x_0} \frac{A_0(s)(s-a)^{-c}}{z(s)^2}\,ds. \tag{7.28}$$

The first term in the right member of (7.28) has limit

$$\lim_{x\to a} \frac{v(x_0)}{z(x_0)}z(x) = \frac{v(x_0)}{z(x_0)}$$

and the second term can be expressed as

$$z(x)\int_x^{x_0} \frac{A_0(s)(s-a)^{-c}}{z(s)^2}\,ds = \frac{z(x)A_0(s_x)}{z(s_x)^2}\int_x^{x_0} (s-a)^{-c}\,ds$$

for some s_x between x and x_0 by the mean value theorem for integrals. Since

$$\int_x^{x_0} (s-a)^{-c}\,ds = \begin{cases} (1-c)^{-1}((x_0-a)^{1-c} - (x-a)^{1-c}) & \text{if } c \ne 1 \\ \ln(x_0-a) - \ln(x-a) & \text{if } c = 1 \end{cases},$$

we have

$$\lim_{x\to a} (x-a)^{c-1}\int_x^{x_0} (s-a)^{-c}\,ds = 1/(c-1) \quad \text{if } c > 1,$$

$$\lim_{x\to a} \frac{1}{\ln(x-a)}\int_x^{x_0} (s-a)^{-c}\,ds = -1 \quad \text{if } c = 1,$$

$$\lim_{x\to a} \int_x^{x_0} (s-a)^{-c}\,ds = \frac{(x_0-a)^{1-c}}{1-c} \quad \text{if } 0 < c < 1.$$

Using these limits and the fact that $x_0 > a$ can be chosen arbitrarily close to a in the foregoing results, it follows from (7.28) that there exists

$$\lim_{x\to a} (x-a)^{c-1}v(x) = -\frac{A_0(a)}{c-1} \quad \text{if } c > 1, \tag{7.29}$$

$$\lim_{x\to a} \frac{v(x)}{\ln(x-a)} = A_0(a) \quad \text{if } c = 1, \tag{7.30}$$

$$\lim_{x\to a} v(x) = \frac{v(x_0)}{z(x_0)} - \frac{A_0(a)(x_0-a)^{1-c}}{1-c} \quad \text{if } 0 < c < 1. \tag{7.31}$$

These preliminaries enable us to establish the following variation of parameters result.

Theorem 193 (Variation of Parameters) *Let $z(x)$ be the unique solution to (7.22), $v(x)$ be a solution to the differential equation in (7.22) that is linearly independent of $z(x)$, and $W_{z,v}(x)$ be their Wronskian. If $g(x)$ is continuous on $[a, b]$, then the initial value problem*

$$\begin{cases} (x-a)w'' + \alpha(x)w' + \beta(x)w = g(x) & \text{for } a < x \leq b, \\ w(a) = 0, \quad w'(a) = g(a)/\alpha(a), \end{cases} \tag{7.32}$$

has the unique solution w that is continuously differentiable function on $[a, b]$. The solution is given explicitly by

$$w(x) = \begin{cases} A(x)z(x) + B(x)v(x) & \text{for } a < x \leq b \\ 0 & \text{for } x = a \end{cases} \tag{7.33}$$

where

$$A(x) = -\int_a^x \frac{v(s)g(s)}{(s-a)\,W_{z,v}(s)}\,ds \quad \text{and} \quad B(x) = \int_a^x \frac{z(s)g(s)}{(s-a)\,W_{z,v}(s)}\,ds.$$

Proof. The initial value problem has a unique solution that is continuously differentiable function on $[a, b]$ by Theorem 162. What is new here is the explicit representation of the solution.

The improper integrals that define $A(x)$ and $B(x)$ both converge. To confirm this, use $W_{z,v}(x) = A_0(x)(x-a)^{-c}$, where $c = 2\nu + 1 > 0$, to express the integrand for $A(x)$ as

$$\frac{v(s)g(s)}{(s-a)\,W_{z,v}(s)} = \frac{v(s)g(s)}{(s-a)A_0(s)(s-a)^{-c}} = \frac{(s-a)^{c-1}v(s)g(s)}{A_0(s)}.$$

The integrand has a finite limit as $x \to a$ when $c > 1$ by (7.29); has a logarithmic singularity as $x \to a$ when $c = 1$ by (7.30); and is weakly singular when $0 < c < 1$ by (7.31). In each case, the improper integral defining $A(x)$ converges. Likewise,

$$\frac{z(s)g(s)}{(s-a)\,W_{z,v}(s)} = \frac{(s-a)^{c-1}z(s)g(s)}{A_0(s)}$$

and the integrand for $B(x)$ has a finite limit as $x \to a$ when $c \geq 1$ and has a weak singularity when $0 < c < 1$. In each case, the improper integral defining $B(x)$ converges. Since both improper integrals converge,

$$\lim_{x \to a} A(x) = 0 \quad \text{and} \quad \lim_{x \to a} B(x) = 0$$

and it is convenient to define $A(a) = 0$ and $B(a) = 0$ so that A and B are continuous on $[a, b]$.

As in the proof of Theorem 189, it is simplest just to check that the expression for $w(x)$ has the required properties. This follows if we establish:

A. There exists $\lim_{x \to a} w(x) = 0$.

B. There exists $\lim_{x \to a} w'(x) = g(a)/\alpha(a)$.

C. $w(x)$ satisfies the differential equation in (7.32).

To prove A note that

$$\lim_{x \to a} w(x) = \lim_{x \to a} A(x)z(x) + \lim_{x \to a} B(x)v(x)$$

because both limits on the right exist: first,

$$\lim_{x \to a} A(x)z(x) = 0 \cdot 1 = 0.$$

Second, by the mean value theorem for integrals, for some s_x is between a and x,

$$\begin{aligned} B(x)v(x) &= \left(\int_a^x \frac{(s-a)^{c-1} z(s)g(s)}{A_0(s)} \, ds \right) v(x) \\ &= \left(\frac{z(s_x)g(s_x)(x-a)^c}{A_0(s_x)c} \right) v(x) \to 0, \end{aligned}$$

as $x \to a$ because $(x-a)^c v(x) \to 0$ as $x \to a$ by the asymptotic results (7.29), (7.30), and (7.31) established for $v(x)$. Combining these results gives

$$\lim_{x \to a} w(x) = 0,$$

which establishes A and the continuity of $w(x)$ at $x = a$. Since $w(x) = A(x)z(x) + B(x)v(x)$ is continuous on $a < x \leq b$, it follows that $w(x)$ is continuous on $[a, b]$. We establish B in a similar way starting with the observation that

$$A'(x)z(x) + B'(x)v(x) = -\frac{v(x)g(x)}{(x-a)\, W_{z,v}(x)} z(x) + \frac{z(x)g(x)}{(x-a)\, W_{z,v}(x)} v(x) = 0$$

for $a < x \leq b$, and, hence,

$$w'(x) = A(x)z'(x) + B(x)v'(x)$$

there. First

$$\lim_{x \to a} A(x)z'(x) = 0 \cdot z'(a) = 0.$$

Second, since $zv' - z'v = W_{z,v}(x) = A_0(x)(x-a)^{-c}$,

$$\begin{aligned} v'(x) &= \frac{z'(x)}{z(x)} v(x) + \frac{A_0(x)}{z(x)} (x-a)^{-c}, \\ (x-a)^c v'(x) &= \frac{z'(x)}{z(x)} (x-a)^c v(x) + \frac{A_0(x)}{z(x)}. \end{aligned}$$

Use the asymptotic properties (7.29), (7.30), and (7.31) of $v(x)$ to find that

$$\lim_{x \to a} (x-a)^c v'(x) = A_0(a).$$

By the mean value theorem for integrals,

$$B(x)v'(x) = \frac{z(s_x)g(s_x)(x-a)^c}{A_0(s_x)c} v'(x)$$

for some s_x is between a and x, and, hence, there exists

$$\lim_{x \to a} B(x)v'(x) = \lim_{x \to a} \frac{z(s_x)g(s_x)}{A_0(s_x)c} (x-a)^c v'(x) = \frac{g(a)}{A_0(a)c} A_0(a) = \frac{g(a)}{c}.$$

Since $c = \alpha(a)$, it follows that

$$\lim_{x \to a} w'(x) = \frac{g(a)}{\alpha(a)}$$

and B is established. The continuity of w on $[a, b]$ and B imply that there exists $w'(a) = g(a)/\alpha(a)$ and that w' is continuous at $x = a$ by Lemma 11. The expression $w' = Az' + Bv'$ shows that w' is continuous on $a < x \leq b$. Thus, w is continuously differentiable on $[a, b]$.

It remains to establish C. From the proof of B

$$w'(x) = A(x)z'(x) + B(x)v'(x).$$

Furthermore,

$$\begin{aligned} &(x-a)(A'(x)z'(x) + B'(x)v'(x)) \\ &= (x-a)\left(-\frac{v(x)g(x)}{(x-a)\,W_{z,v}(x)}z'(x) + \frac{z(x)g(x)}{(x-a)\,W_{z,v}(x)}v'(x)\right) \\ &= \frac{W_{z,v}(x)g(x)}{W_{z,v}(x)} = g(x) \end{aligned}$$

for $a < x \leq b$. Consequently, $w = Az + Bv$ satisfies

$$\begin{aligned} (x-a)w'' + \alpha w' + \beta w &= (x-a)(Az'' + Bv'' + A'z' + B'v') \\ &\quad + \alpha(Az' + Bv') + \beta(Az + Bv) \\ &= A \cdot 0 + B \cdot 0 + g \quad \text{for } a < x \leq b, \end{aligned}$$

which establishes C and the proof is complete. ■

The proof shows that $\lim_{x \to a}(A(x)z(x) + B(x)v(x)) = 0$. So the variation of parameters solution can be expressed simply as $w(x) = A(x)z(x) + B(x)v(x)$ if $A(x)z(x) + B(x)v(x)$ is interpreted at $x = a$ as its limiting value as x tends to a. Furthermore, the proof also shows that $A(x)$ and $B(x)$ are chosen so that

$$w'(x) = A(x)z'(x) + B(x)v'(x),$$

a result that plays an essential role in the convergence analysis in the next two sections.

7.3.2 Bisection Method and Convergence

The convergence analysis that follows parallels closely that in Section 7.1.2 for regular problems and that in Section 7.2.2 for the singular problems in Chapter 5.

Recall that λ is an eigenvalue of the eigenvalue problem (7.19) and y is its corresponding normalized eigenfunction if and only if $y = (x-a)^{\nu}\, z(x, \lambda)$ where $\nu = \sqrt{q_1(a)/\varphi(a)}$ and $z(x, \lambda)$ solves the initial value problem (7.22) and also satisfies $\tilde{F}(\lambda) = 0$, where

$$\tilde{F}(\lambda) = \tilde{\gamma} z(b, \lambda) + \tilde{\delta} z'(b, \lambda), \tag{7.34}$$

$\tilde{\gamma} = \gamma(b-a) + \delta\nu$, and $\tilde{\delta} = \delta(b-a)$.

We establish next that the eigenvalues and eigenfunctions of (7.19) can be found by a shooting method that determines values of λ such that the unique solution $z(x, \lambda)$ to (7.22) satisfies $\tilde{F}(\lambda) = 0$. Such a λ is an eigenvalue of (7.19) and $y = (x-a)^{\nu} z(x, \lambda)$ is its corresponding normalized eigenfunction. As for the other regular and singular eigenvalue problems already discussed, the basic conceptual approach is based on Theorem 193, the variation of parameters formula associated with the singular differential equation in (7.19). We continue to use the notation introduced earlier.

Theorem 194 *Let $z(x,\lambda)$ be the unique solution to (7.22), and $\tilde{F}(\lambda) = \gamma z(b,\lambda) + \delta z'(b,\lambda)$ where $\tilde{\gamma} = \gamma(b-a) + \delta\nu$ and $\tilde{\delta} = \delta(b-a)$. Under the standing assumption (1)–(5), if λ is an eigenvalue of the Sturm-Liouville problem (7.19), then $\tilde{F}(\lambda) = 0$ and*

$$\tilde{F}'(\lambda) = \tilde{\gamma} w(b,\lambda) + \tilde{\delta} w'(b,\lambda) \neq 0$$

where $w = w(x,\lambda) = \partial z(x,\lambda)/\partial\lambda$ and $y = y(x,\lambda) = (x-a)^{\nu}\, z(x,\lambda)$ is the corresponding normalized eigenfunction.

Proof. Let $y(x) = y(x,\lambda)$ be the normalized eigenfunction corresponding to the eigenvalue λ. We established earlier that $y(x) = y(x,\lambda) = (x-a)^{\nu}\, z(x,\lambda)$ where $\nu = \sqrt{q_1(a)/\varphi(a)}$, $z(x,\lambda)$ is the unique solution to (7.22) and also that $\tilde{F}(\lambda) = 0$ because $\gamma y(b,\lambda) + \delta y'(b,\lambda) = 0$.

It remains to show that $\tilde{F}'(\lambda) \neq 0$. Let $v(x) = v(x,\lambda)$ be a solution of the differential equation in (7.22) that is linearly independent of $z(x,\lambda)$ so that the Wronskian $W_{u,v}(x) \neq 0$ on $a < x \leq b$. In particular, $W_{u,v}(b) \neq 0$. Since $\tilde{F}(\lambda) = \tilde{\gamma} z(b,\lambda) + \tilde{\delta} z'(b,\lambda) = 0$, if $\tilde{\gamma} \neq 0$ then $z'(b,\lambda) \neq 0$ because otherwise the boundary condition implies $z(b,\lambda) = 0$ in which case $z = 0$, contradicting the fact that z is nontrivial. Furthermore,

$$W_{z,v}(b) = \begin{vmatrix} z(b) & v(b) \\ z'(b) & v'(b) \end{vmatrix} = \tilde{\gamma}^{-1} \begin{vmatrix} \tilde{\gamma} z(b) + \tilde{\delta} z'(b) & \tilde{\gamma} v(b) + \tilde{\delta} v'(b) \\ z'(b) & v'(b) \end{vmatrix}$$
$$= -\tilde{\gamma}^{-1} z'(b)(\tilde{\gamma} v(b) + \tilde{\delta} v'(b))$$

and $\tilde{\gamma} v(b) + \tilde{\delta} v'(b) \neq 0$. Likewise, if $\tilde{\delta} \neq 0$, then $z(b) \neq 0$,

$$W_{z,v}(b) = \tilde{\delta}^{-1} z(b)(\tilde{\gamma} v(b) + \tilde{\delta} v'(b))$$

and $\tilde{\gamma} v(b) + \tilde{\delta} v'(b) \neq 0$. Since one of $\tilde{\gamma}$ and $\tilde{\delta}$ is nonzero, it follows that

$$\tilde{\gamma} v(b) + \tilde{\delta} v'(b) \neq 0.$$

Under the standing assumptions (1)–(5), we can differentiate the initial value problem for $z = z(x,\lambda)$ with respect to the parameter λ to obtain the variational initial value problem

$$\begin{cases} (x-a)w'' + \alpha(x)w' + \beta(x)w = -r(x)z(x)/\varphi(x) & \text{for } a < x \leq b \\ z(a) = 0,\ z'(a) = -(r(a)/\varphi(a))/\alpha(a) & \end{cases}$$

for $w = w(x) = \partial z(x,\lambda)/\partial\lambda$. Apply Theorem 193 with $g(x) = -r(x)z(x)/\varphi(x)$ to express the solution to the variational initial value problem as

$$w(x) = \begin{cases} A(x)z(x) + B(x)v(x) & \text{for } a < x \leq b \\ 0 & \text{for } x = a \end{cases}$$

where

$$A(x) = \int_a^x \frac{v(s)r(s)z(s)}{(s-a)\varphi(s)\, W_{z,v}(s)}\, ds \quad \text{and} \quad B(x) = -\int_a^x \frac{r(s)z(s)^2}{(s-a)\varphi(s)\, W_{z,v}(s)}\, ds$$

and the dependence on λ is suppressed. The integrand for $B(x)$ is not identically zero and maintains a fixed sign on $a < x \leq b$. ($W_{z,v}(s)$ maintains a fixed sign by (7.27).) Hence, $B(b,\lambda) \neq 0$. Recall that the coefficients in the variation of parameters solution are chosen so that

$$w'(x) = A(x)z'(x) + B(x)v'(x).$$

For an eigenvalue λ of the Sturm-Liouville problem,

$$\tilde{F}(\lambda) = \tilde{\gamma} z(b, \lambda) + \tilde{\delta} z'(b, \lambda) = 0$$

because $y = (x - a)^{\nu}\ z(x, \lambda)$ is the normalized eigenfunction corresponding to λ and

$$\begin{aligned} \tilde{F}'(\lambda) &= \frac{\partial}{\partial \lambda}(\tilde{\gamma} z(b, \lambda) + \tilde{\delta} z'(b, \lambda)) = \tilde{\gamma} w(b, \lambda) + \tilde{\delta} w'(b, \lambda). \\ &= \tilde{\gamma}(A(b, \lambda) z(b, \lambda) + B(b, \lambda) v(b, \lambda)) + \\ &\quad \tilde{\delta}(A(b, \lambda) z'(b, \lambda) + B(b, \lambda) v'(b, \lambda)) \\ &= A(b, \lambda)(\tilde{\gamma} z(b, \lambda) + \tilde{\delta} z'(b, \lambda)) + B(b, \lambda)(\tilde{\gamma} v(b, \lambda) + \tilde{\delta} v'(b, \lambda)) \\ &= B(b, \lambda)(\tilde{\gamma} v(b, \lambda) + \tilde{\delta} v'(b, \lambda)) \neq 0 \end{aligned}$$

and the proof is complete. ■

It follows immediately from Theorem 194 that the bisection method can be used to find each eigenvalue to any desired accuracy. Moreover, the numerically determined solutions to the z-initial value problem (7.22) yield approximate eigenfunctions $y = (x - a)^{\nu}\ z(x, \lambda)$ that converge uniformly to the normalized eigenfunction corresponding to the given eigenvalue.

Theorem 195 *Under the standing assumptions (1)–(5), if λ is an eigenvalue of the Sturm-Liouville eigenvalue problem (7.19) and $y = y(x, \lambda)$ is the corresponding normalized eigenfunction, then the bisection method can be used to generate a sequence of approximate eigenvalues $\lambda^{(n)} \to \lambda$ and corresponding approximate eigenfunctions $y_n(x, \lambda^{(n)}) = (x - a)^{\nu} z(x, \lambda^{(n)})$ where $\nu = \sqrt{q_1(a)/\varphi(a)}$ obtained by solving the z-initial value problem (7.22) such that $y_n(x, \lambda^{(n)}) \to y(x, \lambda)$ and $y_n'(x, \lambda^{(n)}) \to y'(x, \lambda)$ uniformly on $a \leq x \leq b$.*

Proof. By Theorem 194, there is a $\delta_1 > 0$ so that λ is the only zero of $\tilde{F}(\mu)$ in the interval $|\mu - \lambda| < \delta_1$ and $\tilde{F}$ changes sign at λ. Consequently, the bisection method can be used to generate a sequence $\lambda^{(n)}$ of approximate zeros of $\tilde{F}$ with $\lambda^{(n)} \to \lambda$ and corresponding solutions $z(x, \lambda^{(n)})$ to the z-initial value problem with parameter $\lambda^{(n)}$. By Theorem 163, given $\varepsilon > 0$ there is a $\delta_2 > 0$ so that $|\mu - \lambda| < \delta_2$ implies that the solution $z(x, \mu)$ to the z-initial value problem with parameter μ satisfies

$$|z(x, \mu) - z(x, \lambda)| < \varepsilon \quad \text{and} \quad |z'(x, \mu) - z'(x, \lambda)| < \varepsilon$$

for $a \leq x \leq b$. Since $\lambda^{(n)} \to \lambda$ as $n \to \infty$,

$$|z(x, \lambda^{(n)}) - z(x, \lambda)| < \varepsilon \quad \text{and} \quad |z'(x, \lambda^{(n)}) - z'(x, \lambda)| < \varepsilon$$

for $a \leq x \leq b$ provided n is sufficiently large. Since $(x - a)^{\nu}$ is bounded for x in $[a, b]$, $y_n(x, \lambda^{(n)}) \to y(x, \lambda)$ and $y_n'(x, \lambda^{(n)}) \to y'(x, \lambda)$ uniformly on $a \leq x \leq b$. ■

7.3.3 Newton's Method and Convergence

Theorem 194 is also a key step in establishing that the eigenvalues and eigenfunctions of a Sturm-Liouville eigenvalue problem (7.19) can be found using Newton's method as the root finder.

As in the previous section $z = z(x) = z(x, \lambda)$ is the unique solution to the initial value problem (7.22) and $\tilde{F}(\lambda) = \tilde{\gamma} z(b, \lambda) + \tilde{\delta} z'(b, \lambda)$ is continuously differentiable.

Fix an eigenvalue λ of the eigenvalue problem (7.19). Since $\tilde{F}(\lambda) = 0$ and $\tilde{F}'(\lambda) \neq 0$, if $\lambda^{(0)}$ is a sufficiently good initial guess of λ, then all the Newton iterates

$$\lambda^{(n+1)} = \lambda^{(n)} - \frac{\tilde{F}(\lambda^{(n)})}{\tilde{F}'(\lambda^{(n)})}$$

are defined and $\lambda^{(n)} \to \lambda$ as $n \to \infty$. See Theorem 46. Now reasoning in the same way as for the bisection method, one obtains the following result.

Theorem 196 *Under the standing assumptions (1)–(5), if* λ *is an eigenvalue of the Sturm-Liouville eigenvalue problem (7.19) and* $y = y(x, \lambda)$ *is the corresponding normalized eigenfunction, then Newton's method can be used to generate a sequence of approximate eigenvalues* $\lambda^{(n)} \to \lambda$ *and corresponding approximate eigenfunctions* $y_n(x, \lambda^{(n)}) = (x-a)^{\nu} z(x, \lambda^{(n)})$ *where* $\nu = \sqrt{q_1(a)/\varphi(a)}$ *obtained by solving the z-initial value problem (7.22) such that* $y_n(x, \lambda^{(n)}) \to y(x, \lambda)$ *and* $y_n'(x, \lambda^{(n)}) \to y'(x, \lambda)$ *uniformly on* $a \le x \le b$.

7.3.4 Numerical Results

We give several examples of singular eigenvalue problems of the type considered in Chapter 6 to illustrate the convergence results of the previous section. Additional results will be given in Chapter 8.

Since the shooting method described above for singular eigenvalue problems is an interactive method, just like the method used for regular Sturm-Liouville eigenvalue problems, the discussion in Section 7.1.4 about determining good starting values that lead to numerical convergence of the method applies verbatim to the singular problems. We urge the reader to review that discussion.

The following analogues of SL-1 to SL-5 for the singular problems in Section 7.2.4 hold. The singular eigenvalue problem (7.19) has the following properties, under the standing assumptions of Section 7.3:

SL-1. The eigenvalues are real, simple, and can be ordered as

$$\lambda_0 < \lambda_1 < \lambda_2 < \cdots$$

(See Theorem 184.)

SL-2. The eigenfunction y_n corresponding to λ_n has exactly n nodal zeros (interior zeros where a sign change occurs) and no other interior zeros. Moreover, the nodes of y_{n-1} and y_n strictly interlace. (See Theorem 184.)

SL-3. If $q \ge 0$ on $a < x \le b$, then all the eigenvalues are positive. (See Theorem 180.)

SL-4. If $q \ge 0$ on $a < x \le b$, then

$$\lambda_0 \ge \lim_{c \to a} \min_{c \le x \le b} \frac{q(x)}{r(x)} = \min_{a < x \le b} \frac{q(x)}{r(x)}.$$

The inequality in SL-4 satisfied by λ_0 is a consequence of reasoning that parallels the standard argument leading to SL-3 for regular problems. Normalize the eigenfunction y_0 by $\int_a^b y_0(x)^2 r(x)\, dx = 1$ and recall that y_0 is continuous on $[a, b]$ and continuously differentiable on $a < x \leq b$. Consequently, for $a < c < b$,

$$\lambda_0 = \int_a^b \lambda_0 y_0^2 r\, dx = \lim_{c \to a} \int_c^b \lambda_0 y_0^2 r\, dx = \lim_{c \to a} \int_c^b (-(py_0')' y_0 + qy_0^2)\, dx$$

$$= \lim_{c \to a} \left(\int_c^b (p(y_0')^2 + qy_0^2)\, dx - py_0 y_0' |_c^b \right).$$

The integral in the right member of the last equality increases as c decreases to a and by Lemma 160 $\lim_{c \to a} p(c) y_0(c) y_0'(c) = 0$. Hence,

$$\lambda_0 \geq \int_a^b (p(y_0')^2 + qy_0^2)\, dx - p(b) y_0(b) y_0'(b)$$

and

$$\lambda_0 \geq \int_a^b qy_0^2\, dx$$

because $\gamma\delta \geq 0$ implies that $-p(b) y_0(b) y_0'(b) \geq 0$. Now

$$\int_a^b qy_0^2\, dx \geq \int_c^b \frac{q}{r} y_0^2 r\, dx \geq \left(\min_{c \leq x \leq b} \frac{q(x)}{r(x)} \right) \int_c^b y_0^2 r\, dx.$$

Thus,

$$\lambda_0 \geq \left(\min_{c \leq x \leq b} \frac{q(x)}{r(x)} \right) \int_c^b y_0^2 r\, dx.$$

Since $\min_{c \leq x \leq b} q(x)/r(x)$ decreases as c decreases to a, it follows that

$$\lambda_0 \geq \lim_{c \to a} \left(\min_{c \leq x \leq b} \frac{q(x)}{r(x)} \right) \int_a^b y_0^2 r\, dx = \lim_{c \to a} \left(\min_{c \leq x \leq b} \frac{q(x)}{r(x)} \right).$$

Finally, since $q(x) = q_1(x)/(x - a)$ and $q_1(a) > 0$ and $r(x) > 0$ on $a < x \leq b$, it follows that $\min_{c \leq x \leq b} q(x)/r(x)$ is a constant function of c for $c > a$ sufficiently near to a and the limit on the right above is $\min_{a < x \leq b} q(x)/r(x)$.

SL-5. *(Rayleigh Quotient)* If the weight function is $r(x) = (x - a)^m \rho(x)$ where $m \geq 0$ and $\rho(x)$ is positive and continuous on $[a, b]$, then

$$\lambda_0 = \min \frac{\langle Ly, y \rangle}{\langle y, y \rangle_r} = \min R(y),$$

where

$$R(y) = \frac{-p(b) y(b) y'(b) + \int_a^b (py'^2 + qy^2)\, dx}{\int_a^b y^2 r\, dx}$$

and the minimum is over all functions $y \neq 0$ in the domain of L that satisfy the boundary conditions $|y(a)| < \infty$, $\gamma y(b) + \delta y'(b) = 0$, and for which $\lim_{x \to a} Ly(x)/(x - a)^m$ exists and is finite. The minimum is achieved if and only if y is an eigenfunction corresponding to λ_0. (See Theorem 185.)

Let $\tilde{F}(\lambda) = \tilde{\gamma} z(b, \lambda) + \tilde{\delta} z'(b, \lambda)$ be the function introduced in the previous section, where $z(x, \lambda)$ is the solution to (7.22). We established that $\tilde{F}(\lambda)$ is a smooth function of λ for λ real, that it has only simple zeros, and those zeros are the eigenvalues of the Sturm-Liouville eigenvalue problem (7.19). The corresponding normalized eigenfunctions are

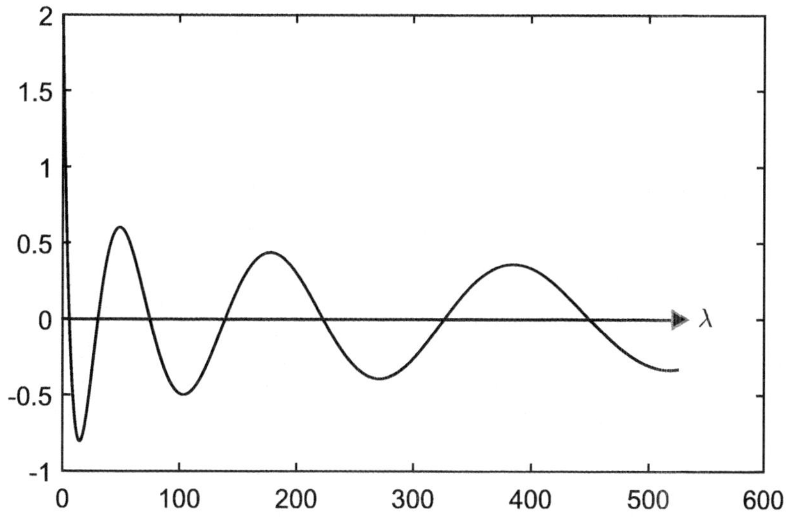

FIGURE 7.9: Graph of $\tilde{F}(\lambda) = 2J_0(\sqrt{\lambda})$

$y(x) = (x-a)^{\nu} z(x, \lambda)$ where $\nu = \sqrt{q_1(a)/\varphi(a)}$. Figure 7.9 illustrates a function $\tilde{F}(\lambda)$, see Example 1 below, and suggests a practical strategy for implementing the shooting method to determine eigenvalues and eigenfunctions of (7.19).

We apply the strategy first to determine numerically the smallest eigenvalue λ_0 and its corresponding eigenfunction y_0 using the shooting method and then using this information to systematically search for any additional eigenvalues and eigenfunctions that may be needed. We assume that Newton's method is used in the update step. As we said earlier and as Figure 7.9 suggests, it is not difficult to find a starting value for Newton's method that gives numerical convergence to some eigenvalue λ (perhaps not λ_0) and corresponding normalized eigenfunction $y(x, \lambda)$. A graph of $y(x, \lambda)$ reveals the number of its nodes. If there are, say, seven nodes, then $\lambda = \lambda_7$. Informed by this information, new starting values for Newton's method can be chosen and the shooting algorithm run again. This trial and error method leads to a determination of λ_0 in reasonably short time, usually a minute or two at most. The same approach can be used to locate other desired eigenvalues and eigenfunctions. See also the advice on page 309.

If $q(x) \geq 0$, the trial and error approach can be refined if finding helpful starting values proves difficult. In this case, from SL-4 and SL-5, the smallest eigenvalue λ_0 of the eigenvalue problem satisfies

$$\min_{a<x\leq b} \frac{q(x)}{r(x)} \leq \lambda_0 \leq R(y) \tag{7.35}$$

for any function $y \neq 0$ in the domain of L that satisfies the boundary conditions $|y(a)| < \infty$, $\gamma y(b) + \delta y'(b) = 0$ and is such that $\lim_{x\to a} Ly(x)/(x-a)^m$ exists and is finite. A routine verification establishes that a function y with these properties is

$$y(x) = (x-a)^{\mu} + c(x-a)^{\mu+1}$$

where

$$\mu = \nu = \sqrt{q_1(a)/\varphi(a)} \text{ if } m \leq \nu \quad \text{and} \quad \mu = m+1 \text{ otherwise}$$

and

$$c = -\frac{\gamma(b-a)+\delta\mu}{(b-a)(\gamma(b-a)+\delta(\mu+1))}.$$

Here $p(x) = (x-a)\varphi(x)$ and $q(x) = q_1(x)/(x-a)$ as in the standing assumptions. The double inequality (7.35) helps to inform a trial and error approach for finding a starting value for the shooting parameter that gives convergence to λ_0. Further help in finding suitable initial guesses for Newton's method in sensitive cases can be found by graphing the function $\tilde{F}(\lambda)$ over some interval with left endpoint at most $\min_{a<x\leq b} q(x)/r(x)$. The standard fourth order ordinary differential equation solvers yield numerical versions of $u(x,\lambda)$ and $u'(x,\lambda)$ for λ at a suitable set of equally spaced points, say, and, hence, a numerical version of $\tilde{F}(\lambda)$. The graph of $\tilde{F}(\lambda)$ can be used to select useful starting values for Newton's method. The same strategies apply when the bisection method is used as the root-finder.

The numerical results in the following examples were obtained with the shooting method, using Newton's method to update the shooting parameter, and following the practical suggestions given earlier for determining starting values. Newton's method was stopped when $|\tilde{F}(\lambda)| < 10^{-6}$ and the change in magnitude of the shooting parameter was less than 10^{-6}. The algorithm was run on a standard desktop computer and convergence was obtained in a matter of seconds, once good starting values were determined.

Example 1. A singular Sturm-Liouville eigenvalue problem of the form

$$\begin{cases} -(xy')' + \frac{1}{x}y = \lambda xy, & 0 < x < 1, \\ |y(0)| < \infty, & y(1) + y'(1) = 0. \end{cases}$$

arises in connection with heat conduction in a circular plate with insulated top and bottom and whose circumference obeys Newton's law of cooling. All thermal coefficients have been set equal to 1 for simplicity. By SL-3 all the eigenvalues are positive. The differential equation is Bessel's differential equation of order 1 and parameter λ. Consequently, the normalized eigenfunctions corresponding to the eigenvalues $\lambda_0, \lambda_1, \ldots$ are nonzero multiples of

$$J_1(\sqrt{\lambda_n}x)$$

for $n = 0, 1, 2, \ldots$ and the eigenvalues are the positive zeros of the function

$$\begin{aligned} F(\lambda) &= J_1(\sqrt{\lambda}) + \sqrt{\lambda}J_1'(\sqrt{\lambda}) \\ &= J_1(\sqrt{\lambda}) + \sqrt{\lambda}J_0(\sqrt{\lambda}) - J_1(\sqrt{\lambda}) = \sqrt{\lambda}J_0(\sqrt{\lambda}), \end{aligned}$$

where the formula $zJ_1'(z) = zJ_0(z) - J_1(z)$ was used.

In this example, $a = 0$, $b = 1$, $\varphi(x) = 1$, $q_1(x) = 1$, $\nu = \sqrt{q_1(0)/\varphi(0)} = 1$, $\gamma = 1$, and $\delta = 1$ so that $\tilde{\gamma} = \gamma(b-a) + \delta\nu = 2$, $\tilde{\delta} = \delta(b-a) = 1$, and the function $\tilde{F}$ of the previous section is

$$\tilde{F}(\lambda) = 2z(1,\lambda) + z'(1,\lambda).$$

It is informative to express $\tilde{F}(\lambda)$ in terms of Bessel functions. By Theorem 164 any nontrivial solution to the differential equation in (7.20) is a nonzero multiple of $(x-a)^{\nu}z(x,\lambda)$, where $z(x,\lambda)$ is the solution of the initial value problem (7.22). For Bessel's equation of order 1 and parameter λ this means that

$$J_1(\sqrt{\lambda}x) = cxz(x,\lambda)$$

for some constant c. Since

$$\sqrt{\lambda}J_1'(\sqrt{\lambda}x) = cz(x,\lambda) + cxz'(x,\lambda),$$

$z(0,\lambda) = 1$ and $J_1'(0) = 1/2$, $c = \sqrt{\lambda}/2$. So

$$z(x,\lambda) = 2\frac{J_1(\sqrt{\lambda}x)}{\sqrt{\lambda}x},$$

$$2J_1'(\sqrt{\lambda}) = z(1,\lambda) + z'(1,\lambda),$$

and

$$\begin{aligned}\tilde{F}(\lambda) &= 2z(1,\lambda) + z'(1,\lambda) = 2z(1,\lambda) + (2J_1'(\sqrt{\lambda}) - z(1,\lambda))\\ &= 2\frac{J_1(\sqrt{\lambda})}{\sqrt{\lambda}} + 2J_1'(\sqrt{\lambda}) = \frac{2}{\sqrt{\lambda}}(J_1(\sqrt{\lambda}) + \sqrt{\lambda}J_1'(\sqrt{\lambda}))\\ &= \frac{2}{\sqrt{\lambda}}F(\sqrt{\lambda}) = 2J_0(\sqrt{\lambda}).\end{aligned}$$

The relation $\tilde{F}(\lambda) = 2\lambda^{-1/2}F(\sqrt{\lambda}) = 2J_0(\sqrt{\lambda})$ and the fact that J_0 has only simple zeros shows that $\tilde{F}$ and F have the same positive simple zeros. Of course, this assertion follows from the general theory developed in Chapters 6 and 7. A graph of $\tilde{F}(\lambda)$ is shown in Figure 7.9.

The double inequality (7.35) for λ_0 applied with the polynomial $y = x - 2x^2/3$ yields the bounds

$$1 \le \lambda_0 \le R(y) = \frac{-xyy'\big|_{x=1} + \int_0^1 (xy'^2 + x^{-1}y^2)\,dx}{\int_0^1 y^2 x\,dx} = \frac{1/9 + 2/9}{31/540} = \frac{180}{31} < 5.9.$$

In the table that follows, the initial guess at λ_0 of 3 was suggested by the bounds above. The first guess at λ_1 was chosen as about twice $R(y)$ and an interactive doubling and halving of previous initial guesses and/or previously found approximate eigenvalues was used after that to get the eigenvalues λ_2, λ_3, and λ_4.

The shooting method produced the following approximations of the first five eigenvalues. The table shows all the initial guesses used, in the order they were used, to find the first five eigenvalues and corresponding eigenfunctions. The first column shows which eigenvalue was found with the corresponding initial guess. The check column are the squares of the zeros of $J_0(z)$ computed in MATLAB. The relative error is calculated using the values in the check column as proxies for the exact eigenvalues.

n	**Initial Guess**	$\lambda_n \approx$	**Check**	**Relative Error**
0	3	5.7832	5.7832	0
0	12	5.7832		
1	24	30.4718	30.4713	1.6×10^{-5}
0	48	5.7832		
2	60	74.8958	74.8870	1.2×10^{-4}
3	120	139.0491	139.0403	6.3×10^{-5}
4	240	222.9380	222.9323	2.6×10^{-5}

The Rayleigh quotient of $y = x - 2x^2/3$ gives a reasonable approximation of λ_0. Graphs of the corresponding normalized eigenfunctions y_0, y_1, y_2, y_3, and y_4 are shown in Figure 7.10.

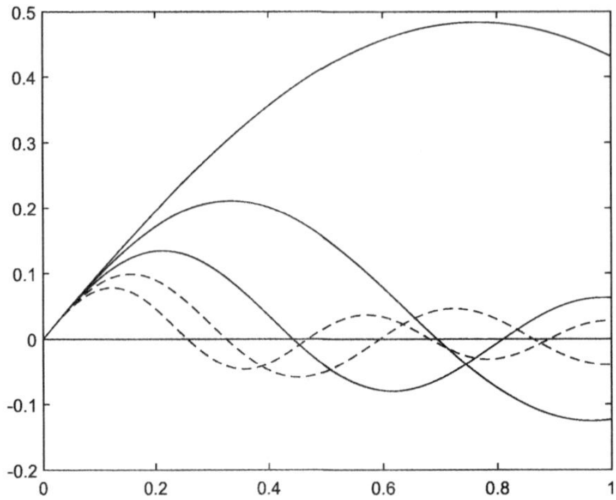

FIGURE 7.10: Eigenfunctions for Example 1

The graph shows the interlacing of the nodes in SL-2. The initial guess 3.5 produces the approximate eigenvalue 5.7832 in the table and a graph of the corresponding eigenfunction with no nodes in (0, 1). We conclude that 5.7832 is an approximate value of the eigenvalue λ_0. The other eigenvalues are identified in the same manner.

Example 2. Use the shooting method to find the first five eigenvalues and eigenfunctions of the singular Sturm-Liouville eigenvalue problem

$$\begin{cases} -(xy')' + \dfrac{\cos x}{4x} y = \lambda(\sin x)y, \quad 0 < x < 1, \\ |y(0)| < \infty, \; y(1) = 0. \end{cases}$$

This example $a = 0$, $b = 1$, $p(x) = x$, $q(x) = (\cos x)/4x$, $r(x) = \sin x$ so $\varphi(x) = 1$, $q_1(x) = 4^{-1}\cos x$, $\rho(x) = (\sin x)/x$, $m = 1$, and $\nu = \sqrt{q_1(0)/\varphi(0)} = 1/2$. Since $m > \nu$, the double inequality (7.35) for λ_0 can be applied with the polynomial $y(x) = x^2 - x^3$ yields the bounds

$$0.16 < \frac{\cos 1}{4\sin 1} \leq \lambda_0 \leq R(y) = \frac{\int_0^1 xy'^2 + ((\cos x)/4x)y^2\,dx}{\int_0^1 y^2 \sin x\,dx} < 18.4.$$

The shooting method with the indicated initial guesses lead to the following approximations of the first five eigenvalues. The initial guesses shown were the first ones tried in the search process. The following table shows all the initial guesses used, in the order they were used, to find the first five eigenvalues and corresponding eigenfunctions. The first column shows which eigenvalue was found with the corresponding initial guess. The strategy for choosing initial guesses was the same as for Example 1.

n	**Initial Guess**	$\lambda_n \approx$
0	9	10.2162
1	36	41.5311
2	72	93.7298
3	144	166.7988
4	288	260.7434

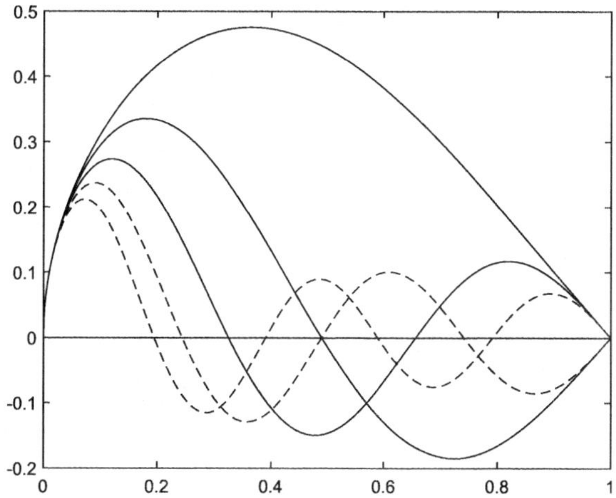

FIGURE 7.11: Eigenfunctions for Example 2

In this example, the Rayleigh quotient for $y = x^2 - x^3$ is not a good approximation of λ_0. Graphs of the corresponding normalized eigenfunctions y_0, y_1, y_2, y_3, and y_4 are shown in Figure 7.11.

Just as for regular problems or the singular problems in Chapter 5, if the shooting method converges numerically to $\tilde{\lambda}$, then an initial value problem solver can be used to evaluate $\tilde{F}(\tilde{\lambda} - \varepsilon)$ and $\tilde{F}(\tilde{\lambda} + \varepsilon)$ for some $\varepsilon > 0$. If ε can be chosen so that $\tilde{F}(\tilde{\lambda} - \varepsilon)$ and $\tilde{F}(\tilde{\lambda} + \varepsilon)$ are of opposite sign, then $\tilde{\lambda}$ approximates an eigenvalue λ of the eigenvalue problem to within an error of at most ε. A plot of $(x - a)^{\nu} z(x, \tilde{\lambda})$ will reveal the number of nodes of the approximate eigenfunction and, therefore, which eigenvalue has been approximated to accuracy ε. Since $\tilde{\lambda}$ is almost certainly not exactly an eigenvalue, $\tilde{F}(\tilde{\lambda}) \neq 0$ and hence $\tilde{F}(\tilde{\lambda} - \varepsilon)$ and $\tilde{F}(\tilde{\lambda} + \varepsilon)$ have the same sign for $\varepsilon > 0$ suitably small. Our experience is that by experimenting with different choices of ε reasonable small a sign change can be detected.

Example 2 (continued) We found $\lambda_2 \approx \tilde{\lambda} = 93.729772$ correctly rounded, where two more digits of the numerical output are shown here. In Example 2, $\tilde{F}(\lambda) = z(1, \lambda)$, where z is the solution of the initial value problem

$$\begin{cases} xz'' + 2z' + \left(\dfrac{1 - \cos x}{4x} + \lambda \sin x\right) z = 0, & 0 < x \leq 1, \\ z(0) = 1, \ z'(0) = -\beta(0)/\alpha(0) = -0/2 = 0, \end{cases}$$

where the coefficients and initial conditions are given by (7.23), (7.24), and (7.25). Numerical experiments with various choices for a possible error bound $\varepsilon > 0$ yield $F(\tilde{\lambda} - 2 \times 10^{-6}) \approx 1.7 \times 10^{-8}$ and $F(\tilde{\lambda} + 2 \times 10^{-6}) \approx -4.8 \times 10^{-9}$. It follows from the intermediate value theorem that there is an eigenvalue that differs from $\tilde{\lambda}$ by at most 2×10^{-6}. Since the approximate normalized eigenfunction $x^{1/2} z(x, \tilde{\lambda})$ has two nodes in $0 < x < 1$, we conclude that $|\tilde{\lambda} - \lambda_2| < 2 \times 10^{-6}$. Since $F(\tilde{\lambda} - 10^{-6}) \approx 1.1 \times 10^{-8}$ and $F(\tilde{\lambda} + 10^{-6}) \approx 6.5 \times 10^{-10}$ we further conclude that $10^{-6} < |\tilde{\lambda} - \lambda_2| < 2 \times 10^{-6}$, which shows that the error bound 2×10^{-6} is reasonably sharp.

Example 3. The damped vibrations (small transverse displacements) u of a circular membrane with radius 1 can be modeled by the damped wave equation $u_{tt} + \alpha u_t = \Delta u$

where $\alpha > 0$ is the damping constant. The change of variables $u = e^{\beta t} v$ with $\beta = -\alpha/2$ transforms the damped wave equation to the wave equation $v_{tt} = \Delta v + (\alpha^2/4)v$. Separation of variables in this equation with $v(t, r, \theta) = T(t)R(r)\Theta(\theta)$ leads to the following family of eigenvalue problems for R

$$\begin{cases} -(rR')' + \left(-\dfrac{r\alpha^2}{4} + \dfrac{n^2}{r}\right)R = \lambda r R, & 0 < x < 1, \\ |R(0)| < \infty, \ \gamma R(1) + \delta R'(1) = 0, \end{cases}$$

where n is a nonnegative integer and γ and δ determine how the circumference of the membrane is supported. If $\alpha = 0$ and there is no damping the differential equation is Bessel's equation of order n and parameter λ. By way of an example and to illustrate Neumann boundary data at $r = 1$, we choose $n = 2$, $\alpha = 2$, $\gamma = 0$, and $\delta = 1$. The choice of boundary conditions means the circumference is unconstrained. (The choices $\gamma = 1$ and $\delta = 0$ correspond to a standard drum head.)

Use the shooting method to find the first five eigenvalues and eigenfunctions of the singular Sturm-Liouville eigenvalue problem

$$\begin{cases} -(xy')' + \dfrac{2^2 - x^2}{x} y = \lambda x y, & 0 < x < 1, \\ |y(0)| < \infty, \ y'(1) = 0, \end{cases}$$

This example $a = 0$, $b = 1$, $p(x) = x$, $q(x) = (4 - x^2)/x$, $r(x) = x$ so $\varphi(x) = 1$, $q_1(x) = 4 - x^2$, $\rho(x) = 1$, $m = 1$, and $\nu = \sqrt{q_1(0)/\varphi(0)} = 2$. Since $m < \nu$, the double inequality (7.8) for λ_0 can be applied with the polynomial $y(x) = x^2 - 2x^3/3$ and yields the bounds

$$3 = \frac{4-1}{1} \le \lambda_0 \le R(y) = \frac{\int_0^1 xy'^2 + ((4 - x^2)/x)y^2\, dx}{\int_0^1 y^2 x\, dx} = \frac{50/189}{2/63} = \frac{25}{3} < 8.4.$$

The shooting method with the indicated initial guesses lead to the following approximations of the first five eigenvalues. The initial guesses shown were the first ones tried in the search process. The following table shows all the initial guesses used, in the order they were used, to find the first five eigenvalues and corresponding eigenfunctions. The first column shows which eigenvalue was found with the corresponding initial guess. The strategy for choosing initial guesses was the same as for Example 1.

n	**Initial Guess**	$\lambda_n \approx$
0	5.7	8.3284
0	17	8.3284
1	34	43.9731
3	68	172.4798
2	88	98.4048
4	264	266.2652

The Rayleigh quotient of $y = x^2 - 2x^3/3$ is a rather good approximation of the smallest eigenvalue, just as we have observed often before. Graphs of the corresponding normalized eigenfunctions y_0, y_1, y_2, y_3, and y_4 are shown in Figure 7.12. The figure shows the interlacing of nodes in SL-2. The approximate eigenvalue 8.3284 is identified as an approximation to λ_0 because the corresponding approximate eigenfunction has no nodes in $(0, 1)$. Similar determinations were made for the other approximate eigenvalues.

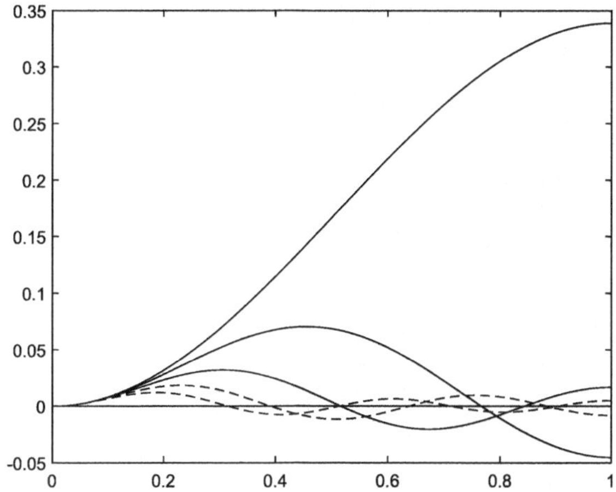

FIGURE 7.12: Eigenfunctions for Example 3

Problems with Neumann boundary conditions are typically the most challenging for finding good initial guesses. The flexible use of the usual strategy of doubling or halving either a previous initial guess or previously found approximate eigenvalue serves well here. The first guess 5.7 at λ_0 is about midway between the bounds $3 \le \lambda_0 \le 8.4$. The second guess of 17 is roughly double 8.4. Since this guess also gives convergence to λ_0, that guess is doubled to 34 and convergence to λ_1 is obtained. The next doubling of the initial guess to 68 gives convergence to λ_3, not to λ_2. With knowledge of that result, we chose the next guess as 88, about double the approximate value of λ_1. That guess gave convergence to λ_2. Doubling 88 to 176 gives an initial guess that would probably give convergence to λ_2 again. So we tripled the previous initial guess to 264 and obtained convergence to λ_4.

Example 3 (continued) We found $\lambda_4 \approx \tilde{\lambda} = 266.265224$ correctly rounded, where two more digits of the numerical output are shown here. In Example 3, $\tilde{F}(\lambda) = 2z(1, \lambda) + z'(1, \lambda)$, where z is the solution of the initial value problem

$$\begin{cases} xz'' + 5z' + x(1+\lambda)z = 0, & 0 < x \le 1, \\ z(0) = 1,\ z'(0) = -\beta(0)/\alpha(0) = -0/5 = 0, \end{cases}$$

and where the data in the initial value problem is given by (7.23), (7.24), and (7.25). Numerical experiments with various choices for a possible error bound $\varepsilon > 0$ yield $F(\tilde{\lambda} - 2 \times 10^{-5}) \approx 5.0 \times 10^{-9}$ and $F(\tilde{\lambda} + 2 \times 10^{-5}) \approx -1.5 \times 10^{-7}$. It follows from the intermediate value theorem that there is an eigenvalue that differs from $\tilde{\lambda}$ by at most 2×10^{-5}. Since the approximate normalized eigenfunction $x^2 z(x, \tilde{\lambda})$ has four nodes in $0 < x < 1$, we conclude that $|\tilde{\lambda} - \lambda_4| < 2 \times 10^{-5}$. Since $F(\tilde{\lambda} - 10^{-5}) \approx -5.4 \times 10^{-8}$ and $F(\tilde{\lambda} + 10^{-5}) \approx -1.0 \times 10^{-7}$ we further conclude that $10^{-5} < |\tilde{\lambda} - \lambda_4| < 2 \times 10^{-5}$, which shows that the error bound 2×10^{-5} is reasonably sharp.

7.3.5 Concluding Remarks

A few concluding remarks are in order about the numerical implementation of the shooting method for finding eigenvalues and eigenfunctions of (7.19) via the initial value problems (7.22) in which λ is the shooting parameter. Since the initial value problem is singular a $x = a$ standard initial value problem solvers do not apply on the full interval $a \le x \le b$.

We use the initial data in (7.19) and an Euler-like method to extend the initial data to $x = a + \varepsilon$, where $\varepsilon > 0$ is fixed suitably small. The initial value problem is regular on $a + \varepsilon \leq x \leq b$ and the solution z to (7.22) can be extended from $a + \varepsilon$ to b by a standard initial value solver. Recall that $p(x) = (x - a)\varphi(x)$ where $\varphi(x)$ is positive and continuous on $[a, b]$. The Euler-like step during the shooting procedure is done as follows. From the initial data in (7.22)

$$z(a + \varepsilon) = 1 + \varepsilon z'(a) = 1 + \varepsilon(-\beta(a)/\alpha(a))$$

to first order accuracy in ε. Since $\lim_{x\to a}(x - a)z''(x) = 0$ by Theorem 162, (7.22) leads to the following approximation,

$$z'(a + \varepsilon) = -\beta(a + \varepsilon)z(a + \varepsilon)/\alpha(a + \varepsilon).$$

A convenient choice for ε in code written for MATLAB and ode45 is $\varepsilon = \varepsilon ps$, the distance from 1.0 to the next larger positive double-precision number (approximately 10^{-16}).

If Newton's method is used in the shooting procedure, then the variational initial value problem of (7.22), its derivative with respect to the shooting parameter λ, is needed. The variational initial value problem also is singular at $x = a$ and first step way from the singularity at $x = a$ can be handled using the same ideas used for z above.

For programming purposes it is convenient to express the initial value problem for z as

$$\begin{aligned} z'' &= -\left(\frac{2\nu + 1}{x - a} + \frac{\varphi'(x)}{\varphi(x)}\right)z' - \left(\frac{q_2(x) + \nu\varphi'(x) + \lambda r(x)}{(x - a)\varphi(x)}\right)z \\ &= A(x)z' + B(x)z \end{aligned}$$

where

$$\begin{aligned} A(x) &= -\left(\frac{2\nu + 1}{x - a} + \frac{\varphi'(x)}{\varphi(x)}\right), \\ B(x) &= -\left(\frac{\tilde{q}_2(x) + \nu\varphi'(x)}{(x - a)\varphi(x)}\right), \end{aligned}$$

and $\tilde{q}_2(x)$ is given by (7.25) and $a < x \leq b$. The corresponding variational problem for $v = \partial z/\partial\lambda$ is

$$\begin{aligned} v'' &= A(x)v' + B(x)v + C(x)z(x), \\ v(a) = 0,\ v'(a) &= -\frac{r(a)}{(2\nu + 1)\varphi(a)}, \end{aligned}$$

and where

$$C(x) = -\frac{r(x)}{(x - a)\varphi(x)}.$$

Chapter 8

Concluding Examples and Observations

In this final chapter, we illustrate the results of the previous chapters with three typical problems in which Sturm-Liouville problems determine the characteristic frequencies and normal modes associated with a physical process and in which the solution can be conveniently represented by an eigenfunction expansion. Approximate eigenvalues and eigenfunctions are easily computable by the shooting methods of Chapter 7 or by similar methods suggested in this chapter.

8.1 Hanging Chains

Daniel Bernoulli (1700–1782) determined the normal modes of a hanging chain in 1732. This was the first use of a Bessel function. Later F. W. Bessel (1784–1846) investigated the functions that now bear his name. The hanging chain was further discussed by Euler in 1781. Bernoulli assumed the chain experienced small transverse oscillations so that a linear model was used. We make the same assumption in this section. Bernoulli assumed the chain was homogeneous; in particular, that it had constant density. We relax that assumption. The wave equation derived for the chain also models the small transverse vibrations of a hanging flexible metal cable, string, slinky, or bungee cord.

See https://www.acs.psu.edu/drussell/Demos/HangChain/HangChain.html for an online demonstration of the first three modes of Bernoulli's hanging chain.

A derivation of the wave equation for the small transverse oscillations of a hanging chain follows. The chain is modeled as a one-dimensional continuum that hangs vertically downward, has a variable mass density, and experiences small transverse displacements. The top of the chain is pinned and the bottom is free.

Basic Assumptions: The chain has length l. Two forces, gravity and tension, act on the chain. The force of gravity has constant gravitational acceleration $\mathbf{g}$. The tension $\mathbf{T}$ in the chain that acts at a given cross section is due to the portion of chain hanging below it and is directed tangentially to the chain.

Set up coordinates as follows: the x-axis is directed vertically upward with origin at the free end of the chain hanging at rest. Let $\mathbf{R} = x\mathbf{i} + u(x,t)\mathbf{j}$ be the position vector to the point on the chain at time t that would occupy position x when at rest in equilibrium. Assume the chain has (linear) mass density $\rho_0(x)$ when in equilibrium and has density $\rho(x,t)$ at time t. Denote arc length along the chain by $s = s(x,t)$.

If C is a segment of chain which would be the segment $[a,b]$ on the x-axis when in equilibrium, then by conservation of mass

$$\int_a^b \rho_0(x)\,dx = \int_C \rho\,ds = \int_a^b \rho(x,t)|\mathbf{R}_x|\,dx,$$

where $\mathbf{R}_x = \partial\mathbf{R}/\partial x$. Since a and b are arbitrary, we conclude that

$$\rho(x, t)|\mathbf{R}_x| = \rho_0(x)$$

for all time t.

By Newton's second and third laws

$$\frac{d}{dt}\int_C \rho\mathbf{R}_t\, ds = \mathbf{T}(a, t) - \mathbf{T}(b, t) + \int_C \rho\mathbf{g}\, ds$$

$$\frac{d}{dt}\int_a^b \rho(x, t)\mathbf{R}_t|\mathbf{R}_x|\, dx = -\int_a^b \mathbf{T}_x(x, t)\, dx + \int_a^b \rho(x, t)\mathbf{g}|\mathbf{R}_x|\, dx$$

$$\frac{d}{dt}\int_a^b \rho_0(x)\mathbf{R}_t\, dx = -\int_a^b \mathbf{T}_x(x, t)\, dx + \int_a^b \rho_0(x)\mathbf{g}\, dx$$

$$\int_a^b \rho_0(x)\mathbf{R}_{tt}\, dx = -\int_a^b \mathbf{T}_x(x, t)\, dx + \int_a^b \rho_0(x)\mathbf{g}\, dx$$

Since a and b can be chosen arbitrarily, it follows that

$$\rho_0(x)\mathbf{R}_{tt}(x, t) = -\mathbf{T}_x(x, t) + \rho_0(x)\mathbf{g} \tag{8.1}$$

is a differential equation that describes the motion.

By our basic assumption on the tension

$$\mathbf{T}(x, t) = T(x, t)\left(-\frac{\mathbf{R}_x}{|\mathbf{R}_x|}\right) = -\frac{T}{|\mathbf{R}_x|}\mathbf{R}_x$$

where $T(x, t)$ is the magnitude of the tension at cross section x at time t. Resolve (8.1) into components to obtain

$$\rho_0(x)u_{tt}\mathbf{j} = -\frac{\partial}{\partial x}\left(-\frac{T}{|\mathbf{R}_x|}(\mathbf{i} + u_x\mathbf{j})\right) - \rho_0(x)g\mathbf{i}$$

and

$$\left(\frac{\partial}{\partial x}\left(\frac{T}{|\mathbf{R}_x|}\right) - \rho_0(x)g\right)\mathbf{i} + \left(\frac{\partial}{\partial x}\left(\frac{T}{|\mathbf{R}_x|}u_x\right) - \rho_0(x)u_{tt}\right)\mathbf{j} = \mathbf{0}.$$

Consequently,

$$\frac{\partial}{\partial x}\left(\frac{T}{|\mathbf{R}_x|}\right) = \rho_0(x)g$$

and

$$\rho_0(x)u_{tt} = \frac{\partial}{\partial x}\left(\frac{T}{|\mathbf{R}_x|}u_x\right).$$

Since $T(0, t) = 0$, integration of the first equation gives

$$\frac{T}{|\mathbf{R}_x|} = \int_0^x \rho_0(\xi)g\, d\xi$$

and from the second

$$\rho_0(x)u_{tt} = \frac{\partial}{\partial x}\left(\left(\int_0^x \rho_0(\xi)g\, d\xi\right)u_x\right) = \left(\left(\int_0^x \rho_0(\xi)g\, d\xi\right)u_x\right)_x,$$

which is the wave equation for the transverse oscillations $u(x, t)$ of the chain.

Under the foregoing assumptions the initial boundary value problem for the chain is

$$\begin{cases} \rho_0(x)u_{tt} = (p(x)u_x)_x, & 0 < x < l, \quad t > 0, \\ |u(0,t)| < \infty, & u(l,t) = 0, \quad t \geq 0, \\ u(x,0) = f(x), & u_t(x,0) = v(x) \quad 0 \leq x \leq l, \end{cases} \tag{8.2}$$

where

$$p(x) = g\int_0^x \rho_0(\xi)\,d\xi$$

$f(x)$ specifies the initial shape of the chain, and $v(x)$ is its initial velocity profile. Observe that the differential equation is singular because $p(0) = 0$. Typically such equations can have both bounded and unbounded solutions. Physically realistic solutions for the displacement $u(x,t)$ must be bounded. This leads to the boundary condition $|u(0,t)| < \infty$ which means that the displacement is bounded for $x > 0$ and near 0 for all time t. It follows that $u(x,t)$ is bounded in space and time. We also note that $p(x) = x\varphi(x)$ where $\varphi(x)$ is continuous on $0 \leq x \leq l$ and $\varphi(0) \neq 0$ provided $\rho_0(0) \neq 0$. Indeed,

$$\varphi(x) = \frac{1}{x}\int_0^x g\rho_0(\xi)\,d\xi$$

for $0 < x \leq l$ and $\varphi(0) = g\rho_0(0)$. Thus, the eigenvalue problems that follow are singular of the type considered in Chapter 5 or Chapter 6 or can be transformed into such problems.

To get the normal modes of the motion, we seek separated solutions

$$u(x,t) = T(t)X(x)$$

to the differential equation in (8.2). Such a solution will satisfy the partial differential equation if and only if

$$\begin{aligned} \rho_0(x)\ddot{T}X &= (p(x)TX')', \\ \rho_0(x)\ddot{T}X &= T(p(x)X')' \\ \frac{\ddot{T}}{T} &= \frac{(p(x)X')'}{\rho_0(x)X} = -\lambda \end{aligned}$$

where $-\lambda$ is the separation constant. The separated solution also will satisfy the boundary conditions in (8.2) if $|X(0)| < \infty$ and $X(l) = 0$. Thus, the normal modes are determined by the singular Sturm-Liouville eigenvalue problem

$$\begin{cases} -(p(x)X')' = \lambda\rho_0(x)X, & 0 < x < l, \\ |X(0)| < \infty, \; X(l) = 0, \end{cases} \tag{8.3}$$

and the equation $\ddot{T} + \lambda T = 0$. By Theorem 151 all the eigenvalues of (8.3) are positive.

Next we determine the eigenvalues and eigenfunctions when the density of the chain is $\rho_0(x) = \bar{\rho}x^n$ with $\bar{\rho} > 0$ a constant and $n \geq 0$. The eigenvalue problem is

$$\begin{cases} -\left(g\dfrac{x^{n+1}}{n+1}X'\right)' = \lambda x^n X, & 0 < x < l, \\ |X(0)| < \infty, \; X(l) = 0. \end{cases} \tag{8.4}$$

In the original Bernoulli's problem $n = 0$ and the eigenvalue problem is

$$\begin{cases} -(gxX')' = \lambda X, & 0 < x < l, \\ |X(0)| < \infty, \; X(l) = 0, \end{cases} \tag{8.5}$$

a singular eigenvalue problem of the type treated in Chapter 5.

The differential equation in the eigenvalue problem (8.4) can be transformed into a Bessel equation: the differential equation for X can be expressed as

$$(x^{n+1}X')' + \left(\frac{n+1}{g}\lambda\right)x^n X = 0$$

where $0 < x < l$. This equation can be transformed into the equation

$$(uy')' - \frac{n^2}{u}y + \mu u y = 0,$$

which is Bessel's equation of order n and parameter $\mu = 4(n+1)\lambda/g$, by the change of variables

$$x = u^2,\ X(x) = u^{-n}y(u)$$

for $0 < u < \sqrt{l}$. Since $y(u) = x^{n/2}X(x)$ and $X(x)$ is bounded for $x > 0$ and near 0, y satisfies the boundary condition $|y(0)| < \infty$ and $y(\sqrt{l}) = 0$. Consequently, the corresponding eigenvalue problem for y is

$$\begin{cases} (uy')' - \dfrac{n^2}{u}y + \mu u y = 0, & 0 < u < \sqrt{l}, \\ |y(0)| < \infty,\ y(\sqrt{l}) = 0, \end{cases}.$$

a singular eigenvalue problem of the type treated in Chapter 6. The differential equation and boundary condition at $u = 0$ imply that $y(u)$ is a multiple of

$$J_n(\sqrt{\mu}u) = J_n(\sqrt{\mu x}) = J_n\left(\sqrt{\frac{4(n+1)\lambda x}{g}}\right)$$

and the second boundary condition is satisfied if and only if

$$J_n\left(\sqrt{\frac{4(n+1)\lambda l}{g}}\right) = 0;$$

that is, if and only if

$$\lambda = \lambda_{n,m} = \frac{g}{4(n+1)l}\zeta_{n,m}^2 \tag{8.6}$$

where $\zeta_{n,m}$ is the m-th positive zero of $J_n(\zeta)$. Thus the eigenvalues of (8.4) are given by (8.6) and the corresponding eigenfunctions are the nonzero multiples of

$$X_{n,m}(x) = x^{-n/2}J_n\left(\zeta_{n,m}\sqrt{\frac{x}{l}}\right) = l^{-n/2}\left(\sqrt{\frac{x}{l}}\right)^{-n} J_n\left(\zeta_{n,m}\sqrt{\frac{x}{l}}\right)$$

for $m = 0, 1, 2, \ldots$. The equation for T with $\lambda = \lambda_{n,m}$ has solutions the multiples of

$$T_{n,m}(t) = \cos\left(\sqrt{\zeta_{n,m}}t - \tau_{n,m}\right),$$

where $\tau_{n,m}$ is a phase angle. Consequently the normal modes are multiples of

$$u_{n,m}(x,t) = \left(\sqrt{\frac{x}{l}}\right)^{-n} J_n\left(\zeta_{n,m}\sqrt{\frac{x}{l}}\right)\cos\left(\sqrt{\zeta_{n,m}}t - \tau_{n,m}\right).$$

At any time t the chain profile is a constant multiple of

$$\left(\sqrt{x/l}\right)^{-n} J_n\left(\zeta_{n,m}\sqrt{x/l}\right).$$

If $n=0$ the eigenvalue problem (8.4) is a singular Sturm-Liouville eigenvalue problem of Type I. The shooting method in Chapter 7 for such singular problems leads to the following table, in which the relative errors where computed in MATLAB using double precision, for a chain of length $l=1$ m and $g = 9.8$ m/s^2

m	guess	λ by shooting	$\lambda = g\zeta_{0,m}^2/4$	Rel Error
0	7	14.1688	14.1688	-2.5×10^{-7}
1	56	74.6548	74.6546	2.4×10^{-6}
2	150	183.4756	183.4732	1.3×10^{-5}

and the plot in Figure 8.1. The initial guess 7 for $m=0$ was chosen because $0 \le \lambda_0 \le 3g/2 = 14.7$. The second inequality follows from use of the Rayleigh quotient $R(y)$ with $y=1-x$. In the figure, the profiles of the first three normal modes of the Bernoulli chain are normalized so that the horizontal deflection is 0.5 m at the free end of the chain.

As a further check on the accuracy of the shooting method, the profiles in Figure 8.1 were plotted in three different colors in MATLAB on a computer screen and then were exactly overwritten one-by-one by plots in black of the functions $0.5\,J_0(\zeta_{0,m}\sqrt{x})$ for $m=0$, 1, 2.

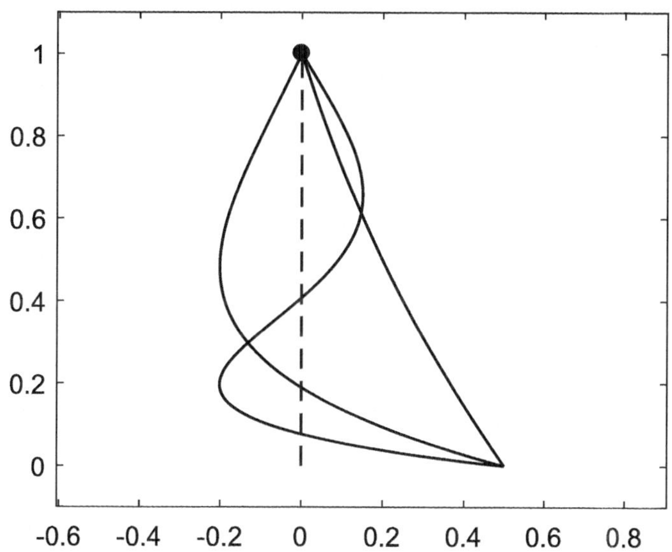

FIGURE 8.1: First three profiles of a Bernoulli chain

If $n>0$ the eigenvalue problem (8.4) is not a singular Sturm-Liouville eigenvalue problem of Type I or II but is equivalent to a singular problem of Type II via the change of variables above that expresses the eigenvalue, eigenfunction pairs λ, X in terms of the eigenvalue, eigenfunction pairs μ, y of the eigenvalue problem

$$\begin{cases} (uy')' - \dfrac{n^2}{u}y + \mu u y = 0, & 0 < u < \sqrt{l}, \\ |y(0)| < \infty, \ y(\sqrt{l}) = 0, \end{cases}$$

where $\lambda = g\mu/4(n+1)$ and $X(x) = u^{-n}y(u)$. When this Type II eigenvalue problem is solved by the shooting method of Chapter 7 for $n = 2$ (a quadratic density), a chain of length $l = 1$ m, and $g = 9.8$ m/s^2 the following data is obtained

m	μ guess	μ by shooting	$\lambda = g\mu/12$	$\mu = \zeta_{2,m}^2$	Rel Error
0	12	26.3746	21.5392	26.3746	-1.6×10^{-6}
1	52	70.8585	57.8678	70.8500	1.2×10^{-4}
2	142	135.0404	110.2830	135.0207	1.6×10^{-4}

as well as the plot in Figure 8.2. The initial guess 12 for $m = 0$ was chosen because $4 \le \lambda_0 \le 28$. The inequalities follow from Properties SL-4 and SL-5 in Section 7.3.4 and use of the Rayleigh quotient $R(y)$ with $y = x^2 - x^3$. The exact μ eigenvalues are $\zeta_{2,m}^2$. In Figure 8.2, the profiles of the first three normal modes of the chain are scaled so that the horizontal deflection is 0.5 m at the free end. The plots were obtained as follows. An X-eigenfunction is determined by $X(x) = u^{-2}y(u)$ where y has the form $y(u) = u^{\nu}z(u)$ where $\nu = 2$ in this case and $z(u)$ is the solution to a singular initial value problem. See (7.22), (7.23), (7.24), and (7.25). Consequently, $X(x) = z(u)$ for $x = u^2$ and $0 \le u \le \sqrt{l} = 1$. A numerical selection of values of z is found by shooting and used to make the plots of the normal modes shown in Figure 8.2.

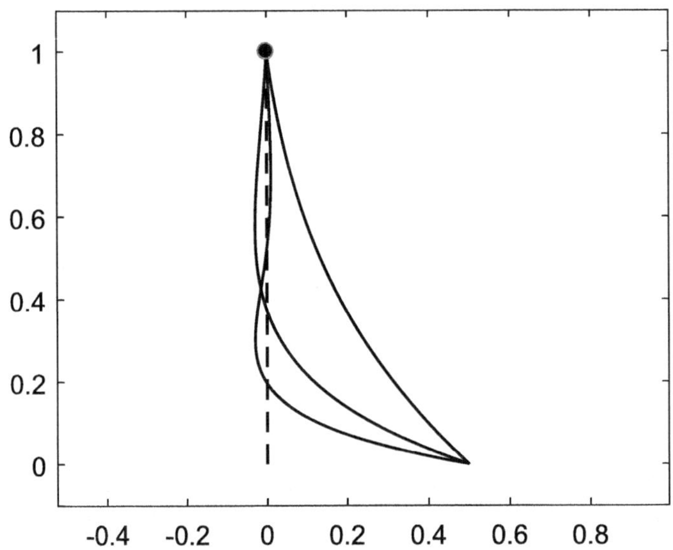

FIGURE 8.2: First three profiles for a quadratic density

Just as for Bernoulli's chain, as a further check on the accuracy of the shooting method applied to a chain with a quadratic density, the profiles in Figure 8.2 were plotted in three different colors in MATLAB on a computer screen via shooting and then were exactly overwritten one-by-one by plots in black of the functions $\left((4/\zeta_{2,m}^2)x^{-1}J_2(\zeta_{2,m}\sqrt{x}),\, x\right)$ for $m = 0, 1, 2$ in which the exact eigenfunctions are normalized to have displacement 0.5 m at their free end.

Next we seek the normal modes $u(x,t) = T(t)X(x)$ of a chain with density

$$\rho_0(x) = \bar{\rho}\cos\left(\frac{\pi(x-c)}{2l}\right)$$

where $\bar{\rho}$ is a positive constant and c is fixed in $0 \le c \le l$. For these densities

$$p(x) = g\int_0^x \bar{\rho}\cos\left(\frac{\pi(\xi - c)}{2l}\right)d\xi = \frac{2lg\bar{\rho}}{\pi}\left(\sin\left(\frac{\pi(x-c)}{2l}\right) + \sin\left(\frac{\pi c}{2l}\right)\right)$$

and the normal modes are determined by the eigenvalue problem

$$\begin{cases} -(p(x)X')' = \lambda\rho_0(x)X, & 0 < x < l, \\ |X(0)| < \infty, \ X(l) = 0, \end{cases} \tag{8.7}$$

and the equation $\ddot{T} + \lambda T = 0$. The eigenvalue problem is singular in the sense of Chapter 5 for $0 \le c < l$ but not for $c = l$. The eigenvalues and eigenfunctions of (8.7) cannot be expressed in terms of standard special functions. Nevertheless the shooting method of Chapter 7 can be used to find accurate numerical approximations to the eigenvalues and eigenfunctions and hence the profiles of the normal modes for any choice of c with $0 \le c < l$. The following table gives the approximate values of the first three eigenvalues for the cases $c = 0$, $l/4$, $l/2$ and $3l/4$ for a chain of length $l = 1$ meter.

$c \backslash m$	0	1	2
0	13.2724	84.0815	217.1227
$l/4$	13.8859	76.8065	192.7343.
$l/2$	14.4802	71.2081	174.7854
$3l/4$	15.3163	65.5182	155.8711

Figures 8.3, 8.4, 8.5, and 8.6 show the spatial factor of the first three normal modes for $c = 0$, $l/4$, $l/2$ and $3l/4$ for a chain of length $l = 1$ meter, respectively. As c increases the chain becomes less dense near its free end, more dense toward its pinned end, and has maximum density at $x = c$. The shooting method of Chapter 7 produces an eigenfunction (spatial factor of a normal mode) normalized to be 1 at $x = 0$. Such spatial profiles are shown in the figures.

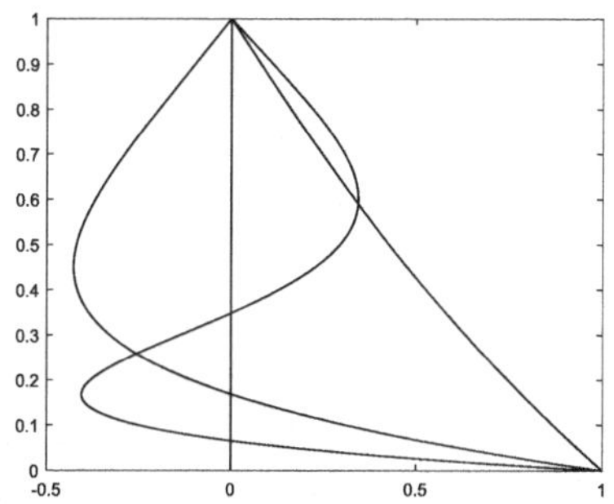

FIGURE 8.3: Normal modes $c = 0$

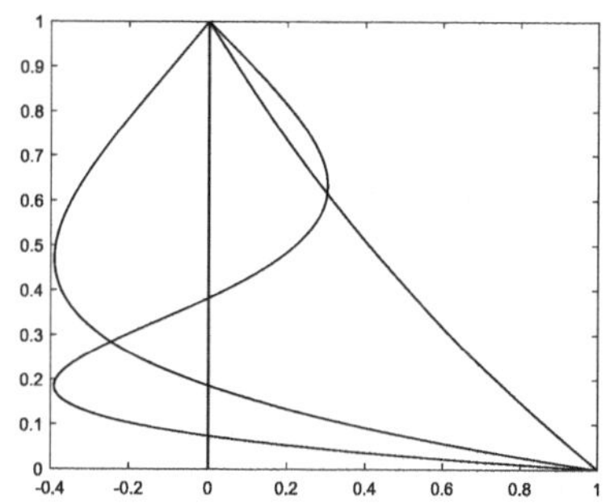

FIGURE 8.4: Normal modes $c = 1/4$

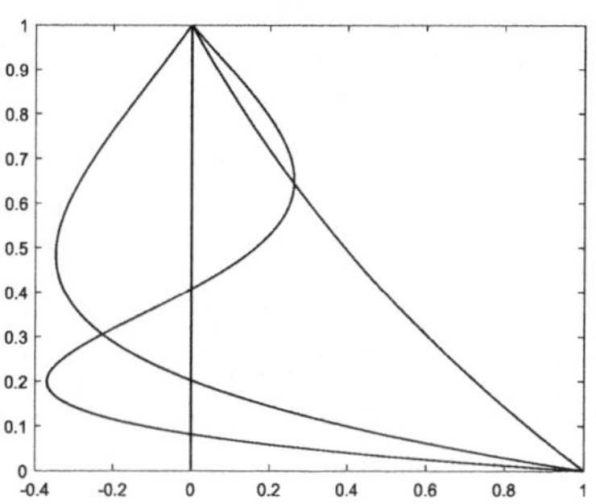

FIGURE 8.5: Normal modes $c = 1/2$

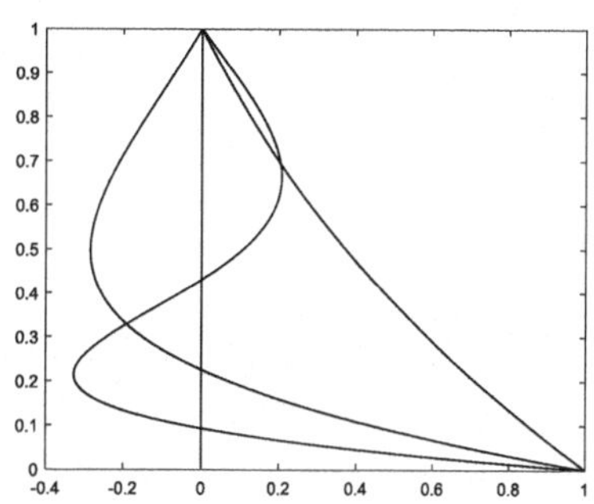

FIGURE 8.6: Normal modes $c = 3/4$

The normal modes of the family of chains of length l and density $\rho_0(x) = \bar{\rho}\exp((x-c)/l)$, where $\bar{\rho} > 0$ and c is a parameter, are surprising. Different choices for c give chains with manifestly different density distributions. Nevertheless, the normal modes of all these chains are the same! Indeed, it is readily confirmed that $p(x)$ for these densities is $p(x) = gl\bar{\rho}(\exp((x-c)/l) - \exp(-c/l))$ and that the eigenvalue problem (8.3) reduces to

$$\begin{cases} -(gl(e^{x/l}-1)X')' = \lambda e^{x/l}X, & 0 < x < l, \\ |X(0)| < \infty, \; X(l) = 0, \end{cases}$$

after cancellation of common factors. Thus, the spatial factor X of each normal mode is independent of c, the eigenvalues λ are independent of c, and the temporal factor determined by $\ddot{T} + \lambda T = 0$ is independent of c. Use of the shooting method of Chapter 7 yields the following results. The first three eigenvalues for a chain of length 1 meter are

$$\lambda = 15.1687, \; 66.5073, \; 158.9361$$

and the corresponding spatial profiles of the first three normal modes are shown in Figure 8.7.

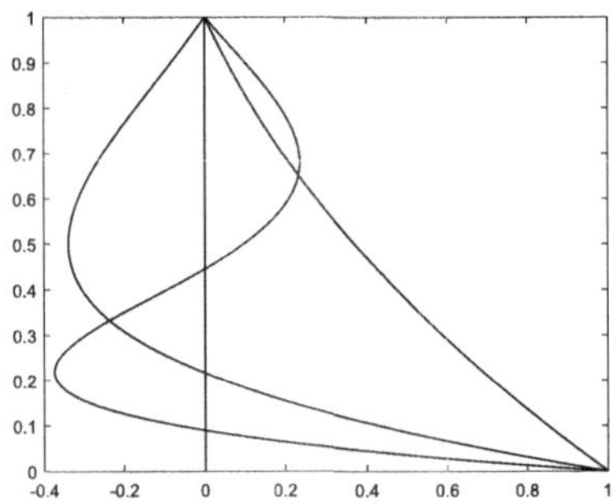

FIGURE 8.7: Normal modes for $\rho_0(x) = \bar{\rho}\exp(x-c)$ and any c

8.2 Vibrating Strings

Given its importance, there are many derivations for the model governing the small transverse vibrations of a one-dimensional elastic continuum, such as a string on a piano, violin or guitar. We shall give two derivations of the model, one based directly on Newton's laws and the second on energy considerations.

Consider a tightly stretched string of length l anchored at its endpoints. We assume the string is taught enough and thin enough to be modeled as a one-dimensional continuum. When at rest, the string is modeled as the segment $0 \le x \le l$ of the x-axis. For ease of discussion we assume the string lies horizontally when at rest. We shall neglect the force of gravity because its effect is assumed to be small compared to the other forces that act and assume that the external forces that act on the string act transversely to it. It is assumed that the external forces produce small transverse displacements of the points of the string from their rest positions. That is, as the string vibrates, the point on the string located at position x when the string is at rest moves perpendicular to the x-axis and occupies, at time t, the position $u(x,t)$, vertically above or below point x on the x-axis. These assumptions imply that the string is so flexible and the displacements from the rest position are so small that the slight displacements from the vertical that the physical string elements experience are negligible.

The fact that the string is modeled as a one-dimensional continuum means that its mass while in equilibrium can be described by a density, mass per unit length, $\rho(x)$ that varies continuously and that the mass of the string when in motion is given by a continuous mass density

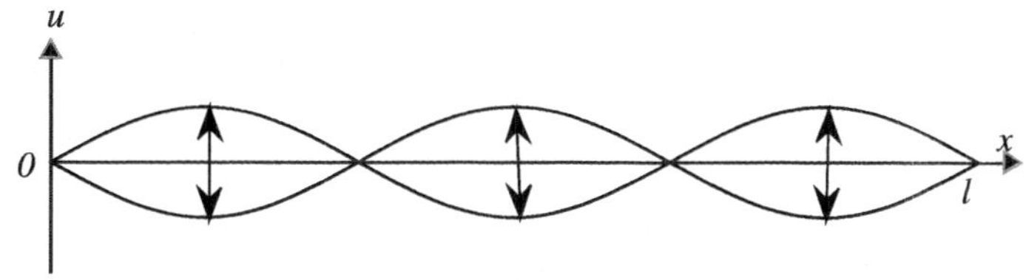

FIGURE 8.8: Transverse vibrations of a string

$\tilde{\rho}(x,t)$. The segment of string between x and $x+\Delta x$ when the string is in equilibrium moves into an arc, say C_t, at time t when the string is in motion. By conservation of mass

$$\int_x^{x+\Delta x} \rho(y)\,dy = \int_{C_t} \tilde{\rho}\,ds = \int_x^{x+\Delta x} \tilde{\rho}(y,t)\sqrt{1+u(y,t)^2}\,dy,$$

where $s = s(y,t)$, $0 \le y \le l$ is arc length along the string at time t measured from its left end. Divide this equation by Δx, let $\Delta x \to 0$, and use l'Hôpital's rule or the definition of a derivative to conclude that

$$\tilde{\rho}(x,t)\sqrt{1+u(x,t)^2} = \rho(x)$$

for $0 \le x \le l$.

Since $u_t(x,t)$ is the velocity of the point (x,t) on the arc of string C_t, the time rate of change of the momentum of the arc is

$$\frac{d}{dt}\int_{C_t} \tilde{\rho}u_t\,ds = \frac{d}{dt}\int_x^{x+\Delta x} \tilde{\rho}(y,t)\sqrt{1+u(y,t)^2}\,u_t(y,t)\,dy = \frac{d}{dt}\int_x^{x+\Delta x} \rho(y)u_t(y,t)\,dy.$$

Thus, by Newton's second law

$$\int_x^{x+\Delta x} \rho(y)u_{tt}(y,t)\,dy = \sum \text{forces}, \tag{8.8}$$

where the sum is over all forces that act on the arc of string C_t. As was mentioned previously, gravitational forces are being neglected and we also will neglect resistance forces from the medium surrounding the string. This last assumption is justified by the fact that the properties of the string we shall study can be determined by the motion of the string over a small time interval. Thus, the only external forces we shall include in our model are the tension forces that act at the ends of the arc C_t. The assumption that the string is flexible means that these forces act tangentially to the string. (Later we shall include some of the forces neglected at this time. See also [18].) A force diagram is shown in Figure 8.9.

Since the string arc C_t only moves vertically there can be no net horizontal force acting on it. Therefore,

$$T(x+\Delta x,t)\cos\alpha(x+\Delta x,t) - T(x,t)\cos\alpha(x,t) = 0,$$

where $T(x,t)$ is the magnitude of the tension at the cross section of the string through the point $(u(x,t),t)$ and α is the angle shown in Figure 8.9. Divide this equation by Δx and let $\Delta x \to 0$ to find that

$$\frac{\partial}{\partial x}(T(x,t)\cos\alpha(x,t)) = 0$$

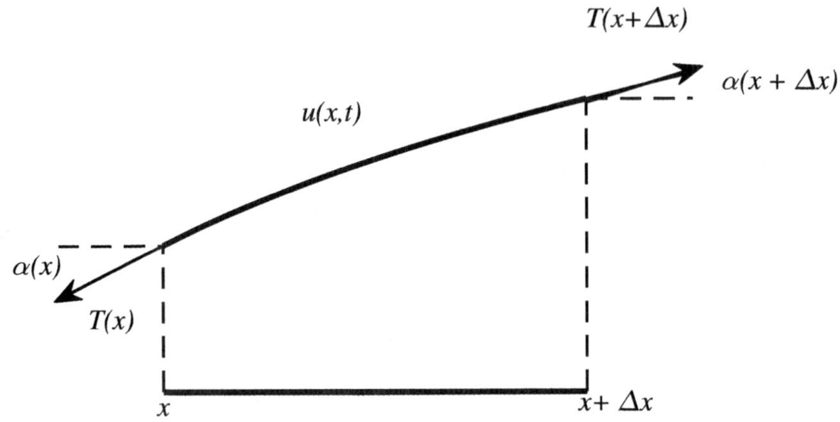

FIGURE 8.9: Force Diagram

and

$$T(x,t)\cos\alpha(x,t) = \tau,$$

where τ is a constant or at most a function of t. We shall assume that the horizontal component of tension τ is constant, unless the contrary is explicitly stated. This means, for example, that we are ignoring any thermal effects that may occur due to the vibrations. This assumption is reasonable because we plan only to study the vibrations over a short time interval.

The net vertical component of force acting on C_t is

$$\begin{aligned} &T(x+\Delta x,t)\sin\alpha(x+\Delta x,t) - T(x,t)\sin\alpha(x,t) \\ &= T(x+\Delta x,t)\cos\alpha(x+\Delta x,t)\frac{\sin\alpha(x+\Delta x,t)}{\cos\alpha(x+\Delta x,t)} - T(x,t)\cos\alpha(x,t)\frac{\sin\alpha(x,t)}{\cos\alpha(x,t)} \\ &= \tau(\tan\alpha(x+\Delta x,t) - \tan\alpha(x,t)) = \tau(u_x(x+\Delta x,t) - u_x(x,t)). \end{aligned}$$

Thus, under our assumptions, Newton's second law (8.8) can be expressed as

$$\int_x^{x+\Delta x} \rho(y) u_{tt}(y,t)\,dy = \tau(u_x(x+\Delta x,t) - u_x(x,t)).$$

Again, divide by Δx and let $\Delta x \to 0$ to find that

$$\rho(x) u_{tt}(x,t) = \tau u_{xx}(x,t)$$

for $0 < x < l$ and all relevant t. This partial differential equation, a basic wave equation, must be combined with boundary and initial data to determine the motion of the string. Thus, we are lead to the following initial boundary value problem for the vibrations of a string:

$$\begin{cases} u_{tt} = c^2 u_{xx}, & 0 < x < l, & t > 0, \\ u(0,t) = 0, & u(l,t) = 0, & t \geq 0, \\ u(x,0) = f(x), & u_t(x,0) = v(x), & 0 \leq x \leq l, \end{cases} \tag{8.9}$$

where $c = \sqrt{\tau/\rho}$, the string is set in motion at time $t = 0$, $f(x)$ is the initial shape of the string, and $v(x)$ is its velocity profile at $t = 0$. Here τ may depend on the time t and ρ may depend on position x but we shall assume that these physical parameters are constant unless explicitly stated to the contrary.

The derivation of the wave equation just given follows directly from first principles, Newton's laws and conservation of mass. The second derivation, which follows, is based on energy considerations and variational methods. It is more abstract but adds insight into our understanding of oscillatory motion of conservative systems and the exchange of energy in such systems. It is based on the principle of least action which states that the action, the integral over any time interval during the motion of the kinetic energy minus the potential energy, must be stationary when compared to all possible (virtual) motions of the physical system. Further explanation of the action integral and a motivation for it follow.

Let C be the position of the string at time t; that is, C is the graph of $u(x, t)$ versus x at time t. The total kinetic energy of the string is

$$K = \int_C \frac{1}{2}\tilde{\rho} u_t^2 \, ds = \frac{1}{2}\int_0^l \rho u_t^2 \, dx$$

because $ds = \sqrt{1 + u_x^2}\, dx$ and $\tilde{\rho}\sqrt{1 + u_x^2} = \rho$, the rest density of the string. The potential energy U of the string is its stored elastic energy due to the stretching of the string as it oscillates. An element of the string at rest of length Δx moves into an element of length $\Delta s = \sqrt{1 + u_x^2}\Delta x$, up to first order terms in Δx, at time t; see Figure 8.10.

The incremental work ΔU done by the tension T during the displacement $\Delta s - \Delta x$ is

$$\Delta U = T(\Delta s - \Delta x) = T\left(\frac{\Delta s}{\Delta x} - 1\right)\Delta x$$

and the elastic potential energy U stored in the string is

$$U = \int_0^l T\left(\frac{ds}{dx} - 1\right) dx = \int_0^l T\left(\sqrt{1 + u_x^2} - 1\right) dx.$$

Since the vibrations are small, we assume $u_x^2 << 1$; consequently, $\cos\alpha(x, t) = 1/\sqrt{1 + u_x^2} \approx 1$, $T = \tau/\cos\alpha(x, t) \approx \tau$, and $\sqrt{1 + u_x^2} - 1 \approx u_x^2/2$. With these approximations, the expression for U reduces to

$$U = \frac{1}{2}\int_0^l \tau u_x^2 \, dx,$$

which we take for the potential energy of the string at time t. The action integral for such a string is

$$I(u) = \int_{t_1}^{t_2} (K - U)\, dt = \frac{1}{2}\int_{t_1}^{t_2}\left(\int_0^l (\rho u_t^2 - \tau u_x^2)\, dx\right) dt, \tag{8.10}$$

where t_1 and t_2 are any two times during the motion. Now comes an important change in point of view. Regard $u = u(x, t)$ as a potential shape for the string at position x and time t.

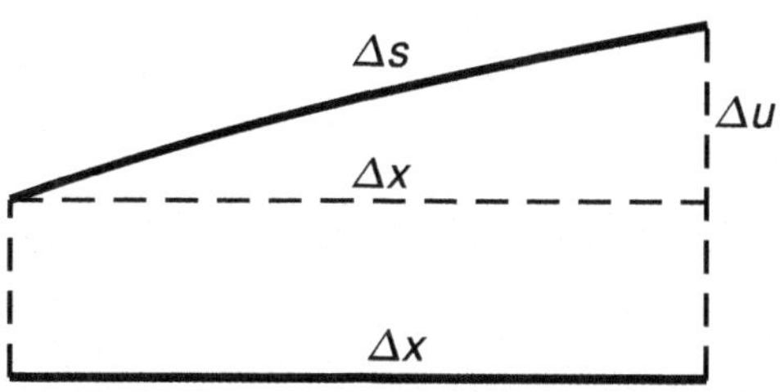

FIGURE 8.10: Element of arc $\Delta s \approx \sqrt{1 + u_x^2(x, t)}\Delta x$

This potential shape is sometimes called a **virtual motion** of the string. For us a virtual motion is any continuously differentiable function of space and time that satisfies the given boundary conditions; here that the string has fixed ends. The **principle of least action** asserts that among all possible virtual motions of the string, the actual motion of the string u makes the action integral stationary. The original statement of the principle replaced "stationary" by "a minimum". It was a fundamental belief of the mathematicians and physicists that developed the consequences of Newton's laws that the processes of the physical world evolved in as economical a way as possible. In the case of a vibrating string, the total energy, kinetic plus potential energy, is conserved (is constant) but during the motion energy is constantly flowing back and forth between kinetic and potential energy. The inner integral in (8.10) is a measure of this ebb and flow of energy. The outer integral averages the ebb and flow over time, apart from a constant factor $1/(t_2 - t_1)$. The early practitioners of mathematical physics asserted that the actual motion of the string minimized the action integral among all virtual motions of the string. To find the minimum, one seeks to set the derivative of $I(u)$ to zero. A virtual motion u for which the derivative of $I(u)$ is zero is called a **stationary point** of $I(u)$. It was realized later by looking at particular conservative systems that the correct formulation of the principle of least action was that the actual motion u makes the action integral $I(u)$ stationary; often the motion u that makes the action stationary does minimize the integral, but not always.

We mentioned above that the total energy of the string is conserved. This should be expected because we have ignored frictional effects in our model. However, a proof is needed, in part to confirm that the model has the properties expected from the physical assumptions we have made. The total (mechanical) energy of the string at time t is

$$E = \frac{1}{2}\int_0^l (\rho u_t^2 + \tau u_x^2)\, dx \tag{8.11}$$

and

$$\begin{aligned} \frac{dE}{dt} &= \int_0^l (\rho u_t u_{tt} + \tau u_x u_{xt})\, dx \\ &= \int_0^l u_t(\rho u_{tt} - \tau u_{xx})\, dx = 0, \end{aligned}$$

upon integration by parts on the second summand in the first integrand and use of the boundary conditions $u(0,t) = 0$ and $u(l,t) = 0$. Thus, the total energy is constant as expected. Energy is conserved for the string even if it is inhomogeneous (as the reasoning above shows) as long as the horizontal component of tension τ, which may depend on time t, is a constant. If $\tau = \tau(t)$, then reasoning above leads to

$$\begin{aligned} \frac{dE}{dt} &= \int_0^l u_t(\rho u_{tt} - \tau u_{xx})\, dx + \int_0^l \tau'(t) u_x^2\, dx, \\ \frac{dE}{dt} &= -\tau'(t)\int_0^l u_x^2\, dx. \end{aligned} \tag{8.12}$$

This result implies a number of properties of a vibrating string that do not strike us as likely to be observed experimentally, except perhaps by targeted experiments suggested by what follows. The second factor on the right of (8.12) equals zero if and only if $u_x(x,t) = 0$ for all $0 \le x \le l$ and all time t during the motion. It follows by integration with respect to x that $u(x,t) = 0$ for all $0 \le x \le l$ and all time t because $u(0,t) = 0$ for all time t during the motion. Consequently, apart from a string at rest, (8.12) implies that (1) energy is conserved if and only if the horizontal component of tension τ is independent of the time t, (2) the total energy decreases in time, if τ increases in time, and (3) the total energy increases in time if τ decreases in time. Of course these properties also hold for a string at rest. Furthermore, if $\tau'(t) \geq 0$, then $dE/dt \leq 0$ and $E(t) \leq E(0)$ for all time t. Since the difference of two solution to (8.9) satisfy that problem with zero initial conditions, $E(0) = 0$ for the difference of two solutions to (8.9), $E(t) = 0$ for the difference, and the two solutions to (8.9) are the same. This establishes the uniqueness of the solution to the initial boundary value problem when τ is constant or increases in time.

Now we apply the principal of least action to give an alternative derivation of the wave equation for the vibrating string. We use a simple but powerful idea of Euler, later refined by Lagrange: the action integral $I(u)$ is a function whose inputs are (other) functions. So the standard calculus of functions of a real variable available to him did not apply directly. Euler finessed this obstacle as follows. He considered virtual motions of the form $u + \varepsilon\zeta$, where u is the actual motion, ε is a real parameter, and ζ is any continuously differentiable function of space and time that satisfies $\zeta(0, t) = 0$ and $\zeta(l, t) = 0$ for all time t. These side conditions on ζ guarantee that $u + \varepsilon\zeta$ is a virtual motion; it is smooth and has the same fixed ends as the actual motion u. Since u is the actual motion of the string, the action integral $I(u + \varepsilon\zeta)$ evaluated at the comparison (test) functions $u + \varepsilon\zeta$ must be stationary when $\varepsilon = 0$; that is

$$\frac{d}{d\varepsilon} I(u + \varepsilon\zeta)\bigg|_{\varepsilon=0} = 0,$$

which is a standard calculus problem. From a more modern perspective, this derivative is the direction derivative of the function I at u in the direction of the function ζ. It is often denoted by

$$\delta I(u)\zeta = \frac{d}{d\varepsilon} I(u + \varepsilon\zeta)\bigg|_{\varepsilon=0}$$

and called the **first variation** of I.

For the string,

$$\begin{aligned}\delta I(u)\zeta &= \frac{d}{d\varepsilon}\left[\frac{1}{2}\int_{t_1}^{t_2}\left(\int_0^l \left(\rho(u_t + \varepsilon\zeta_t)^2 - \tau(u_x + \varepsilon\zeta_x)^2\right) dx\right) dt\right]\bigg|_{\varepsilon=0} \\ &= \int_{t_1}^{t_2}\left(\int_0^l (\rho u_t\zeta_t - \tau u_x\zeta_x)\, dx\right) dt.\end{aligned}$$

Integrate by parts to remove the temporal and spatial derivatives from ζ and use $\zeta(0, t) = 0$ and $\zeta(l, t) = 0$ for all t to obtain

$$\delta I(u)\zeta = -\int_{t_1}^{t_2}\left(\int_0^l (\rho u_{tt}(x, t) - \tau u_{xx}(x, t))\, dx\right)\zeta(x, t)\, dt.$$

If $\rho u_{tt}(x, t) - \tau u_{xx}(x, t)$ were not equal to zero at a point (x, t) with $0 < x < l$ and $t > 0$, say was positive there, then we could choose a test function ζ that has the same sign as $\rho u_{tt}(x, t) - \tau u_{xx}(x, t)$ near the point in question and becomes identically zero before $\rho u_{tt}(x, t) - \tau u_{xx}(x, t)$ changes its sign. For such a ζ, $\delta I(u)\zeta < 0$ but by the principle of least action $\delta I(u)\zeta = 0$ for all test functions ζ because u is the actual motion of the string. This contradiction forces us to conclude that the actual motion of the string u satisfies $\rho u_{tt}(x, t) - \tau u_{xx}(x, t) = 0$ for $0 < x < l$ and $t > 0$. This is the same equation of motion we found before and the motion of the string is modeled by the initial boundary value problem (8.9).

We now take a closer look at properties of a vibrating string when the speed of propagation $c = \sqrt{\tau/\rho}$ is constant. The normal modes of vibration are determined by the separated solutions $u(x, t) = X(x)\,T(t)$ of the wave equation and boundary conditions in (8.9). A nontrivial separated solution $u = XT$ will satisfy the wave equation if and only if

$$XT'' = c^2 X'' T,$$

$$\frac{T''}{c^2 T} = \frac{X''}{X} = -\lambda,$$

for some separation constant $-\lambda$. A nontrivial separated solution will satisfy the boundary conditions if and only if $X(0) = 0$ and $X(l) = 0$. The nontrivial normal modes of the string are given by solutions $u(x, t) = X(x)T(t)$ such that

$$\begin{cases} -X''(x) = \lambda X(x), & 0 < x < l, \\ X(0) = 0, & X(l) = 0, \end{cases} \tag{8.13}$$

and

$$T''(t) + \lambda c^2 T(t) = 0$$

for all time t during the motion.

(8.13) is a regular Sturm-Liouville eigenvalue problem. Hence it has an infinite sequence of positive eigenvalues with corresponding eigenfunctions that are orthogonal (with weight function 1) by the principal results in Section 4.8. Of course, for the homogeneous string the eigenvalues are well-known,

$$\lambda = \lambda_n = (n\pi/l)^2$$

for $n = 1, 2, 3, \dots$ and corresponding eigenfunctions are the nonzero multiples of

$$X_n(x) = \sin(n\pi x/l).$$

The companion solution for the temporal factor of a normal mode is

$$T_n(t) = a_n \cos(cn\pi t/l) + b_n \sin(cn\pi t/l),$$

where a_n and b_n are arbitrary constants. Thus, the normal modes are multiples of

$$u_n(x, t) = (a_n \cos(cn\pi t/l) + b_n \sin(cn\pi t/l)) \sin(n\pi x/l).$$

At each instant of time the wave profile is a multiple of $\sin(n\pi x/l)$.

A formal solution to the wave equation and the boundary conditions in (8.9) is obtained by an infinite superposition of the normal modes,

$$\begin{aligned} u(x, t) &= \sum_{n=1}^{\infty} u_n(x, t) = \sum_{n=1}^{\infty} T_n(t) X_n(x) \\ &= \sum_{n=1}^{\infty} (a_n \cos(cn\pi t/l) + b_n \sin(cn\pi t/l)) \sin(n\pi x/l). \end{aligned} \tag{8.14}$$

Any partial sum of the formal solution satisfies the wave equation and boundary conditions. If the coefficients a_n and b_n converge rapidly enough to zero so that the series for u, u_{tt}, and u_{xx} converge uniformly for $0 \le x \le l$ and t in any bounded time interval, then the full infinite series will satisfy the wave equation and boundary conditions. Let us assume this is the case and inquire if the coefficients a_n and b_n can be chosen to satisfy the initial conditions in (8.9). For simplicity, assume the string is a piano string that is hit by a hammer. Then $f(x) = 0$ for $0 \le x \le l$ and the hammer gives an initial velocity profile $v(x)$ for $0 \le x \le l$ to the string.

The series (8.14) will satisfy the initial condition $u(x, 0) = 0$, that is

$$u(x, 0) = \sum_{n=1}^{\infty} a_n \sin(n\pi x/l) = 0$$

for $0 \le x \le l$, if $a_n = 0$ for all n. To satisfy the initial condition $u_t(x, 0) = v(x)$, the b_n must be chosen to satisfy

$$u_t(x, 0) = \sum_{n=1}^{\infty} b_n(cn\pi/l) \sin(n\pi x/l) = v(x)$$

for $0 \le x \le l$. Since the eigenfunctions $X_n(x) = \sin(n\pi x/l)$ are orthogonal on $0 \le x \le l$, multiplication of the series by X_m and term by term integration, justified by the assumed uniform convergence, yields

$$b_m(cm\pi/l) \int_0^l \sin^2(m\pi x/l)\, dx = \int_0^l v(x) \sin(m\pi x/l)\, dx,$$

$$b_m = \frac{2}{cm\pi} \int_0^l v(x) \sin(m\pi x/l)\, dx$$

for $m = 1, 2, 3, \ldots$.

Thus, the vibrations of the piano string can be expressed as

$$u(x, t) = \sum_{n=1}^{\infty} b_n \sin\left(c\sqrt{\lambda_n} t\right) \sin\left(\sqrt{\lambda_n} x\right)$$

where

$$b_n = \frac{2}{cn\pi} \int_0^l v(x) \sin\left(\sqrt{\lambda_n} x\right) dx.$$

The individual term $u_n(x, t)$ in the series solution (8.14) are the normal modes, usually called harmonics here, and determine the nature of the sound produced. The first harmonic

$$u_1(x, t) = b_1 \sin\left(c\sqrt{\lambda_1} t\right) \sin\left(\sqrt{\lambda_1} x\right)$$

has period $2\pi/c\sqrt{\lambda_1}$ and fundamental frequency

$$\frac{c\sqrt{\lambda_1}}{2\pi} = \frac{\sqrt{\tau/\rho}(\pi/l)}{2\pi} = \frac{1}{2l}\sqrt{\frac{\tau}{\rho}}.$$

Likewise, the n-th harmonic (overtone) has frequency

$$\frac{c\sqrt{\lambda_n}}{2\pi} = \frac{\sqrt{\tau/\rho}(n\pi/l)}{2\pi} = n\frac{1}{2l}\sqrt{\frac{\tau}{\rho}},$$

exactly n times the fundamental frequency. Observe that the fundamental frequency can be increased, that is the pitch made higher, by increasing the tension τ, and/or decreasing the density ρ, and/or decreasing the length l of the string. These precise conclusions about the dependence of the frequencies on the physical parameters τ, ρ, and l can be confirmed qualitatively by looking at a piano keyboard and pressing various keys. These observations are the basis for the so-called Pythagorean rules for tuning.

In the analysis above we have assumed that the series expansion in (8.14) and the expansions for its derivatives converge sufficiently rapidly so that the term-by-term integrations are valid. This validity depends on the assumptions made on the velocity profile $v(x)$ and is part of the theory of eigenfunction expansions. These assumptions in turn determine how rapidly the coefficients b_n tend to zero and, hence, how many overtones one can hear. Moreover, if the piano string is simply stretched between two posts, the sound it produces is so faint one can scarcely hear it. The piano sounding board amplifies the sound so it is easily heard. The sound box on a violin or guitar plays the same role for these stringed instruments.

We have ignored one obvious property of real strings in this discussion. The vibrations die out over a relatively short time interval and energy is lost in the process. The string vibrates in air which resists its motion and the movement of the string causes it to heat up a little. In each case, the more rapidly the string vibrates the more pronounced the damping effects. This leads to a damped wave equation model for the string in which damping effects are modeled by $-\rho k u_t$, where k is a constant with units m/sec^3 and the minus sign occurs because the damping effects oppose the motion. Adding this term to the right member of (8.8) leads to the damped wave equation

$$\rho u_{tt} + \rho k u_t = \tau u_{xx}$$

or

$$u_{tt} + k u_t = c^2 u_{xx},$$

in which we continue to assume the speed of propagation $c = \sqrt{\tau/\rho}$ is constant. The initial boundary value problem for the damped wave equation is

$$\begin{cases} u_{tt} + k u_t = c^2 u_{xx}, & 0 < x < l, & t > 0, \\ u(0,t) = 0, & u(l,t) = 0, & t \geq 0, \\ u(x,0) = f(x), & u_t(x,0) = v(x), & 0 \leq x \leq l. \end{cases} \tag{8.15}$$

We should expect that the energy of a damped string modeled by (8.15) decreases with time. To confirm this, differentiate the total energy E in (8.11) and use the damped wave equation $\rho u_{tt} + \rho k u_t = \tau u_{xx}$ to find that

$$\frac{dE}{dt} = \int_0^l u_t(\rho u_{tt} - \tau u_{xx})\, dx = -\int_0^l \rho k u_t^2\, dx \leq 0.$$

Thus, the total mechanical energy of the damped string decreases.

The separated solutions $u = TX$ for (8.15) are determined by the same eigenvalue problem for X as before and the equation

$$T'' + kT' + \lambda c^2 T = 0.$$

Thus, the eigenvalues are $\lambda_n = (n\pi/l)^2$ for $n = 1, 2, 3, \ldots$, corresponding orthogonal eigenfunctions are

$$X_n(x) = \sin(n\pi x/l),$$

and the temporal factors are

$$T_n(t) = e^{-kt/2}\left[a_n \cos\left(\frac{\sqrt{4\lambda_n c^2 - k^2}}{2}t\right) + b_n \sin\left(\frac{\sqrt{4\lambda_n c^2 - k^2}}{2}t\right)\right],$$

where we have assumed that

$$4\lambda_1 c^2 - k^2 = 4\pi^2 \frac{\tau}{\rho l^2} - k^2 > 0$$

and, hence, that $4\lambda_n c^2 - k^2 > 0$ for all n. Roughly, this means that damping is relatively weak compared to the effects of tension.

Finally, consider an undamped, inhomogeneous string so that $\rho = \rho(x)$ varies with position and assume that the horizontal component of tension τ is constant. The normal modes $u = T(t)X(x)$ in this case are determined by the eigenvalue problem

$$\begin{cases} -X''(x) = \lambda c^2(x)X(x), & 0 < x < l, \\ X(0) = 0, & X(l) = 0, \end{cases} \tag{8.16}$$

where $c(x) = \sqrt{\tau/\rho(x)}$ and the temporal factor satisfies

$$T'' + \lambda T = 0.$$

By the principal results in Chapter 4, the regular eigenvalue problem (8.16) has all positive eigenvalues, say λ_n, and corresponding orthogonal eigenfunctions $X_n(x)$, where X_n has exactly n nodes in $0 < x < l$. At each fixed time t, the spatial profile of a normal mode is a multiple of $X_n(x)$. The following table shows the first three eigenvalues λ and corresponding frequencies $2\pi/\sqrt{\lambda}$ for a string of length 1 meter and for

$$c(x)^2 = \tau/\rho(x) = 1 + 2x^2, \;\; 1/(1+x), \;\; \cos(x - 1/2),$$

respectively.

$c^2(x)$	λ_n, $n = 1, 2, 3$	$2\pi/\sqrt{\lambda_n}$
$1+2x^2$	6.1928	2.5249
$1+2x^2$	24.6132	1.2665
$1+2x^2$	55.1490	0.8461
$1/(1+x)$	14.5112	1.6494
$1/(1+x)$	57.6534	0.8275
$1/(1+x)$	129.5411	0.5520
$\cos(x-1/2)$	10.0320	1.9838
$\cos(x-1/2)$	40.9026	0.9824
$\cos(x-1/2)$	92.3677	0.6538

Figures 8.11–8.13 show profiles of the first three normal modes (graphs of the first three eigenfunctions $X_n(x)$).

The eigenvalues in the table and the graphs were found using the shooting method of Chapter 7. The shooting method normalizes the profiles shown to have slope 1 at $x = 0$. The actual normal modes have the indicated profile at each instant in time but with much smaller vertical displacements.

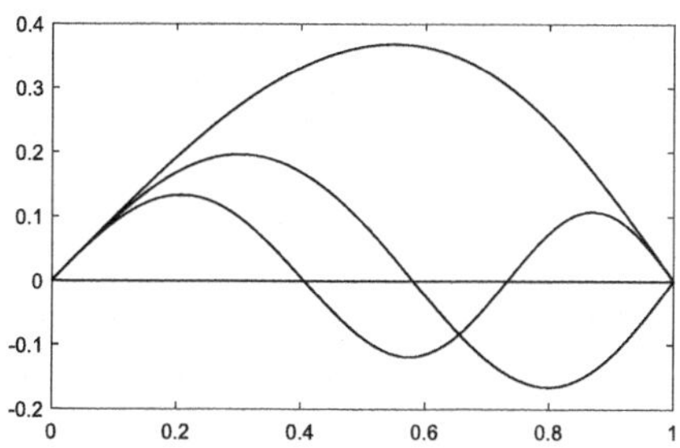

FIGURE 8.11: Profiles for $c^2(x) = 1 + 2x^2$

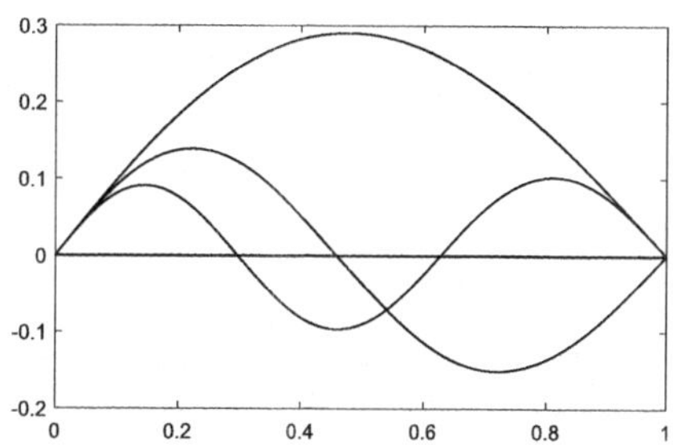

FIGURE 8.12: Profiles for $c^2(x) = 1/(1+x)$

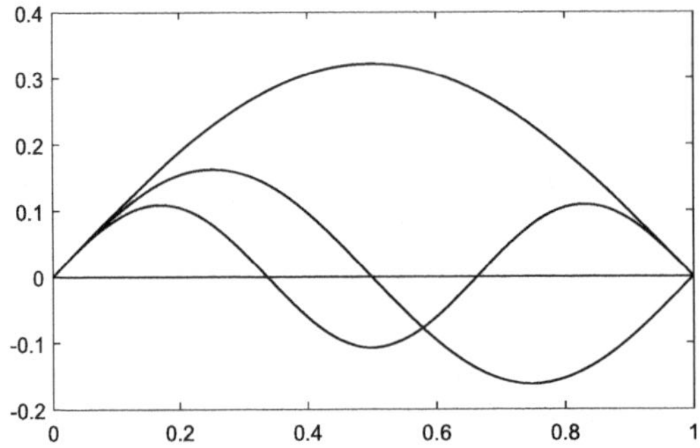

FIGURE 8.13: Profiles for $c^2(x) = \cos(x - 1/2)$

8.3 Vibrating Bars

This section suggests how shooting methods developed earlier to treat Sturm-Liouville eigenvalue problems can be extended to higher order self-adjoint problems. We use the fourth order wave equation for the vibrations of a bar (beam) to motivate ideas and because it is among the most important higher order equations. Although the shooting methods suggested are primarily of interest for inhomogeneous bars, it is instructive to begin with an example for a homogeneous bar. The eigenvalues and eigenfunctions associated with the vibrations of a bar subject to the usual boundary conditions behave qualitatively like those of Sturm-Liouville problems. In particular, the eigenvalues $\{\lambda_n\}_{n=0}^{\infty}$ are all simple and an eigenfunction ϕ_n corresponding to λ_n has exactly n nodal zeros because the Green's functions associated with the standard bar problems are Kellogg kernels. The interested reader can find these results and many more in [16].

8.3.1 Homogeneous Bars

The small transverse vibrations $u(x, t)$ of a homogeneous bar (beam) about its neutral axis (the x-axis) satisfy the biharmonic wave equation

$$u_{tt} + a^2 u_{xxxx} = 0$$

where

$$a^2 = EI/\rho A.$$

Here E is Young's modulus of elasticity, A is the cross sectional area of the beam at x, I is the moment of inertia of the cross section at x about an axis perpendicular to the neutral axis of the bar, ρ is the linear density of the bar, and l is its length. See for example [18]. We assume the beam is uniform and homogeneous so all physical and geometric parameters are positive constants. We also assume no external load is applied to the bar and that gravitational effects are negligible.

In the context above, the small transverse displacements of a bar clamped at both ends are determined by the initial boundary value problem

$$\begin{cases} u_{tt} + a^2 u_{xxxx} = 0, & 0 < x < l, & t > 0, \\ u(0, t) = u_x(0, t) = 0, & u(l, t) = u_x(0, t) = 0, & t \geq 0, \\ u(x, 0) = f(x), & u_t(x, 0) = v(x), & 0 \leq x \leq l. \end{cases}$$

The (nontrivial) normal modes associated with the bar are determined by separated solutions $u(x,t) = X(x)T(t)$ that satisfy the biharmonic wave equation and the homogeneous boundary conditions. Such solutions are determined by the eigenvalue problem

$$\begin{cases} X'''' = \lambda X, & 0 < x < l, \\ X(0) = X'(0) = 0, & X(l) = X'(l) = 0, \end{cases} \tag{8.17}$$

and the equation $\ddot{T} + \lambda a^2 T = 0$, where $-\lambda$ is the separation constant. The equation for T suggests that $\lambda > 0$. This is easy to confirm. Multiply the differential equation for X by $\bar{X}$ and integrate by parts twice to obtain

$$\int_0^l |X''(x)|^2\, dx = \lambda \int_0^l |X(x)|^2\, dx.$$

It follows that $\lambda \geq 0$. If equality were to hold, then $X(x)$ would be linear and hence identically equal to zero. Thus, $\lambda > 0$.

The eigenvalue problem can be solved explicitly in the following sense. A standard approach to this eigenvalue problem is to start with the general solution of the differential equation $X'''' - \lambda X = 0$ and show that the boundary conditions are satisfied by a nontrivial solution X if a certain transcendental equation is satisfied by λ. We prefer to take a variant of the route to these results that can be used when the eigenvalue problem involves variable coefficients.

The solution space of $X'''' - \lambda X = 0$ is four dimensional but the eigenfunctions lie in the two dimensional subspace in which $X(0) = X'(0) = 0$. A basis for this subspace is $v(x,\lambda)$ and $w(x,\lambda)$ where

$$\begin{aligned} v''''(x,\lambda) - \lambda v(x,\lambda) = 0,\ 0 < x < l, \\ v(0,\lambda) = 0,\ v'(0,\lambda) = 0, \\ v''(0,\lambda) = 1,\ v'''(0,\lambda) = 0, \end{aligned}$$

and

$$\begin{aligned} w''''(x,\lambda) - \lambda w(x,\lambda) = 0,\ 0 < x < l, \\ w(0,\lambda) = 0,\ w'(0,\lambda) = 0, \\ w''(0,\lambda) = 0,\ w'''(0,\lambda) = 1. \end{aligned}$$

Consequently, the eigenfunctions of (8.17) have the form

$$X(x) = Av(x,\lambda) + Bw(x,\lambda)$$

where A and B not both zero are chosen to satisfy $X(l) = X'(l) = 0$; that is,

$$\begin{bmatrix} v(l,\lambda) & w(l,\lambda) \\ v'(l,\lambda) & w'(l,\lambda) \end{bmatrix} \begin{bmatrix} A \\ B \end{bmatrix} = \begin{bmatrix} 0 \\ 0 \end{bmatrix}. \tag{8.18}$$

Since A and B are not both zero, the eigenvalues are determined by the equation

$$v(l,\lambda)w'(l,\lambda) - v'(l,\lambda)w(l,\lambda) = 0. \tag{8.19}$$

The initial value problems for v and w have solutions

$$v(x,\lambda) = \frac{1}{2\mu^2}(\cosh \mu x - \cos \mu x) \tag{8.20}$$

and

$$w(x,\lambda) = \frac{1}{2\mu^3}(\sinh \mu x - \sin \mu x), \tag{8.21}$$

respectively where $\mu = \lambda^{1/4}$. Consequently equation (8.19) can be conveniently expressed as

$$\cosh \mu l \cos \mu l = 1,$$

where $\mu = \lambda^{1/4}$. A plot of $1/\cos \mu l$ and $\cosh \mu l$ reveals that the equation has an infinite number of positive roots, μ_n, such that

$$\lim_{n\to\infty} \frac{\mu_{2n}}{(4n+3)\pi/2} = 1 \quad \text{and} \quad \lim_{n\to\infty} \frac{\mu_{2n+1}}{(4n+1)\pi/2} = 1.$$

The eigenvalues of (8.17) are $\lambda_n = \mu_n^4$ and the corresponding eigenfunctions are

$$Av(x,\lambda_n) + Bw(x,\lambda_n),$$

where the constants A and B satisfy (8.18). Since $v(l,\lambda_n) > 0$ and $w(l,\lambda_n) > 0$, it follows that $B = -\big(v(l,\lambda_n)/w(l,\lambda_n)\big)A$ with A an arbitrary constant. Thus, each eigenvalue λ_n is simple and its corresponding eigenfunctions are the nonzero multiples of

$$X_n(x) = v(x,\lambda_n) - \frac{v(l,\lambda_n)}{w(l,\lambda_n)} w(x,\lambda_n), \tag{8.22}$$

where v and w are given by (8.20) and (8.21).

For the record the first three roots of the equation are

$$\mu_0 = 4.7300,\ \mu_1 = 7.8532,\ \text{and}\ \mu_2 = 10.9956.$$

The corresponding eigenvalues of (8.17) are

$$\lambda_0 = 500.5639,\ \lambda_1 = 3,803.5371,\ \ \text{and}\ \lambda_2 = 14,617.6301.$$

Since the temporal factor of a normal mode is a multiple of $T(t) = \cos\big(a\sqrt{\lambda}t - \phi\big)$, where ϕ is an arbitrary phase angle, the vibrational frequency of the first three normal modes is

$$\frac{a\sqrt{\lambda_0}}{2\pi} = 3.5608a,\ \frac{a\sqrt{\lambda_1}}{2\pi} = 9.8155a,\ \text{and}\ \frac{a\sqrt{\lambda_2}}{2\pi} = 19.2424a,$$

where $a = \sqrt{EI/\rho A}$. For example, if the fundamental frequency of the bar is 440 Hz, then $a \approx 123.57$ and the next two frequencies are about 1,213 Hz and 2,378 Hz.

Since the eigenvalues $\lambda = \mu^4$ of (8.17) satisfy $\cosh \mu l \cos \mu l = 1$, accurate numerical approximations to the first few eigenvalues can be found with the aid of a root-finder. Corresponding approximate eigenfunctions are given by (8.22). Such an explicit equation for the eigenvalues is available only if the general solution to the differential equation in (8.17) can be expressed in a convenient closed form. This is normally not the case for an inhomogeneous bar.

8.3.2 Inhomogeneous Bars

The small transverse vibrations $u(x,t)$ of an inhomogeneous bar (beam) about its neutral axis (the x-axis) satisfy the biharmonic wave equation

$$A\rho u_{tt} + (EIu_{xx})_{xx} = 0.$$

Here E is Young's modulus of elasticity, A is the cross sectional area of the bar at x, I is the moment of inertia of the cross section at x about an axis perpendicular to the neutral axis of the bar, ρ is the linear density of the bar at x, and l is its length. All geometric and physical parameters are positive and may vary with x. We also assume no external load is applied to the bar and that gravitational effects are negligible.

In the context above, the small transverse displacements of a bar clamped at both ends are determined by the initial boundary value problem

$$\begin{cases} A\rho u_{tt} + (EIu_{xx})_{xx} = 0, & 0 < x < l, & t > 0, \\ u(0,t) = u_x(0,t) = 0, & u(l,t) = u_x(0,t) = 0, & t \geq 0, \\ u(x,0) = f(x), & u_t(x,0) = v(x) & 0 \leq x \leq l. \end{cases}$$

The (nontrivial) normal modes associated with the bar are determined by separated solutions $u(x,t) = X(x)T(t)$ that satisfy the biharmonic wave equation and the homogeneous boundary conditions. Such solutions are determined by the eigenvalue problem

$$\begin{cases} (EIX'')'' = \lambda A\rho X, & 0 < x < l, \\ X(0) = X'(0) = 0, & X(l) = X'(l) = 0, \end{cases} \tag{8.23}$$

and the equation $\ddot{T} + \lambda T = 0$, where $-\lambda$ is the separation constant.

The equation for T suggests that $\lambda > 0$. This is easy to confirm. Multiply the differential equation for X by $\bar{X}$ and integrate by parts twice to obtain

$$\int_0^l EI|X''(x)|^2\,dx = \lambda \int_0^l A\rho|X(x)|^2\,dx.$$

It follows that $\lambda \geq 0$. If equality were to hold, then $X(x)$ would be linear and hence identically equal to zero. Thus, $\lambda > 0$.

The solution space of $(EIX'')'' = \lambda A\rho X$ is four dimensional but the eigenfunctions lie in the two dimensional subspace in which $X(0) = X'(0) = 0$. A basis for this subspace is $v(x,\lambda)$ and $w(x,\lambda)$ where v and w satisfy

$$\begin{cases} (EIv'')'' - \lambda A\rho v = 0, \ 0 < x < l, \\ v(0,\lambda) = 0, \ v'(0,\lambda) = 0, \\ v''(0,\lambda) = 1, \ v'''(0,\lambda) = 0, \end{cases} \tag{8.24}$$

and

$$\begin{cases} (EIw'')'' - \lambda A\rho w = 0, \ 0 < x < l, \\ w(0,\lambda) = 0, \ w'(0,\lambda) = 0, \\ w''(0,\lambda) = 0, \ w'''(0,\lambda) = 1, \end{cases} \tag{8.25}$$

respectively. Consequently, the eigenfunctions of (8.23) have the form

$$X(x) = Av(x,\lambda) + Bw(x,\lambda)$$

where A and B not both zero are chosen to satisfy $X(l) = X'(l) = 0$; that is,

$$\begin{bmatrix} v(l,\lambda) & w(l,\lambda) \\ v'(l,\lambda) & w'(l,\lambda) \end{bmatrix} \begin{bmatrix} A \\ B \end{bmatrix} = \begin{bmatrix} 0 \\ 0 \end{bmatrix}. \tag{8.26}$$

Since A and B are not both zero, the eigenvalues are determined by the equation

$$v(l,\lambda)w'(l,\lambda) - v'(l,\lambda)w(l,\lambda) = 0. \tag{8.27}$$

The corresponding equation for a homogeneous bar could be expressed as a simple transcendental equation because the corresponding solutions $v(x,\lambda)$ and $w(x,\lambda)$ could be found explicitly in terms of standard function of calculus. For an inhomogeneous clamped bar such explicit solutions for $v(x,\lambda)$ and $w(x,\lambda)$ are not available but a shooting method can be used to find accurate numerical approximations to the eigenvalues and eigenfunctions.

We outline such an approach next. Let

$$D(\lambda) = v(l,\lambda)w'(l,\lambda) - v'(l,\lambda)w(l,\lambda), \tag{8.28}$$

where v and w are the solutions to the initial value problems (8.24) and (8.25). The four functions in (8.28) can be evaluated at any particular value of λ by a standard initial value problem solver. Consequently, a rough plot of $D(\lambda)$ over a suitably chosen interval using a reasonably coarse grid of sample points can be used to find initial estimates for the first few eigenvalues. The essence of an algorithm for solving the eigenvalue problem follows.

Step 1. Determine an initial guess (approximate value) λ of an eigenvalue of interest. Even better, determine an interval that contains the eigenvalue.

Step 2. Solve the initial value problems (8.24) and (8.25) for $v(x,\lambda)$ and $w(x,\lambda)$.

Step 3. If $D(l,\lambda) = 0$ (or is zero to within an acceptable error) STOP; λ is an eigenvalue (approximate eigenvalue) and one of the equations in (8.26) determines a corresponding eigenfunction (approximate eigenfunction). ELSE

Step 4. Use a root-finder to update the current estimate of λ as a root of $D(\lambda) = 0$ and GO TO Step 1 with the updated λ.

Step 3 deserves a small clarification. The 2×2 matrix in (8.26) cannot be the zero matrix; otherwise, v and w would be linearly independent eigenfunctions corresponding to λ, which contradicts the fact that all the eigenvalues are simple. Consequently, at least one of the two equations in (8.26) determines a corresponding eigenfunction.

If Newton's Method is used as the root-finder, the derivative of $D(\lambda)$ will be needed. This calculation requires solving the variational equations associated with the initial value problems that determine $v(x,\lambda)$ and $w(x,\lambda)$. These problems are, respectively,

$$\begin{aligned}(EIv_\lambda'')'' - \lambda A\rho v_\lambda &= A\rho v, \ 0 < x < l,\\ v_\lambda(0,\lambda) = 0,\ & v_\lambda'(0,\lambda) = 0,\\ v_\lambda''(0,\lambda) = 0,\ & v_\lambda'''(0,\lambda) = 0,\end{aligned}$$

and

$$\begin{aligned}(EIw_\lambda'')'' - \lambda A\rho w_\lambda &= A\rho w, \ 0 < x < l,\\ w_\lambda(0,\lambda) = 0,\ & w_\lambda'(0,\lambda) = 0,\\ w_\lambda''(0,\lambda) = 0,\ & w_\lambda'''(0,\lambda) = 0,\end{aligned}$$

where the subscripts indicate differentiation with respect to λ. Once again an initial value problem solver yields numerical approximations for $v_\lambda(l,\lambda)$, $v_\lambda'(l,\lambda)$, $v_\lambda''(l,\lambda)$, $w_\lambda(l,\lambda)$, $w_\lambda'(l,\lambda)$, $w_\lambda''(l,\lambda)$, and, hence, for $D'(\lambda)$.

The algorithm just described is easily implemented in many programming languages, runs rapidly on almost any current laptop or desktop computer, and can be easily adjusted to handle the other standard boundary conditions for a bar:

clamped end	$u(c,t) = 0,\ u_x(c,t) = 0,$
simply supported end	$u(c,t) = 0,\ u_{xx}(c,t) = 0,$
free end	$u_{xx}(c,t) = 0,\ u_{xxx}(c,t) = 0,$

where $c = 0$ or l is an endpoint of the bar. One of the three boundary conditions is applied at each end of the bar. The eigenfunctions lie in the 2-dimensional subspace of solutions to $(EIX'')'' = \lambda A\rho X$ that satisfy one of the chosen boundary conditions, say the boundary condition at the left end of the bar. The functions v and w are chosen as a basis for that subspace.

The same approach can be used to find the eigenvalues and eigenfunctions of other eigenvalue problems arising from initial boundary value problems involving a linear fourth order partial differential equation and separated linear boundary conditions.

Appendix A

Mildly Singular Compound Kernels

If $k(x, s)$ is a continuous kernel on $[a, b] \times [a, b]$, then each of its compound kernels $k_{[n]}(x, s)$ is continuous on the simplex Δ_n, the corresponding integral operator $K_{[n]}$ is a bounded, linear, compact operator on $C(\Delta_n)$, and Jentzsch's theorem extends to compound kernels that satisfy $k_{[n]}(x, s) \geq 0$ on $\Delta_n \times \Delta_n$ with $k_{[n]}(x, x) > 0$ for all $x = (x_1, \ldots, x_n)$ in Δ_n with $a < x_1 < \cdots < x_n < b$. These results are established by the same reasoning used in the proofs when $n = 1$. Here and in what follows the context determines the dimension of the variables x and s. Thus, x and s are real variables in $k(x, s)$ and are elements of $\mathbb{R}^n$ in $k_{[n]}(x, s)$.

The compound kernel versions of the foregoing results are true for the two types of singular kernels (Green's functions) that arise from the singular Sturm-Liouville problems studied in Chapters 5 and 6. These Green's functions are particular instances of the mildly singular kernels $k(x, s)$ that are the subject of this appendix. The proofs of the analogues of Theorems 52 and 54 when $n > 1$ are essentially the same as for the case $n = 1$, once the theorems are properly stated for the higher dimensional situation. The proof that the compound kernels of a mildly singular kernel satisfy the hypotheses of the general theorems when $n > 1$ is more involved. We establish here that they do.

A real-valued kernel $k(x, s)$ with domain $[a, b] \times [a, b] \backslash \{(a, a)\}$ is **mildly singular** if either

$$(i)\ k(x, s) = h(x, s) \ln(\max(x, s) - a)$$

for all (x, s) in its domain and where $h(x, s)$ is a continuous function on $[a, b] \times [a, b]$; or

$$(ii)\ k(x, s) \text{ is bounded and continuous}$$

for all (x, s) in its domain and the kernel does not have a continuous extension to $[a, b] \times [a, b]$. The Green's functions of the singular Sturm-Liouville problems in Chapter 5 are mildly singular of type (i) and the Green's functions of the singular Sturm-Liouville problems in Chapter 6 are mildly singular of type (ii).

Throughout the appendix

$$\Delta_n = \{u = (u_1, \ldots, u_n) : a \leq u_1 \leq \cdots \leq u_n \leq b\},$$

$$\widetilde{\Delta}_n = \{u \in \Delta_n : u_1 > a\},$$

$$F_1 = \{u \in \Delta_n : u_1 = a\},$$

and for any a' with $a < a' < b$,

$$\Delta_n' = \{u \in \Delta_n : u_1 \geq a'\},$$

Thus, F_1 is the face of the simplex Δ_n in the hyperplane perpendicular to the u_1-axis at $u_1 = a$, $\widetilde{\Delta}_n$ is the simplex Δ_n with its face F_1 removed, and Δ_n' is a subsimplex of Δ_n at a positive distance from F_1. It is instructive for the arguments that follow to make sketches of these sets when $n = 2$ and the simplices are solid triangles. In the applications to Green's functions of singular Sturm-Liouville problems, the face F_1 of the simplex Δ_n contains all the singularities of the nth compound kernel of the Green's function.

In Theorem 52, the singular case when $n=1$, the kernel $k(x,s)$ is defined on $[a,b]\times[a,b]/\{(a,a)\}$. Notice that

$$\begin{aligned}[a,b]\times[a,b]/\{(a,a)\} &= ([a,b]\times(a,b])\cup((a,b]\times[a,b]) \\ &= \left(\Delta_1\times\widetilde{\Delta}_1\right)\cup\left(\widetilde{\Delta}_1\times\Delta_1\right).\end{aligned}$$

The analogue of Theorem 52 when $n>1$ is:

Theorem 197 *Let $k_{[n]}(x,s)$ be a continuous real or complex-valued kernel defined on $\left(\Delta_n\times\widetilde{\Delta}_n\right)\cup\left(\widetilde{\Delta}_n\times\Delta_n\right)$. If*
(a) for each f in $C(\Delta_n)$ and x^0 in F_1, $K_{[n]}f(x^0)=\int_{\Delta_n}k_{[n]}(x^0,s)f(s)\,ds$ exists as a convergent improper Riemann integral,
(b) $\int_{\Delta_n}|k_{[n]}(x,s)|\,ds\le M$ for some constant M and all x in Δ_n,
(c) $\int_{\Delta_n}\left|k_{[n]}(x,s)-k_{[n]}(x^0,s)\right|ds\to 0$ as $x\to x^0$ for each x^0 in F_1,
then $K_{[n]}:C(\Delta_n)\to C(\Delta_n)$ and $K_{[n]}$ is a bounded, linear, compact operator on $C(\Delta_n)$ equipped with the maximum norm.

Proof. Given f in $C(\Delta_n)$ and x^0 in F_1, $K_{[n]}f(x^0)$ is defined by (a) and for x in $\widetilde{\Delta}_n$, $K_{[n]}f(x)$ is given by a proper Riemann integral. So $K_{[n]}f$ is a well defined function on Δ_n. We claim that

$$\int_{\Delta_n}\left|k_{[n]}(x,s)-k_{[n]}(x^0,s)\right|ds\to 0 \text{ as } x\to x^0$$

for each x^0 in Δ_n. If x^0 is in F_1, the limit holds by (c). Fix x^0 in $\widetilde{\Delta}_n$ and set $a'=\left(a+x_1^0\right)/2$. Then $a'>a$ and the kernel $k_{[n]}(x,s)$ is continuous on $\Delta'_n\times\Delta_n$ and, hence, uniformly continuous there. Given $\varepsilon>0$ there is a $\delta>0$ such that

$$\left|k_{[n]}(x,s)-k_{[n]}(x^0,s)\right|<\varepsilon \quad \text{for } x \text{ in } \Delta'_n \quad \text{and} \quad s \text{ in } \Delta_n \quad \text{when} \quad \left|x-x^0\right|<\delta.$$

Consequently, if $|\Delta_n|$ is the volume of Δ_n, for x in Δ'_n,

$$\int_{\Delta_n}\left|k_{[n]}(x,s)-k_{[n]}(x^0,s)\right|ds\le\varepsilon|\Delta_n| \quad \text{when} \quad \left|x-x^0\right|<\delta$$

and the claim is established for x^0 in $\widetilde{\Delta}_n$. Thus, for f in $C(\Delta_n)$,

$$\left|K_{[n]}f(x)-K_{[n]}f(x^0)\right|\le\|f\|_{\max}\int_{\Delta_n}\left|k_{[n]}(x,s)-k_{[n]}(x^0,s)\right|ds\to 0$$

as $x\to x^0$, the function $K_{[n]}f$ is continuous on Δ_n, and $K_{[n]}:C(\Delta_n)\to C(\Delta_n)$. By (b) the operator $K_{[n]}$ is bounded because

$$\begin{aligned}\left|K_{[n]}f(x)\right|&\le\|f\|_{\max}\int_{\Delta_n}\left|k_{[n]}(x,s)\right|ds\le M\|f\|_{\max}, \\ \left\|K_{[n]}f\right\|_{\max}&\le M\|f\|_{\max}.\end{aligned}$$

It remains to show that $K_{[n]}$ is a compact operator. If $\{f_m\}$ is a bounded sequence in $C(\Delta_n)$, with $\|f_m\|_{\max}\le M'$ for all m, then $\{K_{[n]}f_m\}$ is uniformly bounded on Δ_n because

$\|K_{[n]}f_m\|_{\max} \leq M\|f_m\|_{\max} \leq MM'$. Applying the inequality above for $|K_{[n]}f(x) - K_{[n]}f(x^0)|$ with $f = f_m$ yields

$$\begin{aligned} |K_{[n]}f_m(x) - K_{[n]}f_m(x^0)| &\leq \|f_m\|_{\max} \int_{\Delta_n} |k_{[n]}(x, s) - k_{[n]}(x^0, s)|\, ds \\ &\leq M' \int_{\Delta_n} |k_{[n]}(x, s) - k_{[n]}(x^0, s)|\, ds \to 0 \end{aligned}$$

as $x \to x^0$. Thus, $\{K_{[n]}f_m\}$ is equicontinuous at x^0 for each x^0 in Δ_n and $\{K_{[n]}f_m\}$ is equicontinuous on Δ_n by Proposition 42. The compactness of $K_{[n]}$ follows from the Arzelà-Ascoli theorem by the same reasoning used in the proof of Theorem 51. ■

The analogue of Theorem 54 when $n > 1$ is formulated in parallel to Theorem 197 and its proof follows along the same lines. We leave both the statement of the theorem and its proof to the reader. In fact, we only used Theorem 197 and the corresponding theorem of Jentzsch for compound kernels in the text.

We establish in the next two sections that the mildly singular kernels of types (i) and (ii) satisfy the conditions of Theorem 197.

A.1 Mildly Singular Kernels of Type (i)

A mildly singular kernel $k(x, s)$ of type (i) has domain $[a, b] \times [a, b] \backslash \{(a, a)\}$ and satisfies

$$k(x, s) = h(x, s) \ln(\max(x, s) - a) \quad \text{for } a \leq x, s \leq b \quad \text{with } (x, s) \neq (a, a)$$

and with $h(x, s)$ continuous on $[a, b] \times [a, b]$. We shall check that (a), (b), and (c) of Theorem 197 hold for $k_{[n]}(x, s)$. The reasoning will be presented in the case $n = 2$, for clarity, but using arguments that extend naturally to a general n.

When $n = 2$,

$$k_{[2]}(x, s) = \begin{vmatrix} k(x_1, s_1) & k(x_1, s_2) \\ k(x_2, s_1) & k(x_2, s_2) \end{vmatrix}$$

and this compound kernel will be defined at (x, s) unless $x_i = s_j = a$ for some i and j, which happens if and only if $a = x_1 = \cdots = x_i$ and $a = s_1 = \cdots = s_j$ for some i and j; that is, if and only $x_1 = s_1 = a$. It follows that the domain of $k_{[2]}(x, s)$ is $\left(\Delta_2 \times \widetilde{\Delta}_2\right) \cup \left(\widetilde{\Delta}_2 \times \Delta_2\right)$. The definition of $k(x, s)$ shows that it is continuous in a neighborhood of any point (x, s) in its domain and, hence, $k_{[2]}(x, s)$ is continuous in a neighborhood of any point (x, s) in its domain:

A. $k_{[2]}(x, s)$ is continuous on its domain $\left(\Delta_2 \times \widetilde{\Delta}_2\right) \cup \left(\widetilde{\Delta}_2 \times \Delta_2\right)$.

For any f in $C(\Delta_2)$, if $x \in \widetilde{\Delta}_2$ then $k_{[2]}(x, s)f(s)$ is continuous for s in Δ_2; hence,

B. For x in $\widetilde{\Delta}_2$, $K_{[2]}f(x) = \int_{\Delta_2} k_{[2]}(x, s)f(s)\, ds$ exists as an ordinary Riemann integral for all f in $C(\Delta_2)$. $K_{[2]}f(x)$ exists as an improper Riemann integral when x is in the face F_1:

$$K_{[2]}f(x) = \int_{\Delta_2} k_{[2]}(x, s)f(s)\, ds = \lim_{a' \to a} \int_{\Delta_2'} k_{[2]}(x, s)f(s)\, ds.$$

That the limit exists is among several consequences of the following observations about the kernel $k_{[2]}(x, s)$.

Lemma 198 *For all x in $[a, b]$ and s in $(a, b]$,*

$$\left|\ln(\max(x, s) - a)\right| \leq \max(|\ln(s-a)|, \ln(b-a)).$$

Hence,

$$\left|\ln(\max(x, s) - a)\right| \leq \left|\ln(s-a)\right| + \left|\ln(b-a)\right|.$$

Proof. The function $\left|\ln(u-a)\right|$ decreases on $a < u \leq a+1$ and increases $a+1 \leq u < \infty$. Let x be in $[a, b]$ and s be in $(a, b]$. If $\max(x, s) = s$ the desired conclusions are clear. If $\max(x, s) = x$ and $b \geq x \geq a+1$, then

$$\left|\ln(\max(x, s) - a)\right| = \ln(x-a) \leq \ln(b-a).$$

If $\max(x, s) = x$ and $a+1 \geq x \geq a$, then $a+1 \geq x \geq s > a$ and

$$\left|\ln(\max(x, s) - a)\right| = \left|\ln(x-a)\right| \leq \left|\ln(s-a)\right|.$$

So, if $\max(x, s) = x$ then

$$\left|\ln(\max(x, s) - a)\right| \leq \max(\left|\ln(s-a)\right|, \ln(b-a)).$$

Thus,

$$\left|\ln(\max(x, s) - a)\right| \leq \max(\left|\ln(s-a)\right|, \ln(b-a))$$

for all x in $[a, b]$ and s in $(a, b]$. ∎

For $(x, s) \in \Delta_2 \times \widetilde{\Delta}_2$, the kernel $k_{[2]}(x, s)$ is continuous and

$$k_{[2]}(x, s) = k(x_1, s_1)k(x_2, s_2) + (-1)k(x_2, s_1)k(x_1, s_2) = \alpha + \beta.$$

From the formula for $k(x, s)$ and Lemma 198

$$|\alpha|, |\beta| \leq \|h\|_{\max}^2 \max(|\ln(s_1-a)|, \ln(b-a)) \max(|\ln(s_2-a)|, \ln(b-a)),$$

where

$$\|h\|_{\max} = \max_{a \leq x, s \leq b} |h(x, s)|.$$

Since $s_1 \leq s_2$,

$$\max(|\ln(s_2-a)|, \ln(b-a)) \leq \max(|\ln(s_1-a)|, \ln(b-a))$$

because if $s_2 \leq a+1$, $|\ln(s_1-a)| \geq |\ln(s_2-a)|$ while if $b \geq s_2 > a+1$, then $|\ln(s_2-a)| \leq \ln(b-a)$. Consequently,

$$\begin{aligned} |\alpha|, |\beta| &\leq \|h\|_{\max}^2 [\max(|\ln(s_1-a)|, \ln(b-a))]^2 \\ &\leq \|h\|_{\max}^2 [(|\ln(s_1-a)| + |\ln(b-a)|)]^2 \end{aligned}$$

and for $(x, s) \in \Delta_2 \times \widetilde{\Delta}_2$

$$|k_{[2]}(x, s)| \leq (2!)\|h\|_{\max}^2 (|\ln(s_1-a)| + |\ln(b-a)|)^2 \tag{A.1}$$

because $\left|k_{[2]}(x, s)\right| \leq |\alpha| + |\beta|$.

In the same way, introducing two more terms α^0 and β^0 corresponding to $k_{[2]}(x^0, s)$, if (x, s), $(x^0, s) \in \Delta_2 \times \widetilde{\Delta}_2$, then

$$\left|k_{[2]}(x, s) - k_{[2]}(x^0, s)\right| \leq 2 \cdot (2!)\|h\|^2_{\max}(|\ln(s_1 - a)| + |\ln(b - a)|)^2. \tag{A.2}$$

In the case of a general n, the $2!$ is replaced by $n!$ and the exponent 2 is replaced by n.

If $a < a'' < a' < b$, then from (A.1)

$$\begin{aligned}
0 &\leq \int_{\Delta_2''} \left|k_{[2]}(x, s)\right| ds - \int_{\Delta_2'} \left|k_{[2]}(x, s)\right| ds \\
&= \int_{\Delta_2''\backslash\Delta_2'} \left|k_{[2]}(x, s)\right| ds = \int_{a''}^{a'} ds_1 \int_{s_1}^{b} \left|k_{[2]}(x, s)\right| ds_2 \\
&\leq \int_{a''}^{a'} ds_1 \int_{s_1}^{b} (2!)\|h\|^2_{\max}\big(|\ln(s_1 - a)| + \left|\ln(b - a)\right|\big)^2 \, ds_2 \\
&= (2!)\|h\|^2_{\max} \int_{a''}^{a'} (b - s_1)\big(|\ln(s_1 - a)| + \left|\ln(b - a)\right|\big)^2 \, ds_1 \\
&\leq (2!)\|h\|^2_{\max}(b - a) \int_{a''}^{a'} \big(|\ln(s_1 - a)| + \left|\ln(b - a)\right|\big)^2 \, ds_1
\end{aligned}$$

and the right member tends to 0 as a'' and a' tend to a because the improper integral $\int_a^b |\ln(s_1 - a)|^p \, ds_1$ converges for all integers $p \geq 1$. It follows from the Cauchy criterion that there exists

$$\lim_{a'' \to a} \int_{\Delta_2''} |k_{[2]}(x, s)| \, ds \stackrel{\text{def}}{=} \int_{\Delta_2} |k_{[2]}(x, s)| \, ds.$$

Let $a'' \to a$ in the foregoing inequalities to obtain

$$\begin{aligned}
0 &\leq \int_{\Delta_2} \left|k_{[2]}(x, s)\right| ds - \int_{\Delta_2'} \left|k_{[2]}(x, s)\right| ds \\
&\leq (2!)\|h\|^2_{\max}(b - a) \int_{a}^{a'} \big(\left|\ln(s_1 - a)\right| + \left|\ln(b - a)\right|\big)^2 \, ds_1
\end{aligned} \tag{A.3}$$

for x in Δ_2. This estimate shows that $\int_{\Delta_2'} \left|k_{[2]}(x, s)\right| ds$ converges uniformly for x in Δ_2 to $\int_{\Delta_2} \left|k_{[2]}(x, s)\right| ds$ as $a' \to a$. Since $k_{[2]}(x, s)$ is continuous on $\Delta_2 \times \Delta_2'$, $\int_{\Delta_2'} \left|k_{[2]}(x, s)\right| ds$ is a continuous function of x in Δ_2 by Proposition 18. Its uniform limit $\int_{\Delta_2} \left|k_{[2]}(x, s)\right| ds$ is continuous for x in Δ_2 by Theorem 23.

C. $\int_{\Delta_2} \left|k_{[2]}(x, s)\right| ds$ is continuous for x in Δ_2.

In the same manner, using (A.2) it follows that there exists

$$\lim_{a'' \to a} \int_{\Delta_2''} \left|k_{[2]}(x, s) - k_{[2]}(x^0, s)\right| ds \stackrel{\text{def}}{=} \int_{\Delta_2} \left|k_{[2]}(x, s) - k_{[2]}(x^0, s)\right| ds$$

and that

$$\begin{aligned}
0 &\leq \int_{\Delta_2} \left|k_{[2]}(x, s) - k_{[2]}(x^0, s)\right| ds - \int_{\Delta_2'} \left|k_{[2]}(x, s) - k_{[2]}(x^0, s)\right| ds \\
&\leq 2(2!)\|h\|^2_{\max}(b - a) \int_{a}^{a'} \big(\left|\ln(s_1 - a)\right| + \left|\ln(b - a)\right|\big)^2 ds_1
\end{aligned} \tag{A.4}$$

for x and x^0 in Δ_2. Since the integral over Δ_2' is a continuous function of x for x in Δ_2, and this integral converges uniformly for x in Δ_2 to the integral over Δ_2 by the foregoing inequality, it follows that

D. $\int_{\Delta_2} \left| k_{[2]}(x, s) - k_{[2]}(x^0, s) \right| ds$ is continuous for x in Δ_2.

We can now prove the last assertion in B above, namely, that for f in $C(\Delta_2)$ and x in F_1, $K_{[2]}f(x)$ is defined by the improper Riemann integral

$$K_{[2]}f(x) = \int_{\Delta_2} k_{[2]}(x, s)f(s)\, ds = \lim_{a' \to a} \int_{\Delta_2'} k_{[2]}(x, s)f(s)\, ds.$$

Indeed, if $a < a'' < a' < b$, then

$$\left| \int_{\Delta_2''} k_{[2]}(x, s)f(s)\, ds - \int_{\Delta_2'} k_{[2]}(x, s)f(s)\, ds \right| \le \|f\|_{\max} \int_{\Delta_2''\setminus\Delta_2'} |k_{[2]}(x, s)|\, ds$$

$$\le \|f\|_{\max}(2!)\|h\|_{\max}^2(b-a)\int_{a''}^{a'} \left(|\ln(s_1 - a)| + |\ln(b-a)|\right)^2 ds_1$$

and the right member tends to 0 as a'' and a' tend to a. Hence, there exists

$$\lim_{a' \to a} \int_{\Delta_2'} k_{[2]}(x, s)f(s)\, ds \stackrel{\text{def}}{=} \int_{\Delta_2} k_{[2]}(x, s)f(s)\, ds.$$

This establishes that (a) in Theorem 197 holds for the compound kernel $k_{[2]}(x, s)$ of a mildly singular kernel $k(x, s)$, and for $k_{[n]}(x, s)$ by the same line of reasoning.

Part (b) in Theorem 197 holds for the compound kernel $k_{[2]}(x, s)$ because M can be chosen as the maximum of the integral in C. Part (c) of Theorem 197 follows directly from D.

In summary, the compound kernels $k_{[n]}(x, s)$ of a mildly singular kernel $k(x, s)$ of type (i) determine compact, bounded, linear, integral operators $K_{[n]}$ on $C(\Delta_n)$ equipped with the maximum norm. Moreover, given the compactness of $K_{[n]}$ and the fact that D implies

$$\lim_{x \to x^0} \int_{\Delta_n} \left| k_{[n]}(x, s) - k_{[n]}(x^0, s) \right| ds = 0$$

for each x^0 in Δ_n, the reasoning used in Chapter 3 to establish Jentzsch's theorem when $n = 1$ carries over without essential change to the compound kernels $k_{[n]}(x, s)$ of a mildly singular kernel $k(x, s)$. Thus, Jentzsch's theorem holds for the compound kernels of a mildly singular kernel $k(x, s)$ of type (i) that satisfy $k_{[n]}(x, s) \ge 0$ on their domains with $k_{[n]}(x, x) > 0$ for all $x = (x_1, \ldots, x_n)$ in Δ_n with $a < x_1 < \cdots < x_n < b$.

A.2 Mildly Singular Kernels of Type (ii)

A mildly singular kernel $k(x, s)$ of type (ii) has domain $[a, b] \times [a, b]\setminus\{(a, a)\}$ and is bounded and continuous there but does not have a continuous extension to $[a, b] \times [a, b]$. So $k(x, s)$ has a singularity at (a, a) that cannot be removed and there is a constant M such that

$$|k(x, s)| \le M \quad \text{for all}(x, s) \text{ in } [a, b] \times [a, b]\setminus\{(a, a)\}.$$

We shall check that (a), (b), and (c) of Theorem 197 hold for the compound kernels $k_{[n]}(x, s)$ of a mildly singular kernel $k(x, s)$ of type (ii), after making a number of preliminary observations. Since the kernel $k(x, s)$ is bounded certain simplifications occur compared to the treatment of type (i) singularities in the last section, but the basic line of reasoning is the same. We continue to use the notations Δ_n, $\widetilde{\Delta}_n$, Δ'_n, and F_1 from the previous section.

The reasoning will be presented in the case $n = 2$, for clarity, but using arguments that extend naturally to a general n.

When $n = 2$,

$$k_{[2]}(x, s) = \begin{vmatrix} k(x_1, s_1) & k(x_1, s_2) \\ k(x_2, s_1) & k(x_2, s_2) \end{vmatrix}$$

and this compound kernel will be defined at (x, s) unless $x_i = s_j = a$ for some i and j, which happens if and only if $a = x_1 = \cdots = x_i$ and $a = s_1 = \cdots = s_j$ for some i and j; that is, if and only $x_1 = s_1 = a$. It follows that the domain of $k_{[2]}(x, s)$ is $\left(\Delta_2 \times \widetilde{\Delta}_2\right) \cup \left(\widetilde{\Delta}_2 \times \Delta_2\right)$. The definition of $k(x, s)$ show that it is continuous in a neighborhood of any point (x, s) in its domain and, hence, $k_{[2]}(x, s)$ is continuous in a neighborhood of any point (x, s) in its domain:

A. $k_{[2]}(x, s)$ is continuous on its domain $\left(\Delta_2 \times \widetilde{\Delta}_2\right) \cup \left(\widetilde{\Delta}_2 \times \Delta_2\right)$.

For any f in $C(\Delta_2)$, if $x \in \widetilde{\Delta}_2$ then $k_{[2]}(x, s)f(s)$ is continuous for s in Δ_2; hence,

B. For x in $\widetilde{\Delta}_2$, $K_{[2]}f(x) = \int_{\Delta_2} k_{[2]}(x, s)f(s)\, ds$ exists as an ordinary Riemann integral for all f in $C(\Delta_2)$. $K_{[2]}f(x)$ exists as an improper Riemann integral when x is in the face F_1:

$$K_{[2]}f(x) = \int_{\Delta_2} k_{[2]}(x, s)f(s)\, ds = \lim_{a' \to a} \int_{\Delta'_2} k_{[2]}(x, s)f(s)\, ds.$$

That the limit exists is among several consequences of the following observations about the kernel $k_{[2]}(x, s)$.

For $(x, s) \in \Delta_2 \times \Delta'_2$, the kernel $k_{[2]}(x, s)$ is continuous and

$$k_{[2]}(x, s) = k(x_1, s_1)k(x_2, s_2) + (-1)k(x_2, s_1)k(x_1, s_2).$$

Since $k(x, s)$ is bounded by M, for $(x, s) \in \Delta_2 \times \Delta'_2$,

$$\left|k_{[2]}(x, s)\right| \leq (2!)M^2. \tag{A.5}$$

In the same way, if (x, s), $(x^0, s) \in \Delta_2 \times \Delta'_2$, then

$$\left|k_{[2]}(x, s) - k_{[2]}(x^0, s)\right| \leq 2 \cdot (2!)M^2. \tag{A.6}$$

In the case of a general n, the 2! is replaced by $n!$ and the exponent 2 is replaced by n.

If $a < a'' < a' < b$, then from (A.5)

$$\begin{aligned} 0 &\leq \int_{\Delta''_2} \left|k_{[2]}(x, s)\right| ds - \int_{\Delta'_2} \left|k_{[2]}(x, s)\right| ds = \int_{\Delta''_2 \setminus \Delta'_2} \left|k_{[2]}(x, s)\right| ds \\ &= \int_{a''}^{a'} ds_1 \int_{s_1}^{b} \left|k_{[2]}(x, s)\right| ds_2 \leq \int_{a''}^{a'} ds_1 \int_{s_1}^{b} (2!)M^2\, ds_2 \\ &= (2!)M^2 \int_{a''}^{a'} (b - s_1)\, ds_1 \leq (2!)M^2(b - a)(a' - a''). \end{aligned}$$

The right member tends to 0 as a'' and a' tend to a and it follows from the Cauchy criterion that there exists

$$\lim_{a''\to a}\int_{\Delta_2''}\left|k_{[2]}(x,s)\right|\,ds \stackrel{\text{def}}{=} \int_{\Delta_2}\left|k_{[2]}(x,s)\right|\,ds.$$

Let $a'' \to a$ in the foregoing inequalities to obtain

$$\int_{\Delta_2}\left|k_{[2]}(x,s)\right|\,ds - \int_{\Delta_2'}\left|k_{[2]}(x,s)\right|\,ds \le (2!)\|h\|_{\max}^2(b-a)\big(a'-a\big) \tag{A.7}$$

for x in Δ_2. This estimate shows that $\int_{\Delta_2'}\left|k_{[2]}(x,s)\right|\,ds$ converges uniformly to $\int_{\Delta_2}\left|k_{[2]}(x,s)\right|\,ds$ for x in $[a,b]$ as $a' \to a$. Since $\int_{\Delta_2'}\left|k_{[2]}(x,s)\right|\,ds$ is a continuous function of x, its uniform limit $\int_{\Delta_2}\left|k_{[2]}(x,s)\right|\,ds$ is continuous for x in Δ_2.

C. $\int_{\Delta_2}\left|k_{[2]}(x,s)\right|\,ds$ is continuous for x in Δ_2.

In the same manner, using (A.6), there exists

$$\lim_{a''\to a}\int_{\Delta_2''}\left|k_{[2]}(x,s)-k_{[2]}(x^0,s)\right|\,ds \stackrel{\text{def}}{=} \int_{\Delta_2}\left|k_{[2]}(x,s)-k_{[2]}(x^0,s)\right|\,ds$$

and

$$\begin{aligned}0 &\le \int_{\Delta_2}\left|k_{[2]}(x,s)-k_{[2]}(x^0,s)\right|\,ds - \int_{\Delta_2'}\left|k_{[2]}(x,s)-k_{[2]}(x^0,s)\right|\,ds\\ &\le 2(2!)\|h\|_{\max}^2(b-a)\big(a'-a\big)\end{aligned} \tag{A.8}$$

for x and x^0 in Δ_2. Since the integral over Δ_2' is a continuous function of x for x in Δ_2, and this integral converges uniformly for x in Δ_2 to the integral over Δ_2 by the foregoing inequality, it follows that

D. $\int_{\Delta_2}\left|k_{[2]}(x,s)-k_{[2]}(x^0,s)\right|\,ds$ is continuous for x in Δ_2.

We can now prove the last assertion in B above, namely, that for f in $C(\Delta_2)$ and x in F_1, $K_{[2]}f(x)$ is defined by the improper Riemann integral

$$K_{[2]}f(x) = \int_{\Delta_2} k_{[2]}(x,s)f(s)\,ds = \lim_{a'\to a}\int_{\Delta_2'} k_{[2]}(x,s)f(s)\,ds.$$

Indeed, if $a < a'' < a' < b$, then

$$\begin{aligned}&\left|\int_{\Delta_2''} k_{[2]}(x,s)f(s)\,ds - \int_{\Delta_2'} k_{[2]}(x,s)f(s)\,ds\right|\\ &\le \|f\|_{\max}\int_{\Delta_2''\setminus\Delta_2'}\left|k_{[2]}(x,s)\right|\,ds \le \|f\|_{\max}(2!)\|h\|_{\max}^2(b-a)\big(a'-a''\big)\end{aligned}$$

and the right member tends to 0 as a'' and a' tend to a. Hence, there exists

$$\lim_{a'\to a}\int_{\Delta_2'} k_{[2]}(x,s)f(s)\,ds \stackrel{\text{def}}{=} \int_{\Delta_2} k_{[2]}(x,s)f(s)\,ds.$$

This establishes that (a) in Theorem 197 holds for the compound kernel $k_{[2]}(x,s)$ of a mildly singular kernel $k(x,s)$, and for $k_{[n]}(x,s)$ by the same line of reasoning.

Part (b) in Theorem 197 holds by C because continuous functions on Δ_2 are bounded. That (c) in Theorem 197 holds follows directly from D.

In summary, the compound kernels $k_{[n]}(x, s)$ of a mildly singular kernel $k(x, s)$ of type (ii) determine bounded, linear, compact integral operators $K_{[n]}$ on $C(\Delta_n)$ with the maximum norm. Moreover, given the compactness of $K_{[n]}$ and the fact that D implies

$$\lim_{x \to x^0} \int_{\Delta_n} \left| k_{[n]}(x, s) - k_{[n]}(x^0, s) \right| ds = 0$$

for each x^0 in Δ_n, the reasoning used in Chapter 3 to establish Jentzsch's theorem when $n = 1$ carries over without essential change to the compound kernels $k_{[n]}(x, s)$. Thus, Jentzsch's theorem holds for the compound kernels of a mildly singular kernel $k(x, s)$ of type (ii) that satisfy $k_{[n]}(x, s) \geq 0$ on their domains with $k_{[n]}(x, x) > 0$ for all $x = (x_1, \ldots, x_n)$ in Δ_n with $a < x_1 < \cdots < x_n < b$.

Appendix B

Iteration of Mildly Singular Kernels

As in Appendix A, a real-valued kernel $k(x, s)$ with domain $[a, b] \times [a, b] \backslash \{(a, a)\}$ is **mildly singular** if either

$$(i)\ k(x, s) = h(x, s) \ln(\max(x, s) - a)$$

for all (x, s) in its domain and where $h(x, s)$ is a continuous function on $[a, b] \times [a, b]$; or

$$(ii)\ k(x, s) \text{ is bounded and continuous}$$

for all (x, s) in its domain and the kernel does not have a continuous extension to $[a, b] \times [a, b]$. The Green's functions of the singular Sturm-Liouville problems in Chapter 5 are mildly singular of type (i) and the Green's functions of the singular Sturm-Liouville problems in Chapter 6 are mildly singular of type (ii).

If $k(x, s)$ and $l(s, t)$ are mildly singular kernels of the same type with corresponding integral operators K and L, then

$$m(x, t) = \int_a^b k(x, s)l(s, t)\, ds \tag{B.1}$$

is the kernel of the integral operator KL, where $KL(f) = K(Lf)$ for f in $C[a, b]$. If $x \neq a$ and $t \neq a$, the integrand in (B.1) is continuous and the integral is a proper Riemann integral. When $x = a$ and/or $t = a$, at least one of $k(x, s)$ and $l(s, t)$ is singular at $s = a$ and the integral in (B.1) is an improper Riemann integral,

$$\int_a^b k(x, s)l(s, t)\, dt = \lim_{a' \to a} \int_{a'}^b k(x, s)l(s, t)\, ds,$$

with $a' > a$ understood. We will establish that the limit exists for all (x, t) in $[a, b] \times [a, b]$ and that $m(x, t)$ is continuous on the full square $[a, b] \times [a, b]$. We also establish the corresponding results for the compound kernels of $k(x, s)$ and $l(s, t)$. Of course, the limit is also the value of the proper Riemann integral when $x \neq a$ and $t \neq a$.

Throughout the appendix, we will refer to the integral defining $m(x, t)$ and corresponding integrals involving the compound kernels of $k(x, s)$ and $l(s, t)$ as improper integrals even in the case when in fact the integrals are proper. No harm will result from this abuse of notation because the limits that define the improper integrals correctly evaluate the integrals when they are proper.

Throughout Appendix B the simplices Δ_n, $\widetilde{\Delta}_n$, and Δ'_n are defined as in Appendix A.

B.1 Mildly Singular Behavior of Type (i)

The main result of this section is

Proposition 199 *If $k(x,s)$ and $l(s,t)$ are mildly singular kernels of type (i) on $[a,b]\times[a,b]\backslash\{(a,a)\}$, then for $n\geq 1$ the improper integral $\int_{\Delta_n} k_{[n]}(x,s)l_{[n]}(s,t)\,ds$ converges for (x,t) in $\Delta_n\times\Delta_n$ and is continuous there.*

Proof. Consider first the case $n=1$ where $\Delta_1=[a,b]$ and

$$\int_{\Delta_1} k_{[1]}(x,s)l_{[1]}(s,t)\,ds=\int_a^b k(x,s)l(s,t)\,ds.$$

For any a' with $a<a'<b$ and (x,t) in $[a,b]\times[a,b]$, the integral $\int_{a'}^b k(x,s)l(s,t)\,ds$ is an ordinary Riemann integral because its integrand is continuous. For fixed (x,t) in $[a,b]\times[a,b]$ and $a<a''<a'<b$,

$$\left|\left(\int_{a''}^b-\int_{a'}^b\right)k(x,s)l(s,t)\,ds\right|=\left|\int_{a''}^{a'}k(x,s)l(s,t)\,ds\right|$$

$$\leq\int_{a''}^{a'}|h_1(x,s)\ln(\max(x,s)-a)h_2(s,t)\ln(\max(s,t)-a)|\,ds$$

$$\leq\|h_1h_2\|_{\max}\int_{a''}^{a'}|\ln(\max(x,s)-a)|\,|\ln(\max(s,t)-a)|\,ds$$

$$\leq\|h_1h_2\|_{\max}\int_{a''}^{a'}(|\ln(s-a)|+|\ln(b-a)|)^2\,ds$$

by Lemma 198, where

$$\|h_1h_2\|_{\max}=\max_{a\leq x,s,t\leq b}|h_1(x,s)h_2(s,t)|.$$

The improper integral

$$\int_a^b(|\ln(s-a)|+|\ln(b-a)|)^2\,ds$$

converges because the improper integrals $\int_a^b|\ln(s-a)|^p\,ds$ converge for all integers $p\geq 1$. Consequently, the right member in the chain of inequalities above tends to 0 as a' and a'' tend to a. By Cauchy's criterion

$$\lim_{a'\to a}\int_{a'}^b k(x,s)l(s,t)\,ds$$

exists and is finite and $\int_a^b k(x,s)l(s,t)\,ds$ is defined as the improper integral

$$\int_a^b k(x,s)l(s,t)\,ds=\lim_{a'\to a}\int_{a'}^b k(x,s)l(s,t)\,ds.$$

Let a'' tend to a in the chain of inequalities to obtain

$$\left|\int_a^b k(x,s)l(s,t)\,ds-\int_{a'}^b k(x,s)l(s,t)\,ds\right|$$

$$\leq\|h_1h_2\|_{\max}\int_a^{a'}(|\ln(s-a)|+|\ln(b-a)|)^2\,ds.$$

The right member tends to 0 as a' tends to a uniformly for (x, t) in $[a, b] \times [a, b]$ because the improper integral $\int_a^b (|\ln(s-a)| + |\ln(b-a)|)^2 \, ds$ converges. The integral $\int_{a'}^b k(x, s)l(s, t)\, ds$ is continuous for (x, t) in $[a, b] \times [a, b]$ by Proposition 18 because its integrand is continuous on $[a, b] \times [a', b] \times [a, b]$. Thus, $\int_a^b k(x, s)l(s, t)\, ds$ is the uniform limit of continuous functions on $[a, b] \times [a, b]$ and, hence, is continuous there by Theorem 23. This establishes the Proposition when $n = 1$.

For clarity we give the proof for $n \geq 2$ for the case $n = 2$ using reasoning that applies to a general n. Recall from Appendix A that $\widetilde{\Delta}_2 = \{s \in \Delta_2 : s_1 > a\}$. The kernel $k_{[2]}(x, s)$ is continuous for (x, s) in $\Delta_2 \times \widetilde{\Delta}_2$, (see Appendix A) and for $(x, s) \in \Delta_2 \times \widetilde{\Delta}_2$

$$\left|k_{[2]}(x, s)\right| \leq (2!)\|h_1\|_{\max}^2(|\ln(s_1 - a)| + |\ln(b-a)|)^2$$

by (A.1) in Appendix A. Likewise, for $(s, t) \in \widetilde{\Delta}_2 \times \Delta_2$,

$$\left|l_{[2]}(s, t)\right| \leq (2!)\|h_2\|_{\max}^2(|\ln(s_1 - a)| + |\ln(b-a)|)^2.$$

Thus,

$$\left|k_{[2]}(x, s)l_{[2]}(s, t)\right| \leq (2!)^2\|h_1\|_{\max}^2\|h_2\|_{\max}^2\left[(|\ln(s_1 - a)| + |\ln(b-a)|)^2\right]^2$$

for (x, s, t) in $\Delta_2 \times \widetilde{\Delta}_2 \times \Delta_2$. By the same reasoning, the corresponding inequality for $k_{[n]}(x, s)l_{[n]}(x, s)$ is

$$\left|k_{[n]}(x, s)l_{[n]}(s, t)\right| \leq (n!)^2\|h_1\|_{\max}^n\|h_2\|_{\max}^n\left[\left(|\ln(s_1 - a)| + |\ln(b-a)|\right)^n\right]^2$$

for (x, s, t) in $\Delta_n \times \widetilde{\Delta}_n \times \Delta_n$. For $a < a'' < a' < b$ and (x, t) in $\Delta_2 \times \Delta_2$,

$$\left|\left(\int_{\Delta_2''} - \int_{\Delta_2'}\right) k_{[2]}(x, s)l_{[2]}(s, t)\, ds\right| = \int_{\Delta_2''\backslash\Delta_2'} \left|k_{[2]}(x, s)l_{[2]}(s, t)\right| ds$$

$$\leq (2!)^2\|h_1\|_{\max}^2\|h_2\|_{\max}^2 \int_{\Delta_2''\backslash\Delta_2'} \left[\left(|\ln(s_1 - a)| + |\ln(b-a)|\right)^2\right]^2 ds,$$

where $\Delta_2' = \{s \in \Delta_2 : s_1 \geq a'\}$ and $\Delta_2'' = \{s \in \Delta_2 : s_1 \geq a''\}$. Since the improper integral $\int_{\Delta_2}[(|\ln(s_1 - a)| + |\ln(b-a)|)^2]^2\, ds$ converges, the last integral tends to 0 as a' and a'' tend to a. By the Cauchy criterion,

$$\lim_{a' \to a} \int_{\Delta_2'} k_{[2]}(x, s)l_{[2]}(s, t)\, ds$$

exists and is finite, and $\int_{\Delta_2} k_{[2]}(x, s)l_{[2]}(s, t)\, ds$ is defined as the improper integral

$$\int_{\Delta_2} k_{[2]}(x, s)l_{[2]}(s, t)\, ds = \lim_{a' \to a} \int_{\Delta_2'} k_{[2]}(x, s)l_{[2]}(s, t)\, ds.$$

Note that the integral on the right is an ordinary Riemann integral because its integrand is continuous on $\Delta_2 \times \Delta_2' \times \Delta_2$.

Let a'' tend to a in the foregoing inequality to obtain

$$\left|\int_{\Delta_2} k_{[2]}(x, s)l_{[2]}(s, t)\, ds - \int_{\Delta_2'} k_{[2]}(x, s)l_{[2]}(s, t)\, ds\right|$$

$$\leq (2!)^2\|h_1\|_{\max}^2\|h_2\|_{\max}^2 \int_{\Delta_2\backslash\Delta_2'} \left[\left(|\ln(s_1 - a)| + |\ln(b-a)|\right)^2\right]^2 ds.$$

The right member tends to 0 as a' tends to a uniformly for (x,t) in $\Delta_2 \times \Delta_2$ because the improper integral $\int_{\Delta_2}[(|\ln(s_1-a)| + |\ln(b-a)|)^2]^2\,ds$ converges. Since the integrand in $\int_{\Delta_2'} k_{[2]}(x,s)l_{[2]}(s,t)\,ds$ is continuous $\Delta_2 \times \Delta_2' \times \Delta_2$, the integral over Δ_2' is continuous on $\Delta_2 \times \Delta_2$ by Proposition 18. Therefore, its uniform limit $\int_{\Delta_2} k_{[2]}(x,s)l_{[2]}(s,t)\,ds$ is continuous on $\Delta_2 \times \Delta_2$ by Theorem 23. ■

B.2 Mildly Singular Behavior of Type (ii)

If $k(x,s)$ and $l(s,t)$ are mildly singular kernels of type (ii), then each kernel is continuous and bounded on $[a,b] \times [a,b]\backslash\{(a,a)\}$ and neither has a continuous extension to the full square. Consequently, there exists $B > 0$ such that

$$|k(x,s)|,\ |l(s,t)| \le B$$

for all (x,s) and (s,t) in $[a,b] \times [a,b]\backslash\{(a,a)\}$. The corresponding compound kernels $k_{[n]}(x,s)$ and $l_{[n]}(s,t)$ are expressible as sums with $n!$ terms each of which is an n-fold product of values of the original kernel. Hence,

$$|k_{[n]}(x,s)|,\ |l_{[n]}(s,t)| \le n!B^n$$

for (x,s) and (s,t) in $\left(\Delta_n \times \widetilde{\Delta}_n\right) \cup \left(\widetilde{\Delta}_n \times \Delta_n\right)$, the domain of the compound kernels.

Proposition 200 *If $k(x,s)$ and $l(s,t)$ are mildly singular kernels of type (ii) on $[a,b] \times [a,b]\backslash\{(a,a)\}$, then for $n \ge 1$ the improper integral $\int_{\Delta_n} k_{[n]}(x,s)l_{[n]}(s,t)\,ds$ converges for (x,t) in $\Delta_n \times \Delta_n$ and is continuous there.*

Proof. Since $k_{[n]}(x,s)$ is continuous on $\Delta_n \times \widetilde{\Delta}_n$ and $l_{[n]}(s,t)$ is continuous on $\widetilde{\Delta}_n \times \Delta_n$, $\int_{\Delta_n'} k_{[n]}(x,s)l_{[n]}(s,t)\,ds$ is an ordinary Riemann integral. Moreover, for $a < a'' < a' < b$,

$$\left|\int_{\Delta_n''} k(x,s)_{[n]}l_{[n]}(s,t)\,ds - \int_{\Delta_n'} k_{[n]}(x,s)l_{[n]}(s,t)\,ds\right|$$

$$= \left|\int_{\Delta_n''\backslash\Delta_n'} k_{[n]}(x,s)l_{[n]}(s,t)\,ds\right| \le (n!B^n)^2\left(|\Delta_n''| - |\Delta_n'|\right),$$

where $|A|$ is the n-dimensional volume of the set A. Since the right member of this inequality tends to 0 as a' and a'' tend to a, given (x,t) in $\Delta_n \times \Delta_n$ there exists

$$\lim_{a' \to a} \int_{\Delta_n'} k_{[n]}(x,s)l_{[n]}(s,t)\,ds$$

by the Cauchy criterion. Thus, $\int_{\Delta_n} k_{[n]}(x,s)l_{[n]}(s,t)\,ds$ is defined as the improper Riemann integral

$$\int_{\Delta_n} k_{[n]}(x,s)l_{[n]}(s,t)\,ds = \lim_{a' \to a} \int_{\Delta_n'} k_{[n]}(x,s)l_{[n]}(s,t)\,ds$$

for all (x,t) in $\Delta_n \times \Delta_n$.

It remains to show that $\int_{\Delta_n} k_{[n]}(x, s) l_{[n]}(s, t)\, ds$ is continuous on $\Delta_n \times \Delta_n$. Let a'' tend to a in the inequality above to obtain

$$\left| \int_{\Delta_n} k_{[n]}(x, s) l_{[n]}(s, t)\, ds - \int_{\Delta'_n} k_{[n]}(x, s) l_{[n]}(s, t)\, ds \right| \leq (n! B^n)^2 (|\Delta_n| - |\Delta'_n|)$$

for all (x, t) in $\Delta_n \times \Delta_n$. The right member tends to 0 uniformly in (x, t) as a' tends to a. Since the integrand of the integral over Δ'_n is continuous for (x, s, t) in $\Delta_n \times \Delta'_n \times \Delta_n$, the integral is a continuous function for (x, t) in $\Delta_n \times \Delta_n$ by Proposition 18. Thus, $\int_{\Delta_n} k(x, s)_{[n]} l_{[n]}(s, t)\, ds$ is the uniform limit of continuous functions on $\Delta_n \times \Delta_n$ and, hence, is continuous there. ■

B.3 Iterated Kernels

If $k(x, s)$ is a mildly singular kernel of type (i) or type (ii), the second iterated kernel

$$k_2(x, t) = \int_a^b k(x, s) k(s, t)\, ds$$

is continuous on $[a, b] \times [a, b]$, by the results of the previous sections. Virtually the same discussion as the one given there shows that

$$m(x, s) = \int_a^b k(x, s) l(s, t)\, ds$$

exists and is continuous on $[a, b] \times [a, b]$ if one of k and l is mildly singular and the other kernel is continuous on $[a, b] \times [a, b]$, and the corresponding conclusions hold for the compound kernels of k and l. Consequently, all iterated kernels $k_m(x, s)$ for $m > 1$ of a mildly singular kernel $k(x, s)$ are continuous on $[a, b] \times [a, b]$ and the iterated compound kernels $\left(k_{[n]}\right)_m(x, s)$ are continuous on $\Delta_n \times \Delta_n$.

Appendix C

The Kellogg Conditions

In Section 1.11.2 we mentioned that Kellogg found the Kellogg conditions K1 and K2,

$$\text{K1.} \quad \det\big[k(x_i, x_j)\big]_{n\times n} > 0 \quad \text{for} \quad 0 < x_1 < \cdots < x_n < 1,$$

$$\text{K2.} \quad \det\big[k(x_i, s_j)\big]_{n\times n} \geq 0 \quad \text{for} \quad \begin{array}{l} 0 \leq x_1 \leq \cdots \leq x_n \leq 1, \\ 0 \leq s_1 \leq \cdots \leq s_n \leq 1, \end{array}$$

by purely mathematical considerations. Here $k(x, s)$ is the influence or Green's function for the Sturm-Liouville problem under consideration. Kellogg considered self-adjoint problems so that the Green's function was symmetric, $k(x, s) = k(s, x)$.

Later Gantmacher and Krein showed that the Kellogg conditions reflect familiar properties of many one-dimensional elastic continua that experience transverse deflections within their elastic limits (the linear regime). We reprise Gantmacher and Krein's reasoning here. It reveals important physical interpretations of the Kellogg conditions.

To be concrete, we assume the one-dimensional elastic continuum is a violin string S pinned at its left end $x = 0$ and its right end $x = 1$. The unforced violin string is modeled as the segment $0 \leq x \leq 1$ of the x-axis, the y-axis is transverse to the equilibrium position of the violin string, and the origin is at the left end of the string. We make the following physical assumptions:

H1 Forces act transversely to the equilibrium position of the string and points of the string experience transverse displacements. All points in the string are movable except its endpoints.

H2 The deflection $k(x, s)$ at x due to unit positive force applied at s is continuous for $0 \leq x, s \leq 1$; moreover, a nonzero force applied to any interior point of the string produces a nonzero displacement in the direction of the applied force. That is, $k(s, s) > 0$ for $0 < s < 1$.

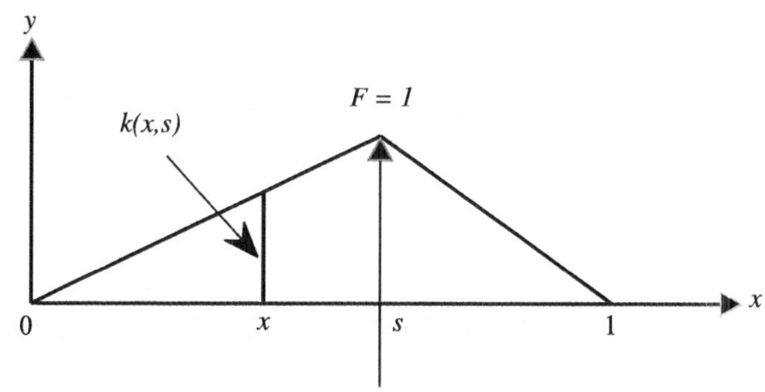

FIGURE C.1: Influence Function

H2* Let S^* be the continuum obtained from the violin string S by placing physical restraints at distinct points $s_1, s_2, \ldots, s_n$ of S that make these points immovable in S^*; that is, if $k^*(x, s)$ is the influence function for S^*, then $k^*(s_i, s_i) = 0$. All points of S^* except its endpoints and the points $s_1, s_2, \ldots, s_n$ are movable and if a force acts at a movable point of S^* that point is displaced in the direction of the force; that is, $k^*(s, s) > 0$ for $s \neq 0, s_1, s_2, \ldots, s_n, 1$ in S^*. (Of course, H2* includes H2 when no physical restraints are imposed at interior points of the string.)

H3 (Superposition Principle – roughly small impressed force assumption) If forces F_1, $F_2, \ldots, F_n$ are applied at $s_1, s_2, \ldots, s_n$, then the displacement at x is $y(x) = \sum_{j=1}^{n} k(x, s_j) F_j$.

H4 If n forces are applied along the string, the resulting deflection $y(x)$ changes its sign (crosses the equilibrium position of the string) at most $n-1$ times.

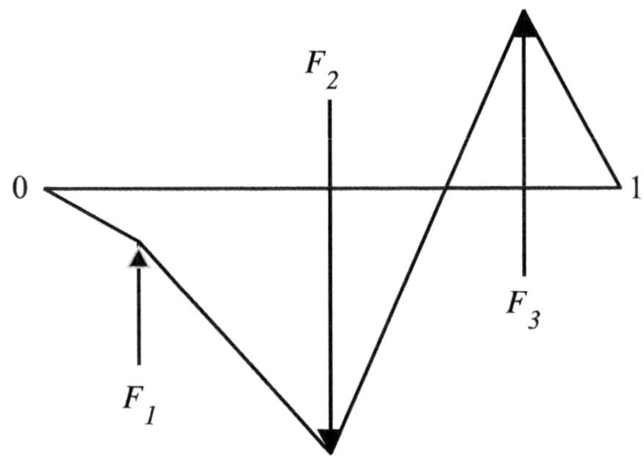

FIGURE C.2: At most two sign changes

H5 (Conservative System) The work W needed to bring the violin string into a given configuration depends only on that configuration; consequently, the work done to achieve the configuration $y(x) = \sum_{j=1}^{n} k(x, s_j) F_j$ depends only upon the forces $F_1, F_2, \ldots, F_n$; that is, $W = W(F_1, F_2, \ldots, F_n)$. We assume the potential energy W is twice continuously differentiable.

H6 The potential energy of the violin string is uniquely minimized (and normalized to 0) when it is in its equilibrium position and no external forces act on it; that is, $W \geq 0$ with equality if and only if no external forces act on S.

We use the notation for compound kernels of $k(x, s)$ introduced in the main text: if $x_1, x_2, \ldots,$ x_n and $s_1, s_2, \ldots, s_n$ are points in S, then

$$k\begin{pmatrix} x_1, x_2, \ldots, x_n \\ s_1, s_2, \ldots, s_n \end{pmatrix} = \det\left[k(x_i, s_j)\right]_{n \times n}$$

where $0 \leq x_1 \leq \cdots \leq x_n \leq 1$, $0 \leq s_1 \leq \cdots \leq s_n \leq 1$.

C.1 Consequences of Conservation of Energy

Fix points $s_1, s_2, \ldots, s_n$ in S with $0 < s_1 < s_2 < \cdots < s_n < 1$ and apply transverse forces $F_1, F_2, \ldots, F_n$ at those points. By H3 the string assumes the shape

$$y(x) = \sum_{j=1}^{n} k(x, s_j) F_j.$$

If each force $F_1, F_2, \ldots, F_n$ receives a increment $dF_1, dF_2, \ldots, dF_n$, then the displacement at s_i, $y_i = y(s_i)$, receives a differential displacement

$$dy_i = \sum_{j=1}^{n} k(s_i, s_j)\, dF_j$$

and the corresponding differential of work is

$$\begin{aligned} dW &= \sum_{i=1}^{n} F_i dy_i \\ &= \sum_{j=1}^{n} \left(\sum_{i=1}^{n} k(s_i, s_j) F_i \right) dF_j. \end{aligned}$$

Hence,

$$\frac{\partial W}{\partial F_j} = \sum_{i=1}^{n} k(s_i, s_j) F_i$$

and

$$\frac{\partial^2 W}{\partial F_i \partial F_j} = k(s_i, s_j).$$

Since

$$\frac{\partial^2 W}{\partial F_i \partial F_j} = \frac{\partial^2 W}{\partial F_j \partial F_i},$$

it follows that

$$k(x, s) = k(s, x).$$

The influence function is symmetric, which is **Maxwell's reciprocity theorem.**

Furthermore, since $W(0, 0, \ldots, 0) = 0$,

$$\begin{aligned} W(F_1, F_2, \ldots, F_n) &= \int_0^1 \frac{d}{dt} W(tF_1, tF_2, \ldots, tF_n)\, dt \\ &= \int_0^1 \sum_{j=1}^{n} \frac{\partial W(tF_1, tF_2, \ldots, tF_n)}{\partial F_j} F_j\, dt \\ &= \sum_{j=1}^{n} \int_0^1 \sum_{i=1}^{n} k(s_i, s_j) tF_i F_j\, dt \\ &= \frac{1}{2} \sum_{i,j=1}^{n} k(s_i, s_j) F_i F_j = \frac{1}{2} \langle \tilde{K} F, F \rangle, \end{aligned}$$

where $\tilde{K} = \left[k(s_i, s_j)\right]_{n\times n}$ and $F = \left(F_1, F_2, \ldots, F_n\right)^T$. By H6 the quadratic form $\langle \tilde{K}F, F\rangle$, which is twice the potential energy, is positive definite: indeed if F is an eigenvector of $\tilde{K}$ with corresponding eigenvalue λ, then

$$0 < W(F) = \langle \tilde{K}F, F\rangle = \lambda\langle F, F\rangle \Rightarrow \lambda > 0.$$

Hence,

$$\det(\tilde{K}) = k\begin{pmatrix} s_1, s_2, \ldots, s_n \\ s_1, s_2, \ldots, s_n \end{pmatrix} = \prod \lambda > 0,$$

where the product is over all the eigenvalues λ of $\tilde{K}$. That is, for any n and any selection of points $0 < s_1 < s_2 \cdots < s_n < 1$ in S

$$k\begin{pmatrix} s_1, s_2, \ldots, s_n \\ s_1, s_2, \ldots, s_n \end{pmatrix} > 0.$$

which is K1. Thus, K1 has the following physical interpretation:

K1 reflects the fact that the violin string S is in stable equilibrium when unforced.

A second interpretation of K1 follows.

C.2 Consequences of H2 and H2*

Let S^* be obtained from the violin string S by introducing restraints at points $s_1, s_2, \ldots, s_n$ interior to S that make these points immovable. The influence function of S is $k(x, s)$ and the influence function of S^* is $k^*(x, s)$. Apply a unit force $F = 1$ at a movable point s in S^*. The displacement at any movable point x in S^* is $k^*(x, s)$ and reaction forces $R_1, R_2, \ldots, R_n$ arise in the constraints at $s_1, s_2, \ldots, s_n$ due to the impressed force at s. The same displacement at x must arise in the original continuum (string) S if, in addition to the unit force applied at s, forces $R_1, R_2, \ldots, R_n$ are applied at the points $s_1, s_2, \ldots, s_n$; hence,

$$k^*(x, s) = k(x, s) + \sum_{j=1}^{n} R_j k(x, s_j).$$

Since the constrained points s_i in S^* cannot move when the unit force is applied at s, $k^*(s_i, s) = 0$; that is,

$$k(s_i, s) + \sum_{j=1}^{n} R_j k(s_i, s_j) = 0$$

for $i = 1, 2, \ldots, n$. This gives $n + 1$ equations for the unknowns $R_0 = 1, R_1, R_2, \ldots, R_n$ which we express as

$$(k(x, s) - k^*(x, s))R_0 + \sum_{j=1}^{n} R_j k(x, s_j) = 0,$$

$$k(s_i, s)R_0 + \sum_{j=1}^{n} R_j k(s_i, s_j) = 0,$$

for $i = 1, 2, \ldots, n$ and x and s any movable points in S^*. Since the homogeneous system has a nontrivial solution,

$$\begin{vmatrix} k(x, s) - k^*(x, s) & k(x, s_1) & \cdots & k(x, s_n) \\ k(s_1, s) - 0 & k(s_1, s_1) & \cdots & k(s_1, s_n) \\ & & \cdots & \\ k(s_n, s) - 0 & k(s_n, s_1) & \cdots & k(s_n, s_n) \end{vmatrix} = 0.$$

Expand the determinant by its first column to get

$$k\begin{pmatrix} x, s_1, \ldots, s_n \\ s, s_1, \ldots, s_n \end{pmatrix} - k^*(x, s)k\begin{pmatrix} s_1, \ldots, s_n \\ s_1, \ldots, s_n \end{pmatrix} = 0.$$

Rename s as s_{n+1} and set $x = s_{n+1}$ to find

$$k\begin{pmatrix} s_1, \ldots, s_n, s_{n+1} \\ s_1, \ldots, s_n, s_{n+1} \end{pmatrix} = k^*(s_{n+1}, s_{n+1})k\begin{pmatrix} s_1, \ldots, s_n \\ s_1, \ldots, s_n \end{pmatrix},$$

where s_{n+1} is any movable point in S^*; that is, $s_{n+1} \neq 0, s_1, \ldots, s_n, 1$.

By H2* and H2, $k^*(s_{n+1}, s_{n+1}) > 0$ and

$$k\begin{pmatrix} s_1 \\ s_1 \end{pmatrix} = k(s_1, s_1) > 0.$$

It follows by an inductive argument that

$$k\begin{pmatrix} s_1, \ldots, s_n \\ s_1, \ldots, s_n \end{pmatrix} > 0$$

for any selection of points $s_1, \ldots, s_n$ with $0 < s_1 < \cdots < s_n < 1$. Consequently, the Kellogg condition K1 is a consequence of the following physical property of the violin string:

> If any number of fixed supports are imposed on S and a single force acts at a movable point of S, then the deflection at that point is nonzero and in the direction of the impressed force.

Now suppose that K1 holds. If fixed supports are imposed at $0 < s_1 < \cdots < s_n < 1$ and s_{n+1} is any movable point in S^*, then, as above,

$$k\begin{pmatrix} x, s_1, \ldots, s_n \\ s, s_1, \ldots, s_n \end{pmatrix} - k^*(x, s)k\begin{pmatrix} s_1, \ldots, s_n \\ s_1, \ldots, s_n \end{pmatrix} = 0$$

with x and s any movable points of S^*. Set $s = x = s_{n+1}$ to obtain

$$k^*(s_{n+1}, s_{n+1}) = \frac{k\begin{pmatrix} s_1, \ldots, s_n, s_{n+1} \\ s_1, \ldots, s_n, s_{n+1} \end{pmatrix}}{k\begin{pmatrix} s_1, \ldots, s_n \\ s_1, \ldots, s_n \end{pmatrix}} > 0.$$

Since $k^*(s_{n+1}, s_{n+1}) > 0$ is just a concise way to express the property of S given above, it follows that K1 is equivalent to that displayed physical property of S.

C.3 Consequences of H4 (H2 and H2*)

Let $k(x, s)$ be the influence function for the violin string S and $s_1, \ldots, s_n$ be distinct points of the string with $0 < s_1, \ldots, s_n < 1$. In this section we show that the Kellogg condition K2 is essentially equivalent to H4 with help from H2 and H2*. The equivalence follows easily from a basic relationship between Tchebycheff systems and weak Tchebycheff systems together with the observation that H3 implies that $k(x, s_1), \ldots, k(x, s_n)$ is a weak Tchebycheff system on $0 \leq x \leq 1$.

Relevant background material on Tchebycheff systems and the kernel

$$l_\sigma(x, s) = \frac{1}{\sqrt{\pi\sigma}} e^{-(x-s)^2/\sigma}, \ \sigma > 0,$$

used by Weierstrass in his original proof of the Weierstrass approximation theorem, are given in Section 2.4.1. It is established there that $l_\sigma(x, s)$ is strictly totally positive on $(-\infty, \infty) \times (-\infty, \infty)$ meaning that

$$l_\sigma\begin{pmatrix} x_1, x_2, \ldots, x_n \\ s_1, s_2, \ldots, s_n \end{pmatrix} > 0$$

for all $-\infty < x_1 < x_2 < \cdots < x_n < \infty$, $-\infty < s_1 < s_2 < \cdots < s_n < \infty$, and $n = 1, 2, 3, \ldots$. In addition, for any function $f(x)$ that is continuous on the closed bounded interval $a \leq x \leq b$,

$$\lim_{\sigma \to 0+} \int_{-\infty}^{\infty} l_\sigma(x, s) f(s)\, ds = f(x)$$

uniformly on $(-\infty, \infty)$, where $f(x)$ is extended to $(-\infty, \infty)$ by setting $f(x) = f(a)$ for $x < a$ and $f(x) = f(b)$ for $x > b$. See Theorem 34. If, in the proof of that theorem, $f(x)$ is extended to be 0 outside the interval $[a, b]$, the reasoning in the proof is easily modified to establish

$$\lim_{\sigma \to 0+} \int_a^b l_\sigma(x, s) f(s)\, ds = f(x)$$

with pointwise convergence on $a < x < b$ and uniform convergence on any closed subinterval of (a, b). We only require the pointwise convergence on $a < x < b$ in what follows.

The next two lemmas establish the basic relationship between Tchebycheff systems and weak Tchebycheff systems that are needed to establish the essential equivalence of H4 and the Kellogg condition K2.

Lemma 201 *If $\varphi(x)$ is a continuous function on a closed bounded interval $I = [a, b]$ that changes sign at most $n - 1$ times on I, then for fixed $\sigma > 0$ the function*

$$\Phi(x, \sigma) = \int_I l_\sigma(x, s)\varphi(s)\, ds$$

has at most $n - 1$ zeros in I.

Proof. We say a continuous function f has m sign changes on I if there exist $m + 1$ points $x_1 < x_2 < \cdots < x_{m+1}$ in I such that

$$f(x_i) f(x_{i+1}) < 0$$

and there is no set of $m+2$ points in I with this property. The hypothesis of the lemma guarantees that there are points

$$a = t_0 < t_1 < \cdots < t_n = b$$

such that $\varphi(x)$ maintains a fixed sign on $I_i = (t_{i-1}, t_i)$ and is nonzero there. Since

$$\begin{aligned}\Phi(x) &= \int_I l_\sigma(x,s)\varphi(s)\,ds \\ &= \sum_{i=1}^{n} \int_{t_{i-1}}^{t_i} l_\sigma(x,s)\varphi(s)\,ds = \sum_{i=1}^{n} \Phi_i(x)\end{aligned}$$

is a polynomial in the functions $\Phi_1, \Phi_2, \ldots, \Phi_n$, where

$$\Phi_i(x) = \int_{t_{i-1}}^{t_i} l_\sigma(x,s)\varphi(s)\,ds$$

for $i = 1, \ldots, n$, $\Phi(x)$ will have at most $n-1$ zeros in I by Proposition 29 if either $\Phi_1, \Phi_2, \ldots, \Phi_{n-1}, \Phi_n$ or $\Phi_1, \Phi_2, \ldots, \Phi_{n-1}, -\Phi_n$ is a Tchebycheff system on I. To see that $\Phi_1, \Phi_2, \ldots, \Phi_{n-1}, \pm\Phi_n$ is a Tchebycheff system use the fact that a determinant is a linear function of each of its columns to obtain

$$\det\,[\Phi_i(x_j)]_{n\times n} =$$

$$\begin{vmatrix} \int_{t_0}^{t_1} l_\sigma(x_1,s_1)\varphi(s_1)\,ds_1 & \cdots & \int_{t_0}^{t_1} l_\sigma(x_n,s_n)\varphi(s_n)\,ds_n \\ \int_{t_1}^{t_2} l_\sigma(x_1,s_1)\varphi(s_1)\,ds_1 & \cdots & \int_{t_1}^{t_2} l_\sigma(x_n,s_n)\varphi(s_n)\,ds_n \\ & \cdots & \\ \int_{t_{n-1}}^{t_n} l_\sigma(x_1,s_1)\varphi(s_1)\,ds_1 & \cdots & \int_{t_{n-1}}^{t_n} l_\sigma(x_n,s_n)\varphi(s_n)\,ds_n \end{vmatrix}$$

$$= \int_{t_0}^{t_1}\int_{t_1}^{t_2}\cdots\int_{t_{n-1}}^{t_n} \begin{vmatrix} l_\sigma(x_1,s_1) & \cdots & l_\sigma(x_n,s_n) \\ l_\sigma(x_1,s_1) & \cdots & l_\sigma(x_n,s_n) \\ & \cdots & \\ l_\sigma(x_1,s_1) & \cdots & l_\sigma(x_n,s_n) \end{vmatrix} \varphi(s_1)\cdots\varphi(s_n)\,ds_1\cdots ds_n$$

$$= \int_{t_0}^{t_1}\int_{t_1}^{t_2}\cdots\int_{t_{n-1}}^{t_n} l_\sigma\begin{pmatrix} x_1,\ldots,x_n \\ s_1,\ldots,s_n \end{pmatrix}\varphi(s_1)\cdots\varphi(s_n)\,ds_1\cdots ds_n.$$

The integrand maintains a fixed sign and is not identically zero; hence, $\det\,[\Phi_i(x_j)]_{n\times n} \neq 0$ and maintains a fixed sign for all $x_1 < x_2 < \cdots < x_n$ in I and either $\Phi_1, \Phi_2, \ldots, \Phi_{n-1}, \Phi_n$ or $\Phi_1, \Phi_2, \ldots, \Phi_{n-1}, -\Phi_n$ is a Tchebycheff system on I. ∎

Lemma 202 *Let $\varphi_1(x), \ldots, \varphi_n(x)$ be continuous and linearly independent on $I = [a,b]$. Then a necessary and sufficient condition that every nontrivial linear combination of these functions changes sign at most $n-1$ times in I is that the determinant*

$$\det\,[\varphi_i(x_j)], \quad x_1 < x_2 < \cdots < x_n \text{ in } I,$$

whenever it is nonzero, maintains the same sign independent of $x_1 < x_2 < \cdots < x_n$ in I.

Proof. ⇒: Suppose every nontrivial linear combination

$$\varphi(x) = \sum_{i=1}^{n} c_i\varphi_i(x)$$

changes sign at most $n-1$ times on I. Let $\sigma > 0$ and

$$\Phi_i(x, \sigma) = \int_I l_\sigma(x, s)\varphi_i(s)\, ds$$

so that

$$\Phi_i(x, \sigma) \to \varphi_i(x) \quad \text{as } \sigma \to 0$$

at each interior point x of I. By Lemma 201 every nontrivial linear combination

$$\Phi(x, \sigma) = \sum_{i=1}^{n} c_i \Phi_i(x, \sigma)$$

has at most $n-1$ zeros. By Proposition 30 $\Phi_1(x, \sigma),\ \Phi_2(x, \sigma), \ldots, \Phi_{n-1}(x, \sigma),\ \pm\ \Phi_n(x, \sigma)$ is a Tchebycheff system on I. Hence, $\det[\Phi_i(x_j, \sigma)] \neq 0$ maintains a fixed sign for all $x_1 < x_2 < \cdots < x_n$ in I. Since

$$\det[\varphi_i(x_j)] = \lim_{\sigma \to 0} \det[\Phi_i(x_j,\ \sigma)],$$

$\det[\varphi_i(x_j)]$, whenever it is nonzero, must maintain the same sign independent of $x_1 < x_2 < \cdots < x_n$ in I.

$\Leftarrow$: Apply Schur's lemma (Lemma 70) with $\phi_i(s) = \varphi_i(s)$ and $\psi_j(s) = l_\sigma(x_j, s)$ to obtain

$$\det[\Phi_i(x_j,\ \sigma)] = \int_{\Delta_n} l_\sigma \begin{pmatrix} x_1, \ldots, x_n \\ s_1, \ldots, s_n \end{pmatrix} \det[\varphi_i(s_j)]\, ds_1 \cdots ds_n$$

where Δ_n is the simplex of points $x_1, \ldots, x_n$ with $x_1 < \cdots < x_n$ in I. The integrand maintains a fixed sign and is not identically zero because the φ_i are linearly independent. (See Lemma 71.) Consequently $\Phi_1(x, \sigma),\ \Phi_2(x, \sigma), \ldots, \Phi_{n-1}(x, \sigma),\ \pm\ \Phi_n(x, \sigma)$ is a Tchebycheff system on I. Since any nontrivial Φ-polynomial has at most $n-1$ zeros and any nontrivial φ-polynomial is the limit of Φ-polynomials, it follows from the intermediate value theorem that any nontrivial φ-polynomial can change sign at most $n-1$ times. ■

Now we are prepared to show the essential equivalence of H4 and the Kellogg condition K2. As before $k(x, s)$ is the influence function of the violin string S.

Lemma 203 *The functions* $k(x, s_1), \ldots, k(x, s_n)$ *are linearly independent on* $0 < x < 1$ *for any choice of* $0 < s_1 < \cdots < s_n < 1$.

Proof. If

$$k(x, s_1)F_1 + \cdots + k(x, s_n)F_n = 0$$

for certain constants $F_1, \ldots, F_n$ and for $0 < x < 1$, then

$$k(s_i, s_1)F_1 + \cdots + k(s_i, s_n)F_n = 0$$

for $i = 1, \ldots, n$; that is $\tilde{K}F = 0$ where $\tilde{K} = [k(s_i, s_n)]_{n\times n}$ and $F = [F_1, \ldots, F_n]^T$. Since $\det(\tilde{K}) > 0$ by H2 and H2*, $\tilde{K}$ is nonsingular, $F = 0$, and $k(x, s_1), \ldots, k(x, s_n)$ are linearly independent on $0 < x < 1$. ■

By H4, for any fixed set of points $s_1 < \cdots < s_n$ in S and any constants $F_1, \ldots, F_n$ the k-polynomial

$$\sum_{j=1}^{n} F_j k(x, s_j)$$

changes sign at most $n - 1$ times. Consequently, by Lemma 202

$$\det\left[k(x_i, s_j)\right] = k\begin{pmatrix} x_1, \ldots, x_n \\ s_1, \ldots, s_n \end{pmatrix}$$

maintains a fixed sign for $x_1 < \cdots < x_n$ in S. Consequently

$$k\begin{pmatrix} x_1, \ldots, x_n \\ s_1, \ldots, s_n \end{pmatrix} \geq 0$$

because

$$k\begin{pmatrix} s_1, \ldots, s_n \\ s_1, \ldots, s_n \end{pmatrix} > 0$$

for $s_1 < \cdots < s_n$ in S by H2 and H2*. Since $k(x, s)$ is continuous on $[0, 1] \times [0, 1]$,

$$k\begin{pmatrix} x_1, \ldots, x_n \\ s_1, \ldots, s_n \end{pmatrix} \geq 0$$

holds for all $0 \leq x_1 \leq \cdots \leq x_n \leq 1$ and $0 \leq s_1 \leq \cdots \leq s_n \leq 1$ in S. Thus, H4 (H2 and H2*) imply K2. Conversely, if K2 holds, Lemma 202 and Lemma 203 imply H4.

Bibliography

[1] Anselone, P. M. and Lee, J. W., *The Heart of Calculus*, The Mathematical Association of America, Washington, DC (2015).

[2] Berezanskii, Ju. M., *Expansions in Eigenfunctions of Selfadjoint Operators*, Vol. 17, Translations of Mathematical Monographs, American Mathematical Society, Providence Rhode Island (1968).

[3] Bergendahl, G., *Convergence and summability of eigenfunction expansions connected with elliptic differential equations*, Medd. Lunds Univ. Mat. Sem. **15,** 1–63 (1959).

[4] Bieberbach, L., *Theorie der gewöhnlichen Differentialgleichungen*, Die Grundlehren der mathematischen Wissenschaften, Springer Verlag, Berlin, Göttingen, Heidelberg (1953).

[5] Birkhoff, G. and Rota, G-C., *Ordinary Differential Equations, 4th ed.*, John Wiley & Sons, Inc., New York (1989).

[6] Brown, J. W. and Churchill, R. V., *Complex Variables and Applications, 9th ed.*, McGraw-Hill, New York (2013).

[7] Collatz, L., *Eigenwertaufgaben mit technischen Anwendungen*, 2. Auflage, Akademische Verlagsgesellschaft, Geest & Portig K.-G., Leipzig (1963).

[8] Collatz, L., *Einschliesungsstaz für die characteristischen Zahlen von Matrizen*, Math. Zeitschr. **48,** 221–226 (1942).

[9] Coddington, E. A. and Levinson, N., *The Theory of Ordinary Differential Equations*, McGraw-Hill Book Company, New York (1955).

[10] Courant, R. and Hilbert, D., *Methods of Mathematical Physics*, Vol. 1, Interscience Publishers, Inc., New York (1953).

[11] Curtis, C., *Linear Algebra: An Introductory Approach*, Springer-Verlag, New York (1984).

[12] Franklin, J., *Matrix Theory*, Dover Publications, Mineola, New York (2000).

[13] Fredholm, I., *Sur une classe d'équations fonctionnelles*, Acta Mathematica, **27,** 365–390 (1903).

[14] Friedberg, S. H., Insel, A. J., and Spence, L. E., *Linear Algebra, 3rd ed.*, Prentice Hall, Inc. (1997).

[15] Fulks, W., *Advanced Calculus: An Introduction to Analysis, 3rd ed.*, John Wiley & Sons, Inc. (1978).

[16] Gantmacher, F. R. and Krein, M. G., *Oszillationsmatrizen, Oszillationskerne und Kleine Schwingungen Mechanischer Systeme*, Academe Verlag, Berlin (1960).

[17] Granas, A., Guenther, R. B., and Lee, J. W., *Nonlinear Boundary Value Problems for Ordinary Differential Equations*, in Dissertationes Mathematicae, CCXLIV, Polska Akademia Nauik. Instytut Matematyczny, Warszawa (1985).

[18] Guenther, R. B. and Lee, J. W., *Partial Differential Equations of Mathematical Physics and Integral Equations*, Dover Publications, Inc., New York (1996).

[19] Hille, H., *Ordinary Differential Equations in the Complex Domain*, Dover Publications Inc., Meola, New York (1997) (Reprint of the 1976 edition published by John Wiley & Sons, Inc.).

[20] Hoffman, K. and Kunze, R., *Linear Algebra, 2nd ed.*, Prentice Hall, Englewood Cliffs, New Jersey (1971).

[21] Isaacson, E. and Keller, H. B., *Analysis of Numerical Methods*, John Wiley & Sons, New York (1966).

[22] Jentzsch, R., *Über Integralgleichungen mit positizvem Kern*, J. Math. Crelle, **141,** 235–244 (1912).

[23] Kamke, E., *Differentialgleichungen*, 4. Auflage, vol. I und II, Akademische Verlagsgesellschaft, Geest & Portig K.-G., Leipzig (1962).

[24] Karlin, S., *Total Positivity, Vol. 1*, Stanford University Press, Palo Alto, California (1968).

[25] Karlin, S. and Studden, W., *Tchebycheff Systems: with applications in analysis and statistics*, Interscience Publishers, New York (1966).

[26] Kellogg, O. D., *The Oscillation of Functions of an Orthogonal Set*, Amer. J. Math. **38,** 1–5 (1916).

[27] Kellogg, O. D., *Orthogonal Function Sets Arising from Integral Equations*, Amer. J. Math. **40,** 145–154 (1918).

[28] Knopp, K., *The Theory of Functions, Part I and II*, Dover Publications, Mineola, New York (1996).

[29] Loomis, L. and Sternberg, S., *Advanced Calculus*, Addison-Wesley, Reading, Massachusetts (1968).

[30] Mangoldt, H. and Knopp, K., *Einführung in die höhere Mathematik*, S. Hirzel Verlag, Stutgart (1958).

[31] Meinardus, G., *Approximation of Functions: Theory and Numerical Methods*, Springer Verlag, New York (1967).

[32] Pincus, Allan, *Spectral Properties of Totally Positive Kernels and Matrices*, in Total Positivity and Its Applications, M. Gasca, C. A. Micchelli (eds.), pp. 477–511, Kluwer Academic Publishers (1996).

[33] Riesz, F. and Nagy, B., *Functional Analysis*, Frederick Ungar Publishing Co., New York (1955).

[34] Ross, K. A., *Elementary Analysis: The Theory of Calculus*, Undergraduate Texts in Mathematics, Springer Verlag, New York (2013).

[35] Royden, H., *Real Analysis, 2nd ed.*, The Macmillan Company, London (1968).

[36] Schur, I., *Über die charakterischen Wurzeln einer linearen Substitution mit einer Anwendung auf die Theorie der Integralgleichungen*, Math. Ann. **66,** 488–510 (1909). Also in *Gesammelte Abhandlungen, Vol. 1*, Eds. A. Brauer and H. Rohrbach, Springer Verlag, Berlin (1973).

[37] Schur, I., *Zur Theorie der linearen homogenen Integralgleichungen*, Math. Ann. **67,** 306–359 (1909) Also in *Gesammelte Abhandlungen, Vol. 1*, Eds. A. Brauer and H. Rohrbach, Springer Verlag, Berlin (1973).

[38] Smirnov, V. I. and Lohwater, A. J., *A Course in Higher Mathematics, 1st ed., Vol. 4*, Elsevier Science, (2014) (Available as ebook).

[39] Smith, K. T., *Primer of Modern Analysis, 1st ed*, Bogden and Quigley Inc., New York (1971). Also, 2nd ed, Springer Verlag (1983).

[40] Sperner, E., *Einführung in die analytische Geometrie und Algebra, I. Teil*, Vandenhoek & Ruprecht, Göttingen (1959).

[41] Stoer, J. and Bulirsch, R., *Introduction to Numerical Analysis, 2nd ed.*, Springer Verlag, New York (1993).

[42] Strang, G. and Fix, G., *Analysis of the Finite Element Method, 2nd ed.*, Wellesley-Cambridge (2008).

[43] Tychonoff, A. N. and Samarski, A. A., *Differentialgleichungen der Mathematischen Physik*, Veb Deutscher Verlag der Wissenschaften, Berlin (1959).

[44] Tychonoff, A. N. and Samarski, A. A., *Partial Differential Equations of Mathematical Physics, Vol. II*, Holden-Day, Inc. San Francisco (1967).

[45] Weyl, H., *Gruppentheorie und Quantenmechanik*, Wissenschaftliche Buchgesellschaft, Darmstadt, (1981). A reprint of the second edition, Leipzig (1931).

Index

V

W